Rebellion in the Province: The Landvolkbewegung and the Rise of National Socialism in Schleswig-Holstein

Kieler Werkstücke

Reihe A:
Beiträge zur schleswig-holsteinischen und skandinavischen Geschichte

Herausgegeben von Oliver Auge und Thomas Riis
Begründet von Erich Hoffmann

Band 36

PETER LANG
EDITION

Alexander Otto-Morris

Rebellion in the Province: The Landvolkbewegung and the Rise of National Socialism in Schleswig-Holstein

PETER LANG
EDITION

Bibliographic Information published by the Deutsche Nationalbibliothek
The Deutsche Nationalbibliothek lists this publication in the Deutsche Nationalbibliografie; detailed bibliographic data is available in the internet at http://dnb.d-nb.de.

Gratefully acknowledging financial support
from Monash University, Clayton/Australia.

Cover illustration: Seal of the Christian-Albrechts-Universität zu Kiel.
Printed with kind permission of the Christian-Albrechts-Universität zu Kiel.

Library of Congress Cataloging-in-Publication Data

Otto-Morris, Alexander, 1974-
Rebellion in the province : the Landvolkbewegung and the rise of national socialism in Schleswig-Holstein / Alexander Otto-Morris.
pages cm. — (Kieler Werkstücke. Reihe A, Beiträge zur schleswig-holsteinischen und skandinavischen Geschichte, ISSN 0936-4005 ; volume 36)
Includes bibliographical references and index.
ISBN 978-3-631-58194-0
1. Farmers—Political activity—Germany—Schleswig-Holstein—History—20th century. 2. Protest movements—Germany—Schleswig-Holstein—History—20th century. 3. Germany—Economic conditions—1918-1945. 4. Schleswig-Holstein (Germany)—Economic conditions—20th century. 5. Agriculture and state—Germany—Schleswig-Holstein—History—20th century. 6. National socialism—Germany—Schleswig-Holstein. 7. Schleswig-Holstein (Germany)—Politics and government—20th century. I. Title.
DD801.S338O89 2013
322.4'2094351209042—dc23
2013014655

ISSN 0936-4005
ISBN 978-3-631-58194-0 (Print)
E-ISBN 978-3-653-03527-8 (E-Book)
DOI 10.3726/978-3-653-03527-8

Peter Lang – Frankfurt am Main · Bern · Bruxelles · New York ·
Oxford · Warszawa · Wien

www.peterlang.de

Acknowledgements

This book has benefited from the assistance of many individuals and institutions in Australia and Germany. In particular, I would like to express my deep thanks to my doctoral supervisors, Dr. Eleanor Hancock and Professor David Garrioch. Their guidance and encouragement over many years were invaluable.

I am indebted to Monash University for a postgraduate scholarship and the School of Historical Studies for assistance in publishing this work as well as to Professor Mark Peel and to Liisa Williams who paved the way for my research in Germany. Acknowledgments are also due to Professor Olaf Mörke at the Christian-Albrechts-Universität zu Kiel and to the Kiel International Centre for making scholarly life in Germany so much easier.

I am grateful to Professors Oliver Auge and Thomas Riis for facilitating the publication of this book and to the latter's colloquium at the University of Kiel for kindly giving me the chance to present and discuss my work. I am also grateful to my doctoral examiners, Professors Jeremy Noakes and John Moses, for their advice. I benefited from discussions on anti-Semitism with Professor Gerhard Paul and Dr. Claudia Prestel and from the advice of Dr. Michael Hau and Michael Spurr.

Thanks are also due to the various institutions at which my research was conducted, in particular, to the extraordinarily helpful staff at Schleswig-Holstein's Landesarchiv in Schleswig and the Landesbibliothek and Zentralbibliothek für Wirtschaftswissenschaften in Kiel. I am indebted to Klaus Tidow at the Stadtarchiv Neumünster and to Kai Illfrich and the Zeitzeichen Archiv in Kiel for transcripts of interviews of the Landvolk leader in Plön County, Peter Petersen, and of Mrs Hamkens, wife of Landvolk leader Wilhelm Hamkens of Tetenbüll. I am grateful to the Peters family and the Archiv der Kirchspielslandgemeinde Lunden for copies from the *Heider Anzeiger*, to Claus Heinrich Bill for allowing me to view copies of *Blut und Boden*, to Kay Dohnke for access to his collection of photographs and to Frank Omland who very kindly produced the map of Schleswig-Holstein. Further, I wish to thank Michael Rücker at Peter Lang for his technical assistance.

I am indebted to my friends, Julia Schmitt-Eliassen, who generously helped me to read Le Bars' French thesis, and Nils Teifke who assisted me in clarifying some of the intricacies of Weimar's legal system. Special thanks are due to my father-in-law, Karl-Ulrich Otto, who took me on my first tour of Schleswig-Holstein's west coast and supplied me with reading material while I was in Australia. I am very grateful to my mother-in-law, Helgard Otto, for reading the footnotes, to Lyndal Turner, who read some of the early drafts, and to my proofreaders: Clare McCausland, my mother, Anna Hall, and especially my father, Peter Morris. Needless to say, the responsibility for any shortcomings is mine alone.

Finally, my biggest debt is to my wife, Christine. Without her patience and loving support, this book could not have been written.

Contents

Abbreviations

BDC	Berlin Document Center (Bundesarchiv)
BdL	Bund der Landwirte
CNBLP	Christlich-Nationale Bauern- und Landvolk-Partei
DDP	Deutsche Demokratische Partei
DNVP	Deutschnationale Volkspartei
DSP	Deutsche Staatspartei
DVP	Deutsche Volkspartei
GStA	Geheimes Staatsarchiv Preußischer Kulturbesitz
KPD	Kommunistische Partei Deutschlands
LAS	Landesarchiv Schleswig-Holstein
NSDAP	Nationalsozialistische Deutsche Arbeiterpartei
NSFB	Nationalsozialistische Freiheitsbewegung
Preuß.	Preußischer
OC	Organisation Consul
RLB	Reichlandbund
RM	Reichsmark
SA	Sturmabteilung
SHLP	Schleswig-Holsteinische Landespartei
SPD	Sozialdemokratische Partei Deutschlands
StaNms	Stadtarchiv Neumünster
USPD	Unabhängige Sozialdemokratische Partei Deutschlands
VSB	Völkisch-Sozialer Block

Introduction

In the late 1920s, Germany's northern province, Schleswig-Holstein, was home to one of the most unusual political movements of the Weimar Republic. Even though Schleswig-Holstein had no recent history of rural unrest, the *Landvolkbewegung* (Rural People's Movement) became the most active and radical farmers' movement of the Weimar period. This book analyses the historical development of the Landvolkbewegung. In short, it examines the transformation of rural discontent about deteriorating economic circumstances into a unique militant movement. It demonstrates how farmers, a group which rarely protested publicly, became increasingly disillusioned, not only turning against the authorities but also actively promoting and directing wider discontent toward them. It reveals how the Landvolkbewegung mobilised, publicised and radicalised rural dissatisfaction with Weimar, and the way it came to reject the Republic completely. The history of the movement is significant for the insight it provides into the failure of German democracy of this era. Furthermore, it helps to explain the early and considerable success of National Socialism in Schleswig-Holstein, particularly in rural areas, as the Landvolkbewegung paved the way for the advancement of the National Socialist German Workers' Party (*Nationalsozialistische Deutsche Arbeiterpartei*, NSDAP) in the province.

The Landvolkbewegung developed in response to the agricultural crisis of the late 1920s which preceded the general economic depression. The damaging impact of the agricultural crisis was strongly felt in Schleswig-Holstein, and on January 28, 1928 farmers converged on the province's towns to hold unprecedented protest rallies. But after the Reich government failed to provide the relief the farmers expected and the existing agricultural associations were unable to unite and establish a more effective means of serving agriculture's interests, the province's farmers adopted alternative means to preserve their property. Embittered by the tax burden, from the autumn of 1928, the *Landvolk* (rural people) took obstructive, direct action to prevent compulsory tax collection. They voiced their anger at countless meetings and later founded their own newspaper, employing a number of nationalist publicists. The farmers rallied in solidarity with their colleagues who faced legal action, leading to clashes and sometimes violent conflict with the police. Bombs exploded in public buildings and after police suppressed a rally in Neumünster, confiscating the farmers' flag, the farmers boycotted the town. Though the Landvolk continued its activities throughout 1930 and 1931, the movement gradually lost support. Again and again the leaders strove to recover momentum but what remained of the movement was soon quashed after the Nazi seizure of power.

Charting the course of the Landvolkbewegung, its origins, development and decline, is the principal object of this study. The sources of rural discontent that led to unrest and the factors that shaped the form of the farmers' protests are also subjected to in-depth examination, as are the symptoms, stages and contributory factors of the movement's decline. Further points of interest are the Landvolk's organisational and leadership structures and the roles played by individuals and organisations from outside agriculture. The main focus of analysis, however, is the process of radicalisation within the movement and its radicalisation of the province's rural population. Central to this are the means by which the Landvolkbewegung stirred discontent, directed it toward the authorities and generated opposition to them as well as the Republic, in particular, at its meetings, through public actions and in the press. Moreover, the character, function and reception of the Landvolk's ideology as well as the way the farmers came to embrace increasingly militant and violent methods require clarification. So too do the nature and efficacy of the movement's aims, tactics and strategies. It is also necessary to determine why the opponents of the movement failed to stop the course of radicalisation in the province. Consequently, the Landvolk's dealings with the authorities and the authorities' responses to the movement are examined, as is the effect of movement's demagogy and defiant undertakings in undermining the authorities' standing and stretching their means of enforcement. While the Landvolkbewegung's significance should not be viewed solely within the context of the history of National Socialism, the movement's role in the NSDAP's advance in the countryside is of particular concern. And although this is not an account of the rise of the NSDAP, it does provide a detailed analysis of the relationship between the two movements and thus seeks to contribute to an explanation of the conditions and developments which facilitated the NSDAP's astonishing success in Schleswig-Holstein.

The Landvolkbewegung has been the subject of numerous fictional and historical works. Most notably, the movement provided the background for Hans Fallada's *A Small Circus* (*Bauern, Bonzen und Bomben*), a novel praised, in particular, for capturing the atmosphere of a small town at the end of the Weimar Republic. Fallada had reported on one of the many Landvolk trials whilst working for a newspaper in Neumünster and parts of his novel are remarkably close to actual events. The accuracy of his portrayal of the Landvolkbewegung has been debated, but there can be no question that the novel, which has since been adapted for theatre and television, has done most to preserve the memory of the farmers' movement.[1]

---

1 H. Fallada, *Bauern, Bonzen und Bomben* (Hamburg, 1997 [1931]); for the English translation, H. Fallada, *A Small Circus* (London, 2012); E. Monk (screenplay), *Bauern, Bonzen und Bomben* (Studio Hamburg, 1972). For Kurt Tucholsky's praise of the novel, see the *Weltbühne*, 27/14, (1931). Martin Sadek pointed to the close relationship between the novel and reality. See M. Sadek, 'Bauern, Bonzen und Bomben – Realität und Roman' in G. Müller-Waldeck and R. Ulrich (eds), *Hans Fallada: Beiträge zu Leben und Werk* (Rostock, 1995), pp. 100-116. Ellis Shookman criticised Fallada for remaking history in *Bauern, Bonzen und Bomben*, for sympathising with the Landvolkbewegung, downplaying its violence and hiding the farmers' susceptibility to Nazism. Thomas Bredohl, on the other hand, insisted that the novel was only ever intended to portray the farmers in so far as they concerned the po-

Ernst von Salomon, whose brother Bruno was the editor of the farmers' newspaper, had more intimate contact with the movement. His little known novel *Die Stadt*, a fictional account of his brother's experiences, is supplemented by a humorous account of his own dealings with the Landvolk in his autobiographical *Der Fragebogen*.[2] Nationalist adventurer and Landvolk activist, Herbert Volck, and the editor of the National Socialist *Schleswig-Holsteinische Tageszeitung*, Bodo Uhse, offered further autobiographical perspectives.[3] Foremost among those who experienced the movement first-hand and wrote with more overt partisan motivation, was the Landvolk's lawyer, Walter Luetgebrune, whose *Neu-Preußens Bauernkrieg* condemned the authorities for political intrusion in the Landvolk's trials. His criticisms were echoed in a chapter of the second volume of Gottfried Zarnow's *Gefesselte Justiz*. Zarnow notoriously stood the history of Weimar's "political justice" on its head, alleging the judiciary indulged the political left and treated the right malevolently.[4]

The first academic study to take note of the Landvolkbewegung was Rudolf Heberle's pioneering regional sociological study *From Democracy to Nazism*.[5] Later published in German in full, it analysed voting behaviour in rural Schleswig-Holstein, examined the ecology, that is the soil types and modes of agriculture, and the social and economic structures of the various regions in the province to explain

---

litical life of a town, that Fallada in fact supported Weimar and was critical of the farmers. E. Shookman, 'Making History in Hans Fallada's Bauern, Bonzen und Bomben: Schleswig-Holstein, Nazism, and the Landvolkbewegung', *German Studies Review*, 13 (1990), pp. 461-480; T. Bredohl, 'Some thoughts on the Political Opinions of Hans Fallada: A Response to Ellis Shookman', *German Studies Review*, 15 (1992), pp. 525-545; E. Shookmann, 'Figuring Out Hans Fallada: A Reply to Thomas Bredohl', *German Studies Review*, 17 (1994), pp. 79-81. Fallada was not the first to fictionalise the farmers' movement. From January 1930, Ernst Benedix sought permission to present his play, "Kampf um der Väter Erbe" (*Struggle for the Fathers' Inheritance*) in Schleswig-Holstein. See the *Flensburger Nachrichten* No. 4, 6.1.1930 and the letter from the Nirvana-Verlag to the Polizeibehörde Itzehoe, Hamburg, 16.1.1930, LAS 309/22783.

2 E. von Salomon, *Die Stadt* (Berlin, 1932) and *Der Fragebogen* (Hamburg, 1961).

3 H. Volck, *Rebellen um Ehre. Mein Kampf für die nationale Erhebung 1918-33* (Gütersloh, 1936); B. Uhse, *Söldner und Soldat* (Paris, 1935).

4 W. Luetgebrune, *Neu-Preußens Bauernkrieg. Entstehung und Kampf der Landvolkbewegung* (Hamburg, 1931); G. Zarnow, *Gefesselte Justiz*, vol. 2 (Munich, 1932). Other notable contemporary accounts include J. Schimmelreiter (pseudonym: Peter Petersen), *Unter der schwarzen Bauernfahne. Die Landvolkbewegung im Kampf für Deutschlands Befreiung* (Munich, 1929) and R. Schapke, *Aufstand der Bauern* (Leipzig, 1933).

5 R. Heberle, *From Democracy to Nazism. A Regional Case Study* (Baton Rouge, 1945) and *Landbevölkerung und Nationalsozialismus. Eine soziologische Untersuchung der politischen Willensbildung in Schleswig-Holstein 1918-1932* (Stuttgart, 1963). Heberle conducted his research from 1932 to 1934 but the National Socialist seizure of power prevented him from publishing all but a summary of his findings in the short term. See R. Heberle 'Die Politische Haltung des Landvolks in Schleswig-Holstein 1918-1932', *Volksspiegel: Zeitschrift für deutsche Soziologie und Volkswissenschaft*, 1 (1934), pp. 166-172.

the conditions and developments that determined their receptiveness to Nazism. Heberle also assessed the role of occupation, social class and the impact of the agricultural crisis. He found that the main support for the NSDAP came from the lower middle class, particularly from the central regions of the province where the social structure of the villages presented little class antagonism and a high degree of community solidarity. Additionally, Heberle sketched the rural organisations and interest groups which helped to form public opinion in the province. In this context, he provided an overview of the Landvolkbewegung, noting that its activities encouraged farmers to believe that Weimar was an enemy from which they could expect little or no protection. He also noted that the NSDAP capitalised on the unrest the Landvolk caused and that Landvolk followers joined the NSDAP after their own movement declined.[6] Yet this account of the movement, though well known, is but a hasty summary. As Heberle acknowledged, in his analysis of voting behaviour, he was not nearly as interested in historical processes as in sociological aspects and correlating the ecological and the resulting economic and social conditions with election results.[7]

Heberle contended that the value of studying Schleswig-Holstein was its "general validity" as a case study because the province was typical of conditions and developments elsewhere in Germany. This view, however, is flawed. As Jürgen Falter's research has since shown, the province did not follow the general pattern of political developments in the Reich in all essential respects as Heberle claimed.[8] Historically at least, Schleswig-Holstein is interesting precisely because of the unusual developments there, because of the Landvolk's unique radicalisation of the farmers as well as the NSDAP's early breakthrough in the province which gave it the status of a forerunner in capturing rural support.[9]

Gerhard Stoltenberg's *Politische Strömungen im schleswig-holsteinischen Landvolk 1918-1933* dealt with the Landvolkbewegung in the wider context of the changing political character of Schleswig-Holstein's rural population from the First

---

6 Heberle, *From Democracy to Nazism*, pp. 65-70 and, with minor additions, *Landbevölkerung und Nationalsozialismus*, pp. 156-160. NB This study cites from the better-known *From Democracy to Nazism*, unless there are additions or revisions in the German edition.

7 Heberle, *Landbevölkerung und Nationalsozialismus*, pp. 12 and 63.

8 Heberle, *From Democracy to Nazism*, pp. 23 and 34. According to Falter, many of Heberle's conclusions in respect of the number of farm employees, farm-size and soil fertility did not apply to the rest of German agriculture. In particular, religion was a much more important factor and the prevalence of these other conditions did not necessarily determine their receptiveness to Nazism where the rural population was Catholic. Moreover, Falter showed that agriculture itself was only a clear factor in the NSDAP's success in all regions of Germany from the Reichstag elections of July 1932. J. Falter, *Hitlers Wähler* (Munich, 1991), pp. 256-266.

9 For this reason Schleswig-Holstein has been described as the NSDAP's model Gau. U. Danker 'Die Zerstörung der Weimarer Demokratie in Schleswig-Holstein' in G. Wewer (ed.), *Demokratie in Schleswig Holstein* (Opladen, 1998), p. 214.

World War to National Socialist *Gleichschaltung* (coordination).[10] This too was Heberle's subject, but Stoltenberg's thesis was a political historical study investigating the rise of National Socialism by explaining the developments within the parties and organisations, including the agricultural and paramilitary associations, and their leaderships from the late imperial period. One of his most arresting findings was the evident weakness of the middle-class parties' organisations in rural areas in contrast to the relative strength of interest groups at the communal level.[11] Stoltenberg also highlighted the Landvolkbewegung's radicalising influence and, although not explicitly drawing the conclusion himself, he demonstrated convincingly that the movement played a significant role in the political shift from left to right in the province during the Weimar Republic.[12] Stoltenberg's overview of the movement's development showed the radicalisation of Landvolkbewegung, but it did not reveal how the farmers actually came to interpret their economic difficulties as a struggle against the Weimar system. Stoltenberg offered almost no analysis of the economic factors behind the farmers' unrest. Nor did he examine in any detail the movement's meetings, ideas and organisation. For this reason, it seems, he judged incorrectly that the farmers' movement was under *völkisch* (racialist) influence from the beginning. Moreover, his assessment concerning the Landvolk's role in the rise of National Socialism did not do justice to the movement's originality. Though noting that the Landvolk did pose competition initially, he wrote that ultimately it was "only .... a ferment of radicalisation to the benefit of the National Socialists".[13] It is perhaps too much to expect Stoltenberg to have offered closer analysis since his was not a study of the movement. Yet although Stoltenberg devoted little more than 30 pages to the Landvolkbewegung, his thesis has remained the authoritative history on the subject.

In a series of essays in the late 1950s and early 1960s, Hans Beyer argued that the Landvolkbewegung could not simply be viewed from the perspective of political radicalisation and he objected to the view that the Landvolkbewegung be considered prehistory of National Socialism.[14] Protesting against Stoltenberg and

---

10 G. Stoltenberg, *Politische Strömungen im schleswig-holsteinischen Landvolk 1918-1933. Ein Beitrag zur politischen Meinungsbildung in der Weimarer Republik* (Düsseldorf, 1962). See especially, pp. 121-141 passim and 172-174.

11 Alongside Stoltenberg, Peter Wulf has corroborated this from the viewpoint of the province's trades. P. Wulf, *Die politische Haltung des schleswig-holsteinischen Handwerks, 1928-1932* (Kiel, 1967). Timothy Alan Tilton also underlined this in *Nazism, Neo-Nazism and the Peasantry* (London, 1975), pp. 35-37.

12 Thomas Nipperdey criticised Stoltenberg's thesis more generally for insufficient reflection and for leaving his readers to make up their own minds. See his review of Heberle and Stoltenberg's studies, T. Nipperdey, 'Die deutsche Landbevölkerung und die Politik 1918-1933', *Neue Politische Literatur*, 9 (1964), pp. 389-400.

13 Stoltenberg, *Politische Strömungen im schleswig-holsteinischen Landvolk*, pp. 147 and 197.

14 H. Beyer, 'Die Landvolkbewegung Schleswig-Holsteins und Niedersachsens 1928-1932', *Jahrbuch Kreis Eckernförde*, 15 (1957), pp. 173-202; H. Beyer, 'Die Agrarkrise und die

Heberle's analyses, Beyer highlighted the Landvolkbewegung's differences to the NSDAP, its independence and opposition. Yet these features did not actually diminish the Landvolk's role in facilitating the NSDAP's rise in Schleswig-Holstein. As Beyer himself conceded, the NSDAP profited from the Landvolkbewegung's "slogans". Beyer portrayed the Landvolk's opposition to the party system and stressed its local patriotism and desire for self-administration, but in doing so, he neglected other essential elements of its extreme right-wing propaganda. He provided an innocuous representation of the movement, insisting that its "core" was not radical or völkisch, and that such elements were only to be found in its newspaper, which was taken over by a small group of national revolutionary publicists. He exaggerated their influence, claiming they succeeded in turning the "farmers' movement" into the instrument of their ideology in the winter of 1928/1929. He accorded the farmers a passive role and portrayed them as disadvantaged by the rise of National Socialism.

Writing from the former German Democratic Republic, Ehrenfried Friedrich provided a Marxist interpretation. He presented remarkable coverage of the movement's activities, given his restricted access to the sources which comprised almost entirely of the reports sent to the Prussian Interior Ministry in Berlin. Friedrich viewed the Landvolkbewegung as a failed farmers' movement, highlighting its anti-capitalist affinities and great potential to cooperate with the Communist Party of Germany (KPD), but also seeing it as misguided, falling victim to anti-Semitism and anti-communism. Friedrich redressed the neglect in earlier studies in respect of Bruno von Salomon and Bodo Uhse's efforts from 1931 to win over former Landvolk followers to the KPD, but overstated their success. The class conflict model he imposed upon the explanation of the movement's origins was strained and the conclusion, that the farmers were bound to fail because they cooperated with reactionary forces instead of allying with urban workers, is unsubstantiated.[15]

Michelle Le Bars' *Le Mouvement Paysan Dans Le Schleswig-Holstein 1928-1932* centred overwhelmingly on literary sources, in particular Fallada's *A Small Circus* and Ernst von Salomon's *Die Stadt*. Le Bars showed how the movement's history was reflected in the literature, using the novels, interspersed with a selection of archival documents, to illustrate its origins, actions, radicalisation and connec-

---

Landvolkbewegung in den Jahren 1928-1932: ein Beitrag zur Geschichte "revolutionärer" Bauernbewegungen zwischen den beiden Weltkriegen', *Archiv für Agrargeschichte der holsteinischen Elbmarschen*, 5 (1983), pp. 156-187 (originally published by the Heimatverband für den Kreis Steinburg in 1962); H. Beyer, 'Das Bauerntum Angelns während der großen Krise 1927 bis 1932', *Jahrbuch des Angler Heimatvereins*, 26 (1962), pp. 131-161; H. Beyer, 'Landbevölkerung und Nationalsozialismus in Schleswig-Holstein', *Zeitschrift für Agrargeschichte und Agrarsoziologie*, 11 (1963), pp. 69-74 and his response to Heberle and Stoltenberg in 'Schleswig-Holsteins Landvolk im Zwielicht', *Die Landwirtschaftliche Berufsschule*, 18 (1968), pp. 13-15.

15 E. Friedrich, *Ursachen, Verlauf und agrarhistorische Bedeutung der Landvolkbewegung in Schleswig-Holstein von 1928-1930* (Leipzig, 1967).

tion with the rise of National Socialism. As her thesis focussed on the literary sources, it did not provide a thorough historical account of the Landvolkbewegung, nor did it follow the movement until its end, closing prematurely with the Reichstag election in 1930. Nevertheless, Le Bars added to the existing scholarship through a survey of the movement's propaganda, her emphasis on the influence of völkisch groups upon it, and importantly her argument that when the movement declined, its base went over to the NSDAP, while its leaders refused to do so.[16]

Nils Werner's more recent legal-historical study, *Die Prozesse gegen die Landvolkbewegung in Schleswig-Holstein 1929-1932,* examining the legal actions taken against the movement, stands out from all other studies on the subject to date for its thorough archival research and considerably closer attention to detail.[17] Werner went beyond the movement's most well-known episodes and analysed some 40 court cases, with particular focus on the "Great Bombers' Trial" (*Großer Bombenlegerprozeß*). He categorised the trials according to the actions from which they originated and also attempted to quantify the volume of legal proceedings. Though prevented from a more precise reckoning by deficiencies in the documentation, he asserted that there were considerably fewer convictions than previously presumed. Further, Werner offered a critical view of the provincial authorities' intervention in the legal process, particularly that of *Regierungspräsident* (Provincial Government President) Abegg, who, Werner argued, used and hid behind the courts for his political ends when a more open approach would have much better promoted his credibility and the state's authority. Finally, he assessed Luetgebrune and Zarnow's accusations of the Prussian authorities' impermissible politically motivated interference in the course of justice, and concluded that their claims were not simply polemical, but partly justified.

Unpublished diploma and *Staatsexamen* theses have provided further overviews or focussed on specific facets of the Landvolkbewegung such as the movement's newspaper.[18] This literature is complemented by numerous essays and articles by local historians, chroniclers and former Landvolk followers as well as spe-

---

16 M. Le Bars, *Le Mouvement Paysan Dans Le Schleswig-Holstein 1928-1932* (Bern, 1986).

17 N. Werner, *Die Prozesse gegen die Landvolkbewegung in Schleswig-Holstein 1929-1932* (Frankfurt, 2001).

18 M. Schäfer, *Die schleswig-holsteinischen Bauernunruhen 1928-1932. Eine Darstellung unter besonderer Berücksichtigung der Bestände der Kieler Bibliotheken* (Hamburg, 1974); G. Coppin, *Landvolkbewegung und Nationalsozialismus im schleswig-holsteinischen Marschgebiet 1928-1932* (Lille, 1975); S. Heim, *Die Landvolkbewegung in Schleswig-Holstein 1928/1929. Eine Analyse ihrer sozioökonomischen Entstehungsbedingungen und politischen Aktionsformen* (Examensarbeit, Hamburg, 1980); F. Stoehr, *Die Landvolkbewegung in Schleswig-Holstein* (Kiel, 1986); A. Walter, *Die Entwicklung der schleswig-holsteinischen Landvolkbewegung im Spiegel ihres Zentralorgans "Das Landvolk" (1928-1932)* (Hamburg, 1995).

cific regional treatment in histories of the rise of National Socialism and the Weimar period.[19]

One work deserving further notice is Timothy A. Tilton's *Nazism, Neo-Nazism and the Peasantry*, which considered the Landvolkbewegung in the wider context of the receptivity to right-wing tendencies in Schleswig-Holstein before and after the Nazi period. Tilton demonstrated how political, economic and social conditions changed in Schleswig-Holstein after the Second World War, making a revival of National Socialism unlikely. In his second chapter, he stressed the Landvolkbewegung's role in politicising economic distress and in spreading anti-Republican and anti-Semitic propaganda which prepared the way for the NSDAP's success.[20] To highlight the deficiencies of simplistic economic determinism, Tilton's analysis also included concise comparisons with Denmark, Norway and Sweden. Crises there did not lead to fascism, not only because earlier progressive reform softened the crises' impact, but also because of greater cooperation between farmers and urban workers, the Social Democrats' greater concern for agricultural affairs, their governments more effective response to the farmers' calls for assistance, as well as the stronger democratic traditions in these countries.[21]

Despite these various studies, there is still a great deal to add to the history of the movement. In fact, the sources allow a far more intimate and extensive history of the Landvolkbewegung. Many questions, neglected or unanswered until now, require closer attention and several others must be reconsidered. According to Uwe

19 These include C. H. Bill, 'Organisation und Finanzierung der Landvolkbewegung in Schleswig-Holstein 1928-1933', *Informationen zur Schleswig-Holsteinischen Zeitgeschichte*, 33/34 (1998), pp. 33-52; U. Danker, '"Von heute ab zahle ich keine Steuern mehr!" Steuerstreik, Stadtboykott und Bomben; die schleswig-holsteinische Landvolkbewegung 1928-1930' in U. Danker (ed.), *Die Jahrhundertstory* (Flensburg, 1998), pp. 48-67; H. W. Gondesen, 'Zur Entstehung der schleswig-holsteinischen Landvolkbewegung', *Jahrbuch für die Schleswigsche Geest*, 34 (1986), pp. 8-32; Gondesen, 'Die Landvolkbewegung 1929', *Jahrbuch für die Schleswigsche Geest*, 35 (1987), pp. 48-73; Gondesen, 'Niedergang der Landvolkbewegung, Aufstieg der NSDAP', *Jahrbuch für die Schleswigsche Geest*, 36 (1988), pp.155-176; Gondesen, 'Ende der Weimarer Republik, Ende des Landvolks', *Jahrbuch für die Schleswigsche Geest*, 37 (1989), pp. 7-27; A. Heggen 'Bauern, Bomben und Bonzen – Hans Fallada und der Landvolkprozess in Neumünster 1929' in A. Heggen (ed.) *Projekt Zeitgeschichte Neumünster* (Neumünster, 2006), pp. 42-66; W. Koops, *Südtondern in der Zeit der Weimarer Republik (1918-1933)* (Neumünster, 1993), pp. 241-258; R. Möller, *Eine Küstenregion im politisch-sozialen Umbruch (1860-1933). Die Folgen der Industrialisierung im Landkreis Steinburg* (Hamburg, 2007), pp. 439-448; C. M. Sörensen, *Politische Entwicklung und Aufstieg der NSDAP in den Kreisen Husum und Eiderstedt 1918-1933* (Neumünster, 1995), pp. 224-243; T. Steensen, *Geschichte Nordfrieslands von 1918 bis in die Gegenwart* (Bredstedt, 2008), pp. 49-53; P. Wulf, 'Die Not hat uns zusammengeschmiedet: die Landvolkbewegung Ende der 20er Jahre' in G. Paul, U. Danker and P. Wulf (eds), *Geschichtsumschlungen Schleswig-Holstein 1848-1948* (Bonn, 1996), pp. 192-199.

20 Tilton, *Nazism, Neo-Nazism and the Peasantry*, pp. 62-70.

21 Tilton, *Nazism, Neo-Nazism and the Peasantry*, pp. 46-51.

Danker, there is no satisfactory contemporary general account of the Landvolkbewegung.[22] Nevertheless, the existing literature provides a foundation on which to build. This applies, in particular, to the exploration in this study of the origins of the movement which draws on the work of Heberle, Stoltenberg and others in respect to agricultural conditions in Schleswig-Holstein and the province's political development.[23] Moreover, Heberle must also be acknowledged for providing the impetus to examine the role the unique histories of the west coast marsh farmers played in shaping those farmers' outlooks and the nature of their protests. Heberle raised this issue, yet barely developed it.[24] More recent advances in scholarship, in conjunction with contemporary data, permit a fuller analysis of the economic background. The movement's decline, on the other hand, is at best sketchy in all previous studies. Thorough investigation is required to uncover its course and to determine the many factors involved.

One of the principal themes in the following chapters is that the phenomenon of the Landvolkbewegung was not solely a result of economic distress, but must also be seen in the context of regional history. Other premises are that the Landvolk turned economic difficulties into political ends by exploiting opposition to taxes and that opposition to the authorities, and particularly to the Republic, became ever more the driving force of the movement. A closer examination of the story of unrest in the countryside from 1928 and of the nature of the farmers' protests from the first calls for government assistance to the planting of bombs, allow a more differentiated view of the phases and changes in the process of radicalisation. The quantity of protests and actions before 1933 also need reconsideration.[25]

One of the main deficiencies of earlier studies is their almost exclusive focus on the Landvolkbewegung's more spectacular actions whilst giving insufficient attention to its meetings. In fact, alongside the movement's newspaper, it was the meetings that kept the farmers together, not to mention their importance as a forum

22 Danker, '"Von heute ab zahle ich keine Steuern mehr!", p. 60

23 The most significant addition to the scholarship is R. Rietzler, "*Kampf in der Nordmark" Das Aufkommen des Nationalsozialismus in Schleswig-Holstein (1919-1928)* (Neumünster, 1982).

24 Heberle mentioned the history of the west coast in accounting for the particular temperaments and "political psychology" of the marsh farmers but did not adequately explain how it helped to shape their later actions beyond suggesting that their thinking of themselves "as a ruling class" exacerbated their anger at police intervention. See Heberle, *Landbevölkerung und Nationalsozialismus*, pp. 53 and 157.

25 For a survey of the quantity and nature of rural protest in the Weimar Republic from 1924 to 1933, differentiating between, and underlining the variations among, Germany's northern, central and eastern, and southern and western regions, see J. Bergmann and K. Megerle, 'Protest und Aufruhr der Landwirtschaft in der Weimarer Republik (1924-1933). Formen und Typen der politischen Agrarbewegung im regionalen Vergleich' in J. Bergmann, J. Brockstedt, R. Fremdling et al., *Regionen im historischen Vergleich. Studien zu Deutschland im 19. und 20. Jahrhundert* (Opladen, 1989), pp. 200-287.

at which farmers gave vent to their grievances. A detailed examination of the meetings is required to understand what it meant to farmers to take part in the movement, and to establish their importance in radicalising the rural population. In addition, the location, frequency and attendance levels at meetings and actions serve as valuable indicators of the relative strength of the movement. Other points of interest concern the organisation of the Landvolk's activities, the extent to which they were spontaneous, how farmers came to participate in them, as well as the manner in which the movement's leaders planned and ran these events to ensure the co-operation of their followers. Moreover, it is necessary to determine whether the Landvolkbewegung's activities were a genuine manifestation of distress or a wilful provocation of the authorities.

The idea that the Landvolkbewegung was a "debtor's rebellion" must also be questioned.[26] A misunderstanding that the targets of the bombings were tax offices has bolstered this view. Similarly, it is an exaggeration to speak of the movement's "sporadic armed resistance" which "nearly led to civil war".[27] The Landvolkbewegung began in response to the agricultural crisis, radicalised when it turned to tax-strikes but then swiftly became focussed on the struggle against the Weimar system. Though taxes remained an important issue, the bombings marked the height of the movement's opposition to the system. The bombings are also significant as a subject of study, not only in respect to their effect on support for the Landvolk, but also in so far as the public responses to them can be read as a reflection of growing opposition to the Republic.

Far too little attention has been paid to the Landvolkbewegung's ideas and, in particular, to its anti-Semitism. As Gerhard Paul pointed out, it has been assumed that during the Weimar Republic anti-Semitic polemic reached a peak in Schleswig-Holstein in the period of rural unrest, but neither its strength nor the role of the Landvolkbewegung in generating anti-Semitic sentiment and expression have yet been determined.[28] Hans Beyer criticised Gerhard Stoltenberg for exaggerating the völkisch-anti-Semitic currents in the countryside.[29] Yet Stoltenberg barely scratched the surface when he noted that the notion of "dark powers" was to be found very often in its vocabulary. Actually, Stoltenberg painted far too moderate a picture of the movement's anti-Semitism by stating, for instance, that Landvolk leader Wilhelm Hamkens was "not a völkisch fanatic" and that he only

26 Heberle, *From Democracy to Nazism*, p. 69.

27 See Anna Bramwell's outline of the Landvolkbewegung in her biography of the National Socialist Minister of Agriculture, R. Walther Darré. A. Bramwell, *Blood and Soil: Richard Walther Darré and Hitler's 'Green Party'* (Bourne End, 1985), p. 82.

28 G. Paul, ' "Die Juden sind unser Unglück!" Zum Antisemitismus in einer frühen Hochburg der NSDAP' in G. Paul and M. Gills-Carlebach (eds), *Menora und Hakenkreuz. Zur Geschichte der Juden in und aus Schleswig-Holstein, Lübeck und Altona (1918-1998)* (Neumünster, 1998), pp. 188-189.

29 Beyer, 'Landbevölkerung und Nationalsozialismus in Schleswig-Holstein', p. 70.

aligned with other right-wing forces for tactical reasons.[30] By contrast, Wilhelm Koops, in his history of the County of South Tondern during the Weimar Republic, noted the extreme prevalence of anti-Semitism in the movement, as did Le Bars, ascertaining with just a sprinkling of examples, that the movement's anti-Semitism was frequently linked with anti-capitalist and anti-democratic sentiment.[31]

However, the most important and difficult questions in respect to the Landvolkbewegung's anti-Semitism and its völkisch thought remain unanswered. By investigating the movement's ideas in its newspaper and at its many meetings and rallies, the kinds of anti-Semitic and völkisch ideas used in the movement can be determined, their function clarified and their extent and intensity gauged. Furthermore, the manner in which anti-Semitism was presented and sometimes even the nature of public response to it can be ascertained. A study of the Landvolkbewegung offers important insight into the circulation of right-wing thought in rural circles, and particularly anti-Semitism, before the NSDAP's rise to power. Absurd and repugnant as these ideas were, it is important to discover their appeal to the farmers and understand how they came to be used and why they were taken seriously. For this reason, an attempt must also be made to trace the origins of these ideas in the province.

Since the Landvolk constituted neither a party nor an association, its organisation, or more accurately, the lack of it, was one of the factors that made it such an extraordinary phenomenon Therefore, it is important to determine on what grounds the farmers avoided conventional organisation and to establish which structures, if any, existed in its stead. This necessitates an examination of the nature of the movement's leadership and of the financing of the farmers' activities and newspaper. A further, albeit secondary, object of interest is the contribution of outside revolutionaries and activists. It would appear that Beyer and others, including Rudolf Heberle, Armin Mohler and Otto-Ernst Schüddekopf, overstated the influence of national revolutionary publicists.[32] Their role within the movement, therefore, must be assessed and the extent to which they were responsible for the movement's radicalisation must also be determined.

---

30 Stoltenberg noted the "fierce anti-Semitic attacks" at only one Landvolk meeting and only provided one specific anti-Semitic example, a paraphrased passage from *Das Landvolk*. Stoltenberg, *Politische Strömungen im schleswig-holsteinischen Landvolk*, pp. 121, 124, 132 and 203.

31 Koops, *Südtondern in der Zeit der Weimarer Republik*, pp. 250-251; Le Bars, *Le Mouvement Paysan*, pp. 75-78.

32 Heberle, *From Democracy to Nazism*, pp. 68-69; O-E. Schüddekopf, *Linke Leute von rechts* (Stuttgart, 1960), pp. 306-312; A. Mohler, *Die konservative Revolution in Deutschland 1918-1932* (Darmstadt, 1972), pp. 130-131, 161-165 and 272-273. Also see the supplement to the third edition, in which Mohler removed the Landvolkbewegung – with little explanation – as a distinct ideological component from his "Conservative Revolution" categorisation: Mohler, *Die konservative Revolution in Deutschland 1918-1932. Ergänzungsband* (Darmstadt, 1989), p. 8.

Another significant theme is the authorities' failure to stop the course of radicalisation in the province. Even though Werner has examined the trials against the Landvolk in depth, little has been written on the authorities' and police responses to the movement. This study investigates their reaction to the crisis and their surveillance and management of the Landvolkbewegung. It offers a critical view of the measures taken by the authorities and assessment of the effectiveness of their responses such as timing, the leniency or heavy handedness in particular cases and their cooperation and coordination in dealing with the farmers.

The impact of the Landvolkbewegung on other groups, including the province's agricultural political organisations, demands particular attention, especially the NSDAP. Nowhere was the rise of National Socialism as early and considerable as in Schleswig-Holstein. It was the first and sole German province to give the NSDAP an outright majority before 1933, 51 percent at the Reichstag elections in July 1932.[33] Early scholars of the rise of National Socialism in the province viewed this success as all the more noteworthy on account of the previous strength of the parties friendly to the Republic. Though the shift in voters' loyalties began earlier, the most critical stage was the period between the Reichstag elections of May 1928 and of September 1930. It was then that the NSDAP made its electoral breakthrough, enjoying, above all, substantial gains in the countryside. There can be no doubt that the Landvolkbewegung's radicalisation of the rural population was one of the major factors facilitating this extraordinary success. The NSDAP's success in Schleswig Holstein was significant not just on a regional level. The NSDAP profited from the advent of the Landvolkbewegung because it demonstrated the potential of support in the countryside. It was only after the NSDAP established itself in rural areas that it became a mass political party.

There is general consensus in the literature that the NSDAP was the chief beneficiary of the Landvolkbewegung and that when the Landvolkbewegung declined, its followers turned to the NSDAP.[34] This view is corroborated here but further insight is provided through a closer investigation of the similarities and differences between the two movements' ideas, methods, organisation and aims. Their interaction, connections and responses to one another, are also scrutinised. Research into the NSDAP membership records in the Berlin Document Center (BDC) tests Le Bars' conclusion that the movement's leaders refused to join the NSDAP. Further, this study enquires into the extent to which the NSDAP took over the Landvolkbewegung.[35]

33 *Statistisches Handbuch für Schleswig-Holstein*, Statistisches Landesamt Schleswig-Holstein (Kiel, 1951), p. 537.

34 In addition to the literature noted above, also see J. E. Farquharson, *The Plough and the Swastika: The NSDAP and Agriculture in Germany 1928-1945* (London and Beverly Hills, 1976), p. 8.

35 Bergmann and Megerle wrote of the "Nazification" of the Landvolkbewegung. Bergmann and Megerle, 'Protest und Aufruhr der Landwirtschaft', p. 245.

Tim Mason once argued that Nazi success in seizing power in 1933 was made possible in large part by the disintegration of "civil society".[36] Put simply, Mason attributed this to the pressures of the economic crisis which limited room for social and political compromises and accentuated the struggles between classes and among interest groups. Perhaps the history of the Landvolkbewegung might be viewed as a case study in the breakdown of "civil society" in the province of Schleswig-Holstein. Studying the Landvolkbewegung and its radicalisation of the rural population suggests that it is not nearly as "incredible" as Heberle reckoned that such large masses of generally sober-minded and freedom-loving North Germans should have trusted Hitler.[37] Rather, the examination of the Landvolkbewegung in light of the farmers' disappointment in and growing hostility towards the Weimar Republic helps to make the NSDAP's rise to power much more explicable and comprehensible.

There are boundaries to this study. The first is geographic. The focus, for the most part, is on the movement's activities in Schleswig-Holstein, where the movement had its centre. Though the Landvolkbewegung spread beyond the province's borders, a detailed examination of the movement, for example in Hanover and East Prussia, has not been attempted. The existing literature has encouraged the setting of further limitations. This study is not as interested in the fiction surrounding the movement as Le Bars, nor nearly as concerned with the legal process in respect of the trials as Werner. While it considers election results, it is not a history of elections, political parties and organisations in Schleswig-Holstein and does not seek to surpass Heberle and Stoltenberg's analyses in this respect. Further, it does not consider the intellectual role of the Landvolkbewegung within the "conservative revolution". A detailed examination of the impact of the movement on the writings of individual right-wing intellectuals is beyond the scope of study. Not only would this merit further analysis, so too would a comparison with other agrarian movements of the 1920s and early 1930s such as the French Greenshirts, the Lapua Movement in Finland and Social Credit in Canada.

This study is based largely on the extensive files kept by Schleswig-Holstein's *Oberpräsident* (Governor) and Regierungspräsident on the movement at the initiative of the Prussian Interior Ministry and which are now to be found in Schleswig-Holstein's State Archives in Schleswig.[38] They contain thousands of documents such as correspondence between the Regierungspräsident, Oberpräsident, Prussian Interior and other ministries including reports and their drafts, memoranda and directives; reports from and directives to the province's county administrators (*Landräte*) and community chairmen (*Gemeindevorsteher*); correspondence between the

36 T. W. Mason 'The Primacy of Politics – Politics and Economics in National Socialist Germany' in H. A. Turner (ed.) *Nazism and the Third Reich* (New York, 1972), pp. 175-200.

37 Heberle, *From Democracy to Nazism*, p. 6.

38 Preuß. Minister des Innern to the Regierungspräsident in Schleswig, Berlin, 27.2.1928, LAS 301/4695.

provincial authorities and the public prosecutors; court transcripts, indictments and issues of verdicts and rulings; as well as reports and meeting transcripts from the State Criminal Police departments (*Landeskriminalpolizeistellen)* in Kiel, Flensburg, Altona and Berlin, and from the town and communal police departments and rural constabulary (*Landjägerei*). Further records include copies of the Landvolk's resolutions, notices, leaflets and posters, letters from the farmers and their organisations to the authorities, numerous copies and lists of confiscated correspondence, copies of witness statements, some extracts from and copies of booklets and manuscripts written by Landvolk followers, as well as hundreds of newspapers cuttings from across Germany. Additional indictments, issues of verdicts and correspondence, particularly from the public prosecutors, are found in the files of the Prussian Ministry of Justice preserved at the Geheimes Staatsarchiv Preußischer Kulturbesitz in Berlin. Hundreds of other reports on the movement are also to be found in newspapers held at various libraries in Schleswig-Holstein, principally the State Library in Kiel, including the movement's own newspapers, *Das Landvolk* and the later *Dusendüwelswarf.* Further input comes in the form of memoirs and autobiographic accounts, a small number of letters located at the Town Archive in Neumünster and transcripts of two interviews conducted by the Zeitzeichen Archiv in Kiel.

Though plentiful and wide-ranging, the sources are nevertheless subject to particular problems and limitations. The authorities' records reflect the particular attention paid to certain aspects of the movement rather than others, as the authorities were often looking for an occasion to take action against the Landvolkbewegung, either by way of bans or legal proceedings. Moreover, as their reports were written with their superiors in mind, they require especially careful reading. This is not to suggest that the police, spies and the provincial authorities were inventive. However, one must differentiate carefully between observations based on intelligence, and the conclusions drawn and predictions made from them, as these were sometimes shaped to satisfy particular hopes and expectations. Reports on the farmers' meetings often include near-verbatim records of their speeches and are almost always complemented by one or several newspaper reports, thereby facilitating reliable reconstructions and thorough analysis. Less evidence is available, on the other hand, for other aspects of the movement, such as the planning and organisation of actions at compulsory auctions despite the authorities' best efforts to penetrate the farmers' secret undertakings. The Landvolkbewegung itself had no central organisation, membership lists or records. It only had its newspaper and other printed matter, which served propaganda purposes. One can turn to memoirs and court testimony, but these too must be treated carefully, for the farmers often had cause either to exaggerate or, more often, play down their roles.[39] Finally, it should be

39 For an example of overstatement, Herbert Volck grossly exaggerated his own importance in his autobiographical account, *Rebellen um Ehre*, claiming to be one of "two or three" men

noted that much of the provincial press was not as free to report as its journalists may have wished. Most newspapers were either tied to particular parties and interest groups, or generally had such small readerships that they were under considerable pressure to please subscribers, advertisers and the authorities in order to survive.[40]

The structure of this study is essentially chronological, yet each chapter raises particular questions about the movement. The first chapter provides the necessary background to understand the Landvolkbewegung's origins and character. It introduces the region, the nature and conditions of its agriculture and the peculiarities of its history, provides a survey of political developments and agricultural representation from the later imperial period until the mid-1920s and then examines the economic background from the First World War to the agricultural crisis. Chapter Two examines the mobilisation of the rural population in response to the crisis, the farmers' first rallies, their calls for assistance and the responses to them as well as the subsequent destabilisation of their organisations. Chapter Three investigates the rise of the Landvolkbewegung in the winter of 1928/1929. Of particular focus are the new movement's first boisterous meetings and obstructive actions. Chapter Four explores the consolidation and continuing radicalisation of the movement in the first months of 1929, with particular attention paid to the foundation of its newspaper and the first trials against the farmers. Chapter Five surveys the meetings held from the spring of 1929, the endeavours to institutionalise tax obstruction and assume authority by the formation of committees throughout the province, as well as the movement's increasing militancy in the form of pranks and violent actions. Chapter Six examines the Landvolkbewegung's clashes with police in the summer of 1929 and the subsequent boycott of the town of Neumünster. Concentrating on the movement's turn to terrorist tactics, Chapter Seven returns to the close of 1928, to the beginning of the bombing campaign, traces its planning and execution, the bombers' arrests and the various responses to these events up to the ensuing trial in 1930. Finally, Chapter Eight examines the movement's struggle for survival from 1930 and seeks to explain the factors behind its decline.

As the development and course of the movement and radicalisation in the countryside are of central concern, they have determined the shape and structure of the narrative, while other important themes are discussed within this context. An-

---

behind the movement, and only came to mention Wilhelm Hamkens (Tetenbüll) in a later chapter. See Volck, *Rebellen um Ehre,* pp. 308 and 342.

40 Hans Fallada offered an interesting insight on the particular pressure on the reporters of the provincial press in a letter to his sister, Elisabeth, in which he told of his fear of losing his job because he had upset "some important people" due to his reporting on a Landvolk trial in Neumünster. Letter from Hans Fallada (Rudolf Ditzen) to "Ibeth" Neumünster, 20.11.1929, StaNms MA 5360 (I). Further insight on small newspapers' struggle for survival is to be found in his novel, *A Small Circus*, in a conversation in which Stuff explains to Wenk why he can publish neither articles supporting nor criticising the farmers. Fallada, *Bauern, Bomben und Bonzen*, p. 27.

alysis of the movement's ideology, or its refusal to adopt customary organisational structure, for example, are taken up in varying levels of detail at several relevant points throughout the study. Discussion of agricultural policy and its effects are largely concentrated at the point where they were of greatest relevance to the movement's development, that is, in Chapter Two, which concerns the farmers' calls for crisis relief. The responses of the authorities, the agricultural organisations and other groups, in particular the NSDAP, their relationships with the Landvolk-bewegung and their effects on one another are the subjects of recurrent attention.

Illustration 1: Map of Schleswig-Holstein, 1929.

Chapter One

Schleswig-Holstein, regional politics and the agricultural crisis

Schleswig-Holstein's Landvolkbewegung developed largely from the extraordinary rural discontent arising from the agricultural crisis that preceded the Great Depression. But the movement was also shaped by the particular characteristics of agriculture in the province and by longer-term social and political historical structures and processes. Attention must therefore be given to the regional and wider historical background before the onset of the crisis, principally, to the farmers' consciousness of their own unique regional histories, developments from the imperial period to the unsettled early years of the Republic and the parties and organisations which represented rural interests.

Aspects of this background to the farmers' protests have been noted by Rudolf Heberle, Gerhard Stoltenberg and Rudolf Rietzler.[1] Theirs are the most thorough and reliable studies available which deal with Schleswig-Holstein in the Weimar period. Each offers valuable insights, however they certainly do not elucidate the background to the farmers' movement comprehensively or satisfactorily. Their aim is to explain the rise and success of National Socialism in the province, not the origins of the Landvolkbewegung. Heberle's ecological-sociological analysis demonstrates the relationship between the various modes and structures of agriculture in the region and support for Nazism. Stoltenberg's chronological narrative focuses on the political and organisational developments in the countryside that led to the rise of the NSDAP. Rietzler's object of enquiry is the history of the party in the province until 1928, considering political, organisational and ideological aspects. This study has found inspiration in their scholarship and builds on the foundation of the respective strengths of their work. However, since the focus of this chapter is the analysis of the Landvolkbewegung's origins and the factors that contributed to the character of its protests, greater priority is given to the examination of the unique history of the province's west coast and to the economic factors that led to unrest. Only Heberle examined the agricultural crisis in any detail but his interest was the correlation of economic factors with the NSDAP's electoral success, while Stoltenberg offered little analysis of the crisis even though it was the crucial starting point of rural protest.[2] The necessary, much closer analysis of the agricultural crisis here draws on agricultural and rural histories, wider Weimar literature as well as reports

---

1 Heberle, *From Democracy to Nazism* and Heberle, *Landbevölkerung und Nationalsozialismus*; Stoltenberg, *Politische Strömungen im schleswig-holsteinischen Landvolk*; Rietzler, *"Kampf in der Nordmark"*.

2 Heberle, *Landbevölkerung und Nationalsozialismus*, pp. 118-133; Stoltenberg, *Politische Strömungen im schleswig-holsteinischen Landvolk*, pp. 95 and fn. 3, 107.

of contemporary agricultural experts and Schleswig-Holstein's Chamber of Agriculture. The first part of this chapter introduces the region and concentrates on the ecological, socio-historical and political and organisational background that explain the form and characteristics of the Landvolkbewegung. The second part focuses on the economic factors that contributed to the farmers' discontent and the analysis of the agricultural crisis.

Schleswig-Holstein, between the Baltic and North Seas and bordered by Denmark in the north and Hamburg and the Elbe river in the south, was divided by three distinct geological zones running roughly in strips from north to south. The differences between these zones were more marked in Schleswig-Holstein than in most other areas of northern Germany.[3] In the west was the marsh, in the middle the "Geest" and in the east, the hill zone. Each zone differed in fertility and lent itself to different types of farming.

The loam soil of the hills along the Baltic Sea was well suited to crops, principally cereals. There, noble landlords on large estates had long dominated the counties of Eckernförde, Plön and Oldenburg. During the Weimar period, the owners of these estates generally employed farm labourers to work their land or leased what they could not work themselves to small tenant farmers. The region of Angeln in the north was an exception. There, relatively prosperous farmers on farms of various sizes, engaged in diversified agriculture: dairy, livestock-fattening and crops.

Running through the middle of Schleswig-Holstein, the Geest was composed of sandy, heath-covered soil and moor. Small villages and family farms dotted the landscape. The Geest peasants had long been freeholders. Farmers and their farmhands often lived together. Although the introduction of artificial fertilisers in the latter half of the nineteenth century improved productivity, the Geest peasants remained relatively poor. In hard times they toiled merely for their own subsistence. Rye, oats and vegetables were grown on a small-scale. Pig farming was important in the southern Geest counties and there was some dairy closer to the larger towns. Of particular significance was the breeding and raising of cattle, especially for sale to marsh farmers. Rudolf Heberle argued convincingly that little social stratification or class antagonism, strong community ties and village solidarity as well as a

3 The preceding analysis of the province's geographical conditions and agriculture follows the *Statistisches Handbuch für Schleswig-Holstein,* Statistisches Landesamt Schleswig-Holstein (Kiel, 1951), pp. 5-8 and particularly for the modes of farming in Schleswig-Holstein, Heberle, *From Democracy to Nazism*, esp. pp. 34-39 and 101-106 and E. Danker, *Die Verschuldung der schleswig-holsteinischen Landwirtschaft in ihrer regionalen Bedingtheit unter besonderer Berücksichtigung der bäuerlichen Bezirke* (PhD diss, Kiel, 1931), esp. pp. 28-36. For a rudimentary, yet extensive, guide to land-use and the agricultural practices of the period, including worldwide comparisons, see F. Aereboe (ed.), *Wirtschaftslehre des Landbaues* (Berlin, 1930). An additional sketch of land-use in Schleswig-Holstein is to be found in R. E. Dickinson, *The Regions of Germany* (London, 1945), pp. 141-142.

lack of a tradition of political allegiance and participation explain why, of all three geological zones, support for the NSDAP was strongest on the Geest.[4]

Stretching along the North Sea coast was the marsh, the geographical home of the Landvolkbewegung. The marsh plains constituted approximately ten percent of Schleswig-Holstein's total area. The west coast counties – South Tondern, Husum, Eiderstedt, North and South Dithmarschen, Steinburg and Pinneberg – comprised the marsh and parts of the Geest. There was much cultivation of wheat and cabbage, especially in Dithmarschen and further to the south sugar beet, but above all, the rich marsh soil provided pasture ideal for livestock grazing. On smaller farms some dairy was to be found, toward the north and especially in Eiderstedt, cattle grazing dominated, and in the south, particularly in Steinburg and Pinneberg but also in parts of South Dithmarschen, pig-fattening was the typical mode of agriculture.

Schleswig-Holstein's west coast has traditionally been a commercial agricultural area. Progressive and receptive to innovation, the region had already become home to a commercial livestock industry by the sixteenth century. Moreover, the close proximity to the north German and Dutch cities made intensive commercial dairy production possible.[5] The farmers were compelled to specialise in grazing because much of their land was simply not arable, but as their imposing old farmhouses testify, they enjoyed long periods of prosperity.[6]

The second half of the nineteenth century saw a revolution in agriculture. Innovations in machinery, fertilisers, breeding techniques and means of transport led to increased milk and meat production, drastically reduced labour costs, opened previously unsuitable land to pasture, significantly increased fodder supplies and opened new beef markets, first to England and then within Germany itself.[7] Until the Great War, increased production, rising domestic meat and dairy consumption and expanding markets encouraged farmers to specialise ever more in commercial livestock farming. At the same time high transport costs, from the United States for

---

4 Heberle, *From Democracy to Nazism.*

5 R. Hopcroft, *Regions, Institutions, and Agrarian Change in European History* (Michigan, 1999), pp. 157-195; W. Abel, *Geschichte der deutschen Landwirtschaft vom frühen Mittelalter bis zum 19. Jahrhundert* (Stuttgart, 1978), p. 182.

6 A. Mayhew, *Rural Settlement and Farming in Germany* (London, 1973), p. 171; L. Fischer, *Haubarge: Eine Bauernhausform hat abgewirtschaftet?* (Bredstedt, 1987), pp. 6 and 32.

7 F-W. Henning, *Landwirtschaft und ländliche Gesellschaft in Deutschland: Band 2: 1750-1976* (Paderborn, 1978), p. 131; K-J. Lorenzen-Schmidt, 'Neuorientierung auf den deutschen Wirtschaftsraum – Wirtschaftliche Entwicklung 1864-1918' in U. Lange (ed.), *Geschichte Schleswig-Holsteins* (Neumünster, 1996), pp. 389-390; N. Nissen, *Landwirtschaft im Wandel* (Heide, 1989), pp. 19-58 passim; J. Brockstedt, 'Konjunkturen in der Wirtschaftsentwicklung Schleswig-Holsteins 1864-1913, mit einem Überblick der Periode 1773-1864' in J. Brockstedt (ed.), *Wirtschaftliche Wechsellagen in Schleswig-Holstein vom Mittelalter bis zur Gegenwart* (Neumünster, 1991), p. 247.

instance, hindered oversupply from abroad, so that the province's farmers continued to be rewarded with high levels of income.[8]

Marsh farmers administered and generally worked their own farms. Many of these were large properties and although the farmers employed farm labourers, these were only in small numbers and usually on a seasonal or short-term, contractual basis.[9] Socially, the marsh was highly stratified. In Dithmarschen and Eiderstedt, especially, farmers and their families lived on large farmsteads in the middle of their properties while the labourers lived separately in villages or small settlements. Further, pasture requires relatively little labour and the largest farms in particular afforded their owners a comfortable, gentrified existence.[10] Livestock-grazing was not only profitable but also by its nature allowed long periods of free-time, giving the marsh farmers the freedom and independence to pursue other interests, such as politics.

Schleswig-Holstein's livestock farmers long enjoyed prosperity but specialising in a single product, as they so often did on the west coast, made them particularly susceptible to crises. They had no compensation if their crop failed, pestilence afflicted their herds or prices for their product fell. Furthermore, marsh farming was capital-intensive, highly speculative and exposed to debt: in the spring, the marsh farmers invested heavily to buy young cattle. They grazed the animals on their rich pastures during summer and then sold the fattened animals in autumn.[11] Keeping the animals over winter was feasible only if sufficient stables and fodder supplies were available.[12] The marsh farmers were therefore very much subject to the power of the market.

Perishable produce, lack of diversity and substantial capital investments meant that Schleswig-Holstein's west coast farmers were to feel acutely the agricultural crisis which struck Germany in the late 1920s. It was to cause the farmers all the more distress because of their long history of prosperity. This helps to explain their particularly extreme response to the crisis.

Of course farmers elsewhere were vulnerable too. On the Geest, however, there were not such large capital outlays as there were on the marsh, while in Schleswig-

---

8 Henning, *Landwirtschaft und ländliche Gesellschaft*, pp. 114, 117-120 and 156-157; Brockstedt, 'Konjunkturen in der Wirtschaftsentwicklung', pp. 246-248 and 256-258.

9 Schleswig-Holstein had a higher proportion of large farms (20-100 ha) compared with any other significant rural area in Germany. 30.14 percent were between 20 and 100 ha. compared with 11.59 percent in Prussia and 9.66 percent in Germany overall. *Statistisches Jahrbuch für das Deutsche Reich*, Statistisches Reichsamt (Berlin, 1931), pp. 50-51.

10 Aereboe, *Wirtschaftslehre des Landbaues*, pp. 196-203; Heberle, *Landbevölkerung und Nationalsozialismus.*, pp. 14-15, 49-53 and 115.

11 M. Sering, *Die deutsche Landwirtschaft unter volks- und weltwirtschaftlichen Gesichtspunkten* (Berlin, 1932), pp. 481 ff. See also Heberle, *From Democracy to Nazism*, pp. 101-109; Danker, *Die Verschuldung der schleswig-holsteinischen Landwirtschaft*, pp. 29-36.

12 For the difficulties of feeding cattle throughout winter, see F. Aereboe, *Allgemeine landwirtschaftliche Betriebslehre* (Berlin, 1918), pp. 347-349.

Holstein's east, the estate owners weathered crises better not only on account of the large scale of their farms but also because the comparative durability of cereals meant that they were not compelled to sell in the event of short-term price falls. In Angeln farmers endured better because of the greater diversity of their products.

The nature of west coast agriculture is crucial in accounting for the farmers' response to the crisis but it is also important to consider the unique histories of independence and traditions of self-administration on the west coast. Rudolf Heberle noted the history of the west coast in accounting for the particular temperaments and "political psychology" of the marsh farmers.[13] Yet their histories help not only to account for their particular rebelliousness but also to explain the shape of their later actions since the farmers' traditions of independence and self-administration were well preserved in their collective consciousness.

Since settlement the marsh farmers had worked cooperatively to protect their farms with dikes and gain land from the sea.[14] They were never subjected to serfdom. In particular, the histories of Dithmarschen and Eiderstedt deserve special mention because there the farmers also retained control of local affairs. In Dithmarschen, the natural barriers of the marshland and the canals and dikes facilitated centuries of successful resistance against the Wends in the eleventh century, against the Counts of Stade and Holstein, and most famously against the combined forces of the Danish king and Gottorf Duke at a battle near Hemmingstedt in 1500.[15] Effectively free since the thirteenth century, Dithmarschen's farmer republic was finally defeated in 1559, but even then still managed to retain significant freedoms.[16]

Eiderstedt, the peninsula to the north, was never a free state but rather a dominion of the dukes or the Danish crown. It used its considerable economic strength to preserve local administration until Prussian annexation in 1867. From the fourteenth century, the wealthiest farmers, who constituted the assembly of this farmers' patriarchy, ensured that locals supervised the justice system, oversaw the collection of the ruler's taxes and administered the dikes.[17] In fact, even after Prussian annexation, which brought considerable changes to taxation, administration and the legal system, certain regional peculiarities survived, both in Dithmarschen and

---

13 Heberle, *Landbevölkerung und Nationalsozialismus*, p. 53.

14 A. Dopsch, 'Agrarian Institutions of the Germanic Kingdoms from the Fifth to the Ninth Century', p. 191; M. Bloch, 'The Rise of Dependent Cultivation and Seigniorial Institutions' in M. Postan (ed.), *The Cambridge Economic History of Europe, vol.1: The Agrarian Life of the Middle Ages* (Cambridge, 1966), p. 282.

15 W. L. Urban, *Dithmarschen: A Medieval Peasant Republic* (New York, 1991). For an account of the battle near Hemmingstedt, see W. Lammers, *Die Schlacht bei Hemmingstedt. Freies Bauerntum und Fürstenmacht im Nordseeraum* (Heide, 1982), esp. pp. 171-184.

16 A. Scharff, *Geschichte Schleswig-Holsteins* (Freiburg, 1991), p. 48.

17 R. Kuschert, *Landesherrschaft und Selbstverwaltung in der Landschaft Eiderstedt unter den Gottorfern (1544 bis 1713)* (Neumünster, 1981); V. Pauls, *Landesherrschaft und Selbstverwaltung in Eiderstedt* (Garding, 1932).

Eiderstedt, at least temporarily, most notably Eiderstedt's unique civil law until 1900.[18]

The farmers were deeply conscious of their histories. If the Dithmarscher, especially, have a reputation for pronounced wilfulness and obstinacy,[19] this is the product of their strong consciousness of the past. Conveniently ignoring their defeat in 1559, they remembered most fervently their earlier freedom and their victory in 1500 in art, literature and song.[20] In the 1920s, Dithmarschen's farmers would have been familiar with Adolf Bartels' epic, *Die Dithmarscher*, as well as constantly reminded of their history when passing the protuberant obelisk, the Dusenddüwelswarf, erected at the supposed site of the battle on its 400th anniversary in 1900.[21] This keen awareness of their traditions of managing their own affairs and of the struggles to preserve their independence cannot be ignored in accounting for the particular shape of the farmers' protests toward the close of the Weimar Republic. These histories gave the Dithmarscher and Eiderstedter a considerable measure of self-respect and self-assurance as well as a mistrust of strangers and unwillingness to respect the dictates of authorities from far away, that later encouraged their defiance.[22]

Of further significance in shaping the character of the farmers' protests in the late 1920s was German nationalism. Nationalism had long been a potent force in Schleswig-Holstein, surfacing in the uprising against the Danes in 1848-1849.[23] Although the majority of Schleswig-Holsteiners had desired an independent Schleswig-Holstein within the German Federation, after annexation the victorious Franco-Prussian War of 1870-1871 and unification promoted a Prussian-German

18 For an overview of the changes see O. Brandt, *Geschichte Schleswig-Holsteins* (Kiel, 1981), pp. 284-287 and Scharff, *Geschichte Schleswig-Holsteins*, pp. 81-83. Also see A. Bantelmann, A. Panten, R. Kuschert and T. Steensen, *Geschichte Nordfrieslands* (Heide, 1995), p. 253.

19 For descriptions of the Dithmarscher as "pugnacious", "individualistic" and "obstinate", see Urban, *Dithmarschen: A Medieval Peasant Republic*, pp. 99 and 142; Tilton, *Nazism, Neo-Nazism and the Peasantry*, p. 20; Stoltenberg, *Politische Strömungen im schleswig-holsteinischen Landvolk*, p. 10; Heberle, *Landbevölkerung und Nationalsozialismus*, p. 53.

20 H. Haasis, *Spuren der Besiegten* (Reinbek, 1984), p. 308 and for an overview of the art, literature and song on the subject, including works by Friedrich Hebbel, Theodor Fontane and the low German poet Klaus Groth, see F. Trende, *Die Schlacht bei Hemmingstedt: Ein deutscher Mythos zwischen Politik, Poesie und Propaganda* (Heide, 2000).

21 A. Bartels, *Die Dithmarscher: historischer Roman in vier Büchern* (Hamburg, 1927). Highlighting the persisting fascination with the battle of Hemmingstedt during the Weimar period, in 1925 the teacher and later Landvolk leader in Plön, Peter Petersen, although not a Dithmarscher himself, wrote a kitsch, völkisch novel on the subject. See P. Petersen, *Telse Kampen: Aus Dithmarschens großer Zeit* (Neumünster, 1925). For more on the "Dusenddüwelswarf" monument, see Trende, *Die Schlacht bei Hemmingstedt*, pp. 82-100.

22 E. v. Salomon, *Deutschland deine Schleswig-Holsteiner* (Hamburg, 1975), p. 90.

23 For an account of the growth of German nationalism in Schleswig-Holstein and the conflict with Denmark, see W. Carr, *Schleswig-Holstein 1815-1848* (Manchester, 1963).

national consciousness.[24] Though a dislike for Prussia's centralised administrative system remained below the surface, strong and sustained economic growth abetted the province's integration as did the symbolically important marriage of future Emperor Wilhelm to Auguste Viktoria, whose Augustenburg family had stood for an independent Schleswig-Holstein.[25] Moreover, nationalism gained further impetus from Prussian efforts to Germanise the border regions and also from Kiel's status as a naval base which gave the province a sense of national importance.[26]

Like they did throughout Germany, both regional patriotism and nationalism flourished under Prussian rule.[27] Toward the turn of the century, just as nationalism was asserting itself, interest in regional history and culture – the *Heimat* movement – blossomed. Regional poetry and painting prospered, several regional museums opened, regional languages, such as Friesian, found their way into the columns of local newspapers and efforts were made to maintain historically significant buildings and to preserve the region's landscape (*Heimatschutz*).[28] Regional patriotism continued to thrive during the Weimar Republic. In the countryside, most notably the *Wandervogel*-like Young Farmers' Movement (*Jungbauernbewegung*) celebrated local culture with songs, games and dances.[29]

The Danish minority, of course, was an exception, but for German Schleswig-Holsteiners regional patriotism had long been more than merely compatible with German nationalism: it championed it. Popular since 1844, Schleswig-Holstein's hymn proclaims, "Schleswig-Holstein, embraced by the seas, protector of German custom!"[30] From the 1890s prominent writers from the province, foremost Julius Langbehn and Adolf Bartels, propagated this view too. Also notable for their racial understanding of *Volk* and romantic idealised depictions of rural life, they distinguished Schleswig-Holstein as the bulwark of German renewal.[31]

---

24 G. Reinhardt, *Preußen im Spiegel der öffentlichen Meinung Schleswig-Holsteins 1866-1870* (Neumünster, 1954), pp. 82-86 and 129; Brandt, *Geschichte Schleswig-Holsteins*, pp. 290-291; O. Hauser, *Preußische Staatsräson und nationaler Gedanke* (Neumünster, 1960), p. 30; Scharff, *Geschichte Schleswig-Holsteins*, pp. 78 and 85. For Dithmarschen, see E. Opitz, 'Dithmarschen 1773-1867' in M. Gietzelt (ed.), *Geschichte Dithmarschens* (Heide, 2000), p. 254.

25 E. Topf, *Die Grüne Front* (Berlin, 1933), p. 15; Rietzler, "*Kampf in der Nordmark*", p. 47.

26 Rietzler, "*Kampf in der Nordmark*", p. 44.

27 Celia Applegate emphasised the conjunction of national and provincial identity in her study of the Palatinate. See C. Applegate, *A Nation of Provincials: The German Idea of Heimat* (Berkeley, Los Angeles and Oxford, 1990).

28 Bantelmann, Panten, Kuschert and Steensen, *Geschichte Nordfrieslands*, pp. 298-301; Brandt, *Geschichte Schleswig-Holsteins*, p. 306; Hansen, 'Aufbruch in eine neue Zeit: Dithmarschen 1864-1918', p. 292.

29 Heberle, *From Democracy to Nazism*, pp. 54-56.

30 "Schleswig-Holstein meerumschlungen, deutscher Sitte hohe Wacht!"

31 Rietzler, "*Kampf in der Nordmark*", pp. 36-41. Karl Dietrich Bracher also highlighted Langbehn's wide influence in the development of the idea of race as the decisive factor in history and of Germanic Low-German superiority. See K. D. Bracher, *The German Dicta-*

Alongside German nationalism and local patriotism, earlier political developments in Schleswig-Holstein greatly influenced the nature of the farmers' later protests. Not only did they largely determine the direction the farmers would take, but the character of the parties and organisations which represented farming interests was also of considerable subsequent significance in shaping the Landvolkbewegung's form and ideology.

Having said that, the end of the imperial period witnessed increasing support for the oppositional liberals and Social Democrats in Schleswig-Holstein, a reflection of growing demands for greater civil rights and broader political participation.[32] Though it polled more votes than any other party in the province after 1890, support for the Social Democratic Party of Germany (SPD) was concentrated in the cities, while the Progressive People's Party managed to win more mandates because of tactical alliances in run-off elections, joining with the National Liberals and Conservatives in rural areas.[33] The Catholic Centre Party (*Zentrum*) was and remained insignificant in the almost entirely Lutheran province, leaving greater scope for the National Liberal and Progressive Parties.[34] They gained many votes in rural areas, including the west coast, although some wealthier marsh farmers favoured the Free Conservative Party.[35] Together with the German Conservative Party, the electoral strength of the Conservatives was small, generally around ten percent of the vote. With few candidates even running on the west coast, their support came mostly from eastern Holstein, but also from the north where border tensions encouraged nationalist opinion.[36]

The apparent strength of the SPD and liberal parties before the First World War led Rudolf Heberle to interpret the rise of the NSDAP in Schleswig-Holstein as astonishing and a contradiction of political tradition.[37] Though less emphatic, Ger-

---

torship: The Origins, Structure and Consequences of National Socialism (London, 1970), p. 50.

32 Heberle, *From Democracy to Nazism*, pp. 25 and 27; Stoltenberg, *Politische Strömungen im schleswig-holsteinischen Landvolk*, p. 7.

33 Stoltenberg, *Politische Strömungen im schleswig-holsteinischen Landvolk*, pp. 15 and 18; Heberle, *From Democracy to Nazism*, pp. 26 and 30; Rietzler, "*Kampf in der Nordmark*", pp. 46-47.

34 According to the census of June 16, 1925, 93.2 percent of the total and 96.6 percent of the rural population in Schleswig-Holstein was Lutheran. See the *Gemeindelexikon für den Freistaat Preußen*, vol. 9: Provinz Schleswig-Holstein, Preußisches Statistisches Landesamt (Berlin, 1930), p. 18.

35 Heberle, *From Democracy to Nazism*, p. 102 and *Landbevölkerung und Nationalsozialismus*, p. 54; Stoltenberg, *Politische Strömungen im schleswig-holsteinischen Landvolk*, pp. 14 and 17; Rietzler, "*Kampf in der Nordmark*", pp. 50-51.

36 Rietzler, "*Kampf in der Nordmark*", pp. 53-54; Stoltenberg, *Politische Strömungen im schleswig-holsteinischen Landvolk*, p. 17.

37 Heberle, *From Democracy to Nazism*, pp. 32 and 119. Heberle also referred to Schleswig-Holstein as a "stronghold of liberalism" before the war. See Heberle *Landbevölkerung und Nationalsozialismus*, p. 20.

hard Stoltenberg also pointed to the high percentage of left-wing oppositional votes in the province and wrote of the rural population's "political reorientation" (*Neuorientierung*) thereafter.[38] Objecting, however, to the resultant perception of Schleswig-Holstein as a stronghold of liberalism before the war and the following swing to the right as an historical discontinuity, Rudolf Rietzler criticised Heberle and Stoltenberg for overstating liberalism's strength and for not revealing its true nature in the province.[39] Stoltenberg and to a lesser extent Heberle did in fact offer various qualifications. Yet Rietzler gave greater weight to various aspects of the province's liberalism which later promoted the growth of the radical right.

Liberalism in the province was intensely nationalistic, even among the Progressives. This was because it had arisen alongside the awakening of a national identity with the uprising against Denmark in 1848, which had united the population by a common political goal. Indeed, national sentiment was so strong that the liberal parties refused to cooperate with the Danish minority throughout the imperial period.[40]

The Progressives, in particular, opposed the Conservatives, the remnants of feudalism, centralism and Prussian bureaucracy. Opposition on this side of the political spectrum, however, also extended to industrial cartels and department stores which put pressure on smaller middle-class businesses. These were standpoints consistent with later anti-capitalist perspectives. Moreover, on the other side of the political spectrum, the liberal parties opposed socialism.[41] The fact that the liberal parties allied with the Conservatives against SPD candidates at run-off elections, which inflated the Progressives' strength in terms of the number of mandates they won, reflected broadly-held misgivings about socialism in the countryside.

Of further significance to later developments, liberalism in Schleswig-Holstein's countryside was closely tied to the successful promotion of particular economic interests. Livestock farmers tended to support the liberal parties because they advocated free trade, including the import of cheap feed. Tariff policies protecting conservative grain-producing estate owners in the province's east drove up

38 Stoltenberg, *Politische Strömungen im schleswig-holsteinischen Landvolk*, pp. 5-8, 87 and 195. Writing in East Germany, Ehrenfried Friedrich – with unintentional irony – attributed Stoltenberg's perceptions to political motivations, dismissing his thesis as one of many West German historical studies which invented their own kind of liberal-democratic traditions in order to serve their imperialist politics. Friedrich, *Ursachen, Verlauf und agrarhistorische Bedeutung der Landvolkbewegung*, pp. 164-165.

39 Rietzler, "*Kampf in der Nordmark*", p. 48. Also see Tilton's reservations to Heberle's interpretation of election results in respect of the distortion of voting strength in run off elections. See Tilton, *Nazism, Neo-Nazism and the Peasantry*, pp. 9-11.

40 Rietzler, "*Kampf in der Nordmark*", pp. 48-51; Stoltenberg, *Politische Strömungen im schleswig-holsteinischen Landvolk*, p. 13.

41 Heberle, *From Democracy to Nazism*, p. 29; Rietzler, "*Kampf in der Nordmark*", pp. 48-49.

feed prices and hindered livestock exports.[42] Support for the liberal parties in the countryside, however, was likely to weaken if liberal policies no longer accorded with livestock farmers' interests.

Further, the middle-class parties lacked local organisation. In fact, with the exception of the SPD, no party in the province boasted a notable membership and their administrative bodies in both town and countryside were negligible. Only after the SPD's electoral success in 1903 did the Progressives even employ a full-time secretary in the province.[43] Yet afterwards, the absence of farmers among the Progressive candidates in the Reich elections of 1912 marked the continuing dearth of active rural participation in liberal politics.[44] Schleswig-Holstein lacked political structures upon which voter loyalties could be cemented and this did not bode well for a stable political future.

The initial enthusiasm for the First World War, which promoted national unity and later served as an ideal for the extreme right's vision of a *Volksgemeinschaft*, a "national community", gave way as the war continued and social and class tensions resurfaced. However one may evaluate the various underlying causes, it was the sailors' mutiny in Kiel which triggered the unrest and founding of soldiers and workers' councils that sealed the collapse of the monarchy, bringing the liberal and socialist forces to power. Of consequence for the later turn to reaction was that the revolution divided public opinion. Its bloody course in Kiel and some of the province's smaller cities spread fear of violence and for property among much of the middle class. Similarly important for later nationalist propaganda, it was the SPD, just as much as the radical left, which was identified with the revolution. This helped to reinforce a psychological division between the SPD and the middle classes despite much of the old bureaucratic and judicial structures remaining and even though the SPD had actually checked the extreme left's revolutionary endeavours and favoured parliamentary-democracy in line with the liberal parties' aspirations. Few offices beyond the appointment of Heinrich Kürbis as Oberpräsident in Kiel went to the SPD.[45] Until 1928 the office of Regierungspräsident of the *Regier-*

---

42 Heberle, *From Democracy to Nazism*, pp. 23-31 and *Landbevölkerung und Nationalsozialismus*, p. 28; Rietzler, "*Kampf in der Nordmark*", pp. 49-51; Scharff, *Geschichte Schleswig-Holsteins*, p. 86; Stoltenberg, *Politische Strömungen im schleswig-holsteinischen Landvolk*, pp. 17-18.

43 Stoltenberg, *Politische Strömungen im schleswig-holsteinischen Landvolk*, pp. 18-21; Rietzler, "*Kampf in der Nordmark*", pp. 52-53.

44 Stoltenberg interpreted these circumstances as showing that the rural population was largely "unpolitical", an unsubstantiated conclusion if read to include political opinion, rather than just political organisation and active participation. Stoltenberg, *Politische Strömungen im schleswig-holsteinischen Landvolk*, p. 23.

45 Heinrich Kürbis (1873-1951) was a metalworker and town councillor for the SPD in Altona from 1909, district secretary of the SPD for Schleswig-Holstein from 1913 to 1919 and Oberpräsident of Schleswig-Holstein from 1919 to July 1932. See E. R. Huber, *Deutsche*

ungsbezirk (administrative district), in Schleswig, responsible to the Prussian Interior Ministry, was occupied first by a conservative public servant and then a former county administrator loyal to the German People's Party (DVP).[46]

It should be noted that Schleswig-Holstein's farmers' councils were an SPD initiative to secure the delivery of foodstuffs. Their appeal to farmers lay in their hopes of retaining administrative control over agriculture and terminating wartime economic controls. Quickly under the influence of conservative agricultural leaders, strong participation, therefore, was not an expression of rural revolutionary zeal.[47]

The years 1918 to 1924 were characterised by weak voter loyalties and a significant political shift to the right in the countryside. This was clearly reflected in the voting trends in the province at Reich elections. At first, the elections to the national assembly in January 1919 resulted in a continuation of the pre-war voting tendencies in the province.[48] With the SPD and the Progressives' successor, the German Democratic Party (DDP), polling almost three quarters of all votes cast, the Weimar coalition obtained widespread support. Both, however, soon lost votes, especially in rural areas.[49] At the Reichstag elections in February 1921 the DDP's share of the vote plunged, while in May 1924 the SPD suffered its worst election result in the province since 1887.

The fortunes of the national-liberal DVP were an unmistakable sign of the weak rural political allegiances in Schleswig-Holstein. The DVP fared poorly in 1919, made great gains in 1921, but then lost much of that support in May 1924. The success of the German National People's Party (DNVP) highlighted the shift to the right more clearly. Its gains in the countryside were pronounced, particularly in areas where the liberal vote had previously been strong.[50] A fusion of the old Conservative and anti-Semitic parties, the DNVP was self-avowedly reactionary, nationalistic and Christian, and safeguarded private property. In part, its success was a

---

Verfassungsgeschichte seit 1789 Band V: Weltkrieg, Revolution und Reichserneuerung 1914-1919 (Stuttgart, Berlin, Cologne and Mainz, 1978), p. 651.

46 Rietzler, "*Kampf in der Nordmark*", pp. 73-89; Stoltenberg, *Politische Strömungen im schleswig-holsteinischen Landvolk*, pp. 24-27. Heinrich Pauli, a career public servant, was replaced by Adolf Johannsen after the Kapp Putsch.

47 Rietzler, "*Kampf in der Nordmark*", pp. 85-86; Stoltenberg, *Politische Strömungen im schleswig-holsteinischen Landvolk*, pp. 29-30.

48 Heberle, *Landbevölkerung und Nationalsozialismus*, p. 29; Rietzler, "*Kampf in der Nordmark*", p. 91; Stoltenberg, *Politische Strömungen im schleswig-holsteinischen Landvolk*, p. 31.

49 For the shift away from the SPD and DDP in the countryside, see Heberle, *Landbevölkerung und Nationalsozialismus.*, pp. 55-59; Rietzler, "*Kampf in der Nordmark*", pp. 91-92, 106 and 115; Stoltenberg, *Politische Strömungen im schleswig-holsteinischen Landvolk*, pp. 31-33 and 53-54.

50 For the fortunes of the DVP and DNVP in the countryside, see Heberle, *Landbevölkerung und Nationalsozialismus*, p. 42; Rietzler, "*Kampf in der Nordmark*", p. 227; Stoltenberg, *Politische Strömungen im schleswig-holsteinischen Landvolk*, pp. 53-54.

product of the growth of its membership and organisation across the countryside after 1919 – as opposed to the DDP and DVP which remained concentrated in the cities. Yet its success also reflected the party's appeal to the commercial middle classes and farmers with its calls for the removal of wartime economic controls, its racial glorification of farmers, condemnation of socialism, harnessing of nationalism in attacks on the Republic, including blaming the SPD for defeat, as well as intense anti-Semitism.[51]

Table 1: Reich Election Results in Schleswig-Holstein from 1919 to 1924 in Percent

	1919	**1921**	**1924 (I)**	**1924(II)**
DNVP	7.7 (8.6)	20.5 (15.1)	31 (19.5)	33 (20.5)
VSB/NSFB*	-	-	7.4 (6.5)	2.7 (3)
DVP	7.8 (4.4)	18.4 (14)	12.1 (9.2)	14.6 (10.1)
Zentrum	1 (18.8)	0.8 (13.1)	1 (13.4)	1.1 (13.6)
DDP	27.2 (18.1)	9.4 (8.4)	8.1 (5.7)	8.7 (6.3)
SPD	45.7 (37.9)	37.3 (22.1)	24.9 (20.5)	30.3 (26)
USPD	3.4 (7.6)	3 (18)	-	-
KPD	-	6.1 (2)	10.2 (12.6)	6.7 (9.3)
SHLP	7.2	3.8	0.7	-

Reich results in brackets where applicable. *Völkisch-Sozialer Block in May 1924, Nationalsozialistische Freiheitsbewegung in December 1924.

Source: *Statistisches Jahrbuch für das Deutsche Reich* (Berlin, 1919), pp. 236-237, (1921), pp. 356-357, (1924), pp. 390-393.

Fluctuations in support for the middle-class parties were considerable in Schleswig-Holstein where the stabilising effect of the Centre Party was missing.[52] Bringing additional division was the development of a new rural party, the *Schleswig-Holsteinische Landespartei* (The Schleswig-Holstein Country Party, SHLP), a product of growing disappointment in rural circles over the liberal parties' failure to represent farming interests. As a liberal party, the Landespartei demanded equality for all Germans, democracy, free trade and an end to state economic intervention in the form of wartime controls.[53] Its liberalism, however, was

51 Rietzler, "*Kampf in der Nordmark*", pp. 101-104 and 229-234 passim; Stoltenberg, *Politische Strömungen im schleswig-holsteinischen Landvolk*, pp. 31, 38-40, 77-80 and 84; Koops, *Südtondern in der Zeit der Weimarer Republik*, pp. 199-205.

52 Heberle, *Landbevölkerung und Nationalsozialismus*, pp. 34-35; Rietzler "*Kampf in der Nordmark*", p. 93.

53 Surveys of the Landespartei's ideas are to be found in Heberle, *From Democracy to Nazism*, pp. 43-54, Rietzler, "*Kampf in der Nordmark*", pp. 93-98; Stoltenberg, *Politische*

not only opposed to socialism, but also bore a great mistrust of its urban counterparts' alliances with industry and capital. It differentiated between the "golden democracy" of industry and commerce, and a healthy rural democracy of Schleswig-Holstein. Moreover, it employed anti-Semitism to stigmatise those who failed to support its interests and also claimed to combat the predominance of Jews in business and government.[54] Nationalistic, regionally patriotic and opposed to centralism and bureaucracy, the Landespartei underlined tribe and homeland as the sources of the nation's strength, claimed to be above class and party interests and to support a Volksgemeinschaft and a constitution which deferred to occupational interests instead. Some of these ideas would resurface later in the Landvolkbewegung. Moreover, their similarity to National Socialist ideology too led Rudolf Heberle to point to the Landespartei as significant in paving the way for National Socialism in the province.[55] Among its speakers, and for a time its general secretary, was the later NSDAP *Gauleiter*, Hinrich Lohse. Chiefly appealing to smaller Geest farmers, the Landespartei enjoyed strong support in some counties but as a small, regional force, it was compelled to enter an alliance.[56] Joining with the DDP, the very party it had initially rejected, the Landespartei thereby rendered itself meaningless and support for it dwindled thereafter.[57]

National Socialism gained its first members in Schleswig-Holstein in 1922, mostly from various völkisch groups, some of which were loosely affiliated with the party in Munich. Although the party itself was prohibited, a small core of supporters participated in elections for the first time in May 1924 as part of the *Völkisch-Sozialer Block* (VSB), which was composed chiefly of the *Deutschvölkische Freiheitspartei* (DVFP). The VSB polled well in May 1924. However, following a period of internal discord, the extreme right declined into insignificance as the *Nationalsozialistische Freiheitsbewegung* (NSFB) in December.[58]

The extreme left was also weak. Only in 1921 did the Communists (KPD) manage to gain more votes in Schleswig-Holstein than the Reich average. This illu-

---

Strömungen im schleswig-holsteinischen Landvolk, pp. 33-35. Also see T. Thyssen, *Bauer und Standesvertretung. Werden und Wirkung des Bauerntums in Schleswig-Holstein seit der Agrarreform* (Neumünster, 1958), pp. 350-355.

54 Heberle, *From Democracy to Nazism*, p. 46; P. Wulf,' "Jüdische Weltherrschaftspläne" Anti-Semitismus in bürgerlichen und bäuerlichen Parteien und Verbänden in Schleswig-Holstein zu Beginn der Weimarer Republik', *Zeitschrift der Gesellschaft für Schleswig-Holsteinische Geschichte*, 128 (2003), pp. 152-161.

55 Heberle, *From Democracy to Nazism*, p. 45.

56 Heberle, *From Democracy to Nazism*, p. 44; Rietzler, *"Kampf in der Nordmark"*, p. 94.

57 Rietzler, *"Kampf in der Nordmark"*, p. 99; Stoltenberg, *Politische Strömungen im schleswig-holsteinischen Landvolk*, p. 36.

58 For the early history of the NSDAP in Schleswig-Holstein, analysis of the VSB's elections results and the struggle among the leaders of the völkisch parties, see Rietzler, *"Kampf in der Nordmark"*, pp. 187-215, 230-233, 240-251, 254, 343-362. Also see Stoltenberg, *Politische Strömungen im schleswig-holsteinischen Landvolk*, p. 86.

sory success, however, came at the cost of the Independent Socialists (USPD). It was, therefore, consistent with the comparatively poor results in the predominantly rural province in the years to come.

The great swing to the parties of opposition, chiefly to those on the right, in the Reich elections of 1924 followed the turbulence of inflation, the occupation of the Ruhr, communist uprisings and the Hitler putsch. Significant for the extent of this shift in the countryside, where foreign competition for agricultural producers was mounting, was that the DNVP, DVP and völkisch parties advocated agricultural tariffs unlike the SPD and DDP, which stood for international cooperation and free trade.[59] Yet two earlier events of particular consequence in Schleswig-Holstein had already rallied support for the right: the loss of North Schleswig and the Kapp Putsch.

The Treaty of Versailles was perceived as a humiliation across Germany, undermining support for the parties associated with defeat and the Republic. The loss of North Schleswig meant the treaty's effects were felt very tangibly in Schleswig-Holstein and this exacerbated the shift to the right in the province. In the first of two plebiscites, on February 10, 1920, 74.9 percent of those eligible to vote in the northern zone supported joining Denmark. On March 13, 80 percent in the southern zone voted for Germany.[60]

Germans all over Schleswig-Holstein opposed the secession of the northern zone and all parties, including the SPD, called for amendments at the very least. Nevertheless, it was those with the strongest nationalist propaganda, the DNVP and DVP, which stood to profit, while the Republican parties were blamed for betraying the nation.[61] The danger of tariff-free competition quickly reduced the actual appeal of calls for the return of North Schleswig among farmers.[62] However, as Rudolf Rietzler pointed out, the belief that the Reich government failed to understand the border question and provided insufficient support in the endeavours to retain North Schleswig encouraged resentments against Weimar. Moreover, border propaganda obscured social differences and cultivated nationalist and völkisch sentiment, making Schleswig-Holstein more susceptible to nationalist and anti-democratic movements.[63]

Attempts to disband *Freikorps* units in accordance with the Versailles Treaty prompted the Kapp Putsch on March 13, 1920, leading to the government's flight to Dresden and a general strike. In Schleswig-Holstein, *Reichswehr* officers and the conservative Regierungspräsident sided with Kapp, declared a state of emergency and imprisoned SPD leaders. Elsewhere in Germany the putsch broke down in the face of the strikes after four days, but in Schleswig-Holstein turmoil and bloodshed

59 Stoltenberg, *Politische Strömungen im schleswig-holsteinischen Landvolk*, pp. 68-75.

60 *Statistisches Handbuch für Schleswig-Holstein* (1951), p. 536.

61 Rietzler, "*Kampf in der Nordmark*", pp. 110-119.

62 T. Fink, *Geschichte des schleswigschen Grenzlandes* (Copenhagen, 1958), pp. 254-256.

63 Rietzler, "*Kampf in der Nordmark*", pp. 120-121.

continued for several days with almost 50 deaths in shootings and struggles between workers and the Freikorps and military in Kiel.[64]

Highlighting the strength of anti-democratic and anti-Republican forces in the countryside, though they officially rejected the coup afterwards, the agricultural organisations, that is the Chamber of Agriculture and the province's farming associations, publicly supported Kapp.[65] So too did the self-defence organisations. The militia (*Einwohnerwehr*) and "self-defence" (*Selbstschutz*) organisations, which had sprung up in 1919 to protect Germany's borders, defend private property and uphold order, had swiftly grown into anti-revolutionary activism.[66] Though nominally non-political, the privately funded self-defence units in Schleswig-Holstein were strongly nationalist and opposed to communism. Tolerated by the authorities for fear of the radical left, they continued to grow even after the Kapp Putsch, reaching between 25,000 and 30,000 members in November 1920, including a group 10,000 strong in Dithmarschen.[67]

These organisations joined with others across Germany, forming a Reich-wide paramilitary network, the *Organisation Escherich*, and established ties to other nationalist and veterans organisations, including Hermann Ehrhardt's underground *Organisation Consul* (OC).[68] The resultant personal contacts continued after 1921

---

64 Rietzler, "*Kampf in der Nordmark*", pp. 123-126.

65 Stoltenberg, *Politische Strömungen im schleswig-holsteinischen Landvolk*, pp. 49 and 61.

66 R. Bessel, *Germany After the First World War* (Oxford, 1993), pp. 215-217; F. Bösch, 'Von der Einwohnerwehr zur Volkspartei' in D. Münkel (ed.), *Der lange Abschied vom Agrarland. Agrarpolitik, Landwirtschaft und ländliche Gesellschaft zwischen Weimar und Bonn* (Göttingen, 2000), pp. 229-232; J. Flemming, *Landwirtschaftliche Interessen und Demokratie: Ländliche Gesellschaft, Agrarverbände und Staat 1890-1925* (Bonn, 1978), p. 125; J. Flemming, 'Die Bewaffnung des "Landvolks". Ländliche Schutzwehren und agrarischer Konservatismus in der Anfangsphase der Weimarer Republik', *Militärgeschichtliche Mitteilungen*, 26 (1979), pp. 7-36; R. G. Moeller, *German Peasants and Agrarian Politics, 1914-1924: The Rhineland and Westphalia* (Chapel Hill and London, 1986), pp. 69-72; B. Theine, *Westfälische Landwirtschaft in der Weimarer Republik* (Paderborn, 1991), pp. 205-207. For the militias in Schleswig-Holstein, see J. Bracker, 'Die Einwohnerwehren in Schleswig-Holstein', *Zeitschrift der Gesellschaft für Schleswig-Holsteinische Geschichte,* 99 (1974), pp. 255-269 and H. Zimmermann, 'Die Einwohnerwehren. Selbstschutzorganisationen oder konterrevolutionäre Kampforgane?', *Zeitschrift der Gesellschaft für Schleswig-Holsteinische Geschichte,* 128 (2003), pp. 184-211. For the role of these organisations in the development of the Landvolkbewegung, see chapter 3 below.

67 Rietzler, "*Kampf in der Nordmark*", pp. 135-136; Stoltenberg, *Politische Strömungen im schleswig-holsteinischen Landvolk*, p. 50.

68 Led by Hermann Ehrhardt, the Organisation Consul was a secret right-wing organisation founded in 1920, following the dissolution of the "Freikorps Marine-Brigade Ehrhardt." The OC was involved in almost all of the political murders in the early years of the Republic, most notably for the deaths of Erzberger and Rathenau but also the attempt on Scheidemann. Prohibited in 1922, its successor was Bund Wiking. See H. Stern, 'The Organisation Consul', *The Journal of Modern History,* 35 (1963), pp. 20-32. Also see J. Diehl, *Paramili-*

when the Organisation Escherich was officially banned. Many of the old members on the west coast regrouped in 1922 in the *Frontkämpferbund Westküste* and later in the *Stahlhelmwestküste*. Deemed a particularly radical variant of the Stahlhelm veteran organisation, in 1926 it temporarily entered an alliance with the paramilitary groups *Bund Wiking* and *Wehrwolf*.[69]

Reviewing developments after the First World War, the remarkable rise of the DNVP was an early indication of the potential of right-wing opposition to Weimar in Schleswig-Holstein. Yet just as perturbing on the whole was the weakness of middle-class and particularly of rural voter loyalties. Although the elections to the national assembly in 1919 reinforced pre-war trends, the SPD and especially the DDP quickly lost ground to the right, the Landespartei was but a fleeting existence and the DVP failed to retain the surge of support it enjoyed in 1921. Significantly, traditional adjuncts of Schleswig-Holstein's liberalism, nationalism and opposition to socialism, facilitated the exceptional swing to the DNVP in 1924. Further, the period of relative calm that followed concealed the fragmentation of the middle-class parties. They lacked organisational structure and influence in rural areas, particularly at the local level, where economic interest groups proliferated at communal elections.[70] As they did across Germany, these splinter groups, indicative of the growing disaffection among the middle class, eroded support from the larger parties.[71] Essentially, the shifting fortunes of the political parties, division and the

---

tary Politics in Weimar Germany (Bloomington and London, 1977), pp. 107-109, 112-113 and, for Organisation Escherich, pp. 78-93.

69 Rietzler, "*Kampf in der Nordmark*", pp. 133-139, 265-270 and 273-275; Stoltenberg, *Politische Strömungen im schleswig-holsteinischen Landvolk*, pp. 48-51, 80-81 and 97-98. The Wehrwolf, formed in 1923 by Fritz Kloppe, offered military training to men, too young to have served during the war. Increasingly it became political, emphasising its opposition to the Weimar Republic and adversity to international capitalism. Later it pushed for its own candidates' inclusion in right-wing parties' lists at elections. With its members already drifting into the NSDAP, in March 1933 it was absorbed into the SA. Bund Wiking, successor to Organisation Consul, was a small, elitist and tightly organised conspiratorial combat group, which aimed to infiltrate and cooperate with other organisations in its aim to destroy the Republic by provoking a leftist uprising, which would then have to be suppressed by the military and paramilitary. Banned after the discovery of its plans, it cooperated with the Stahlhelm in 1927, before Ehrhardt dissolved the organisation in May 1928. See Diehl, *Paramilitary Politics in Weimar Germany*, pp. 109, 119-120, 171, 217-221, 227-230, 237-241, 261-262 and 295.

70 Stoltenberg, *Politische Strömungen im schleswig-holsteinischen Landvolk*, pp. 48 and 85 and the manuscript of his address 'Der politische Umbruch in Schleswig Holstein, 1928-1930' Neumünster, 13.11.1979, p. 7, StaNms MA 5360 (III).

71 Larry Eugen Jones and Thomas Childers underlined the destabilisation of party loyalties in the aftermath of inflation and stabilisation, pointing to the rise of interest group parties in the mid 1920s. L. E. Jones, *German Liberalism and the Dissolution of the Weimar Party System 1918-1933* (Chapel Hill and London, 1988), esp. pp. 251-290; T. Childers, 'Interest and Ideology: Anti-System Politics in the Era of Stabilization 1924-1928' in G. Feldman (ed.), *Die Nachwirkungen der Inflation auf die deutsche Geschichte 1924-1933* (Munich,

lack of political allegiances and party organisation outside the cities meant that party-political stability was missing in Schleswig-Holstein well before the onset of the agricultural crisis.

The organisational weakness of the middle-class political parties amplified the importance of Schleswig-Holstein's agricultural political associations in representing and shaping rural political opinion. But like the many political parties, these organisations too were fragmented. Though essentially conservative as they were across Germany, the agricultural-political associations encompassed a variety of produce and modes of production and these were reflected in their different programmes and political allegiances.[72]

The provincial Chamber of Agriculture served farmers' commercial and professional interests.[73] Its chief concern was the improvement of the quality and quantity of agricultural production. The chamber spoke for farmers' business interests in the public domain and communicated these to the provincial authorities. It did not fulfil the farmers' need for political representation.

The League of Agrarians (*Bund der Landwirte*, BdL), the oldest of the significant national political agricultural organisations, was formed in Berlin in 1893 and established its provincial branch in Schleswig-Holstein in the same year.[74] In Schleswig-Holstein the BdL always remained the organisation of the politically conservative estate owners in the east of the province. Beyond pockets of support on the west coast, it always found it difficult to attain broader appeal, especially among more liberal circles. Just a fifth of Schleswig-Holstein's farmers were members at the close of the imperial period. Yet the BdL's influence was greatly magnified by its lack of a rival before the war and by way of its national organisation, which boasted over one and a half million members from its support bases across Lutheran northern and eastern Germany.[75]

Regrouping under the Reich Rural League (*Reichslandbund*, RLB) after the war, the BdL also adopted this name in Schleswig-Holstein. It continued to act in the interests of grain producers, advocating tariffs to safeguard their produce and advancing their political concerns at elections by influencing the choice of candi-

---

1985), pp. 1-20; T. Childers 'Inflation, Stabilization, and Political Realignment in Germany 1924 to 1928' in G. Feldman, C.-L. Holtfrerich, G. Ritter and P. Witt (eds), *Die deutsche Inflation: Eine Zwischenbilanz* (Berlin and New York, 1982), pp. 409-431.

72 Report from the Regierungspräsident to the Oberpräsident in Kiel concerning "Entwicklung des landwirtschaftlichen Berufsorganisationswesens in Schleswig-Holstein" Schleswig, 22.1.1930, LAS 309/22765; Gessner, 'The Dilemma of German Agriculture', p. 140.

73 For the history of Schleswig-Holstein's professional agricultural organisations from the 1830s and the province's Chamber of Agriculture (*Landwirtschaftskammer*) during the Weimar period, see Thyssen, *Bauer und Standesvertretung*, pp. 22-209 and 243-265.

74 For the history of the BdL and Rural League in Schleswig-Holstein, see Thyssen, *Bauer und Standesvertretung,* pp. 304-326 and 331-341.

75 Thyssen, *Bauer und Standesvertretung,* pp. 332 and 338.

dates and campaigning for them. Schleswig-Holstein's Rural League was closely allied with the DNVP and disapproved both of the DDP and DVP. It tended to the right of the national organisation and strongly opposed Weimar from the beginning, voicing particularly fierce criticism in its bi-weekly *Landbote für Schleswig-Holstein*, whose editor also often published anti-Semitic and völkisch views.[76]

The fact that just 0.3 percent of Schleswig-Holstein's total population and a mere 0.05 percent of the province's rural population was Jewish would suggest that anti-Semitism in the countryside could not have been much affected by farmers' experiences of Jewish people.[77] However, the farmers had long been exposed to anti-Semitic fears and prejudices as these were entrenched in the province's rural political culture. Of the various parties and organisations which transmitted anti-Semitism among rural circles in Schleswig-Holstein, the Rural League was the most significant. Of the others, the Landespartei has been mentioned already.[78] Its anti-Semitism was amorphous and mild compared with that of the veteran and paramilitary associations such as the Stahlhelm, which expressly called for the combating of the "Jewish spirit" in its statutes, proscribed Jewish membership in 1924 and increasingly called for the boycott of Jewish shops as it moved toward national opposition.[79] Nevertheless, it would be difficult to overstate the consequence of the Rural League in this context. Together with its predecessor, the BdL, it requires particular attention for its central role in the history of anti-Semitism in Schleswig-Holstein, indeed in Germany, in amalgamating different types of anti-Semitic reasoning, integrating it into conservative-nationalist ideology and spreading it to broader circles from the late nineteenth century.[80]

Building on clichéd conceptions of Jews as power-loving profiteers, to be met in the countryside in the form of money-lenders and cattle-traders, BdL ideology drew on opposition to emancipation, liberalism and free trade in conservative circles as well as the association of Jews with capitalism and stock market speculation, which had grown popular among the middle classes during the period of economic uncertainty from 1873 to 1894. Increasingly, the BdL also incorporated the

---

76 Rietzler, "*Kampf in der Nordmark*", p. 104; Stoltenberg, *Politische Strömungen im schleswig-holsteinischen Landvolk*, pp. 45, 64 and 94.

77 According to the results of the census of 16 June 1925: *Gemeindelexikon für den Freistaat Preußen. Band IX: Provinz Schleswig-Holstein*, Preußisches Statistisches Landesamt, Berlin 1930, pp. XVIII, 36, 40 and 48.

78 See p. 28 above.

79 The Stahlhelm had as many as 20,000 members in the province at its zenith in 1927. See V. Berghahn, *Der Stahlhelm Bund der Frontsoldaten 1918-1935* (Düsseldorf, 1966), pp. 64-67; Rietzler, "*Kampf in der Nordmark*", p. 261; Stoltenberg, *Politische Strömungen im schleswig-holsteinischen Landvolk*, p. 97.

80 H.- J. Puhle, *Agrarische Interessenpolitik und preußischer Konservatismus im wilhelminischen Reich (1893-1914)* (Hanover, 1967), pp. 111-140. Also see S. Merkenich, *Grüne Front gegen Weimar: Reichs-Landbund und agrarischer Lobbyismus 1918-1933* (Düsseldorf, 1998), p. 123.

racial tenets of the extreme anti-Semitic organisations in its argumentation, which, when blended with anti-liberal sentiment and fears of growing Jewish influence in society, fuelled the notion that all Jews were exponents of an international conspiracy. Anti-Semitism was a persistent element in BdL ideology and political agitation. However, as a "correlative" of its conservative-nationalist politics, rather than in the form of the concrete political demands of the radical organisations, the BdL's anti-Semitism appeared moderate and was consequently palatable to the Conservatives.[81] For this reason, in addition to its size, organisation and political influence, the BdL proved much more effective in spreading and popularising anti-Semitism than the fanatical anti-Semitic organisations.

It was precisely because of the BdL's assistance in Schleswig-Holstein that one of these organisations, the *Deutsch-Sozialer Reformverein*, enjoyed a rare election success. At the Reichstag elections of 1898 the BdL backed its candidate, Friedrich Raab, in the Flensburg-Apenrade district.[82] Campaigning on the basis of common enmity toward economic liberalism and international capital, the BdL's support in the countryside facilitated Raab's victory in a run-off election against an SPD candidate. Though this result was never replicated, collaboration helped to anchor anti-Semitism in the BdL's ideological conception, fostering the view of Jews as the embodiment of liberalism, capitalism and social democracy.[83]

It would be imprudent to overstate the Reformverein's direct role in conditioning the farmers' later receptiveness to anti-Semitism in the Landvolkbewegung, not least because its activities were limited regionally.[84] Rather it was the BdL and Conservatives, and then after the war, their successors, the Rural League and DNVP, which ensured anti-Semitism continued to spread in the countryside.[85] Though not quite as fanatical, nor placing the same emphasis on racial consciousness as the *Deutschvölkischer Schutz- und Trutzbund*, the umbrella organisation for all anti-Semitic organisations set up on the initiative of the Pan-German League (*Alldeutscher Verband*), they were the principal forces fostering anti-Semitic fears

---

81 Puhle, *Agrarische Interessenpolitik und preußischer Konservatismus im wilhelminischen Reich*, p. 138. Also see S. Volkov, 'Anti-Semitism as A Cultural Code: Reflections on the History and Historiography of Anti-Semitism in Imperial Germany' *Leo Baeck Institute Year Book,* 23 (1978), pp. 25-46.

82 K-G. Riquarts, *Der Anti-Semitismus als politische Partei in Schleswig-Holstein und Hamburg 1871-1914* (Kiel, 1975), pp. 232-245 and 307-312.

83 Pockets of support for anti-Semitism persisted thereafter in those regions where it was backed by the BdL. The Deutsch-Sozialer Reformverein's successor, the Deutschsoziale Partei, received 9 percent of the vote in 1907 and 13.7 percent in 1912 in Flensburg-Apenrade. In Lauenburg it received as much as 32 percent in 1907 and 17 percent in 1912. Rietzler, "*Kampf in der Nordmark*", pp. 55-63; Riquarts, *Der Anti-Semitismus als politische Partei*, pp. 245-249.

84 Riquarts, *Der Anti-Semitismus als politische Partei*, p. 245.

85 Puhle, *Agrarische Interessenpolitik und preußischer Konservatismus im wilhelminischen Reich*, p. 140; Riquarts, *Der Anti-Semitismus als politische Partei*, pp. 302-303.

in rural Schleswig-Holstein.[86] Associating anti-Semitism with the prolongation of wartime controls, defeat, revolution and democracy, the Rural League and closely allied DNVP adapted the old anti-Semitic conceptions to suit the new post-war circumstances. They perpetuated fears of Jewish plans for world domination and claims that the press, art and literature, the Republic and democratic parties were all in Jewish hands. Anti-Semitism was a deliberate and central element of their propaganda, prevalent in the first years of the Republic, which they then revived in connection with the currency stabilisation measures in 1924, when it was linked to price, credit and tax issues.[87] The Rural League and DNVP's anti-Semitism abated from 1925 after they had taken a course of political cooperation.[88] Yet the essentially political and economic anti-Semitic account of Germany's ills was largely developed by that time, ready for reactivation during the agricultural crisis.[89]

After the spring of 1918, the Rural League had to compete with the Schleswig-Holstein Farmers' Association (*Schleswig-Holsteinischer Bauernverein*) for the favour of the province's farmers. Established in response to the "controlled economy" (*Zwangswirtschaft*), the Farmers' Association grew rapidly as it drew its members from small and mid-sized properties, and obtained strong support particularly in the livestock breeding Geest but also in the livestock fattening marsh areas.[90] Since livestock farming constituted the largest sector within Schleswig-

86 The Deutschvölkischer Schutz- und Trutzbund's following in Schleswig-Holstein was small and largely urban: Rietzler, "*Kampf in der Nordmark*", pp. 144-156; Riquarts, *Der Anti-Semitismus als politische Partei*, p. 303; Wulf, ' "Jüdische Weltherrschaftspläne" ', pp. 172-176.

87 Flemming, *Landwirtschaftliche Interessen und Demokratie*, pp. 135-139, 177-178 and 184-185; W. Jochmann, *Gesellschaftskrise und Judenfeindschaft in Deutschland 1870-1945* (Hamburg, 1988), pp. 99-170; J. Kocka, *Klassengesellschaft im Krieg* (Frankfurt/M, 1988), pp. 138-140; H. Reif, 'Antisemitismus in den Agrarverbänden Ostelbiens während der Weimarer Republik' in H. Reif (ed.), *Ostelbische Agrargesellschaft im Kaiserreich und in der Weimarer Republik* (Berlin, 1994), pp. 386-392, 401 and 410; Wulf, ' "Jüdische Weltherrschaftspläne" ', pp. 162-172. For anti-Semitism in the Rural League and DNVP in Schleswig-Holstein, see Koops, *Südtondern in der Zeit der Weimarer Republik*, pp. 200-202; Rietzler, "*Kampf in der Nordmark*", pp. 234 and 253; Sörensen, *Politische Entwicklung und Aufstieg der NSDAP*, pp. 141 and 166, and Stoltenberg, *Politische Strömungen im schleswig-holsteinischen Landvolk*, p. 64.

88 Reif, 'Antisemitismus in den Agrarverbänden Ostelbiens', pp. 388 and 401; Merkenich, *Grüne Front gegen Weimar*, p. 126; Paul, ' "Die Juden sind unser Unglück!" ', p. 189.

89 In brief, Jews were said to have used their influence over the press, international capital and the democratic parties to engineer defeat, revolution and impose the Weimar constitution, to have profited from the controlled economy, reparation payments and inflation, and planned to enslave the German economy and socialise agriculture by exploiting the farmers' need for credit, unfavourable loan conditions, mortgages and the *Rentenbank* debt. See Reif, 'Antisemitismus in den Agrarverbänden Ostelbiens', pp. 393-395.

90 Report from the Regierungspräsident to the Oberpräsident in Kiel concerning "Entwicklung des landwirtschaftlichen Berufsorganisationswesens in Schleswig-Holstein" Schleswig, 22.1.1930 and Regierungspräsident Abegg to the Preuß. Minister des Innern concerning

Holstein's agriculture, the Farmers' Association soon dwarfed the Rural League's membership, reaching 40,000 by the eve of the agricultural crisis.[91]

The Farmers' Association was a driving force in the creation of the aforementioned Schleswig-Holsteinische Landespartei but because the Landespartei was never able to attract a broader support base and quickly declined, the Farmers' Association shifted its alliances. First it bolstered its connections and advocated the DDP and DVP, and then, as the calls for agricultural tariffs increased in 1924, with the DVP and DNVP, thereby reinforcing the swing to the right.[92]

The swift growth of the Farmers' Association, its prominent activity, but especially its progress into counties where the Rural League also had supporters, including Steinburg and Dithmarschen, provoked Rural League opposition.[93] The Rural League attacked the "Catholic spirit" of the Organisation of the German Farmers' Associations (*Vereinigung der Deutschen Bauernvereine*), the largely southern German dominated national organisation to which the Schleswig-Holstein Farmers' Association belonged.[94] Rivalry between the two associations was manifest in the Chamber of Agriculture in which, due to a system of half-elections, the Rural League maintained its supremacy until 1927. Although its influence continued as its universally respected Count Rantzau and Jensen (Ausacker) retained their positions of president and vice-president thereafter, the Rural League's resentment over the Farmers' Association's success was so great, it remembered its election defeat that year as its "black day".[95]

Competition in the province increased further in 1927 when a number of disparate small farmer and settler associations united under the name of Small Farmers' Association (*Kleinbauernverein*) as a provincial branch of the smallest of the national farmers' organisations, the German Peasantry (*Deutsche Bauernschaft*). Unlike the Reich Rural League, which merely tolerated parties and parliaments and viewed its participation as a necessary evil to be used for agriculture's ends, the German Peasantry was politically neutral and receptive to Weimar's institutions.[96]

"die landwirtschaftliche Bewegung in Schleswig-Holstein" Schleswig, 24.1.1930, both in LAS 309/22765; Heberle, *From Democracy to Nazism*, pp. 42-43; Stoltenberg, *Politische Strömungen im schleswig-holsteinischen Landvolk*, pp. 40-44; Thyssen, *Bauer und Standesvertretung*, p. 347.

91 Thyssen, *Bauer und Standesvertretung*, pp. 349-359 and p. 368.

92 Heberle, *From Democracy to Nazism*, pp. 61-62; Stoltenberg, *Politische Strömungen im schleswig-holsteinischen Landvolk*, p. 75. For the Landespartei, see p. 28 above.

93 For examples of the Farmers' Association's activity, see Thyssen, *Bauer und Standesvertretung*, pp. 357-358.

94 Heberle, *From Democracy to Nazism*, p. 43; Thyssen, *Bauer und Standesvertretung*, p. 361.

95 Thyssen, *Bauer und Standesvertretung*, pp. 367-368.

96 For the Reich Rural League see Flemming, *Landwirtschaftliche Interessen und Demokratie*. For the German Peasantry, which was the product of a fusion of three small associations – the Reich Association for Small and Medium-sized Farms (*Reichsverband landwirtschaftlicher Klein- und Mittelbetriebe*), the German Farmers' Association (*Deutscher*

It emphasised the differing interests of estate owners and farmers, of grain growers and livestock farmers, even though it was willing to work with the other agricultural organisations.[97] For the Rural League and the Farmers' Association, the Small Farmers' Association constituted an additional impediment in their attempts to appeal to all farmers. Though it might be considered a reflection of the diversity of Schleswig-Holstein's agriculture, it also amounted to yet another political division in the province in the mid 1920s.

The key immediate factor behind the rise of the Landvolkbewegung was the worldwide agricultural crisis preceding the Great Depression which struck Schleswig-Holstein with particular severity. The causes of the agricultural crisis as it affected the province were both short and long-term, local and international. The First World War, the early years of the Weimar Republic and inflation had a significant and mostly detrimental effect on agricultural development. Moreover, the early years of discontent and uncertainty produced great resentment against the Republic in the countryside.

The war, in particular, led to a considerable decline in agricultural stocks, production and quality.[98] Partly, the blame can be attributed to expectations that the war would be short and end in German victory. The prolongation of the war meant a shortage of labour and draught horses, the enforced slaughter of pigs, a decline in the manufacture and servicing of agricultural machinery, fertiliser and fodder shortages, disturbances in crop rotations and arable land left to fallow. This led to inferior crop yields and degraded harvests for many years to come, as well as dramatic falls in stock numbers and the average weight of animals in the livestock sector.[99] The British naval blockade, not lifted until the summer of 1919, hindered the

---

Bauernbund) and the Westphalian Farmers' Association (*Westfälischer Bauernbund*) – see D. Gessner, *Agrarverbände in der Weimarer Republik* (Düsseldorf, 1976), pp. 101-102; L. E. Jones, 'Crisis and Realignment: Agrarian Splinter Parties in the Late Weimar Republic, 1928-1933' in R. G. Moeller (ed.), *Peasants and Lords in Modern Germany* (London, 1986), pp. 201 and 203.

97 Thyssen, *Bauer und Standesvertretung*, pp. 329-330.

98 For the effects of the war on agricultural production, see F. Aereboe, *Der Einfluß des Krieges auf die landwirtschaftliche Produktion in Deutschland* (Stuttgart, Berlin and Leipzig, 1927), pp. 32-99; H. Haushofer, *Die deutsche Landwirtschaft im technischen Zeitalter* (Stuttgart, 1963), pp. 225-229; Henning, *Landwirtschaft und ländliche Gesellschaft*, pp. 175-192; Sering, *Die deutsche Landwirtschaft*, pp. 14-17. For an overview focusing on circumstances in Bavaria, the Palatinate and the Rhineland, see J. Osmond, 'Peasant Farming in South and West Germany during the War and Inflation 1914 to 1924: Stability or Stagnation' in Feldman, Holtfrerich, Ritter and Witt (eds), *Die deutsche Inflation*, esp. pp. 290-294; J. Osmond, *Rural Protest in the Weimar Republic: The Free Peasantry in the Rhineland and Bavaria* (New York, 1993), pp. 10-12.

99 Aereboe, *Der Einfluß des Krieges*, pp. 88 and 93; Haushofer, *Die deutsche Landwirtschaft*, p. 227; Henning, *Landwirtschaft und ländliche Gesellschaft*, pp. 177-178. For the enforced

import of alternative supplies of artificial fertilisers and fodder. Exports suffered too. While the German engineering and chemical industries improved their positions in the inter-war period, agriculture failed to recapture its old markets: exports in 1928 lagged at just half of their 1913 levels.[100]

It took some ten years for agriculture in Schleswig-Holstein to recover from the war. Not until 1927 did livestock numbers reach levels comparable with 1913/1914 and dairy herds did not fully recover until 1933.[101] However, the loss of North Schleswig, it should be noted, does not appear to have caused lasting disadvantages, say, by way of cutting off significant markets or impeding traditional production practices.[102]

The introduction of a "controlled economy" during the war was a particular source of bitterness for farmers and all the more so when it was not removed at the war's end.[103] The extension of wartime controls contributed to antagonism between producers, consumers and the government, and between country and city. While price controls, distribution measures and requisitioning secured consumers' immediate needs, they were ultimately counter-productive, encouraging farmers to switch to products not under price ceilings, dampening their entrepreneurial spirit and creating opportunity for a flourishing black market.

Controls on the prices of pork and beef were only removed in October 1920, on butter and milk in May-June 1921 and on bread not until the end of 1922. Farmers' resistance to the "controlled economy" and widespread black market trade led to the implementation of a dual system (*Umlagewirtschaft*) in which farmers were required to deliver predetermined produce quotas, but were able to sell the surplus themselves. Nevertheless, farmers were greatly angered that they were unable to

---

slaughter of pigs (*Schweinemord*) in 1915 see A. Skalweit, *Die deutsche Kriegsernährungswirtschaft* (Stuttgart, Berlin and Leipzig, 1927), pp. 92-98 and 113.

100 D. Petzina, 'Was There a Crisis Before the Crisis? The State of the German Economy in the 1920s' in J. v. Kruedener (ed.), *Economic Crisis and Political Collapse: the Weimar Republic, 1924-1933* (New York, Oxford and Munich, 1990), p. 15.

101 *Statistisches Handbuch für Schleswig-Holstein* (1951), pp. 176-178. NB the comparisons here only concern the post-war territory.

102 In the Chamber of Agriculture's extensive reporting on agricultural-economic problems, North-Schleswig only featured in isolated complaints of a political nature. See, for example, the *Geschäftsbericht der Landwirtschaftskammer für die Provinz Schleswig-Holstein für das Rechnungsjahr 1921-1922*, p. 5.

103 See Bessel, *Germany After the First World War*, esp. pp. 212-214; C.-L. Holtfrerich, *Die deutsche Inflation 1914-1923* (Berlin, 1980), pp. 76-92; Kocka, *Klassengesellschaft im Krieg*, pp. 133-138; R. G. Moeller, 'Winners as Losers in the German Inflation: Peasant Protest over the Controlled Economy, 1920-1923' in Feldman, Holtfrerich, Ritter and Witt (eds), *Die deutsche Inflation*, pp. 255-288; Osmond, *Rural Protest in the Weimar Republic*, pp. 22-28 and 57-90; Sering, *Die deutsche Landwirtschaft*, pp. 39-40 and for an account of the "controlled economy" during the war, Skalweit, *Die deutsche Kriegsernährungswirtschaft*.

take advantage of food shortages and world market prices several times higher than those enforced domestically.[104]

The "controlled economy" did little to encourage support of the Republic amongst farmers.[105] Its protraction, as well as the problems of demobilization, Richard Bessel argues, convinced farmers that Berlin "... could not be trusted to give rural interests the priority they deserved".[106] One such problem was the acute shortage of rural labour, exacerbated by the loss of hundreds of thousands of foreign workers and prisoners of war. Germany's returning soldiers proved most unwilling to undertake agricultural work because wages and working conditions were unattractive. Even unemployment was sometimes viewed as preferable.[107] Schleswig-Holstein retained a relatively strong small town and rural character during the Weimar Republic.[108] Yet the shortage of rural labour, in the context of the longer trend of demographic shift from country to city and the growth of the secondary and tertiary sectors as opposed to agriculture, was perceived as a threat to the rural way of life and consequently fostered farmers' dislike of the cities.[109] Moreover, the forced collections and the theft of produce by ravening city-dwellers after the war intensified the farmers' aversion further still.[110]

Inflation had commenced in 1914, but its deferred effects proved slighter than expected with the end of price ceilings in the "controlled economy."[111] The onset of hyperinflation in 1922 and 1923 brought great advantages to Germany's farmers. Quite simply, farmers benefited because they controlled the food supply, that is, indispensable commodities which commanded a relatively strong exchange. Not only were farmers able to pay off debts with devalued currency, they were also effectively freed of all international competition because the rate of exchange acted like an insurmountable tariff.[112] During the hyperinflation, prices for agricultural

104 Moeller, 'Winners as Losers in the German Inflation', pp. 256-257, 274 and 283.

105 Moeller, 'Winners as Losers in the German Inflation', p. 287.

106 Bessel, *Germany After the First World War*, p. 219.

107 Bessel, *Germany After the First World War*, pp. 197-209.

108 A declining portion of province's population was working in agriculture: 70 percent in 1867, 35 percent in 1907 and 29.5 percent in 1925. This figure compared with some 33.5 percent in industry and crafts and 20.4 percent in trade and transport in 1925. *Statistisches Handbuch für Schleswig-Holstein*, 1951, p. 324; *Statistisches Jahrbuch für das Deutsche Reich 1929*, p. 23. Only the populations of the province's five cities exceeded 20,000 and five others 10,000, while a significant portion of the population, 49.05 percent continued to live in the countryside, in villages and small towns. This compared with 44.87 percent for Prussia overall. See the *Statistisches Jahrbuch für den Freistaat Preußen*, Preußisches Statistisches Landesamt Berlin (1929), p. 25.

109 K. J. Bade, 'Arbeitsmarkt, Bevölkerung und Wanderung in der Weimarer Republik' in M. Stürmer (ed.), *Die Weimarer Republik* (Königstein, 1980), pp. 160-162.

110 Bessel, *Germany After the First World War*, pp. 215-217. These thefts prompted farmers to form militias to defend their farms after the war. See p. 30 above.

111 See Holtfrerich, *Die deutsche Inflation*, pp. 90-91.

112 See Holtfrerich, *Die deutsche Inflation*, pp. 76-91.

produce rose faster than production costs and for most other goods: the "price scissors" opened in agriculture's favour. Agricultural specialists, such as Friedrich Aereboe, subsequently encouraged farmers to take the opportunity to update equipment and purchase machinery, improve their soil, restore their inventories and renovate buildings.[113]

Despite its benefits, inflation did not improve the state's standing in the farmers' eyes. Indeed, Robert G. Moeller insists that farmers saw themselves ultimately as losers in the inflationary period. Whilst he acknowledges that inflation helped to wipe out agriculture's debts, he argues that its gains were only relative to other groups. Although they were winners in a sense, farmers nonetheless felt greatly disadvantaged because of the state's regulation of production and supply in the period before the hyperinflation. Furthermore, Moeller underlines the continuity of economic uncertainty for farmers after 1914 and, therefore, the appeal of the National Socialists' promise of an end to economic instability.[114]

Currency stabilisation came at an inopportune time for many German farmers. By November 1923 they had sold their harvest for near-worthless paper money, which did not suffice to cover the cost of inventory for the coming year. Inflation had wiped out debts but it had also wiped out capital and savings, and in this respect, farmers shared the fate of all the German *Mittelstand*. The destruction of Germany's capital necessitated foreign credit and farmers made extensive use of the credit which subsequently flowed into Germany. They were compelled to take out loans not only to cover outlays for the following year's harvest and to replenish stocks, but also to continue the expansion and intensification of their farms to improve their soil and upgrade buildings and machinery.[115] Effortless repayment in the inflationary period, however, had encouraged imprudent investments. Much of the credit was expended poorly, used to cover deficits or for consumptive purposes rather than for new investments.[116]

In order to attract foreign capital, the Reich bank kept interest rates high, higher than all other European industrial nations and the United States. The Reich bank's

---

113 To this very purpose, Aereboe addressed the Chamber of Agriculture in Schleswig-Holstein on January 10, 1922. See Aereboe, *Der Einfluß des Krieges*, pp. 114-116.

114 Moeller, 'Winners as Losers in the German Inflation', pp. 255-288; R. G. Moeller, 'Economic dimensions of Peasant Protest in the Transition from Kaiserreich to Weimar' in R. G. Moeller (ed.), *Peasants and Lords in Modern Germany: Recent studies in Agricultural History* (Boston, London and Sydney, 1986), pp. 140-167; R. G. Moeller, *German Peasants and Agrarian Politics, 1914-1924: The Rhineland and Westphalia* (Chapel Hill and London, 1986), pp. 95-115 and 139-143.

115 Gessner, 'The Dilemma of German Agriculture', p. 136.

116 Aereboe, *Der Einfluß des Krieges*, p. 184; A. Gerschenkron, *Bread and Democracy in Germany* (Berkeley and Los Angeles, 1943), p. 149; H. James, *The German Slump: Politics and Economics 1924-1936* (New York, 1988), p. 254; W. C. McNeil, *American Money and the Weimar Republic. Economics and Politics on the Eve of the Great Depression* (New York, 1986), p. 216.

discount rate fluctuated between seven and ten percent from 1924 to 1930, compared with 4.7 and 4.5 percent in the United States.[117] Consequently, Germany received the largest amount of foreign capital in Europe.[118] Agriculture received a substantial portion of this credit, passed on by way of the *Rentenbank*, *Preußenkasse* and regional banks.[119] Yet governments did not seem concerned that farmers may have overburdened themselves, but instead encouraged agriculture to take up credits.[120]

Agricultural debt was not unusual in itself. However, just seven years after inflation, it had accumulated to pre-war levels.[121] This was remarkable, even in view of the reinstatement of 25 percent of debts paid off in the period from 14 June 1922 to 14 February 1924. Before the First World War interest rates had generally been between 3.5 and 4.5 percent.[122] However, credit available after inflation was offered at much higher rates of interest and mostly on a short-term basis. In 1924 farmers were accepting excessive interest rates, sometimes up to 24 percent for short-term loans. While the rates did fall soon afterwards, they still remained well above their pre-war levels.[123] The average interest rate for the total debt burden on German agriculture at the end of 1928 was estimated at 8.1 percent. In Schleswig-Holstein the average interest burden of more than RM 33 per hectare was among the highest in Germany.[124] Moreover, the burden became all the more pressing when fresh credits began to dry up in the second half of 1927, compelling some farmers to sell off parts of their properties.[125]

Schleswig-Holstein was a particularly debt-ridden province. On March 31, 1928, Schleswig-Holstein's farmers had accumulated secured (mortgage) credit and personal credit debts of RM 398.6 million and by March 31, 1929, RM 468.5 million.[126] Though there was great variation from farm to farm, the west coast coun-

---

117 D. Petzina, *Die deutsche Wirtschaft in der Zwischenkriegzeit* (Wiesbaden, 1977), p. 95.

118 P. Clavin, *The Great Depression in Europe, 1929-1939* (New York, 2000), p. 89.

119 R. Kuczynski, *Deutsche Anleihen im Ausland 1924-1928* (Berlin, 1929), pp. 64 and 79; McNeil, *American Money and the Weimar Republic*, p. 119.

120 James, *The German Slump*, p. 255.

121 Holtfrerich, *Die deutsche Inflation*, pp. 315-327; Sering, *Die deutsche Landwirtschaft*, pp. 49-52.

122 Aereboe, *Der Einfluß des Krieges*, p. 180.

123 J. B. Holt, *German Agricultural Policy 1918-1934* (Chapel Hill, 1936), p. 128; Sering, *Die deutsche Landwirtschaft*, pp. 108, 114 and 128.

124 These figures do not include Rentenbank interest. Schleswig-Holstein's average interest rate burden was the fourth highest of all 23 State Finance Office districts in 1928. See *Die Verschuldungs- und Kreditlage der deutschen Landwirtschaft in ihrer Entwicklung von der Währungsbefestigung bis Ende 1928*. Enquete-Ausschuß, Verhandlungen und Berichte des Unterausschusses für Landwirtschaft (II. Unterausschuß), vol. 12 (Berlin, 1930), pp. 28-29.

125 *Geschäftsbericht der Landwirtschaftskammer für die Provinz Schleswig-Holstein für das Rechnungsjahr 1927-1928*, p. 70. Also see James, *The German Slump*, p. 258.

126 *Geschäftsbericht der Landwirtschaftskammer für die Provinz Schleswig-Holstein für das Rechnungsjahr 1927-1928*, p. 80; *Geschäftsbericht der Landwirtschaftskammer für die*

ties, where agriculture was especially capital-intensive and the level of risk high, were among the most indebted: Eiderstedt RM 631 per hectare, North Dithmarschen RM 673 per hectare, South Dithmarschen RM 717 and Steinburg RM 610 per hectare. Husum and South Tondern were significantly lower, at 422 and 414 per hectare respectively. In other counties in which noteworthy Landvolkbewegung centres developed, debt was also considerable: Segeberg at RM 584 per hectare and Plön RM 525 per hectare.[127] These appeared all the more disturbing when compared with the Reich average of RM 304 per hectare.[128]

Psychologically disturbing for the farmers as these high levels were, the fertile west coast should nevertheless have been able to support a relatively large amount of debt. The rateable values (*Einheitswerte*), based on calculations of yields and soil quality – which were assessed a year in advance for taxation purposes – indicate that the expected returns per hectare were very high. But despite this, the measure of the burden on expected returns in Schleswig-Holstein was still much higher than the Reich average of 29.9 percent.[129] In Eiderstedt, debt measured 35 percent of the rateable value, in North Dithmarschen 39.1 percent, in South Dithmarschen 47.1, Steinburg 42.4, Husum 38.6, South Tondern 49.6, Segeberg 60.9 and Plön 38.5.[130] Bearing in mind that not all farms were in debt, the burden for individual indebted farms was actually much higher. Yet significant as the relative levels of debt were, absolute levels of debt became ever more critical as agricultural prices began to fall.[131] In other words, projected returns were meaningless when agricultural prices faltered and farm profits were destroyed.

The agricultural crisis that struck Germany in the second half of the 1920s was part of a worldwide agricultural depression. The most dramatic falls in prices occurred after the stock market crash, but the agricultural crisis began well before November 1929. For Schleswig-Holstein's farmers the "Golden Twenties", if one may use the term at all, ended in 1927 at the latest. Indeed, even the expression "relative stability" which is sometimes applied to the period of Weimar history from 1924 to 1929 appears quite inappropriate.[132]

Agricultural historians agree on the primacy of worldwide overproduction in explaining price falls and the agricultural crisis, most commonly pointing to the

---

Provinz Schleswig-Holstein für das Rechnungsjahr 1928-1929, p. 53. The 1928 figure is confirmed in Danker, *Die Verschuldung der schleswig-holsteinischen Landwirtschaft*, p. 9.

127 Danker, *Die Verschuldung der schleswig-holsteinischen Landwirtschaft*, p. 11; *Geschäftsbericht der Landwirtschaftskammer für die Provinz Schleswig-Holstein für das Rechnungsjahr 1928-1929*, p. 55.

128 See *Die Verschuldungs- und Kreditlage der deutschen Landwirtschaft*, Enquete-Ausschuß, p. 22.

129 *Die Verschuldungs- und Kreditlage der deutschen Landwirtschaft*, p. 22.

130 Danker, *Die Verschuldung der schleswig-holsteinischen Landwirtschaft*, p. 12.

131 Danker, *Die Verschuldung der schleswig-holsteinischen Landwirtschaft*, p. 19.

132 "Relative stability" is the term Detlev Peukert used. See D. Peukert, *The Weimar Republic. The Crisis of Classical Modernity* (New York, 1993), p. 207.

protracted decline in wheat prices to differentiate it from customary cyclical developments.[133] The German livestock sector was coming under increased pressure as production slowly returned to its pre-war level, whilst at the same time insufficient market rationalisation burdened producers with excessive transport and processing costs. The elastic demand for meat and dairy products meant under-consumption also played a growing role, especially after the onset of the general crisis.[134]

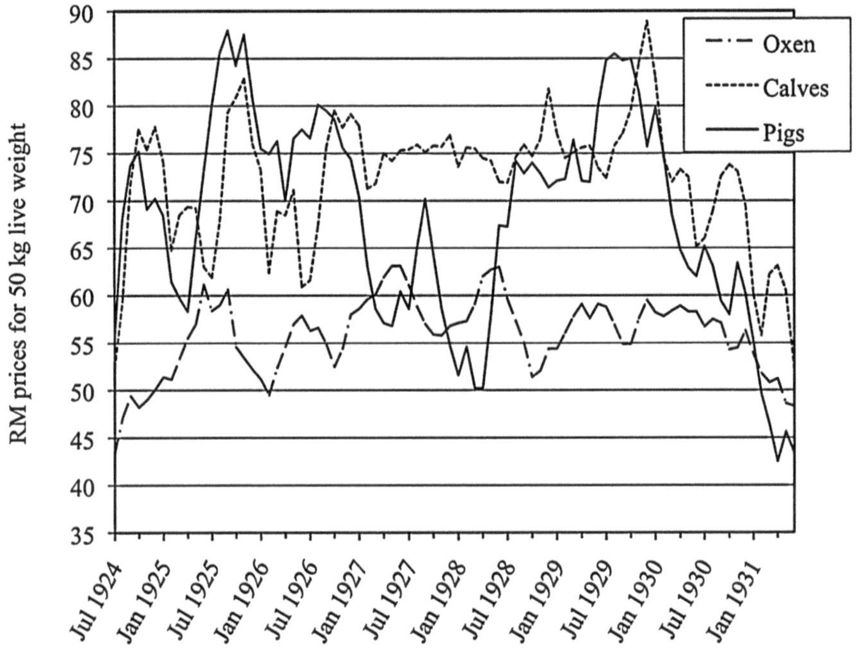

Graph 1: Livestock prices (Hamburg) from July 1924 - June 1931

Source: *Statistisches Jahrbuch für das Deutsche Reich*, Statistisches Reichsamt (Berlin, 1926), pp. 270-271; (1927), pp. 308-309; (1928), pp. 352-353; (1929), pp. 243-244; (1930), pp. 283-284; (1931), pp. 268-269.

The first signs of crisis in Schleswig-Holstein came as early as 1927 when pig prices slumped (See Graph 1). In part, the fall was due to the excessively rapid ex-

133 W. Abel, *Agrarkrisen und Agrarkonjunktur: Eine Geschichte der Land- und Ernährungswirtschaft Mitteleuropas seit dem hohen Mittelalter* (Hamburg and Berlin, 1978), pp. 281-284; Henning, *Landwirtschaft und ländliche Gesellschaft*, p. 194; Sering, *Die deutsche Landwirtschaft*, p. 78.

134 Gessner, *Agrarverbände in der Weimarer Republik*, pp. 87-89.

pansion of production prompted by over-optimistic forecasts.[135] Cattle attained relatively good prices in 1925 but then consistently suffered considerable autumn price falls on account of seasonal oversupply and pressure from Danish and Dutch imports. Some two thirds of the German cattle sold in Hamburg and Husum went to market in autumn, not only because the animals reached their prime weight after grazing throughout the summer, but also because stables for the winter were scarce.[136] Calf prices were much less volatile toward the end of the decade, even increasing in the winter of 1929/1930. However, this was not to the benefit of the cattle graziers, since the purchase of calves was their major outlay.[137]

Table 2: Disparity between Agricultural Prices and Production Costs

	1913/14	1924/25	1925/26	1926/27	1927/28	1928/29	1929/30
Agr. prices	100	121	114	132	132	124	117
Prod. costs	100	136	143	146	152	154	153
Index		89	80	84	87	80	76

Source: Sering, Die deutsche Landwirtschaft, p. 41.

Falls in prices alone did not destroy profitability. Production costs must also be taken into account. Throughout the 1920s production costs increased much faster than the prices attained for agricultural products. Higher costs for machinery and equipment, fuel, general services, wages and social security contributions, and to a lesser extent fertilisers, were all to the farmers' disadvantage. Only the cost of feed fell slightly.[138] The disparity between agricultural prices and production costs eliminated profits and Schleswig-Holstein's Chamber of Agriculture remonstrated its effects, alongside high taxes and the scarcity of capital, to the Prussian Minister of Agriculture from as early as 1925.[139]

---

135 Sering, *Die deutsche Landwirtschaft*, pp. 86-87.

136 See the *Geschäftsbericht der Landwirtschaftskammer für die Provinz Schleswig-Holstein für das Rechnungsjahr 1928-1929*, pp. 26-27; *Geschäftsbericht der Landwirtschaftskammer für die Provinz Schleswig-Holstein für das Rechnungsjahr 1929-1930*, pp. 26-27. For the seasonal cattle price falls as a result of fluctuations in supply, see Aereboe, *Wirtschaftslehre des Landbaues*, pp. 865-867 and Sering, *Die deutsche Landwirtschaft*, p. 488.

137 Danker, *Die Verschuldung der schleswig-holsteinischen Landwirtschaft*, p. 45.

138 *Landwirtschaftliche Buchführungsergebnisse,* Enquete-Ausschuß, Verhandlungen und Berichte des Unterausschusses für Landwirtschaft (II. Unterausschuß), vol. 4 (Berlin, 1929), pp. 11-16.

139 Landwirtschaftskammer für die Provinz Schleswig-Holstein to the Minister der Landwirtschaft, Domänen und Forsten in Berlin concerning "Feststellung der Kreditverhältnisse der schlesw.-holst. Landwirtschaft nach dem Stande von November 1925" Kiel, 20.11.1925, LAS 301/4125 and for a further example see the "Bericht der Landwirtschaftskammer über die Lage der Landwirtschaft" May, 1928, LAS 301/3518.

Farm incomes rose above their pre-war level in the 1920s but farmers were compelled to pay much higher prices for all non-agricultural consumer goods and services following currency stabilisation.[140] The disparity in prices between industrial and agricultural goods, which opened to agriculture's disadvantage after currency stabilisation is evidence of a decline in farmers' relative purchasing power and a rise in the costs of living (See Graph 2). These factors, together with the associated decline in profitability, threatened farmers' living standards and for Schleswig-Holstein's west coast farmers who had long been accustomed to prosperity, adjusting to lower levels of consumption was even more difficult.[141]

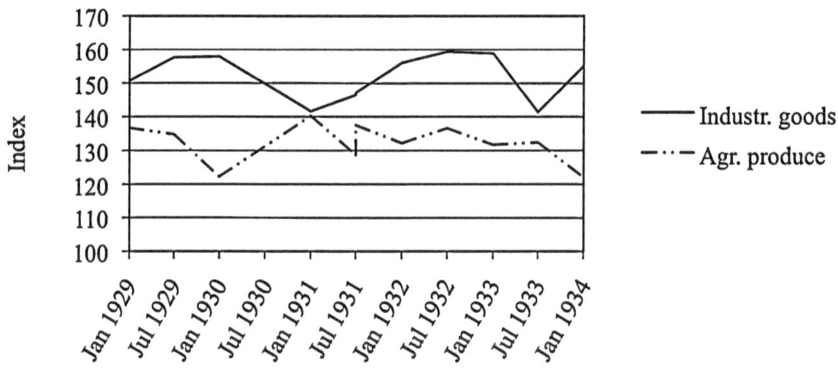

Graph 2: Index for industrial goods and agricultural produce (100 = 1913)

Source: *Geschäftsbericht der Landwirtschaftskammer für die Provinz Schleswig-Holstein für das Rechnungsjahr 1929-1930, p. 29.*

Diminishing fortunes were not just a matter of income. Changes to the taxation system were also important, and particularly so as a cause of discontent because they were a perceptible consequence of the Weimar Republic. Before the First World War, agriculture enjoyed a privileged position in regard to taxation. But under Weimar's taxation system, farmers had to pay a significantly greater share. During the inflationary period, however, they did not feel the effects of the change because, unlike wage earners, they were able to defer payments until after assess-

140 Sering, *Die deutsche Landwirtschaft*, esp. pp. 88-95.

141 Danker, *Die Verschuldung der schleswig-holsteinischen Landwirtschaft*, p. 27; Heberle, *From Democracy to Nazism*, p. 36; Topf, *Die Grüne Front*, p. 25.

ment and pay in devalued currency. Therefore, agriculture began to feel the consequences of tax reform only after currency stabilisation.[142]

Schleswig-Holstein's farmers had to pay the Reich's income, turnover and estate taxes, the Prussian real-estate tax, the communities' property and land tax, and interest on the Rentenbank land debt, as well as various other charges, including social insurance contributions and the Chamber of Agriculture levy. At the end of 1923 and the beginning of 1924, emergency tax measures and pre-payments to restore the state's revenue increased the burden further. Not only were the modes of assessment questionable, such as the application of the anachronistic Defence Tax Value of 1913, but the payment dates compelled farmers to sell at inopportune moments and take up credits, or as Schleswig-Holstein's Chamber of Agriculture complained, to sell their working capital, in order to avoid confiscation for arrears.[143] The transitional regulations thus had grim consequences for agriculture in 1924 and 1925. Indeed, according to Heinrich Becker, the state's liquidity came at the cost of agriculture's liquidity and drove farmers into debt.[144]

Max Sering also argued that farmers opted to take up new loans to avoid the yet more costly interest on arrears, and thus brought themselves further into debt.[145] According to Sering, Germany's farmers were required to pay 370 percent more tax under the new system than they did before the war.[146] Compared with the rateable values, the tax burden appeared tolerable: in 1928 German farms with a rateable value of RM 1,000 paid an average level of RM 25.5 tax per hectare. However, the rateable values were assessed in advance and so circumstances became critical as farm earnings dwindled. Overall, tax levels remained stable, consuming an ever-greater proportion of a farm's gross earnings, and increasingly meant the difference between net profit and loss.[147]

Burkhard Theine has called for a more regionally differentiated assessment of the situation of German agriculture in the Weimar Republic. He argues that in Westphalia, taxes cannot be held accountable for the farmers' debts, for although they did pay more tax than many other occupational groups, farmers' higher incomes meant that they were relatively wealthy. He argues further that the Reich governments showed consideration for agriculture from as early as November 1924

---

142 H. Becker, *Handlungsspielräume der Agrarpolitik in der Weimarer Republik zwischen 1923 und 1929* (Stuttgart, 1990), pp. 114-125; Holt, *German Agricultural Policy 1918-1934*, pp. 72-80.

143 *Geschäftsbericht der Landwirtschaftskammer für die Provinz Schleswig-Holstein für das Rechnungsjahr 1924-1925*, pp. 41, 43 and 50-51.

144 Becker, *Handlungsspielräume der Agrarpolitik*, pp. 212-233. See also Holt, *German Agricultural Policy 1918-1934*, pp. 145-148.

145 Sering, *Die deutsche Landwirtschaft*, p. 176.

146 Sering applied this figure even after the important taxation reforms of 1925. Sering, *Die deutsche Landwirtschaft*, p. 42.

147 *Landwirtschaftliche Buchführungsergebnisse*, Enquete-Ausschuß, p. 17; Sering, *Die deutsche Landwirtschaft*, esp. pp. 181-187.

by lowering income tax rates and increasing family concessions, cutting the estate tax, the turnover tax and, in 1929, the Rentenbank land debt interest payments.[148] Theine's study shows that Westphalian farmers remained relatively prosperous during the Weimar Republic, even after the onset of the crisis. Accordingly, Theine challenges the assumption that German agriculture, in its entirety, belonged to the losers of the Weimar Republic. The complaints of the Westphalian farmers and their organisations, he believes, gave an overly calamitous depiction of their situation.[149]

Theine's argument serves as a valuable warning of the potential perils of accepting the claims of agricultural organisations at face value and raises the question whether it was in fact farmers' perceptions, more than reality, that produced discontent. Former prosperity was bound to magnify the farmers' perception of the tax burden. Moreover, the manifestation of expanding tax collection on Schleswig-Holstein's west coast in the form of the construction of new finance offices, in Husum and in Leck (South Tondern), aroused great resentment in the countryside. Similarly, the rising salaries of civil servants in the mid 1920s appeared unmerited to farmers when in their eyes their own fortunes were deteriorating.[150]

Nonetheless, Theine's assessment of Westphalia cannot be applied to Schleswig-Holstein. Arguably, the province did benefit from the disproportionate transfer of income tax revenue.[151] Yet Reich taxes constituted just 35 percent of the total burden on agriculture by 1928. With income, wealth and sales taxes the domain of the Reich, the states and communities were obliged to raise real estate and land taxes to cover revenue shortfalls. Revenue from the flat-rate land tax, by far the most important agricultural tax, increased from RM 313.7 billion in 1925/1926 to RM 454.8 billion in 1929/1930.[152] Furthermore, the Schleswig-Holstein State Finance Office's assessment of average projected profits (*Durchschnittssätze*) for 1928/1929 moderated the Reich's rate cuts for income and turnover taxes. The increases did follow an improved harvest but were very high considering the widening disparity between production costs and agricultural prices.[153] So while the total

---

148 Theine, *Westfälische Landwirtschaft*, pp. 70-83.

149 Theine, *Westfälische Landwirtschaft*, pp. 43-103.

150 Civil service incomes rose by almost a third in 1927. H. James, 'Economic Reasons for the Collapse of the Weimar Republic' in I. Kershaw (ed.), *Weimar: Why did German Democracy Fail?* (New York, 1990), p. 39. Leck and Husum's finance offices moved into their new building in November 1928. Koops, *Südtondern in der Zeit der Weimarer Republik*, pp. 244-245; Sörensen, *Politische Entwicklung und Aufstieg der NSDAP in den Kreisen Husum und Eiderstedt*, pp. 364 and 367. Also see P. Heinacher, *Der Aufstieg der NSDAP im Stadt- und Landkreis Flensburg* (Flensburg, 1986), p. 270. For the Landvolk's views on civil servants and the new finance office in Husum, see chapter 3.

151 Holt, *German Agricultural Policy 1918-1934*, pp. 150-154.

152 Becker, *Handlungsspielräume der Agrarpolitik*, esp. pp. 243-248.

153 *Geschäftsbericht der Landwirtschaftskammer für die Provinz Schleswig-Holstein für das Rechnungsjahr 1929-1930*, pp. 43-45.

tax burden on agriculture appeared stable, a decline in farming income meant the relative tax burden increased. Though subject to regional variation, taxes were significantly higher than average on Schleswig-Holstein's west coast, and were highest in Dithmarschen and Eiderstedt at between RM 35 and 40 per hectare compared with the average Reich level of RM 25.5 tax per hectare.[154]

Throughout the 1920s, reports of improvements in Schleswig-Holstein's agriculture were short-lived and always qualified. Signs of recovery in the early months of 1926 were tempered by the spread of foot and mouth disease and swine fever, and then by bad weather and storm damage, leading to a particularly poor harvest in 1927.[155] Debt and the interest burden, unsatisfactory prices as well as the discrepancy between production costs and earnings threatened profitability. Very few farms specialising in pig-fattening in Schleswig-Holstein made a net profit in 1927/1928.[156]

Table 3: Compulsory Auctions of Rural Properties in Schleswig-Holstein

1924	**1925**	**1926**	**1927**	**1928**	**1929**	**1930**
1	16	45	34	95	84	86

Source: *Statistisches Jahrbuch für den Freistaat Preußen*, Statistisches Landesamt Berlin, vol. 22 (1926) p. 64; vol. 23 (1927), p. 60; vol. 24 (1928), p. 98; vol. 25 (1929), p. 93; vol. 26 (1930), p. 77; vol. 27 (1931), p. 130.

Debts and losses beleaguered increasing numbers of farmers, so much so that many could no longer endure. The consequence was defaults on interest and tax payments, leading to the confiscation and compulsory auction of stock and inventory and, in some cases, foreclosures. The rise in the number of compulsory auctions of rural properties was the strongest indication of the farmers' deteriorating situation, and marked Schleswig-Holstein as an early casualty of the agricultural crisis. Certainly there were many more compulsory auctions in East Prussia and slightly more in other eastern provinces such as Pomerania, even in 1928, but of

154 Danker, *Die Verschuldung der schleswig-holsteinischen Landwirtschaft*, p. 50.

155 *Geschäftsbericht der Landwirtschaftskammer für die Provinz Schleswig-Holstein für das Rechnungsjahr 1925-1926*, p. 146; *Geschäftsbericht der Landwirtschaftskammer für die Provinz Schleswig-Holstein für das Rechnungsjahr 1926-1927*, p. 29; *Geschäftsbericht der Landwirtschaftskammer für die Provinz Schleswig-Holstein für das Rechnungsjahr 1927-1928*, pp. 54-55.

156 *Geschäftsbericht der Landwirtschaftskammer für die Provinz Schleswig-Holstein für das Rechnungsjahr 1927-1928*, p. 53; Danker, *Die Verschuldung der schleswig-holsteinischen Landwirtschaft*, p. 43.

western German regions, only neighbouring Hanover could be compared with Schleswig-Holstein by 1929 and other regions not until 1930.[157]

The agricultural crisis that preceded the Great Depression struck the farmers on Schleswig-Holstein's west coast cruelly. Having weathered the misfortunes of the First World War and then inflation, they quickly accumulated perilous levels of debts at high levels of interest after currency stabilisation. The nature of agriculture in the region, the lack of diversity, perishable produce and substantial capital investments, meant they were particularly vulnerable, not only to debt, but to the falling prices and rising production costs which increasingly affected profitability negatively from 1927. Higher tax obligations under the Weimar system, which were well above average on the west coast, jeopardised profitability further still. It was the unprecedented severity of the crisis, afflicting farmers who had long been accustomed to prosperity, that was to trigger the unrest that developed into the most radical and vehement farmers' movement of the Weimar Republic.

Alongside economic factors, important regional peculiarities account for the potency of rural mobilisation in Schleswig-Holstein. These included the west coast farmers' confidence, independence and defiance, which were products of the affluence, social conditions and unique histories of the region. These histories, moreover, were to become a source of inspiration for the vocabulary and symbols of the farmers' later protests.

The lack of stability and the fragmentation of political parties and other organisations representing agricultural interests in the province must also be noted in explaining how rural distress could grow into an extraordinary movement. Party organisation in rural areas was weak and though Schleswig-Holstein's west coast farmers were accustomed to caring for local affairs, like the majority of their fellows throughout the province, most had little active experience in party politics. Many farmers were organised in agricultural-political associations and veteran or paramilitary groups, yet only the Rural League had a history preceding the First World War. The short-lived and fragmented nature of the parties and agricultural organisations meant that the farmers' ties to them were fragile.

The mobilisation of the rural population on a course independent of the parties and organisations is the subject of the following chapter, but the factors which paved the farmers' path to the right were in place long before the onset of the crisis. The forces of democracy were not consolidated after the revolution in rural Schleswig-Holstein. The middle-class electorate quickly abandoned liberalism. It might be argued that the Weimar Republic weathered the crises of the Kapp Putsch and inflation. However, it did not do so on account of support from the province's farmers. On the contrary, their swift siding with the putschists revealed their lack of commitment to the Republic as early as March 1920. The belief that agricultural

157 *Statistisches Jahrbuch für den Freistaat Preußen*, Statistisches Landesamt Berlin, vol. 26 (1930), pp. 78-79.

interests would be better served by tariffs rather than free trade bolstered the shift to the right, but so too did elements previously tied to Schleswig-Holstein's liberalism, foremost, regional patriotism, nationalism and the rejection of socialism. They were to help shape not only the ideology but also the form of the Landvolkbewegung together with the anti-Semitism and denunciations of international capital and democracy spread by the Rural League, the DNVP and veteran and paramilitary organisations.

Chapter Two

The mobilisation of the rural people

In the face of rising debt levels and endangered profitability, aggravated by plummeting pig prices and a harvest ravaged by bad weather, Schleswig-Holstein's rural population mobilised from the autumn of 1927. At first, farmers looked to their agricultural associations to secure government assistance but, with discontent growing rapidly, they soon turned to more dynamic methods of protest, holding the largest rural rallies in the history of the Weimar Republic. The rallies launched a movement that destroyed the structure of the agricultural organisations. However, the view that the rallies signified the early radicalisation of the farmers demands reassessment.[1] With their appeals to Berlin for assistance and continuing endeavours to force their organisations to create a more effective agricultural lobby, it is clear that the province's farmers were still looking for a solution to their difficulties within the Weimar system in 1928. Nonetheless, the outcomes of their demands for assistance and for agricultural unity ultimately combined to produce the unique and volatile conditions in which the radical Landvolkbewegung later developed, and which the NSDAP soon strove to use to its advantage. In early 1928 it was not yet clear what form future protest might take, though in other areas of Germany, as well as across the border in North Schleswig, there had been indications that discontent with deteriorating economic conditions might coalesce around the tax issue. However, by 1929 and 1930 agriculture's successful cooperation at the national level and its achievements in influencing policy came too late to curb unrest in the province.

Historians now agree that worldwide overproduction was the chief cause of the falls in agricultural commodity prices in the 1920s. But at the time, experts tended to attribute the falls to the prevailing social, tax and credit policies. They held governments and the social and political changes after the war, rather than deficiencies in market organisation and the state of agricultural modernisation, responsible for the crisis.[2] From this reasoning, it followed that the solution to agriculture's problems lay in government policy and this led farmers to call for government intervention. Furthermore, the introduction of tariffs and their initial successes in the 1880s had also fostered the belief that politics offered solutions to agricultural-economic problems.[3]

The idea that a unified agricultural lobby might better advance rural interests came to the fore in Schleswig-Holstein when the Farmers' Association held a rous-

1 Stoltenberg, *Politische Strömungen im schleswig-holsteinischen Landvolk*, p. 111.

2 Gessner, *Agrarverbände in der Weimarer Republik*, pp. 83-85; D. Orlow, *Weimar Prussia 1925-1933. The Illusion of Strength* (Pittsburgh, 1991), p. 173.

3 James, *The German Slump*, pp. 247-249.

ing meeting in Rendsburg at the end of November 1927. It was the climax of a series of smaller farmers' meetings held over the preceding weeks voicing anger about the crisis.[4] One hundred delegates and one thousand of their farming colleagues heard the chairman, Heinrich Stamerjohann warn that Schleswig-Holstein's "acclaimed circumspection and composure"[5] were at their end. The authorities were informed of the crisis first-hand as the Regierungspräsident, Adolf Johannsen, the Oberpräsident, Heinrich Kürbis, and representatives of the State Finance Office were in attendance, alongside the president of the Chamber of Agriculture, Count Rantzau. Resolutions sent to the Reich chancellor and Reich and Prussian ministers made demands which were repeated frequently over the coming months: extensive tax cuts and cheaper credit, thriftiness in public spending, a simplification of the taxation system and agricultural tariffs at the same level as those of industry.

Regierungspräsident Johannsen showed great empathy for the farmers' plight. In a provincially patriotic speech, he led the call for unification among the farmers – "in the spirit of the 1848 uprising" – to overcome the crisis. Despite the associations' bitter rivalry, the idea of unifying them had persisted throughout the 1920's.[6] However, Johannsen's call in a time of pending crisis provided the required impetus to bring the associations to negotiation.

Given the seemingly irreconcilable differences between the associations, one might think that greater cooperation, rather than unification, would have sufficed to provide a more effectual representation of agricultural interests. But here, concern about the greatly diminished power of the agricultural lobby relative to its privileged pre-war position should not be overlooked. The number of farmers in parliament had been declining since 1893 and agriculture was not accustomed to competing with industry and organised labour.[7] Not only did insecurity replace farmers' traditional feelings of superiority over these other groups during the Weimar Republic, but also because of their generally conservative outlook, they were suddenly at odds with the new form of state.[8] This was all the more true for agriculture in Schleswig-Holstein, as the state of Prussia had been governed by SPD dominated middle and left coalitions since 1919.

Further, the farmers' wish for organisational unification also had particular significance in the context of Weimar's recurrent elections and ever-changing coalitions. The rejection of party-politics and coalitions, frustration with the divisions in German society, and a desire for unity and harmony and against incompleteness,

---

4 O. Clausen, 'Der Aufmarsch des Landvolkes 1928' Schleswig-Holstein, 1 (1978), p. 18.

5 All subsequent quotations here from the rally in the *Kieler Neueste Nachrichten* No. 274, 23.11.1927, pp. 5-6

6 Thyssen, *Bauer und Standesvertretung*, p. 368. For the rivalry between the agricultural associations, also see chapter 1 above.

7 A. Panzer, 'Parteipolitische Ansätze der deutschen Bauernbewegung bis 1933' in H. Gollwitzer (ed.), *Europäische Bauernparteien im 20. Jahrhundert* (Stuttgart and New York, 1977), p. 530.

8 Gessner, 'The Dilemma of German Agriculture', pp. 140-141.

have been dubbed by Peter Gay as a "hunger for wholeness".[9] Schleswig-Holstein's farmers' demand for unification in the winter of 1927/1928 might well be seen as part of this broader phenomenon across German society.

On entering negotiations, Schleswig-Holstein's three agricultural organisations all agreed in principle to unification. The Regierungspräsident had warned the organisations against "half-measures" and the Rural League called for a complete fusion, but the Small Farmers' Association deemed a union with the Rural League premature.[10] Consequently, initial efforts were only partially successful, the Small Farmers' Association joining with the Farmers' Association on January 18. They settled debate about the preference for one or the other's national representative body by allying with both the German Peasantry and the Organisation of the German Farmers' Associations.[11]

On January 25 the Regierungspräsident chaired a meeting between the leaders of the Farmers' Association and the Rural League, hoping to induce them to action. Interestingly, matters of politics and administration appeared more pressing than finding a common programme. It was not so much that the Farmers' Association demanded that the Rural League give greater priority to animal products in trade policy and that it pursue the removal of tariffs on fodder which had caused difficulties. Central to the associations' disagreement was the Farmers' Association's avowal of party-political neutrality and of "respecting" the constitution as well as the question of the national leadership. It was not clear which national organisation a fused Schleswig-Holstein association should join. Despite their declared readiness, the Farmers' Association and the Rural League were unable to put their political differences aside. Instead they agreed only to form a working committee, binding for two years, to deal with economic and political questions for the purpose of a fusion in the future.[12]

Following the Farmers' Association meeting in November 1927, its chairman, Stamerjohann, and the chairman of the Chamber of Agriculture, Thygge Thyssen, met with Reich Chancellor Marx and Reich Agriculture Minister Schiele. The purpose of the meeting was to inform them of the agricultural crisis and consequent dejection among Schleswig-Holstein's farmers, and to request that the cabinet consider how it might help.[13] But impatient for assistance to arrive, and increasingly

9 P. Gay, *Weimar Culture* (London, 1974), pp. 73-106.

10 *Husumer Nachrichten* No. 3, 4.1.1928, p. 1; *Itzehoer Nachrichten* No. 9, 11.1.1928, p. 9.

11 *Itzehoer Nachrichten* No. 17, 20.1.1928, p. 5; Thyssen, *Bauer und Standesvertretung*, pp. 369-371 and 415.

12 *Husumer Nachrichten* No. 22, 26.1.1928, pp. 2-3; *Itzehoer Nachrichten* No. 23, 27.1.1928, p. 1.

13 See the "Vermerk des Ministerialrats Feßler über einen Empfang von Vertretern des Schleswig-Holsteinischen Bauernvereins beim Reichskanzler am 30. November 1927" in G. Abramowski (ed.), *Akten der Reichskanzlei. Die Kabinette Marx III und IV* (Boppard, 1988), pp. 1130-1131.

despondent over the associations' sluggishness to unite, the province's farmers turned to the idea of rallies to draw attention to their plight.

Led by a farmer from the west coast, Otto Johannsen (from Büsum, North Dithmarschen), approximately one hundred representatives met independently of their associations in Neumünster on January 7, 1928. They resolved not only to raise awareness of the crisis and to voice demands by means of large protest rallies on January 28, but also to use the occasion to reinforce the calls for unification made at the Farmers' Association's November meeting. To this end, they issued the organisations an ultimatum: to fuse by January 25, or else the farmers would fuse "over their heads".[14]

The agricultural associations publicised the fact that they played no part in the rallies' organisation. Yet at the same time they expressed their belief that their demands would not be heard in Berlin unless the farmers took action.[15] Rumours circulated in the Farmers' Association that the rallies were a conspiratorial manoeuvre of the Rural League aimed to oust the Farmers' Association from its dominant position in the province. Nevertheless, leaders of both associations were to march with their colleagues on January 28.[16]

The rally organisers did not enjoy the active assistance of the agricultural associations. The rallies were spontaneous, however, only in the sense that they were organised swiftly, not because they were improvised. In this regard, Gerhard Stoltenberg's account is misleading in asserting that Otto Johannsen spread his idea without the assistance of an organisation or apparatus, but by way of a network of private, family, professional and political connections.[17] A respected horse-breeder and the son of a community chairman and member of the board of the Chamber of Agriculture, Otto Johannsen was well connected and presumably used these connections to great effect.[18] Undoubtedly, local networks facilitated the organisation of the rallies, and did so all the more effectively due to the integrative structure of rural villages.[19] However, extensive coverage and publication of arrangements in the press during the lead-up to the rallies indicate that the planning was carried out primarily by publication.

The rallies were coordinated by the farmers who had met in Neumünster on January 7. They spread the idea at other agricultural association meetings and held further preparatory local meetings while continuing to meet in Neumünster.[20] These meetings were attended not only by farmers, but also by local tradespeople and

---

14 *Husumer Nachrichten* No. 9, 11.1.1928, p. 3 and No. 13, 16.1.1928, p. 4.

15 *Husumer Nachrichten* No. 13, 16.1.1928, p. 4 and No. 24, 28.1.1928, p. 6; *Itzehoer Nachrichten* No. 9, 11.1.1928, p. 9 and No. 17, 20.1.1929, p. 7.

16 *Itzehoer Nachrichten* No. 25, 29.1.1928, p. 5.

17 Stoltenberg, *Politische Strömungen im schleswig-holsteinischen Landvolk*, p. 110.

18 Thyssen, *Bauer und Standesvertretung*, pp. 374-375.

19 W. Pyta, *Dorfgemeinschaft und Parteipolitik 1918-1933* (Düsseldorf, 1996), pp. 197-198.

20 *Husumer Nachrichten* No. 16, 19.1.1928, p. 2; *Kieler Neueste Nachrichten* No. 21, 25.1.1928, p. 3; *Nordischer Kurier* No. 13, 15.1.1928, p. 4.

shopkeepers, who agreed to join the rallies in solidarity.[21] The farmers enjoyed their sympathy as a result of the close interdependence between agriculture and their businesses.[22] The county associations of tradesmen, citizens, proprietors and commerce all placed advertisements in the local newspapers calling on their members to support the rallies.[23]

The rallies were so well prepared that the *Itzehoer Nachrichten* deemed their organisation "worthy of admiration".[24] According to the many reports and advertisements in the provincial press in the days before January 28, county leaders were responsible for running individual rallies. Marshals marked with blue armbands would direct the participants, who were instructed to wear white armbands, to assemble in community groups and refrain from carrying flags to display political neutrality. Marches were to follow appointed routes. Speechmakers were chosen and allotted time-slots. Shops were requested to close and in Heide, the police administration announced that the marketplace would be closed to traffic for the benefit of the rally, warning that action would be taken against bystanders thinking to disrupt proceedings.[25] Stoltenberg's depiction of the farmers making their way to the cities "by foot, bicycle or horse and cart" is appealingly romantic but inaccurate.[26] Special trains and ferries – departure times and fares were advertised in the provincial press – had been arranged to bring farmers to the towns.[27]

January 28, 1928 was a crisp and sunny, late winter's day. People came from all over the province, even from the islands, to attend the rallies in almost all counties.[28] Unprecedented in scale, one speaker claimed that the farmers had not taken to the streets like this since the uprisings in the Middle Ages.[29] With many townspeople swelling the farmers' ranks, up to 140,000 people were said to have participated.[30] On the west coast, speakers addressed an 8,000 strong crowd at the

21 *Husumer Nachrichten* No. 16, 19.1.1928, p. 2 and No. 17, 20.1.1928, p. 3; *Itzehoer Nachrichten* No. 25, 29.1.1928, p. 5.

22 Wulf, *Die politische Haltung des schleswig-holsteinischen Handwerks*, pp. 15 and 59-60.

23 *Heider Anzeiger* No. 22, 16.1.1928, p. 4; *Husumer Nachrichten* No. 22, 26.1.1928, p. 4 and No. 23, 27.1.1928, pp. 2-4.

24 *Itzehoer Nachrichten* No. 25, 29.1.1928, p. 5.

25 *Heider Anzeiger* No. 22, 26.1.1928, p. 7; *Itzehoer Nachrichten* No. 24, 28.1.1928, p. 1.

26 Stoltenberg, *Politische Strömungen im schleswig-holsteinischen Landvolk*, p. 111.

27 *Heider Anzeiger* No. 22, 26.1.1928, p. 7; *Husumer Nachrichten* No. 21, 25.1.1928, p. 2, No. 21, 25.1.1928, p. 8, No. 22, 26.1.1928, p. 2 and No. 22, 26.1.1928, p. 4; *Kieler Neueste Nachrichten* No. 23, 27.1.1928, p. 8 and No. 24, 28.1.1928, p.11; *Nordischer Kurier* No. 23, 27.1.1928, p. 8 and No. 24, 28.1.1928, p. 8.

28 *Husumer Nachrichten* No. 24, 28.1.1928, p. 1.

29 *Kieler Neueste Nachrichten* No. 25, 29.1.1928, p. 13

30 Estimated by the farmers themselves, this number was almost certainly an exaggeration and was first given by the *Landbote für Schleswig-Holstein* No. 9, 2.2.1928, p. 1 and by Otto Johannsen during meetings with Reich ministers in Berlin, see the *Husumer Nachrichten* No. 31, 6.2.1928, p. 1. Estimates of the numbers at the individual rallies, in fact, varied among the local newspapers: *Eiderstedter Nachrichten* No. 8050, 31.1.1928, p. 1; *Heider*

marketplace in Husum, and in excess of 10,000 in Heide.[31] Incidentally, a number of the speakers who had come to call attention to the crisis "for one last time", later became active in the Landvolkbewegung.[32]

Rally in Tönning, Eiderstedt, January 28, 1928[33]

The speakers voiced a number of demands which were also articulated in resolutions. Stronger in intensity, the demands were virtually unaltered from those that the agricultural associations had voiced already.[34] First and foremost were the calls for (a) a change in German trade policy to free Germany from the need to import foodstuffs from abroad and to prevent all dispensable food imports; and (b) the assumption of the Rentenbank interest debt and excessive interest liabilities by the Reich, the capping of secured interest rate credit at six percent and the prohibition

Anzeiger No. 24, 28.1.1928, p. 1; *Husumer Nachrichten* No. 25, 30.1.1928, p. 1; *Itzehoer Nachrichten* No. 26, 31.1.1928 pp. 5-6; *Kieler Neueste Nachrichten* No. 25, 29.1.1928, p. 13 and No. 26, 31.1.1928, p. 5; *Nordischer Kurier* No. 25, 29.1.1928, p. 3.

31 The *Heider Anzeiger* estimated that 8,000 participants arrived by special train alone. See the *Heider Anzeiger* No. 24, 28.1.1928, p. 1.

32 These included not just Wilhelm Hamkens (Tetenbüll) and Claus Heim (Ostenfelde) but from the Husum meeting, for example, Harm Jensen (Wittbek) and August Lorenzen (Ostenfeld), in Itzehoe, Max Soth (Lockstedt), in Heide, Paul Adam Ross (Flederwürth) and Heinrich Frauen (Hemme), in Rendsburg, Markus Wulf (Eisendorf) in Bordesholm, Franz Schwarzloh (Neumünster) and in Tönning, Wilhelm Hönck (Harblek).

33 Photo: Dohnke collection.

34 See p. 54 above.

of short-term acceptance credits.[35] Further, the farmers called for greater government control over credit institutions, much greater thriftiness in public expenditure, a review of tax assessment and intriguingly, that communities receive limited rights to levy income tax. This was a remarkable demand because it demonstrated the farmers' faith in the local administration as opposed to the Reich, as income tax was the Reich's domain. The farmers also demanded that the authorities prevent the abuse of social security, insisted on a greater number of candidates from "practical" employment at future elections, demanded that the press "avoid foreign influences" and called for a "strengthening of German feeling" (*Stärkung des Deutschgefühls*). Finally, the farmers called for the immediate fusion of Schleswig-Holstein's agricultural associations.

As further testimony to the careful planning of the rallies, the content of the speeches and resolutions was essentially the same across the province. Priority was given to the first two demands: fundamental reform of trade policy and the provision of low-interest long-term credits.[36] In general, speakers everywhere remained matter-of-fact and rationalised their plight as a consequence of economic factors, arguing for instance, that inflation had deprived the farmers of capital, that foreign products were flooding the German markets, that harvests had suffered as a result of bad weather and that farmers were overburdened by taxes and debts. While there were occasional emotive assertions, such as "the purchase of dispensable foreign goods is a sin against the fatherland"[37] or that farmers were compelled to "murder" their piglets because they were unable to sell the grown animals, the speeches were generally quite measured when compared with later Landvolkbewegung outbursts. Demands for protection may have stemmed from the Rural League but patriotic-völkisch arguments were not such an important feature of the rallies. Calls for the refutation of Germany's sole war guilt and the demand for a revision of reparation payments which were "binding the German people to slavery" were made only at some of the rallies.[38]

In highlighting these demands as illustrative of the influence of völkisch associations on the rallies' leaders, Stoltenberg overstated the importance of völkisch ideas at the January 28, 1928 meetings. Völkisch ideas were characteristic not so much of these early rallies, but of the later Landvolkbewegung. Moreover, by sin-

---

35 For the resolution see the *Husumer Nachrichten* No. 24, 28.1.1928, p. 1 and *Itzehoer Nachrichten* No. 25, 29.1.1928, p. 6. For the speeches, see the *Husumer Nachrichten* No. 25, 30.1.1928, pp. 1-2; *Itzehoer Nachrichten* No. 25, 29.1.1928, pp. 5-6 and No. 26, 31.1.1928, pp. 5-6 and *Nordischer Kurier* No. 25, 29.1.1928, p. 3.

36 Not just at the rallies but also at the subsequent deputation in Berlin. See the *Nordischer Kurier* No. 32, 7.2.1928, p. 2.

37 *Husumer Nachrichten* No. 25, 30.1.1928, p. 1.

38 The calls were made in Heide and Tönning, for instance, but not in Itzehoe and Husum: *Eiderstedter Nachrichten* No. 8050, 31.1.1928, p. 1; *Heider Anzeiger* No. 24, 28.1.1928, p. 2; *Husumer Nachrichten* No. 24, 28.1.1928, p. 1; *Itzehoer Nachrichten* No. 25, 29.1.1928, p. 6.

gling out these demands, Stoltenberg's use of them as examples appears all the more misleading given the meagre coverage of the Landvolk's ideology in his thesis. As a consequence, he obscured a significant difference between the rallies and the later Landvolk movement.[39]

The rallies represented a unique and unprecedented upsurge of rural action: Schleswig-Holstein's farmers expressed their grievances in a manner and on a scale which was astounding. Mass rallies were not the characteristic tactics of farmers, but of the left, particularly of the trade unions.[40] The rallies of January 28, 1928 were a call to the governments for assistance and for protectionism but most significantly, they marked the beginning of the mobilisation of Schleswig-Holstein's rural population.

Stoltenberg's general appraisal is correct: the rallies launched a movement which destroyed the structures of the professional organisations and then, by way of increasing radicalisation and political homelessness, led the rural people to National Socialism.[41] One must differentiate much more clearly, however, between the rallies of January 28, 1928 and the later actions and protests of the Landvolkbewegung from the autumn of that year. Significantly, for the first time, activist expression was given to the rural population's grievances, independent of party and professional association politics. Nevertheless, the rallies were distinct from the Landvolkbewegung's protests because they were conducted in a manner largely cooperative with the authorities. Unlike the later, radical protests of the Landvolkbewegung, they were conducted in a spirit of cooperation with the authorities. The rallies of January 28, 1928 were not directed against the Weimar Republic. Stolten-

---

39 Stoltenberg's mistake appears to be the result of his focussing only on the rally in Heide. Stoltenberg *Politische Strömungen im schleswig-holsteinischen Landvolk,* p. 111. Bergmann and Megerle also mistakenly concluded that protests in Schleswig-Holstein were characterised by the rejection of Weimar and calls for radical change from the beginning. Where they trusted in their own research, however, they more accurately noted the increase in militant and radical actions at the end of 1928. See Bergmann and Megerle, 'Protest und Aufruhr der Landwirtschaft in der Weimarer Republik', pp. 242, 245 and 285, and compare p. 228. For a further example of this mistake, see Merkenich, *Grüne Front gegen Weimar*, p. 250. Beyer's depiction, by contrast, that the meetings were "free of extreme tendencies" was based solely on the meeting in Itzehoe. See Beyer, 'Die Agrarkrise und die Landvolkbewegung in den Jahren 1928-1932', pp. 165-166. Heberle also only noted the economic demands at the rallies: Heberle, *From Democracy to Nazism*, p. 63. In his history of the rise of the NSDAP in Flensburg, Peter Heinacher was more careful to differentiate between the rally in Flensburg, where economic demands were the focus, and Stoltenberg's depiction of the rally in Heide. See Heinacher, *Der Aufstieg der NSDAP im Stadt- und Landkreis Flensburg*, pp. 271-272.

40 This point has previously been made, both by Heberle, *From Democracy to Nazism*, p. 64 and Stoltenberg, *Politische Strömungen im schleswig-holsteinischen Landvolk*, p. 111.

41 Stoltenberg, *Politische Strömungen im schleswig-holsteinischen Landvolk*, p. 111.

berg very much overestimated the level of opposition expressed.[42] Moreover, it is too early to speak of the radicalisation of Schleswig-Holstein's rural population.

Regierungspräsident Johannsen had given the rallies his blessing and expressed hope that they would be used to unify agriculture.[43] Extremely popular, especially among farmers, Johannsen announced his retirement just days after the rallies. As the province's newspapers emphasised, he had been the first native "Schleswig-Holsteiner" appointed as Regierungspräsident since Schleswig-Holstein had become a Prussian province.[44] He was credited with ardent commitment to his province, not least because he was among those recognised for the rallies' success. Though he had reached retirement age, the *Hamburger Fremdenblatt* speculated that Johannsen may have retired because he had been at odds with Berlin, suggesting that his superiors had not approved of the rallies.[45] Prussian Agricultural Minister Steiger had warned of the consequences of illegal actions like those faced by the rioting viticulturists in Bernkastel on the Mosel, indicating that the Prussian authorities were concerned that the rallies had the potential to become disorderly.[46] However, the Interior Ministry had made no public objections. Indeed, local authorities and other public figures took part in the rallies. In Bordesholm, for instance, the Landtag Deputy Milberg gave a speech, and in other counties, so too did a number of community chairmen, including a later opponent of the Landvolkbewegung, Community Chairman Augustin (from Borsfleth, Steinburg County).[47]

The rallies in Schleswig-Holstein were part of a wave of rural protest which swept through northern Germany and across the nation, reaching Upper Silesia and northern Bavaria by mid-March.[48] A demonstration of some 20,000 people in the city of Oldenburg on January 26 was the most prominent in a series of meetings held in Oldenburg and Lower Saxony in January and February 1928.[49]

---

42 Stoltenberg, *Politische Strömungen im schleswig-holsteinischen Landvolk*, pp. 111 and 134, and the manuscript of his address 'Der Politische Umbruch in Schleswig Holstein, 1928-1930' Neumünster, 13.11.1979, p. 8, StaNms MA 5360 (III).

43 *Husumer Nachrichten* No. 24, 28.1.1928, pp. 5-6; *Itzehoer Nachrichten* No. 24, 28.1.1928, p. 6.

44 *Heider Anzeiger* No. 27, 1.2.1928, p. 2; *Kieler Neueste Nachrichten* No. 28, 2.2.1928, p. 5.

45 Cited in the *Husumer Nachrichten* No. 27, 1.2.1928, p. 1.

46 *Nordischer Kurier* No. 20, 24.1.1928, p. 1. For the viticulturists' protests, see p. 75 below.

47 *Itzehoer Nachrichten* No. 25, 29.1.1928, p. 5; No. 26, 31.1.1928, p. 6.

48 Bergmann and Megerle, 'Protest und Aufruhr der Landwirtschaft', pp. 221-228. Also see the *Landbote für Schleswig-Holstein* No. 20, 11.3.1928, p. 1.

49 O. Böckmann, *Landwirtschaft im Oldenburger Münsterland 1919-1933* (Vechta, 2000), pp. 195-196; P. Fritzsche, *Rehearsals for Fascism: Populism and Political Mobilization in Weimar Germany* (New York and Oxford, 1990), pp. 115-118; J. Noakes, *The Nazi Party in Lower Saxony 1921-1933* (London, 1971), pp. 108-112, 118-120; K. Schaap, *Oldenburgs Weg ins "Dritte Reich": Quellen zur Regionalgeschichte Nordwest-Niedersachsens* (Oldenburg, 1983), p. 39; D. Stegmann, *Politische Radikalisierung in der Provinz. Lageberichte und Stärkemeldungen der Politischen Polizei und der Regierungspräsidenten für Osthannover 1922-1933* (Hanover, 1999), pp. 69, 142-151.

As in Schleswig-Holstein, the rallies in Oldenburg had enjoyed the support of the towns, but town support for the rural people rapidly declined in both regions. In Oldenburg, tradespeople's sympathy for the rural people waned significantly after the Rural League called for tax-strikes.[50] In Schleswig-Holstein too, accusations of excessive prices for trade work and assertions of agriculture's primacy swiftly sullied relations between rural and townspeople.[51]

While the mobilisation of the rural population was to be seen across Germany, nowhere did the movement escape the control of the established organisations as it did in Schleswig-Holstein. In Lower Saxony and Oldenburg the rallies were organised by the agricultural organisations, especially the Rural League.[52] Similarly, in Pomerania, the rallies were also a Rural League initiative and the rural protest movement there continued to remain under its influence.[53] Only in Schleswig-Holstein did the rallies allow the development of a unified farmers' movement outside the control of agriculture's elites. Yet this ensued only after the hopes of immediate government assistance were disappointed.

The rallies of January 28, 1928 were part of a wider campaign of lobbying and pressure for state relief. In the weeks leading up to the rallies, the agricultural associations and Chamber of Agriculture, political parties, the Regierungspräsident and Oberpräsident all demanded assistance for agriculture. Regierungspräsident Johannsen was followed by representatives of the agricultural organisations who met with Reich and Prussian ministers as well as with Reich President Hindenburg in Berlin.[54] The DNVP requested in the Landtag that Schleswig-Holstein be declared

---

50 Fritzsche, *Rehearsals for Fascism*, p. 116. For the calls to tax-strikes, see p. 76 below.

51 Wulf, *Die politische Haltung des schleswig-holsteinischen Handwerks*, p. 61.

52 Bergmann and Megerle, 'Protest und Aufruhr der Landwirtschaft', pp. 225, 228, 245-267 passim and 286.

53 See S. Baranowski, *The Sanctity of Rural Life. Nobility, Protestantism, and Nazism in Weimar Prussia* (New York and Oxford, 1995), p. 130.

54 For the meetings of Regierungspräsident Johannsen and then representatives of Schleswig-Holstein's agricultural organisations with Reich and Prussian ministers in Berlin in mid-January, see the *Husumer Nachrichten* No. 16, 19.1.1928, p. 1 and No. 19, 23.1.1928, pp. 1-2. For telegrams sent by the Chamber of Agriculture together with the Oberpräsident to the Reich chancellor and other ministers, see the *Husumer Nachrichten* No. 19, 23.1.1928, p. 2 and *Kieler Neueste Nachrichten* No. 20, 24.1.1928, p. 5. For Stamerjohann's participation in a deputation of the "Organisation of the German Farmers' Associations" with leading Reich ministers, see the *Husumer Nachrichten* No. 28, 2.2.1928, p. 3 and for Friedrich Wilhelm Lübke's (Augaard), participation in a deputation of all three national agricultural organisations to Reich President Hindenburg at the beginning of February see the "Aufzeichnung des Staatssekretärs Meissner über den Empfang von Vertretern der Landwirtschaft beim Reichspräsidenten am 3. Februar 1928, 12 Uhr" in Abramowski, *Akten der Reichskanzlei*, pp. 1284-1286; *Husumer Nachrichten* No. 31, 6.2.1928, p. 2 and *Nordischer Kurier* No. 31, 5.2.1928, p. 1.

a crisis zone, the Rural League released a proposal for an emergency programme, and in the Reichstag there was debate over trade policy and the Reich shouldering the Rentenbank interest debt.[55]

The climax of this campaign took place on February 1 when Otto Johannsen and other rally leaders from Schleswig-Holstein joined further farming representatives from Oldenburg and the Emsland in Berlin to meet with Reich Ministers Hergt, Schiele, Brauns and Köhler.[56] Before departing for Berlin, Johannsen had declared that the measures for agriculture hitherto announced were completely inadequate. Raising expectations of a decisive outcome, he vowed to dispense with commiseration; to secure "deeds" and not "words".[57]

It must be emphasised here that Weimar Germany was overburdened by expectations. Farmers were not alone in their failure to recognise the much weakened condition of the post-war economy, nor to recognise that the state was in no position to meet the multitude of demands imposed upon it from all quarters, even before the Depression.[58] Johannsen's Berlin deputation was no exception.

Johannsen clearly believed it was a matter of the government simply choosing to assist agriculture, and on meeting with the ministers told them he would not return to Schleswig-Holstein empty handed, asking them directly if they wanted to restore agriculture's profitability.[59] He and his fellow representatives deemed profitability the key to solving the crisis and that this was contingent on the realisation of the rallies' two chief demands. The ministers did not – indeed could not – simply agree to these demands, at least not without terminating various bilateral trade treaties. Reich Agriculture Minister Schiele avowed his hope of closing the borders to imports in future, and this boded well with farmers for the long-term. His plan to expand the loan conversion scheme was certainly a step towards the farmers' demand for low-interest rate, long-term credits. However, the other immediate promises of assistance he made – increased veterinary controls, limitations on frozen meat imports, subsidies for evening out market fluctuations and rationalising production – were measures which had long been in discussion, and Johannsen expressed his doubts that these would restore profitability. Schiele was unwilling to

55 *Husumer Nachrichten* No. 14, 17.1.1929, p. 1, No. 16, 19.1.1928, p. 2 and No. 19, 23.1.1929, p. 6. The Rural League's programme was printed in the *Heider Anzeiger* No. 28, 2.2.1928, p. 9 and *Landbote für Schleswig-Holstein* No. 10, 5.2.1928, p. 2.

56 *Heider Anzeiger* No. 27, 1.2.1928, p. 1.

57 *Itzehoer Nachrichten* No. 26, 31.1.1928, p. 5. See also *Nordischer Kurier* No. 25, 29.1.1928, p. 3 and *Heider Anzeiger* No. 28, 2.2.1928, p. 2.

58 See here the essays of Ian Kershaw, Harold James, Dick Geary and Richard Bessel which strongly underline that Weimar was overburdened with expectations in Kershaw, *Weimar: Why did German Democracy Fail?*

59 See the reports and part-transcripts of the meeting in the *Heider Anzeiger* No. 30, 4.2.1928, p. 1 and No. 31, 6.2.1928, pp. 1-2; *Husumer Nachrichten* No. 30, 4.2.1928, p. 3 and No. 31, 6.2.1928, p. 1; *Itzehoer Nachrichten* No. 31, 5.2.1928, p. 5; *Kieler Neueste Nachrichten* No. 31, 5.2.1929, p. 10 and *Nordischer Kurier* No. 32, 7.2.1928, p. 2.

consent to the delegates' demand for Schleswig-Holstein to receive the same measure of help as East Prussia on the grounds that East Prussia required special consideration because of its isolation. However, he agreed that appropriate measures for Schleswig-Holstein were also required.

The farmers' expectations would have been fuelled by Vice-Chancellor Hergt's widely published remark that only as a result of the meeting had he become aware of the full extent of the crisis.[60] This may have mollified Johannsen and his fellow delegates, giving them the impression their efforts had propagated greater understanding for agriculture, but it did not lead to the immediate, fundamental change in trade policy for which they had hoped.

Though they did not meet expectations and only effected a temporary reprieve, the achievements of the farmers' demands for assistance must be acknowledged. Regierungspräsident Adolf Johannsen was instrumental in initiating and securing tax relief measures and other crisis assistance. On account of the weather damage to the 1927 harvest, he directed the land registry offices (*Katasterämter*) to process all applications for postponement and reductions of the Prussian real-estate tax quickly and benevolently, deferring payment until the end of September.[61] Further, he appealed to Prussian Agriculture Minister Steiger, who promised to ensure that credit institutions be as considerate as possible in recalling credits and who agreed to provide emergency aid for those areas of Schleswig-Holstein affected by storm and flood damage.[62]

Weather damage, more than any consideration of indebtedness and profitability, also appeared to be the decisive factor in the Schleswig-Holstein's State Finance Office's decision to grant tax relief. It announced its willingness to accept applications for postponement and tax reductions, designating farms in areas affected by poor weather as worthy of special consideration.[63] At the Reich level, the government parties agreed that farmers should only pay taxes according to their existing incomes, and resolved to allow further postponement of payment and to refrain from imposing arrears. They rejected the idea of shouldering the Rentenbank interest payments, but agreed that the repayment of the previous year's RM 100 million fertiliser credits could be postponed until after the 1928 harvest.[64]

60 *Heider Anzeiger* No. 31, 6.2.1928, pp. 1-2; *Husumer Nachrichten* No. 31, 6.2.1928, p. 1; *Nordischer Kurier* No. 32, 7.2.1928, p. 2. For evidence that Hergt had, in fact, been well aware of the extent of the crisis beforehand, see the "Kabinettssitzung vom 20. Januar 1928, 17 Uhr im Reichstagsgebäude" in Abramowski, *Akten der Reichskanzlei*, p. 1258.

61 *Nordischer Kurier* No. 13, 15.1.1928, p. 1.

62 *Heider Anzeiger* No. 15, 18.1.1928, p. 1 and No. 27, 1.2.1927, p. 2; *Husumer Nachrichten* No. 15, 18.01.1928, p. 1; *Itzehoer Nachrichten* No. 10, 12.1.1928, p. 1 and, among the few reports on agriculture in the SPD press, *Schleswig-Holsteinische Volkszeitung* No. 24, 28.1.1928, p. 7. For a report on the weather damage to agriculture in 1927, see the *Itzehoer Nachrichten* No. 10, 12.1.1928, p. 5.

63 *Heider Anzeiger* No. 25, 28.1.1928, p. 1; *Husumer Nachrichten* No. 13, 16.1.1928, p. 4.

64 *Heider Anzeiger* No. 25, 28.1.1928, p. 1.

After the rallies, Finance Minister Köhler directed the finance offices to be benevolent in accepting applications for the postponement and reduction of taxes, and he also called on credit institutes to relax the recalling of loans.[65] At the Otto Johannsen-led deputation, Köhler confirmed that farmers without income would not be compelled to pay taxes and that the compulsory collection of taxes and confiscation of farmers' inventory for arrears would be avoided. He subsequently repeated these instructions to the finance offices.[66] In the short-term, at least, these measures resulted in a considerable fall in the tax burden.[67]

The Reich government had identified the principal problems of the crisis: Germany's farmers were heavily in debt and their farms unprofitable. On January 20, 1928, Schiele informed the cabinet that agriculture was saddled with an RM 800 to 900 million annual interest burden and was making a total yearly loss of between RM one and one and a quarter billion.[68] The Reich government attempted to address these problems with its "Emergency Programme for Agriculture" at the end of March. Following a meeting of Reich party factions on February 18, the programme was approved within six weeks. This was possible because, with Reich elections imminent, all parties were eager to acknowledge the agricultural crisis in the hope of securing farmers' votes.[69] Yet many of the programme's key points had been promised to the farmers months beforehand.

Schiele had told Stamerjohann and Thyssen at the end of November 1927 that his ministry would soon provide assistance for agricultural indebtedness. However, beyond the special assistance for East Prussia, the cabinet had proved indecisive.[70] But as the farmers and their representatives lobbied throughout January, Schiele's programme took shape. Loan conversion to reduce farm debt and subsidies to even out market fluctuations and improve distribution appeared certain before the rallies.[71] The limitations on trade policy had also been made clear. In mid-January, Schiele promised Regierungspräsident Johannsen that agriculture would not be

---

65 *Heider Anzeiger* No. 27, 1.2.1927, p. 2; *Schleswig-Holsteinische Volkszeitung* No. 24, 28.1.1928, p. 7.

66 *Itzehoer Nachrichten* No. 34, 9.2.1928, p. 7; Becker, *Handlungsspielräume der Agrarpolitik*, p. 248.

67 By April 1928 only 36 percent of the tax due in Schleswig-Holstein had in fact been collected. See James, *The German Slump*, p. 259.

68 "Kabinettssitzung vom 20. Januar 1928, 17 Uhr im Reichstagsgebäude" in Abramowski, *Akten der Reichskanzlei*, p. 1257.

69 A. Panzer, *Das Ringen um die deutsche Agrarpolitik* (Kiel, 1970), pp. 118-121. For the Reich Rural League's similar assessment, see the *Landbote für Schleswig-Holstein* No. 30, 15.4.1928, p. 1.

70 "Vermerk des Ministerialrats Feßler über einen Empfang von Vertretern des Schleswig-Holsteinischen Bauernvereins beim Reichskanzler am 30. November 1927" in Abramowski, *Akten der Reichskanzlei*, pp. 1130-1131; Panzer, *Das Ringen um die deutsche Agrarpolitik*, pp. 116-117.

71 *Husumer Nachrichten* No. 23, 27.1.1928, p. 1; *Itzehoer Nachrichten* No. 24, 28.1.1928, p. 1.

burdened by further meat imports, but regretted that there could be no revision of existing trade agreements.[72] Yet there was soon talk of restricting frozen meat imports, and Schiele also committed to this shortly after meeting with Schleswig-Holstein's rally leaders.[73]

Schiele believed that the agricultural crisis could only be solved by fundamental changes to Germany's economic policy aims. His emergency programme contained many of the elements of his later policies under Brüning, foreshadowing, for instance, the expanded Eastern Aid (*Osthilfe*) programme, which also included loan conversion and other measures to reduce distribution costs.[74] It did not, however, include the much-desired fundamental changes to tariff policy.

The programme was part of the Reich government's and parliament's pre-election spending spree, but was modest in comparison to earlier subsidies, amounting to less than one percent of the Reich budget.[75] The emergency programme of some RM 170 million, which also encompassed a social programme and industry export assistance, provided a total of RM 64 million for agriculture. It granted eight million – subsequently topped up to 22 million – for the reorganisation of livestock and meat production, distribution and marketing to stabilise prices, as well as 30 million for other products, 25 million for the reorganisation of agricultural cooperatives, and one million for poultry breeding development. RM 716,000 were made available to crisis areas for securities for fertiliser credits.[76]

In further measures to buoy livestock and meat prices, the government also expanded the import licence system and limited the tariff-free contingent of frozen meat to half a million tonnes. These measures did indeed lead to a fall in meat imports, but unfortunately there was only a temporary stabilisation of prices.[77] Similarly, the subsidies also failed to bring enduring improvement for agriculture.[78]

RM 200 million were made available to agriculture for loan conversion. This was as a result of Reich and State assistance and a US $25 million loan. The government defrayed the costs of conversion and provided the security for the loan, and the money was distributed by individual credit institutions under public law to be used to consolidate short-term loans and pressing debts under more favourable

72 *Husumer Nachrichten* No. 16, 19.1.1928, p. 1; *Itzehoer Nachrichten* No. 17, 20.1.1928, p. 5.

73 *Heider Anzeiger* No. 25, 28.1.1928, p. 1; *Husumer Nachrichten* No. 30, 4.2.1928, p. 2.

74 Gessner, *Agrarverbände in der Weimarer Republik*, pp. 155-156.

75 Becker, *Handlungsspielräume der Agrarpolitik*, p. 260; Panzer, *Das Ringen um die deutsche Agrarpolitik*, p. 121.

76 Minutes of the ministerial and party-leader discussions concerning the "Arbeitsnotprogramm" from 17.2.1928 to 27.2.1928 in Abramowski, *Akten der Reichskanzlei*, pp. 1319-1341; Holt, *German Agricultural Policy 1918-1934*, pp. 110-111; Panzer, *Das Ringen um die deutsche Agrarpolitik*, pp. 119-120.

77 Panzer, *Das Ringen um die deutsche Agrarpolitik*, p. 122.

78 Becker, *Handlungsspielräume der Agrarpolitik*, p. 268.

rates of interest.[79] Yet, given the substantial amount of short-term debt, RM 200 million was no vast sum, and after a revival of investment in the first half of 1928, long-term loans were subsequently scarce.[80]

After the government crisis of 1928, the Reichstag was dissolved before further measures could be instituted. At the elections in May, consistent with the trend across the Reich, the SPD was the greatest beneficiary and the DNVP the greatest casualty. The SPD captured 35.3 percent of the vote as opposed to 30.3 percent in December 1924, while the DNVP's share fell from 33 to 23 percent. The DDP's share of the vote continued to plummet, falling a further three percent to just 5.7 percent and the DVP's share fell from 14.6 to 13.7 percent. The NSDAP, it should be noted, remained a minor party with just four percent of the vote.[81] The failure of the middle-class parties, the DNVP, DVP and DDP, which the farmers' organisations advocated, ultimately can only have encouraged the view that farmers needed to look to alternatives in the search for a solution to their difficulties.

Of consequence to the emergency programme for agriculture, a new government was formed only after long negotiations. Notwithstanding the delays, the new government, headed by the SPD, did ultimately support the promised loan conversion scheme, the marketing apparatus and distribution measures and the rationalisation of the cooperatives, committing RM 100 million over five years.[82] Beyond this, however, little was done to institute an agricultural programme until April 1929.[83]

Schleswig-Holstein's farmers continued to work towards the unification of their associations rather than simply wait inactively for Reich assistance. Initially, however, there was a setback. In fact, just days after the rallies, it appeared that the merger of the province's agricultural associations had failed. Following a provincial leadership meeting, the Rural League withdrew from the newly-formed working committee, returning to its former standpoint that only complete fusion was acceptable.[84] Although the Farmers' Association had declared its adherence to the principles of private property, its unity programme required party-political neutrality, whereas the Rural League declared that it would never cease combating political parties based on Marxist principles. At the same time, and quite incongruously, the Rural League also declared, "The rural people are fed up being the play-

79 Holt, *German Agricultural Policy 1918-1934*, pp. 138-139; Kuczynski, *Deutsche Anleihen im Ausland 1924-1928*, p. 88.

80 Clavin, *The Great Depression in Europe*, pp. 92-94.

81 *Statistisches Jahrbuch für das Deutsche Reich* (Berlin, 1928), pp. 580-581.

82 Becker, *Handlungsspielräume der Agrarpolitik*, p. 262; Gessner, *Agrarverbände in der Weimarer Republik*, p. 162

83 Gessner, *Agrarverbände in der Weimarer Republik*, p. 165.

84 *Heider Anzeiger* No. 28, 2.2.1928, p. 2; report from the Regierungspräsident to the Oberpräsident in Kiel concerning "Entwicklung des landwirtschaftlichen Berufsorganisationswesens in Schleswig-Holstein" Schleswig, 22.1.1930, LAS 309/22765.

things of political parties."[85] In other words, the Rural League demanded "complete fusion" but without the Farmers' Association's national organisation, the German Peasantry.

Public relations between the Rural League and the Farmers' Association swiftly deteriorated, each blaming the other for failing to show consideration for the others' requirements.[86] The Farmers' Association accused the Rural League of simply trying to exploit the rallies to capture greater support in the province; in fact it was even rumoured that the Rural League had instigated the rallies for this very purpose.[87] The Rural League called the working committee a half-measure and insisted that it was simply following the resolve of the rallies' reported 140,000 participants in calling for a "complete fusion".[88]

But the rallies had given the unity endeavours fresh impulse. On his return from Berlin, Otto Johannsen led a meeting of county representatives in Neumünster at which it was decided that mere cooperation was inadequate, that all counties must fuse to form a "united front".[89] Even before the rallies, some farmers had already recognised the impracticality of a union incorporating the national organisations and had advised a union without them. Likewise at the county level, some farmers' organisations had expressed their intent of fusing without waiting for the provincial leaders' decisions.[90] In Eiderstedt, under the leadership of Wilhelm Hamkens (from Tetenbüll), the associations hesitated no longer and Johannsen called on all the other counties to follow its example.[91] By mid-March, North Dithmarschen, Husum and Stormarn had also merged and joined the "Johannsen" or what was now known as the "Landvolkbewegung". In Oldenburg and Bordesholm most farmers also joined and in almost all other counties, fusion committees were established.[92]

The movement was driven from below but soon the Rural League put itself at the movement's disposal, and Johannsen brought its and the Farmers' Association's provincial leaders together. A "Committee of Six" (*Sechserausschuß*) was formed

---

85 *Husumer Nachrichten* No. 30, 4.2.1928, p. 2; *Kieler Neueste Nachrichten* No. 30, 4.2.1928, p. 5; *Landbote für Schleswig-Holstein* No. 10, 5.2.1928, p. 1.

86 *Itzehoer Nachrichten* No. 36, 11.2.1928, p. 5.

87 See, for example, the resolution of a meeting of Farmers' Association representatives in Neumünster, reported in the *Heider Anzeiger* No. 33, 8.2.1928, p. 1 and *Kieler Neueste Nachrichten* No. 34, 9.2.1928, p. 5. Also see the *Nordischer Kurier* No. 32, 7.2.1928, p. 3 and Stoltenberg, *Politische Strömungen im schleswig-holsteinischen Landvolk*, p. 112.

88 *Heider Anzeiger* No. 34, 9.2.1928, p. 1 and No. 35, 10.2.1928, p. 1; *Kieler Neueste Nachrichten* No. 35, 10.2.1928, p. 5 and No. 36, 11.2.1928, p. 5.

89 *Heider Anzeiger* No. 31, 6.2.1928, p. 1.

90 In Rendsburg and Segeberg: *Itzehoer Nachrichten* No. 18, 21.1.1928, pp. 7-8; *Kieler Neueste Nachrichten* No. 20, 24.1.1928, p. 5 and No. 23, 27.1.1928, p. 5.

91 *Heider Anzeiger* No. 33, 8.2.1928, p. 1, No. 34, 9.2.1928, p. 1 and No. 35, 10.2.1928, p. 1; *Itzehoer Nachrichten* No. 35, 10.2.1928, p. 7 and No. 36, 11.2.1928, p. 5.

92 *Heider Anzeiger* No. 67, 19.3.1928, p. 2.

with three leaders from both the Farmers' Association and Rural League to oversee fusion negotiations at the provincial level.[93] That Johannsen's movement returned to the idea of a working committee proved that it was not a Rural League puppet but a genuine attempt to appease both sides. Nevertheless, negotiations once again focussed on questions of political direction and national leadership for a united organisation. The Committee of Six managed to work out an emergency programme but disputes soon erupted, for instance, over the Farmers' Association's inclusion of the German Democratic Party among the parties it supported at the May Reichstag elections.[94] Above all, the committee was unable to resolve the national leadership question: the Rural League wanted only to be associated with the Reich Rural League and the "Organisation of the German Farmers' Associations" whereas the Farmers' Association remained unwilling to leave the German Peasantry.

After an unproductive provincial conference, the Committee of Six conceded defeat in July. Four members of the committee, Otto Johannsen and the Farmers' Associations' representatives, called for a new organisation to join all three national associations. Criticised by reactionaries in his own Rural League, however, and disillusioned by the apparent failure of his efforts, Johannsen subsequently withdrew from the unity endeavours.[95] Yet fusion continued at the county level. By August the associations had merged in at least 11 counties and these resolved to join in a "Unity Organisation" (*Einheitsorganisation*) under the leadership of the Farmers' Association's Max Tönnsen (from Schaalby) and the Rural League's Otto Köhler (from Bühnsdorf).

Most farmers joined the new organisation, making it by far the strongest in the province, but they also retained their membership in one of the old organisations. The new "Unity Organisation" or "Farmers' League" (*Bauernbund*) agreed to accept all three Reich associations until a ballot was held on April 1, 1929 to determine whether to remain in the German Peasantry permanently. Nonetheless, any such decision was contingent upon the agreement of the national organisations.

Unfortunately, hopes were dashed in September, when at the national level, both the Rural League and the new president of the Organisation of the German Farmers' Associations, Andreas Hermes, declared fusion an impossibility. According to Hermes, the organisations could not surrender their independence, as ideological, cultural and economic differences could not simply be ignored.[96] The new

93 Unless noted otherwise, the following overview of provincial developments follows the report from the Regierungspräsident to the Oberpräsident in Kiel concerning "Entwicklung des landwirtschaftlichen Berufsorganisationswesens in Schleswig-Holstein" Schleswig, 22.1.1930 in LAS 309/22765; Stoltenberg, *Politische Strömungen im schleswig-holsteinischen Landvolk*, pp. 112-114 and 119-121 and Thyssen, *Bauer und Standesvertretung*, pp. 371-386.

94 See, for example, the *Landbote für Schleswig-Holstein* No. 44, 24.5.1928, p. 2.

95 Thyssen, *Bauer und Standesvertretung*, pp. 377 and 384.

96 Thyssen, *Bauer und Standesvertretung*, pp. 385-386.

Farmers' League remained, but this ended its prospects of working with the Reich Organisations. Consequently, national representation remained the domain of the old organisations.[97] Unable to supplant the old organisations, the Farmers' League failed to unify completely Schleswig-Holstein's agriculture. Instead, it too became a further provincial organisation. For their members, the organisations' lack of progress after months of negotiation could only result in increasing frustration and disappointment.

The unification endeavours in the province awakened similar aspirations across Germany in 1928.[98] In Schleswig-Holstein, the rallies set the agricultural lobby on an uncompromising and troublesome course to fusion. Once upon this course, Schleswig-Holstein's farmers perhaps missed the chance to explore alternative forms of cooperation. Elsewhere, however, the agricultural lobby sought to pursue its interests by other means. The formation of an independent farmers' party – the Christian-National Farmers and Rural People's Party (*Christlich-Nationale Bauern- und Landvolk-Partei*, CNBLP) – and of a national coalition of farmers' organisations – the Green Front (*Grüne Front*) – sought to provide farmers with better representation in Berlin.

German farming organisations had generally favoured collaboration with broader based political parties in preference to putting up separate farming lists at elections.[99] With the disintegration of the agricultural-political structures from 1928, however, the CNBLP developed as a new party-political force. Though known as the *Landvolkpartei*, it was not associated with Schleswig-Holstein's Landvolkbewegung, but with the Reich Rural League. The CNBLP never gained significance in Schleswig-Holstein after the Unity Movement discouraged farmers from supporting it in 1928.[100] Elsewhere it was more successful in attracting disenchanted leaders of the DNVP and DVP, seeking to gain concessions for agriculture with pragmatic politics. Contrary to its purpose, however, it did not consolidate the rural vote, but rather split the Rural League's support from the DNVP while putting

---

97 Report from the Regierungspräsident to the Oberpräsident in Kiel concerning "Entwicklung des landwirtschaftlichen Berufsorganisationswesens in Schleswig-Holstein" Schleswig, 22.1.1930, LAS 309/22765.

98 H. - J. Puhle, *Politische Agrarbewegungen in kapitalistischen Industriegesellschaften* (Göttingen, 1975), p. 90; Topf, *Die Grüne Front*, pp. 5 and 14.

99 The main exception was the party-political arm of the Bavarian Farmers' League (*Bayerischer Bauernbund*), the German Farmer's Party (*Deutsche Bauernpartei*). See Jones, 'Crisis and Realignment: Agrarian Splinter Parties', pp. 203, 214 and 216 and Panzer, 'Parteipolitische Ansätze der deutschen Bauernbewegung bis 1933', pp. 533 and 538.

100 The CNBLP received just 0.3 percent of the vote in Schleswig-Holstein in the 1928 Reichstag elections and 3.8 percent in 1930. Heberle, *From Democracy to Nazism*, p. 94. Also see M. Müller, *Die Christlich-Nationale Bauern- und Landvolkpartei, 1928-1933* (Düsseldorf, 2001), pp. 451-455 and Thyssen, *Bauer und Standesvertretung*, p. 378.

pressure on the Rural League as well.[101] The CNBLP's moderate success in the 1928 Reich election came largely at the expense of the DNVP and although it continued to grow, it nevertheless finished well behind the DNVP and the NSDAP in the 1930 elections. Too closely associated with Schiele's policies which failed to alleviate the agricultural crisis, fluctuating between opposition to and support for the Brüning government, weakened by leadership struggles and too dependent on the Rural League for its organisation, the CNBLP suffered massive defeats in the elections of 1932. Like the German Farmers' Party, the CNBLP attracted votes away from the older parties but failed to build a genuine nation-wide voter-base and therefore ultimately served as a "conduit for the transmission of disaffected rural voters … to … the NSDAP."[102]

Beyond Schleswig-Holstein's borders the desire for greater cooperation was articulated in a move to form alliances rather than to strive for unification. The president of the Organisation of the German Farmers' Associations, Andreas Hermes, and the leader of the Bavarian Christian Farmers' Association (*Bayerischer Christlicher Bauernverein*), Georg Heim, deemed organisational fusion an illusion which would not gain greater parliamentary influence on agricultural policy. Instead, they accepted the different political interests and allegiances of the associations and the particular economic interests of the grain producing eastern estate owners as well as the western livestock and garden farmers.[103] Together with the Rural League's Schiele, they began to work on the coalition in 1928, but it was not until January 1929 that the organisations' rank and file were informed of the new Reich Farmers' Front or Green Front.[104]

Where the negotiations in Schleswig-Holstein had failed to settle the question of the inclusion of the German Peasantry, the Green Front managed to involve all three national agricultural organisations in a coalition. It provided the basis for common political activity, enabled the agricultural organisations to exercise influence in parties and parliaments and did so without the organisations forfeiting their autonomy, which to the satisfaction of the Rural League did not necessitate identification with the Weimar system.[105]

The formation of the Green Front demonstrated that agricultural associations did not have to fuse entirely to be effective. Its efforts came too late to curb the spread of unrest in Schleswig-Holstein but reveal, nonetheless, what an alliance could achieve. In March 1929 the Green Front publicised its programme, demanding the unilateral termination of existing trade treaties, the exclusion of foreign

101 Gessner, *Agrarverbände in der Weimarer Republik*, pp. 108-117; Jones, 'Crisis and Realignment: Agrarian Splinter Parties', pp. 198-232; Panzer, 'Parteipolitische Ansätze der deutschen Bauernbewegung', pp. 524-561.

102 Jones, 'Crisis and Realignment: Agrarian Splinter Parties', pp. 198-232.

103 Gessner, *Agrarverbände in der Weimarer Republik*, pp. 120-126.

104 Merkenich, *Grüne Front gegen Weimar*, p. 259.

105 Gessner, *Agrarverbände in der Weimarer Republik*, pp. 127-128.

agricultural produce, subsidies for agriculture, guaranteed prices and tax reductions.[106] It proceeded to apply considerable pressure on the Reich government which at last began to consider changes to tariff policy.[107] There was much reluctance to do so, however, not only because changes to tariffs would jeopardise numerous trade agreements but also because the Social Democrats in particular wished to ensure inexpensive foodstuffs for consumers.

A compromise was reached at the end of June 1929 and the Reich government passed tariff increases for various products, including livestock. The Green Front, however, demanded comprehensive increases and, according to Schiele, livestock farmers were still not protected because frozen meat imports continued to put pressure on livestock and meat prices.[108] The Müller/Dietrich government made further concessions by introducing other protective measures, such as the wheat milling requirement of July 1929, which compelled German millers to mill German surpluses before imports, and it terminated the Swedish trade agreement in order to ease pressure on meat and grains.[109]

Ultimately, the Green Front's continued pressure was effective. In December 1929 a "sliding scale" tariff was introduced as a temporary measure. The "sliding scale" tariff signified a fundamental change to German trade policy for it ultimately allowed increasing insulation from the fluctuations of world production and markets.[110] The SPD accepted this, hoping for domestic price falls, but international price falls meant that tariffs were raised further and further as market prices plummeted.[111]

In March 1930 the Green Front went so far as to demand an enabling law to institute all necessary measures to support prices, a clear step toward autarky. In effect, agriculture wanted the power to circumvent parliament in order to introduce the complete regulation of imports.[112] Despite the fact that the enigmatic DNVP rejected the measures because they did not go far enough, a further compromise was reached with a general increase in tariffs.[113]

The Rural League's enthusiasm for the Green Front declined as its hopes rose in the new Brüning government, particularly because Schiele was again Agriculture Minister. The Green Front then split when the German Peasantry departed over the

---

106 Gessner, 'The Dilemma of German Agriculture', p. 142; Merkenich, *Grüne Front gegen Weimar*, pp. 260-261.

107 Gessner, *Agrarverbände in der Weimarer Republik*, p. 169.

108 Gessner, *Agrarverbände in der Weimarer Republik*, pp. 169-174.

109 Holt, *German Agricultural Policy*, pp. 103 and 115-116.

110 Becker, *Handlungsspielräume der Agrarpolitik*, pp. 354 and 356; Gessner, *Agrarverbände in der Weimarer Republik*, p. 175.

111 Holt, *German Agricultural Policy*, pp. 110 and 123.

112 Gessner, *Agrarverbände in der Weimarer Republik,* p. 175; Merkenich, *Grüne Front gegen Weimar*, p. 265.

113 Gessner, *Agrarverbände in der Weimarer Republik*, p. 176

issue of higher feed tariffs.[114] Never had agriculture gained so many concessions than under the Green Front but these did not effect a lasting improvement. Tariffs buoyed agricultural prices for a time.[115] They could not long prevent the fall of agricultural prices in the face of the deteriorating general economic situation.

Remarkably, it was the parliamentary Müller government with the liberal Agriculture Minister, Dietrich, that responded to the pressure from the agricultural lobby and took the decisive step to protectionism, a step given even greater priority under Brüning and Schiele.[116] Dietrich's efforts, however, were only a partial success. Farm incomes increased in 1928/1929 but most measures favoured Germany's eastern estate owners.[117] Moreover, an assessment of the Eastern Aid programme – that it proved costly for the state but paltry for the agrarian recipients and was not appreciated in the form of political support as the crisis worsened[118] – might be extended, in part, to incorporate the agricultural policies of the preceding governments. That is to say, subsidies and increasing protectionism failed to alleviate the crisis and appease the farmers. What small and transitory inroads government intervention did make came too late to still the mobilisation of Schleswig-Holstein's rural population. Nor could it preclude its radicalisation.

The radicalisation of the farmers' movement from the autumn of 1928 is the subject of the following chapter. Yet the first indications that farmers were considering more severe forms of protest followed soon after the rallies. Before tax relief measures were settled, it swiftly, albeit briefly, became clear that taxes presented a particular focus of resentment. At Rural League meetings in Steinburg County the associations' pleas for assistance were said to have come to nothing. Beyond further calls for thorough trade, tax and social policy reform, speakers warned that the farmers could only pay taxes and interest from their income and not from their "substance", that is, from their stock or working capital.[119] Going further, an unsigned letter to the Schleswig-Holstein's Farmers' Association expressed astonishment that the leaders of the province's agricultural associations had not yet called for "radical self-help". The letter's author declared that nothing had yet come of the rally leaders' deputation to Berlin, that nothing would come of it and demanded a strike and boycott measures "until the gentlemen come to us".[120]

---

114 See Holt, *German Agricultural Policy*, pp. 117 and 123; Merkenich, *Grüne Front gegen Weimar*, p. 261.

115 Henning, *Landwirtschaft und ländliche Gesellschaft*, p. 202.

116 Gessner, *Agrarverbände in der Weimarer Republik*, p. 181.

117 James, *The German Slump*, pp. 263-268.

118 Gessner, 'The Dilemma of German Agriculture' p. 143; James, *The German Slump*, pp. 273-274.

119 *Itzehoer Nachrichten* No. 33, 8.2.1928, p. 7 and No. 34, 9.2.1928, p. 6.

120 Copy of an unsigned letter addressed to the Schleswig-Holsteinischer Bauernverein, 1.3.1928 appended to the report from the Regierungspräsident to the Minister des Innern in

Following the rallies, the Prussian Interior Ministry requested its subordinate provincial authorities to report on developments in the countryside, on the "situation, mood and stance" of the rural population, and in particular whether the situation was "tense or dangerous."[121] The fact that the request was printed in the provincial press only aggravated the mood in rural areas, as rumours circulated that it indicated the authorities intended to suppress protests with police force.[122]

But such concerns proved redundant, at least for the time being. The office of the Regierungspräsident responded to the Interior Ministry's request, reporting that the rural population everywhere was "very irritable and very dejected" but a model of composure and not dangerous.[123] However, the report also warned that circumstances were deteriorating and expressed hope that the farmers' demands would soon be met. While some farmers expected assistance to come, for others – and here the report echoed the unsigned letter to the Farmers' Association – discord and disappointment were replacing hopes, for it was supposed that the promises would not be kept, "that all the organisations' negotiations had been useless". The farmers' dejection, the report noted, was encouraging their receptiveness to radical influences.

Tax relief measures and the hope of government assistance temporarily subdued more radical protest.[124] Nevertheless, there were sufficient precedents to fuel the authorities' fears and upon which Schleswig-Holstein's farmers could draw inspiration.[125] In the Weimar period, no region of Germany experienced more intense rural unrest than Schleswig-Holstein. Yet rural unrest was not unique to the province, nor did it first arise there. From 1924 to 1926, centres of agrarian protest were to be found in central and eastern Germany. In fact, at that time, in Brandenburg, the Rhineland and the Palatinate there was already talk of an agricultural crisis.[126] Furthermore, across the border, in North Schleswig, farmers also protested in response to deteriorating economic conditions.

Berlin concerning "Stimmung der landwirtschaftlichen Bevölkerung" Schleswig, 9.3.1928, LAS 301/4695.

121 See the copy of the memorandum from the Minister des Innern in Berlin to all Regierungspräsidenten, Berlin, 27.2.1928, LAS 301/4695.

122 *Flensburger Nachrichten* No. 60, 10.3.1928, p. 1; *Kieler Neueste Nachrichten* No. 61, 11.3.1928, p. 2 responding to assertions in the *Deutsche Tageszeitung*.

123 Copy of a report from the Regierungspräsident to the Minister des Innern in Berlin concerning "Stimmung der landwirtschaftlichen Bevölkerung" Schleswig, 9.3.1928, LAS 301/4695.

124 *Landbote für Schleswig-Holstein* No. 36, 6.5.1928, p. 1.

125 Reports in the province's press suggest that Schleswig-Holstein's farmers were alert, for example, to the Mosel viticulturists' protest at Bernkastel in 1926: *Heider Anzeiger* No. 50, 1.3.1926, p. 4; *Husumer Nachrichten* No. 48, 26.2.1926, p. 3, No. 49, 27.2.1926, p. 3 and No. 54, 6.3.1926, p. 1. For further evidence of farmers sympathising with their fellows in other parts of Germany, see Bergmann and Megerle, 'Protest und Aufruhr der Landwirtschaft', p. 216 and Osmond, *Rural Protest in the Weimar Republic*, p. 131.

126 Bergmann and Megerle, 'Protest und Aufruhr der Landwirtschaft', pp. 210-219.

The most dramatic instance of rural militancy prior to the radicalisation of the rural population in Schleswig-Holstein was the storming of the tax and customs offices in Bernkastel on the Mosel in February 1926. It was the climax of protests over the wine tax and the trade treaty with Spain which regulated wine imports.[127] The Mosel viticulturists' activism and demands for the end of compulsory tax collection invite parallels with the Landvolkbewegung's later protests. The Bernkastel protest, however, was extraordinary in that it achieved a swift and tangible result – the abolition of the wine tax. In explaining its success, Fritz Blaich underlined the importance of the weakness of Luther's minority cabinet: the abolition of the tax was a compromise between protectionism and subvention, between the Centre Party and the DNVP.[128] Schleswig-Holstein's farmers, by contrast, could not draw on such support from the parties, as they had no comparable tradition of ties as the viticulturists did to the Catholic parties.

Other prominent radical protest came from the Free Peasantry (*Freie Bauernschaft*) and other agricultural organisations in southern Germany in the early 1920s.[129] Almost foreshadowing the Landvolkbewegung's actions, the Free Peasantry's newspaper encouraged farmers to protest over confiscations and one of its leaders was prosecuted for urging farmers to withhold taxes. Similarly, the Free Peasantry took up symbols of peasant history such as the "Peasant boot" (*Bundschuh*) standard of the peasant uprisings from 1493 to 1517 and it commemorated the 400th anniversary of the 1525 peasant battles with a 30,000 strong rally. The parallels to the Landvolkbewegung point to a wider tendency across rural Germany to glorify past triumphs as a means of finding inspiration for confronting present difficulties.

Nearby but less militant inspiration was to be found in North Schleswig, where Danish and German farmers rallied under the *Bondens Selvstyre* (Self-administration Movement) from 1925. It blamed the Danish government for agriculture's difficulties there, and condemned the banks and parliamentary democracy, both in the press and at numerous mass meetings. Avoiding national loyalties, its leader, Cornelius Petersen, a farmer with roots in Eiderstedt, drew on that region's traditions, calling for a self-administered Schleswig with Denmark responsible only for law enforcement, defence and foreign policy. Landowners hoped to govern their communities with executive power invested at the provincial level, and infrastructure and education entrusted to the private sector. Though it continued to hold successful meetings into 1928, the Bondens Selvstyre failed to entice voters, who re-

---

127 F. Blaich, 'Der "Winzersturm von Bernkastel" Ursachen und Auswirkungen eines Steuerstreiks in der Weimarer Republik', *Zeitschrift für Agrargeschichte und Agrarsoziologie,* 33 (1985), pp. 2-26; Osmond, *Rural Protest in the Weimar Republic,* p. 133.

128 Blaich, 'Der "Winzersturm von Bernkastel", pp. 12-17.

129 For the Free Peasantry's later protests in the Rhineland, Palatinate and Bavaria, see Osmond, *Rural Protest in the Weimar Republic*, pp. 132-135.

mained loyal to the established parties, nor did its appeals to the King have any prospect of achieving its aims of self-administration.[130]

In Germany, the Rural League expressed its resentment over taxes after the rallies of January 1928. In fact, in Brandenburg in particular, it had protested against taxes on various occasions throughout the mid 1920s.[131] Then, on March 12, 1928, it lost control of a rally in Kyritz when farmers bombarded the finance office with rocks, and in Langenöls in Silesia, farmers attempted to prevent the compulsory auction of a pig leading to a scuffle with the police.[132] At the beginning of March, the Reich Rural League had issued a declaration to be read at county protest meetings, that its members would no longer pay taxes nor would they tolerate their compulsory collection.[133] But within weeks, on account of disunity in its own ranks as well as pressure from the Reich Government, the Rural League withdrew its summons.[134] The Rural League's failure to organise tax-strikes, according to Dieter Gessner, showed that beyond the desire to participate in government, agriculture could conceive of no viable concept until 1928 as regards inner political tactics, leadership and strategies in the struggle for greater protectionism.[135]

On the eve of the agricultural crisis the NSDAP was still very much a minor party. Since his release from prison in December 1924, Hitler had reorganised the party with himself as undisputed leader, developed an impressive centralised bureaucratic apparatus and determined that the party would seek to acquire power by legal means. However, the NSDAP still seemed a long way from becoming a mass political movement.

The NSDAP neither contributed significantly to the mobilisation of the countryside in Schleswig-Holstein, nor had it threatened to become a notable source of radicalisation before the crisis. As the authorities noted, however, the crisis and resultant unrest facilitated a sudden surge in National Socialist agitation in South Dithmarschen and the formation of local groups in areas where the party's presence had not been noticed previously.[136]

---

130 H. Beyer, 'Der Friese Cornelius Petersen und "Bondens Selvstyre" ', *Zeitschrift für Agrargeschichte und Agrarsoziologie*, 10 (1962), pp. 212-230; Fink, *Geschichte des schleswigschen Grenzlandes*, pp. 212-216 and 246.

131 Merkenich, *Grüne Front gegen Weimar*, pp. 223-225.

132 Bergmann and Megerle, 'Protest und Aufruhr der Landwirtschaft', p. 226. The incident in Kyritz was reported in the *Landbote für Schleswig-Holstein* No. 24, 25.3.1928, p. 1.

133 "Kabinettsitzung vom 16. März, 18 Uhr im Reichstagsgebäude" in Abramowski, *Akten der Reichskanzlei*, pp. 1378-1380; Merkenich, *Grüne Front gegen Weimar*, pp. 225-227.

134 "Kabinettsitzung vom 19. März, 12.30 Uhr" in Abramowski, *Akten der Reichskanzlei*, pp. 1380-1382; Becker, *Handlungsspielräume der Agrarpolitik*, p. 248; Gessner, *Agrarverbände in der Weimarer Republik,* pp. 98-99.

135 Gessner, *Agrarverbände in der Weimarer Republik*, p. 103.

136 Copy of a report from the Regierungspräsident to the Minister des Innern in Berlin concerning "Stimmung der landwirtschaftlichen Bevölkerung" Schleswig, 9.3.1928, LAS 301/4695.

Until late 1927 National Socialist activity focussed on the cities. This was reflected in the party's propaganda and predominantly urban membership.[137] It is difficult to determine when the party's leadership began to pursue rural support, and scholars have emphasised various developments within this shift.[138] Even so, the party's campaign efforts still remained focussed on winning over the industrial working classes.[139] Not until 1930, and in particular after Richard Walther Darré was appointed to advise the Reich leadership on agricultural matters in May, did the party develop a comprehensive ideological and organisational agricultural-political concept and begin a systematic rural campaign, that is, after the radical potential in the countryside had been revealed.[140]

Hitler long neglected agriculture but local and regional NSDAP leaders across Germany had begun to appeal to farmers.[141] Among them, Schleswig-Holstein's Gauleiter Hinrich Lohse stood apart for his more concentrated campaigning and

---

137 D. Orlow, *The History of the Nazi Party, 1919-1933* (Trowbridge and London, 1971), pp. 87-95 and 104-138. Also see Farquharson, *The Plough and the Swastika*, pp. 1-3; J. E. Farquharson 'The Agrarian Policy of National Socialist Germany' in Moeller, *Peasants and Lords in Modern Germany*, pp. 233-238; G. Corni, *Hitler and the Peasants. Agrarian Policy of the Third Reich 1930-1939* (Oxford, 1990), pp. 18-20 and M. H. Kater, *The Nazi Party: A Social Profile of Members and Leaders, 1919-1945* (Oxford, 1983), pp. 39-41, 57-59, 244-245 and 250.

138 The most important turning points identified within this shift are: a *Gauleiter* meeting in Weimar at the end of November 1927 as marking a general change in the party's strategy; Hitler's first ever address to a farmers' meeting in Hamburg in December 1927; and Hitler's "clarification" in early 1928 of article 17 of the party's programme concerning the expropriation of land for the commonwealth. See Noakes, *The Nazi Party in Lower Saxony 1921-1933,* p. 105; J. H. Grill ,'The Nazi Party's Rural Propaganda Before 1928', *Central European History*, 15 (1982), pp.163-164 and 185; H. Gies, *R. Walther Darré und die nationalsozialistische Bauernpolitik in den Jahren 1930 bis 1933* (Frankfurt/M, 1966), p. 28 and Corni, *Hitler and the Peasants*, pp. 18-21. For the party programme, see G. Feder, *Das Programm der NSDAP und seine weltanschaulichen Grundgedanken* (Munich, 1930), p. 9.

139 P. D. Stachura, 'Der kritische Wendepunkt? Die NSDAP und die Reichstagswahlen von 1928', *Vierteljahrshefte für Zeitgeschichte*, 26 (1978) pp. 79-80 and 96; G. Paul, *Aufstand der Bilder. Die NS-Propaganda vor 1933* (Bonn, 1990), pp. 90-93.

140 K. Bergmann, *Agrarromantik und Großstadtfeindschaft* (Meisenheim/Glan, 1970), pp. 279 and 329; Corni, *Hitler and the Peasants*, p. 21; Gies, *R. Walther Darré*, pp. 28 and 35; H. Gies, 'NSDAP und landwirtschaftliche Organisationen in der Endphase der Weimarer Republik', *Vierteljahrshefte für Zeitgeschichte*, 15 (1967), pp. 341-343. For an English translation of the last, H. Gies, 'The NSDAP and Agrarian Organizations in the Final Phase of the Weimar Republic' in H.A. Turner (ed.) *Nazism and the Third Reich* (NY, 1972), pp. 45-88.

141 Grill, 'The Nazi Party's Rural Propaganda Before 1928', pp. 149-185. For the importance of regional initiative in respect of rural propaganda, also see Zdenek Zofka's study of Günzberg, in Bavarian Swabia: Z. Zofka, *Die Ausbreitung des Nationalsozialismus auf dem Lande: Eine regionale Fallstudie zur politischen Einstellung der Landbevölkerung in der Zeit des Aufstiegs und der Machtergreifung der NSDAP 1928-1936* (Munich, 1979), pp. 73-80.

early efforts to build up a rural party structure. A bank clerk with rural roots in Steinburg County, Lohse had been secretary of the Landespartei before joining the NSDAP. He was therefore familiar with agricultural concerns and had excellent contacts in rural circles.[142] Not only did he swiftly direct the party's propaganda toward the farmers, beginning as early as the summer of 1925, he also began to establish a network of contacts and to found local groups, concentrating his efforts in Dithmarschen. The local leaders were typically lower middle-class and likewise most of the new members were not farmers, but young rural traders and craftspeople.[143] Signs of growth, however, encouraged Lohse to focus on the countryside. It should be noted that the party's gains were modest at first. Even by the summer of 1927, the party leadership did not consider Schleswig-Holstein among the regions in which it had made exceptional progress. The sole major function the *Gau* had organised was a meeting of party representatives in Eutin (Lübeck) in May 1926 at which Hitler addressed an audience of approximately 200 members from across the province.[144]

Despite the fact the NSDAP turned to the farmers earlier in Schleswig-Holstein than anywhere else in Germany – and thereby laid the foundations for the party's later success[145] – only after the onset of the agricultural crisis did the NSDAP's fortunes begin to change. It was not until December 1927 that Lohse extended the party's propaganda campaign from Dithmarschen to the neighbouring counties of Steinburg and Rendsburg. By the end of the year, he had managed to set up a few local groups around Lockstedter Lager in Steinburg, an area disposed to right-wing agitation as it had been settled by former Freikorps troops and was then hard hit by the fall in pig prices.[146] The highlight of Lohse's endeavours was Hitler's appearance at a special meeting, specifically held for the province's farmers in Hamburg, which attracted a crowd of some 3,200 people, many of them members of the farming organisations from South Dithmarschen. It was the first time Hitler ever ad-

---

142 For Lohse's biographic details, see P. Hüttenberger, *Die Gauleiter. Studie zum Wandel des Machtgefüges in der NSDAP* (Stuttgart, 1969), p. 216. For Lohse's essay in the *National Socialist Yearbook* in 1927, marking his early programmatic interest in agriculture, see the *Nationalsozialistisches Jahrbuch 1927* (Munich, 1927), pp. 140-145. For his unique early efforts to build up a rural party structure, see Rietzler, *"Kampf in der Nordmark"*, pp. 412 and 434 and Grill, 'The Nazi Party's Rural Propaganda Before 1928', pp. 168-170.

143 Rietzler, *"Kampf in der Nordmark"*, pp. 402-410; Heberle, *From Democracy to Nazism*, pp. 84-87 and 123; Stoltenberg, *Politische Strömungen im schleswig-holsteinischen Landvolk*, p. 149.

144 Rietzler, *"Kampf in der Nordmark"*, pp. 399-400; L. D. Stokes, *Kleinstadt und Nationalsozialismus: ausgewählte Dokumente zur Geschichte von Eutin, 1918-1945* (Neumünster, 1984), pp. 51-56; C. Vollnhals (ed.), *Hitler: Reden, Schriften, Anordnungen: Februar 1925-Januar 1933*, vol. 1 (Munich, London, New York and Paris, 1992), pp. 423-427.

145 Rietzler, *"Kampf in der Nordmark"*, p. 434.

146 Rietzler, "*Kampf in der Nordmark"*, pp. 410-411; Stoltenberg, *Politische Strömungen im schleswig-holsteinischen Landvolk*, pp. 142-143.

dressed a farmers' meeting.[147] The meeting, however, probably did not have much more than a regional impact and should not lead to a miscalculation of the party's appeal in the countryside before the crisis.[148]

The NSDAP's involvement in the rallies of January 28 outside South Dithmarschen was limited to the distribution of fliers.[149] Afterwards, however, Lohse used the agitated atmosphere in the countryside, intensifying party propaganda on the west coast before the May elections and astutely ensuring that local farmers were made candidates for both the Reichstag and Prussian Landtag lists. As a whole, the party's results in Schleswig-Holstein in 1928 were just as unexceptional as they were across the Reich. Reflecting the success of the local rural campaign, however, it obtained 17.7 and 17.2 percent in North and South Dithmarschen respectively and 10.2 percent in Steinburg.[150] These results pointed to the party's potential in the countryside. Moreover, due to Lohse's preparatory work, the party was in a good position to capitalise on the growing unrest in the province. It must be emphasised once again, however, that the party's growth was local before the onset of the crisis and that it did not expand until after the farmers had begun to mobilise themselves.

The rallies of January 28, 1928 mobilised Schleswig-Holstein's rural people, who were hopeful that their solidarity could gain crisis relief and unify the province's divided agricultural associations. They were to be disappointed in both respects. The authorities at all levels were quick to grant short-term tax breaks, but the Reich government's "Emergency Programme for Agriculture" did not satisfy the farmers' calls for fundamental change to tariff policy and the measures to relieve debts also ultimately proved inadequate. The endeavours of Johannsen and his fellow rally leaders' Landvolkbewegung to fuse the agricultural associations from below appeared promising at first, but the farmers' unity depended on avoiding matters of political direction and affiliating with any particular party. The associations stumbled again and again over the questions of political allegiance and national repre-

---

147 For Hitler's speech in Hamburg, see B. Dusik (ed.), *Hitler: Reden, Schriften, Anordnungen: Februar 1925-Januar 1933*, vol. 2 (Munich, London, New York and Paris, 1992), pp. 570-582.

148 See Orlow, *The History of the Nazi Party*, fn. 166, p. 118. Tilton mistakenly assessed the meeting as evidence of the farmers' great interest in Nazism before the crisis, believing it to have taken place in 1925. The error can be attributed to his source, Thyssen, whose dating of the meeting as "in the first year after [Hitler's] release from the Landsberg prison (1923)" was, in fact, doubly incorrect as Hitler did not leave Landsberg until December 1924: Tilton, *Nazism, Neo-Nazism and the Peasantry*, pp. 55-56; Thyssen, *Bauer und Standesvertretung*, p. 296.

149 Fliers were distributed in Itzehoe and Flensburg but to little effect. See the *Nordischer Kurier* No. 25, 29.1.1928, p. 3; Heinacher, *Der Aufstieg der NSDAP im Stadt- und Landkreis Flensburg.*, p. 273 and Rietzler, "*Kampf in der Nordmark*", pp. 414-415.

150 Heberle, *From Democracy to Nazism.*, p. 103; Rietzler, *"Kampf in der Nordmark"*, pp. 415-424; Stoltenberg, *Politische Strömungen im schleswig-holsteinischen Landvolk*, p. 145. Also see p. 67 above.

sentation. Though many agricultural leaders, including Johannsen, resigned in failure, most of the province's farmers finally came together in the new Farmers' League. Its creation, however, left an organisational constellation no less fragmented than before the rallies. The Green Front coalition subsequently showed that uncompromising fusion was not required for an effective agricultural lobby but the rallies of January 1928 elevated the priority given to unity in Schleswig-Holstein and its allure remained thereafter.

Schleswig-Holstein's farmers were still looking for a solution to their difficulties within the Weimar system. The rallies of January 28, 1928 were not radical but largely cooperative. Rather than directed against the authorities, they voiced calls for assistance from them. Fears that the rallies might become riotous like the protests of the Mosel viticulturists proved unwarranted. Moreover, the influence of the völkisch organisations at the rallies should not be overstated. Nor, it should be added, did the NSDAP play a prominent role. Indeed, the party was still an insignificant factor in the province at that time.

The farmers were in distress but calm persisted as long as hope in government assistance and the postponement of tax-payments remained. However, the farmers' fortunes continued to deteriorate, and disappointment and bitterness grew. Only much later, at the end of 1929 when it raised livestock tariffs, did the Reich government take steps along the lines of those the farmers had expected. The early failure to deliver assistance undermined confidence in the state, the associations and their established leaders, just as the associations' failure to meet their members' demands to unite weakened their influence. Schleswig-Holstein's farmers – to recall the previous chapter – were not organised in political parties and now their loyalties to their agricultural associations were shaken also. As autumn approached, rural politics were thoroughly destabilised and tax relief was soon to end. The stage was set for the development of a radical new movement ready to harness the spirit of unity in the Landvolkbewegung.

Chapter Three

The Landvolkbewegung emerges

Until the autumn of 1928 there was little indication that Schleswig-Holstein's rural population would radicalise so swiftly. Since the Prussian Interior Minister's request in February to observe the "situation, mood and stance" of the rural population, the provincial authorities' files had essentially remained empty.[1] But as the agricultural crisis continued and the inflated hopes in the government's promised assistance subsided, discontent resurfaced and, from the west coast once again, a fresh wave of protest commenced that crystallised in the Landvolkbewegung. While the Landvolkbewegung's origins were obscure, by the end of the year it was of major concern to the authorities. Within just a few months, it had organised a number of large meetings and initiated direct action that appears to have attracted significant support in the countryside. These protests in the autumn and winter of 1928 were characterised by a shift from the farmers' pursuit of a remedy for the agricultural crisis to increasingly extreme opposition to the Weimar system.

Various factors led to the beginning of the new protests. These included: the disappointment in the government's promises of assistance, the growing frustration at the failure of the agricultural organisations to form a more effective lobby, and the continued deterioration of the farmers' financial circumstances. Despite an improved harvest and recuperating pig prices in 1928, the disparity between production costs and agricultural prices worsened, the province's agricultural debt increased by a further RM 69.9 million to RM 468.5 million in the year to March 31, 1929 and the number of compulsory auctions of agricultural properties increased.[2] But ultimately the decisive, additional factor that triggered the new protests and the creation of the Landvolkbewegung was the tax issue.

Discontent over taxes, as discussed in Chapter One, stemmed from an indisputably high tax burden which the farmers perceived as all the more considerable due to the loss of the privileges and prosperity they had enjoyed before the First World War. The construction of new finance offices on the west coast and rising civil servant salaries had intensified their resentment. The tax burden, however, had only become critical with the onset of the crisis, since advanced tax assessment meant the relative burden increased as farm earnings declined.[3] In early 1928, before relief measures had been settled, taxes had already become a source of anger in the province. Farmers, it should be recalled, had protested over taxation at the rallies of January 28, 1928, and the agricultural organisations, including the Rural League

1 In fact, the Oberpräsidium's file in which this request was located remained all but bare until November 1928. See LAS 301/4695.

2 See chapter 1, pp. 42 ff.

3 See chapter 1, p. 51.

and the Unity Organisation, had condemned the payment of taxes when it required farmers to sell assets or borrow money at high levels of interest.[4] The subsequent tax relief measures had given farmers a reprieve, but as the postponement of payments was to expire in October, the tax issue resurfaced with renewed vigour.[5]

The earliest radical manifestations of the mounting anger on the west coast came in the form of a series of letters published towards the end of the September in various provincial newspapers such as the *Eiderstedter Nachrichten*, *Dithmarscher Zeitung* and the Rural League's *Landbote*. These letters proclaimed the primary importance of agriculture for the economy and people, discussed the imminent threat of communism, criticised reparation payments and underlined the prevailing difficulties citizens faced in meeting their taxation obligations.[6] Signed simply by a "representative of the rural people" (*ein Vertreter des Landvolks*), the letters presumably stemmed from the same source.[7] Significantly, this apparently self-appointed "representative" maintained that only the elimination of "the system" – meaning the Weimar Republic and everything associated with it – would truly bring an end to the people's suffering.[8]

That such views were spreading on the west coast became apparent at a meeting of 300 graziers and cattle breeders in Husum on October 17. The *Husumer Nachrichten* remarked on the unprecedented level of agitation at the meeting and reported that the farmers had given expression to their "disappointment and bitterness, despair and hatred" in response to the "utterly inadequate protection of do-

---

4 For the Rural League's criticisms, see chapter 2, pp. 73-76, and for examples of the Unity Organisation's complaints, see the *Mitteilungsblatt der Kreisbauernschaft Segeberg: Organ der landwirtschaftlichen Einigungsbewegung* No. 8, Bad Segeberg, 30.6.1928 and the letter from the Schleswig-Holsteinischer Bauernbund e.V. to the Regierungspräsident in Schleswig, Bad Segeberg, 4.12.1928, both in LAS 309/22921. Also see the *Heider Anzeiger* No. 287, 6.12.1928, p. 1.

5 *Geschäftsbericht der Landwirtschaftskammer für die Provinz Schleswig-Holstein für das Rechnungsjahr 1928-1929*, p. 46; report by the Landrat to the Regierungspräsident in Schleswig concerning "Völkische Treibereien" Rendsburg, 26.11.1928, LAS 309/22921; report from Abteilung IA, Außendienst II. G. St., Berlin, 27.12.1928, appended to the report concerning "Ergebnis der Ermittelungen der Beamten des Polizeipräsidiums Berlin" from the Regierungspräsident to the Oberpräsident in Kiel, Schleswig, 9.1.1929, LAS 301/4695. A copy of the report is also to be found in LAS 301/4589.

6 *Eiderstedter Nachrichten* No. 8150, 20.9.1928, pp. 2-3, No. 8152, 25.9.1928, p. 2, No. 8158, 9.10.1928, p. 3.

7 Note in the Regierungspräsident's files titled "Vermerk über eine Besprechung mit Landrat Kracht am 28.10.1928" Schleswig, 29.10.1928, 309/22569; Regierungspräsident to the Oberpräsident in Kiel, Schleswig, 27.11.1928, in both LAS 301/4695 and 309/22569; report from the Landrat to the Regierungspräsident in Schleswig concerning "Völkische Treibereien" Rendsburg, 26.11.1928, LAS 309/22921; report from the Landrat to the Regierungspräsident in Schleswig concerning "Pfandbruch" Heide, 30.11.1928, LAS 309/22569.

8 *Eiderstedter Nachrichten* No. 8165, 25.10.1928, p. 3.

mestic livestock production and the continuing disastrous market situation". It warned that drastic measures had to be taken if acts of despair were to be avoided.[9]

The county administrators of Eiderstedt and Husum, representatives of the Chamber of Agriculture and various agricultural organisations, as well as the DVP Reichstag deputy, Ernst Hamkens (Hanredder) attended the meeting.[10] Hamkens complained about the high level of livestock imports and called for the creation of a new graziers' association to aid him in his efforts in parliament to change tariff policy. Apart from the county administrators, however, the great majority of speakers rejected the proposal, arguing that a new association would only lead to greater division, that sending further commissions to Berlin was useless and that negotiation was futile. According to Paul Adam Ross (Flederwurth), a future Landvolk-bewegung leader, the state had done nothing but lead the farmers to the "edge of the abyss". To great applause, he declared "We have to take things into our own hands in this time of crisis. We do not want to fight with words like Deputy Hamkens, but finally see deeds. If necessary we must secure profitability with violence." Underlining his readiness for confrontation with the authorities, Ross closed his speech, calling, "*Lewwer duad üs Slaav*!" (Rather dead than a slave!)[11] The significance of this Friesian battle-cry, its allusions to the region's history of independence and to resisting the payment of taxes, would have resonated with farmers. It was particularly well known as the refrain of Detlev Freiherr von Liliencron's ballad, relating the tale of Pidder Lüng, a native of the island of Sylt, who rather than submit to paying Danish taxes killed the bailiff, only to be killed by the bailiff's men.

At the meeting in Husum, numerous other speakers also voiced criticism of the government and opposition to the parliamentary system in general. Yet perhaps one moment best illustrated the proliferation of the rebellious mood: When Deputy Hamkens defended his proposal, explaining the need to work within the system since it could not be abolished immediately, he was interrupted by boisterous disapproval.[12]

This meeting marked a turning point in the radicalisation of the countryside, not just because the farmers' discontent was so vociferous and vehement, but also because the anger directed against Weimar was beginning to turn into the outspoken and categorical rejection of it. At this stage, however, unwillingness to work within the Weimar system remained theoretical. The farmers had not given up

---

9 All citations from the *Husumer Nachrichten* No. 246, 18.10.1928, p. 5. Also see the *Rendsburger Tageblatt* No. 256, 18.10.1928, p. 5.

10 Ernst Hamkens (Hanredder) is not to be confused with Landvolk leader, Wilhelm Hamkens (Tetenbüll), with Wilhelm Hamkens (Hoyerswort), the Regierungspräsident of Schleswig-Holstein from 1938-1944, nor with Otto Hamkens, the younger brother of the latter and later county administrator of Eiderstedt.

11 All citations taken from the *Husumer Nachrichten* No. 246, 18.10.1928, p. 5.

12 See the *Heider Anzeiger* No. 246, 18.10.1928, p. 2; *Husumer Nachrichten* No. 246, 18.10.1928, p. 5.

all hope of government intervention. This was clear from the resolution they passed at the end of the meeting, calling on the Unity Organisation to appeal for deferment of payment until the spring, remission of taxes and the immediate end to all meat and livestock imports.[13]

The meeting proved to be no solitary outburst. Two weeks later, on October 30, the Unity Organisation of Eiderstedt, led by Wilhelm Hamkens (Tetenbüll), held a further meeting that echoed much of the mood of the previous meeting. On this occasion not one of the speakers advocated cooperation with the system.[14] Further, Wilhelm Hamkens announced the Unity Organisation's failure to merge at the national leadership level even though, officially, the Unity Organisation had resolved to wait until April to settle the issue. Hamkens called on all farmers to continue to unite at the county level. On the one hand his determination to drive the farmers to unity from below remained in the spirit of the demonstrations of January 28, but on the other it constituted a break from the associations. This division heralded the transition to the new movement: the Landvolkbewegung.

Together with Claus Heim (1884-1965) from St. Annen-Österfeld in North Dithmarschen, Wilhelm Hamkens (1896-1955) from Tetenbüll in Eiderstedt was instrumental in the development of the new movement. Both descended from wealthy farming families dating to the sixteenth century. Typical of their generation, they were veterans. More remarkable was that both were well travelled. Heim had owned a property in Paraguay which for a time he tended after the war. While he stood out for his imposing stature and character, in the Landvolkbewegung he was mainly to play an organisational role. Hamkens, on the other hand, returned from the front a second lieutenant, travelled extensively and managed an estate in East Prussia until 1924, before settling on his grandfather's farm in Tetenbüll. Described in a police report as slender with dark blond hair and twinkling little brown eyes, he proved a talented speaker and was long the public face of the movement.[15]

Prior to January 1928 neither Hamkens nor Heim had been politically active. In the attempts to unite the agricultural associations following the rallies, Hamkens became chairman of the Unity Organisation in Eiderstedt, and both he and Heim attended monthly meetings in Neumünster, which were held among the leaders of

---

13 *Heider Anzeiger* No. 246, 18.10.1928, p. 2; *Husumer Nachrichten* No. 246, 18.10.1928, p. 5.

14 *Husumer Nachrichten* No. 257, 31.10.1928, pp. 1-2.

15 See the transcripts of examinations of Claus Heim, Altona, 11.9.1929 and Altona, 13.9.1929, both in LAS 309/22701 and the undated extracts from them titled "Klaus Heim, 24.3.1884 in Oesterfeld", LAS 309/22784. Also see G. Hamkens, *Die Familie Hamkens aus Eiderstedt: eine Chronik* (Lübeck, 1972), esp. pp. 46 and 67-69; F. Johannsen, 'Zur Erinnerung an Wilhelm Hamkens, Tetenbüll', *Zwischen Eider und Wiedau, Heimatkalender für Nordfriesland* (Husum, 1986), pp. 179-182; the police profiles of both Hamkens and Heim from the Landeskriminalpolizeiamt, Abteilung IA, Außendienst VII., Berlin, 5.6.1930, LAS 309/22701 and the description of Hamkens: report from the Abteilung IA, Außendienst, concerning "Observation des Hamkens, Wilhelm" 9.9.1929, LAS 209/22701.

the rallies of January 28 during the first half of 1928. Abandoning hope in the government's promises of assistance, in May they met with a number of their colleagues and resolved to "enlighten and shake up" their fellow farmers.[16] Together, the two men embarked upon a series of lecture tours in Schleswig-Holstein and elsewhere in Germany.[17]

Wilhelm Hamkens (left) and Claus Heim (right)[18]

Hamkens and Heim turned their attention to taxes as the relief measures came to their end in October. Reasoning that the sustained reduction of a farm's capital could only result in ruin, they went beyond the complaints of the province's agricultural organisations and took a more belligerent approach.[19] Hamkens and Heim may have got the idea for a tax-strike from the Reich Rural League's aborted plans in March 1928 and they may also have been inspired by protests in other areas of Germany, in particular Brandenburg, at that time.[20] Hamkens and Heims' own pro-

16 According to Stoltenberg's interviews of Heim and Hamkens. No such reports, however, are to be found in LAS 309/22696. Compare Stoltenberg, *Politische Strömungen im schleswig-holsteinischen Landvolk*, pp. 121-122.

17 See Wilhelm Hamkens' examination transcript, Altona, 15.9.1929, LAS 309/22701, also to be found in undated extracts from it titled "Kurzer Auszug aus der Vernehmungsniederschrift des Hofbesitzers Wilhelm Cornelius Hamkens...", LAS 309/22784 and C. Heim 'Claus Heim berichtet. Erstveröffentlichung seiner Erinnerungen' in *Bauernruf. Land-Rundschau für Politik und Agrarwirtschaft* No. 23, 26.11.1965 – No. 15, 5.8.1966, LAS EI/4434, p. 2.

18 Photos of Heim and Hamkens: Luetgebrune, *Neu-Preußens Bauernkrieg*, p. 32.

19 For this reasoning, see the report on Hamkens' speech in Garding on 21.12.1928 in *Das Landvolk* No. 1, 3.1.1929, pp. 4-5.

20 See chapter 2, p. 76.

tests commenced with the publication of notices in local newspapers, rejecting the payment of taxes, which precipitated direct action against confiscations and compulsory auctions.

By making his intentions clear in the local press, Heim made his default on tax-payments a genuine tax-strike. After losing his property in Paraguay, he had then had the misfortune of restructuring his family farm in St. Annen on the eve of the crisis and had debts consuming no less than three quarters of his farm's rateable value.[21] On October 16, in the *Heider Anzeiger*, he published a reminder the communal authorities had sent him to pay outstanding Chamber of Agriculture dues (*Landwirtschaftskammerbeiträge*) and occupational society dues (*Beiträge der Berufsgenossenschaft*), i.e., accident insurance. Beneath, he responded that he would not pay taxes including these dues until his farm's profitability was restored.[22] According to Heim, these organisations were self-serving and took insufficient consideration of the agricultural crisis.

Heim's letter subsequently put pressure on the local administration in Lunden, provoking an animated discussion at a community representatives' meeting, which affirmed that a general reduction in taxes could not be granted, but that remission could be permitted in especially dire circumstances.[23] This did not appease Heim, who continued to agitate for tax-relief by publishing letters in the press.[24] Responding to a further reminder from his community chairman to pay outstanding Chamber of Agriculture dues in December, he declared that to support such an organisation was to fund "a gravedigger of one's own economy".[25]

Hamkens' letter, published in November in numerous west coast newspapers, did not concern his own financial circumstances but was rather a more general call for tax-strikes. He called on the community chairmen to inform their superiors that citizens were no longer able to pay tax and that subsidies would therefore be required to cover the communal budgets.[26] Most important was the concluding sentence, "Anyone who continues to pay taxes or encourages [others] to do so, acts deceitfully against himself, his family and the state and makes himself guilty (*mitschuldig*) for the downfall and enslavement of the German people."[27] Ultimately, this letter led to charges against Hamkens for inciting tax-strikes.[28]

---

21 Report from the Amtsvorsteher to the Landrat in Heide concerning "Wirtschaftliche Lage von Führern der Landvolkbewegung" Lunden, 23.7.1929, LAS 309/22784. Also see Heim, 'Claus Heim berichtet', LAS EI/4434, p. 2 and Topf, *Die Grüne Front*, p. 39.

22 *Heider Anzeiger* No. 244, 16.10.1928, p. 6.

23 *Heider Anzeiger* No. 271, 16.11.1928, p. 6.

24 See, for example, *Das Landvolk* No. 73, 4.6.1929, p. 8.

25 *Heider Anzeiger* No. 302, 24.12.1928, p. 3.

26 Landrat des Kreises Eiderstedt to the Regierungspräsident in Schleswig concerning "Vorgehen der Landvolkvereinigung" Tönning, 17.11.1928 and Regierungspräsident to the Oberstaatsanwalt in Flensburg, Schleswig 25.11.1928, both in LAS 309/22569.

27 *Eiderstedter Nachrichten* No. 8147, 15.11.1928, p. 3.

28 For the trial against Hamkens, see chapter 4.

Hamkens and Heim's letters set an insubordinate example.[29] However, they were not alone in moving the farmers to direct action. A further notable impulse came from Hans Siehl, a member of the nationalist German Völkisch Freedom Movement (*Deutschvölkische Freiheitsbewegung*), who travelled the villages of the west coast urging audiences not to pay taxes and interest on loans.[30] Apart from fostering the farmers' distrust of public servants – according to Siehl, the clerks at the finance office simply sat around smoking and talking about girls – he also promoted the idea of impeding compulsory auctions by preventing participants from bidding.[31]

Hamkens and Heim were part of a network that extended along the West Coast marshes and beyond, a network that subsequently enabled them to organise and coordinate direct action. North Dithmarschen's administrator noted that farmers in his county maintained a lively exchange.[32] Doubtless, this included neighbourly interaction, but private, professional and political connections were also of considerable importance, particularly in establishing links between the emerging movement's leaders. Heim, Paul Adam Ross and Otto Johannsen were linked via familial ties.[33] The brother of the last was the founder of the *Stahlhelmwestküste*, a regional variant of the Stahlhelm veteran association, in which Hamkens, Heim and Ross were members.[34] From the beginning of October 1928 Hamkens began to tour and speak throughout the province and on these tours he visited Itzehoe and found allies in Weschke, Soth and Knudsen.[35] Guido Weschke was the Steinburg County's Rural League secretary and Wilhelm Knudsen (Büttel) and the DNVP Reichstag Deputy Max Soth led the Rural League's "National Unity Movement" (*Nationale Einheitsbewegung*) in Steinburg.[36]

---

29 Report from the Landrat to the Regierungspräsident in Schleswig concerning "Vorgehen der Landvolkvereinigung" Tönning, 19.11.1928, LAS 309/22569.

30 Landrat to the Regierungspräsident in Schleswig concerning "Steuerversammlung" Heide, 8.11.1928, LAS 309/22696.

31 Report from the Landrat to the Regierungspräsident in Schleswig concerning "Steuerversammlungen" Heide, 7.11.1928, LAS 309/22696; *Nordischer Kurier* No. 271, 16.11.1928, p. 4.

32 Report from the Landrat in Heide to the Regierungspräsident in Schleswig concerning "Anschlag auf dem Grundstück des Amtsvorstehers Ross in Lunden" Heide, 29.11.1928, LAS 301/4704.

33 Report from the Landeskriminalpolizeiamt, Abteilung IA, Außendienst VII, Berlin, 5.6.1930, LAS 309/22701.

34 Undated document titled "Stahlhelmwestküste" appended to the report from the Regierungspräsident to the Preuß. Minister des Innern concerning "Die Landvolkbewegung und die Sprengstoffanschläge" Schleswig, 17.9.1929, LAS 309/22784.

35 Undated extract from the examination transcript of Wilhelm Hamkens titled "Kurzer Auszug aus der Vernehmungsniederschrift des Hofbesitzers Wilhelm Cornelius Hamkens ...", LAS 309/22784.

36 The Nationale Einheitsbewegung was an initiative which arose among various provincial Rural Leagues. It sought the unity of the right-wing camp and aspired to place the farmers'

Appointed the task of reorganising the Stahlhelmwestküste, Herbert Volck, a former Freikorps leader, arrived in the province in the second half of October 1928.[37] According to Johannes Kühl, a member of the old Stahlhelm, party secretary of the DNVP in Husum and later editor of the Landvolk newspaper, Volck subsequently met with Hamkens, Heim and Ross.[38] Kühl himself had already met Volck when Volck held a series of lectures for the Pan-German League on the west coast the previous year. So too had Hans Nickels, a further member of the Stahlhelmwestküste, who had established a small security firm in Husum after moving there from Mülheim an der Ruhr where he had been active in the resistance efforts against the French occupation. In fact, Nickels had known Volck since he had been a speaker for the German Völkisch Freedom Movement from 1923 to 1925.[39] When plans for the reorganisation of the Stahlhelmwestküste fell through, Heim arranged for Volck and Nickels to establish a new organisation, the "Defence Association for Town and Country" (*Wachvereinigung für Stadt und Land*).[40]

Established as a registered association on October 26, 1928, the Defence Association was a farmers' vigilante network to protect farmers from the "red peril" (*rote Gefahr*).[41] They were inspired by the militias that farmers had formed to defend their properties when hungry city dwellers raided the countryside after the

---

movement in the service of the nationalist front. In Schleswig-Holstein it tried to counter the Farmers' League's acceptance of the German Peasantry by pushing for an immediate vote on the question of national representation. It proved too late and some of its leaders subsequently threw in their lot with the Landvolkbewegung. See Thyssen, *Bauer und Standesvertretung*, p. 384 and also Bergmann and Megerle, 'Protest und Aufruhr der Landwirtschaft', p. 251.

37 Volck grew up in the German minority in Tartu (*Dorpat*) in Estonia, where his father had led a German vigilante group. During the war he was taken prisoner in Russia after his plane crashed. He later escaped captivity, joined the German Freikorps and took part in the Whites' activities in the Caucasus. Mohler, *Die konservative Revolution*, p. 447.

38 Undated extract from the examination transcript of Johannes Kühl titled "Redakteur bei der Landvolkzeitung G.m.b.H. Johann Kühl, ...", LAS 309/22784; report from the Landeskriminalpolizeiamt, Abteilung IA, Außendienst VII, Berlin, 5.6.1930, LAS 309/22701.

39 Undated extract from the examination transcript of Hans Nickels titled "Kurzer Auszug aus der Vernehmungsniederschrift des Polizeihauptmanns a.D. Hans Nickels, ...", LAS 309/22784.

40 The Defence Association was otherwise known as the "Defence Assistance" (*Wachhilfe*) and in some areas as the "Young Northern Lands Defence" (*Jungnordmarkwehr*). For evidence of Heim and Hamkens' leading roles in the association, although their names did not appear in the statutes, see the report from the Regierungspräsident to the Oberpräsident in Kiel concerning "Wachvereinigung für Stadt und Land" Schleswig, 7.2.1929, LAS 301/4589.

41 Landrat to the Oberpräsident concerning "Gründung eines Vereins 'Wachvereinigung für Stadt und Land' " Husum, 8.11.1928 and the appended copy of the statutes titled "Satzung" Husum, 26.10.1928, LAS 301/4589; report from the Landeskriminalpolizeiamt, Berlin, 2.4.1929, sent to the Oberpräsident in Kiel from the Polizeipräsident, Landeskriminalpolizeiamt, Abteilung IA, Berlin, 24.4.1929, in both LAS 301/4695 and 309/22668.

war.[42] Hamkens, Heim and their colleague in South Tondern, a Rural League leader, Max Petersen (Hogelund), promoted the organisation with fanciful stories of imminent communist raids.[43] Members were to arm themselves with truncheons in order to help defend fellow members in need of assistance. Though numerous local groups were quickly established on the west coast, not enough farmers joined, suggesting that most gave little credence to their claims.[44] Signs designed to identify members were never installed.[45] Volck departed in January, fleeing the country temporarily for fear that the authorities were pursuing him on account of seditious speeches, and Nickels resigned soon afterwards.[46] It seems the undertaking was not sufficiently lucrative to sustain his employment.[47] Nickels himself attributed its failure to attract greater support to the competition of the National Socialists' *Sturmabteilung* (SA).[48]

---

42 See also chapter 1, pp. 31 and 40, fn. 110.

43 Report from the Kriminal- und Grenzkommissariat, Niebüll, 20.12.1928, sent by the Landrat in Niebüll to the Regierungspräsident in Schleswig, 22.12.1928, LAS 309/22696; report from the Landeskriminalpolizeistelle Flensburg concerning "Bauernbewegung in der Provinz Schleswig-Holstein" Flensburg, 24.1.1929, in both LAS 301/4589 and 309/22696. Max Petersen and his colleague, Ernst Bossen (Neulandshof), had long been active in the Rural League and had played leading roles in organising the rallies of January 28, 1928 in South Tondern. See Koops, *Südtondern in der Zeit der Weimarer Republik*, pp. 230, 234 and 258.

44 Reports from the Regierungspräsident to the Oberpräsident in Kiel concerning "Jungnordmarkwehr" Schleswig, 3.11.1928 and Schleswig, 25.11.1928, and from the Oberpräsident to the Preuß. Minister des Innern in Berlin concerning "Jungnordmarkwehr" Schleswig, 28.11.1928, all in LAS 301/4589; report from the Landeskriminalpolizeiamt Berlin, Abteilung IA, Heide, 21.3.1929, LAS 309/22668; report from the Kriminal- und Grenzkommissariat, Niebüll 20.12.1928, sent by the Landrat in Niebüll to the Regierungspräsident in Schleswig, 22.12.1928, LAS 309/22696; report from the Landeskriminalpolizeistelle Flensburg to the Regierungspräsident in Schleswig concerning "Gründung von sogen. Wachvereinen" Flensburg, 8.1.1929, LAS 301/4589.

45 Regierungspräsident to the Preuß. Minister des Innern concerning "Unruhe in der Landwirtschaft" Schleswig, 18.2.1929, in both LAS 309/22696 and 301/4695.

46 Regierungspräsident to the Oberpräsident in Kiel concerning "Wachvereinigung für Stadt und Land" Schleswig, 12.2.1929, LAS 301/4589. According to Nickels, Volck departed in December. See the undated extract from Nickel's examination transcript titled "Kurzer Auszug aus der Vernehmungsniederschrift des Polizeihauptmanns a.D. Hans Nickels, ...", LAS 309/22784.

47 Nickels was only once paid his income in full. See the undated document titled "Wachvereinigung für Stadt und Land e.V." appended to the report from the Regierungspräsident to the Preuß. Minister des Innern concerning "Die Landvolkbewegung und die Sprengstoffanschläge" Schleswig, 17.9.1929 and the undated extract from the examination transcript of Hans Nickels titled "Kurzer Auszug aus der Vernehmungsniederschrift des Polizeihauptmanns a.D. Hans Nickels, ...", both in LAS 309/22784.

48 According to Claus Heim, see the undated extracts from the examination transcripts of Claus Heim titled "Klaus Heim, 24.3.1884 in Oesterfeld" and also of Johannes Kühl titled "Redakteur bei der Landvolkzeitung G.m.b.H. Johann Kühl, ...", both in LAS 309/22784.

Absorbed by the Landvolkbewegung by January 1929, the Defence Association was officially dissolved on March 1, although there is evidence it continued underground, particularly in the province's northwest, in South Tondern, Husum and North Dithmarschen.[49] Though difficult to confirm due to its clandestine nature, it seems the Defence Association was, or at least became, involved in resisting confiscations and compulsory auctions as part of the Landvolkbewegung.[50] Yet its principal significance overall lay in the evolution of the movement, and it is the links it helped to establish and cement which must be underlined here. The Defence Association bolstered ties among farmers and right-wing activists, building a network which facilitated the movement's subsequent operations. It brought together the rural members of the former Stahlhelmwestküste with Nickels and Volck, who stayed on in the Landvolkbewegung after the Defence Association's decline, and others too who would later be of consequence.[51] Most notably, Hartmut Plaas, secretary to the leader of Organisation Consul and Bund Wiking, Hermann Ehrhardt, was also offered employment but Plaas declined, recommending Walther Muthmann in his place. Muthmann, a former Freikorps soldier, member of the Wehrwolf and agronomist (*Diplomlandwirt*), did not commence work until December 1928 but remained with the Landvolkbewegung after promoting the Defence Association in the Husum and Plön counties.[52] Even the later business director of the Landvolk's newspaper, John Johnsen, did a few weeks of paper work for the Defence Association.[53]

---

49 Report from the Landeskriminalpolizeistelle Flensburg concerning "Bauernbewegung in der Provinz Schleswig-Holstein" Flensburg, 24.1.1929, in both LAS 301/4589 and 309/22696; report from the Regierungspräsident to the Preuß. Minister des Innern concerning "Unruhe in der Landwirtschaft" Schleswig, 18.2.1929, in both LAS 301/4695 and 309/22696; copy of a report from the Polizeipräsident concerning "Die Nothilfe" Kiel, 30.4.1929, LAS 301/4695; undated extract from the examination transcript of Bruno von Salomon titled "Bruno von Salomon, 7.2.1900 in Stettin, Hauptschriftleiter "Landvolk" in Itzehoe...", LAS 309/22784; report from the Landeskriminalpolizeistelle Flensburg concerning "die im Kreise Husum geführten Ermittelungen" Flensburg, 24.3.1929, both in LAS 309/22668.

50 Report from the Landeskriminalpolizeistelle Flensburg concerning "Bauernbewegung in der Provinz Schleswig-Holstein" Flensburg, 24.1.1929, in both LAS 301/4589 and 309/22696.

51 Copy of the report from the Polizeipräsident concerning "Die Nothilfe" Kiel, 30.4.1929, LAS 301/4695.

52 For Muthmann's activities in Husum and Plön, see the report from the Landeskriminalpolizeistelle Flensburg concerning "Bauernbewegung in der Provinz Schleswig-Holstein" Flensburg, 24.1.1929, in both LAS 301/4589 and 309/22696 and the undated extracts from Muthmann and Hamkens' examination transcripts titled "Geschäftsführer des Landbundes Walther Paul Muthmann, 22.4.1902 in Elberfeld, wohnhaft Plöhn [sic], Markt 3." and "Kurzer Auszug aus der Vernehmungsniederschrift des Hofbesitzers Wilhelm Cornelius Hamkens", both in LAS 309/22784.

53 Undated extract from the examination transcript of John Johnsen titled "John Johnsen, 8.10.1898 in Husum, wohnhaft Itzehoe", LAS 309/22784.

Even while these networks were still forming, radical discussion in Schleswig-Holstein was moving ever closer to action. The final harbinger was a Farmers' League commission to North Dithmarschen's county administrator, headed by Claus Heim, in mid-November. The commission requested the reduction of all outstanding, deferred community taxes, complaining that government assistance had been completely unsatisfactory, that livestock prices were simply too low and that most farmers were paying from their capital (*Substanz*). In his report on this meeting, the county administrator attributed the "particular unrest and commotion" in Eiderstedt and Dithmarschen to a small number of individuals. However, he also reported that there could be no doubt that many small farmers' incomes had fallen below those of the average skilled worker and that farms specialising in single products were unable to improve their lot, despite a relatively good harvest. He compared the loan conversion measures to a drop of water on a hot stone and reported that one community chairman had already resigned on the grounds that he did not feel he could sanction the collection of taxes at the specified level. Signalling the coming conflict between farmers and the authorities, Heim told the county administrator that the mood was the same all over the province, though most farmers had not yet gathered the courage to stop paying taxes which would compel the state to confiscate instead.[54]

The finance office had not moved against tax-strikes because enforcement proceedings had been halted until October 1928.[55] When the authorities subsequently sought to appropriate farmers' property to meet outstanding sums, a previously rare occurrence, farmers moved to block them.[56] In Lunden in North Dithmarschen and in Beidenfleth in Steinburg, farmers obstructed the confiscation of oxen.

On November 19, some 15 to 20 farmers obstructed the transport of six confiscated oxen bound for the station in Lunden, from where they were to be sent to the Hamburg cattle markets to recover outstanding taxes owed by Hinrich Haalck, a farmer in St. Annen-Österfeld. Police later found the animals on the property of a community chairman, Johann Mahmens (Preil). Paul Adam Ross had led a meeting in Lunden at which the farmers had formed the plan to assist Haalck. Haalck

54 Report from the Landrat to the Regierungspräsident in Schleswig concerning "Steuern der Landwirtschaft" Heide, 16.11.1928, LAS 309/22569.

55 *Geschäftsbericht der Landwirtschaftskammer für die Provinz Schleswig-Holstein für das Rechnungsjahr 1928-1929*, p. 46; report by the Landrat to the Regierungspräsident in Schleswig concerning "Völkische Treibereien" Rendsburg, 26.11.1928, LAS 309/22921; report from the Abteilung IA, Außendienst II. G. St., Berlin, 27.12.1928, appended to the report from the Regierungspräsident to the Oberpräsident in Kiel concerning "Ergebnis der Ermittelungen der Beamten des Polizeipräsidiums Berlin" Schleswig, 9.1.1929, LAS 301/4695.

56 According to the county administrator, Göppert, Beidenfleth was the first confiscation to take place in Steinburg County in his seven years in office. See County Administrator Göppert's testimony, "Zeugenvernehmung Landrat Göppert" 8.10.1930, sent by the Landrat to the Regierungspräsident, Itzehoe 9.10.1930, LAS 309/22764.

claimed ignorance but witnesses identified nine of those involved, including Mahmens and another community chairman, Johannes Karstens (Nesserdeich), who were all charged with impeding a confiscation (*Pfandbruch*).[57]

That this incident marked the emergence of the Landvolkbewegung in North Dithmarschen became increasingly clear when farmers demonstrated support for the action at a Farmers' League meeting in Lunden three days later. There, much like Hamkens had called on farmers to continue to unite at the local level at the meeting in Husum, Paul Adam Ross called for the creation of a new Unity Organisation. Moreover, the rebellious tone was evident again. Ross roused the audience's anger when he reported that Haalck's cattle had been auctioned in spite of the farmers' efforts. He also criticised the county administrators for pursuing taxes too rigorously, condemned wasteful expenditure on extravagant public buildings and accused the authorities of deficient understanding of the agricultural crisis. His brother, Max Ross, the district chairman (*Amtsvorsteher*), defended his own work in office, but was interrupted by persistent booing. A further speaker, another community chairman, declared that he would welcome a farmers' revolt so that the government would finally see how bad the situation was.[58]

Important as the Lunden incident was for the evolution of the Landvolkbewegung in North Dithmarschen, the incident in Beidenfleth was much more sensational and far-reaching in its consequences, and therefore merits a more detailed analysis. Beidenfleth farmers Heinrich Kock and Albert Kühl both had sizeable outstanding communal taxes, Kock RM 380 and Kühl RM 524, in November 1928.[59] Both also had substantial debts, and Kühl, especially, had suffered great misfortune as a consequence of epizootics, poor harvests and family illness.[60] However, this did not secure them sufficient sympathy from their community representatives (*Gemeindevertretung*) who either did not believe Kock and Kühl's distress was so severe, or were unmoved due to an overriding concern for the communal budget or by a sense of obligation to fulfilling the duties of office. Whatever the case may be, the community representatives were at odds with an increasing number of farmers in their district and incited the anger of many who believed they

---

57 Indictment of the Oberstaatsanwalt, Flensburg, 7.12.1928, GStA PK, I. HA Rep. 84a, No. 55359; copy of a report from the Landrat to the Regierungspräsident in Schleswig concerning "Störung der öffentlichen Ordnung" Heide, 22.11.1928, in both LAS 309/22569 and 301/4695; note in the Regierungspräsident's files titled "Vermerk über ein Ferngespräch mit Landrat Kracht in Heide" Schleswig, 22.11.1928, LAS 309/22569; *Heider Anzeiger* No. 276, 23.11.1928, p. 1.

58 Report from the Landeskriminalpolizeistelle Flensburg concerning "Bauernbundversammlung am 22.11.28 in Lunden" Flensburg, 26.11.1928, LAS 301/4695; *Heider Anzeiger* No. 277, 24.11.1928, pp. 2-3.

59 Unless otherwise noted, the following account of the disturbance in Beidenfleth is based on the issue of the verdict of the Schöffengericht in Itzehoe from 9-17 April 1929, sent by the Oberstaatsanwalt to the Regierungspräsident in Schleswig, 28.5.1929, LAS 309/22569.

60 *Husumer Nachrichten* No. 83, 10.4.1929, p. 5.

should have been lenient rather than authorise their chairman, Mahlstedt, to confiscate one ox from each of Kock and Kühl's herds. On November 13, Mahlstedt visited Kock and Kühl and marked out the two animals to be fetched on the morning of November 19 and taken to the Hamburg livestock market for auction.

Some days beforehand, Kock and Kühl held a meeting with seven or eight of their friends at a cafe in Itzehoe and conspired to prevent the confiscations. Although it is neither clear who conceived the idea, nor how far the plans developed there, it should be noted that the emerging Landvolk leaders, Guido Weschke, Wilhelm Hamkens and possibly Herbert Volck, attended the meeting.[61] Their presence suggests that more than local forces were at work. As Steinburg's county administrator commented afterwards, the farmers' movement had branches spread far and wide.[62]

Farmers with burning hay-bales in Beidenfleth, November 19, 1928[63]

The planning of the particulars was made locally. Kock and Kühl arranged to thwart the confiscations with a substantial gathering of rural people. By telephone

61 See the note in the Regierungspräsident's files titled "Über eine Besprechung des Sachbearbeiters mit Staatsanwaltschaftsrat Dr. Junker" Schleswig, 3.2.1929, LAS 309/22569.

62 Report from the Landrat to the Regierungspräsident in Schleswig concerning "Steuerverweigerung in der Landwirtschaft" Itzehoe, 24.11.1928, LAS 309/22668.

63 Photo: Johannes Böge, Itzehoe.

and by word of mouth, they summoned their friends and neighbours, some to Kühl's farm, some to Kock's and others to the neighbouring property of Johann Holler. Kock arranged for his son, and Kühl his farmhand, to set a fire with hay-bales on the road outside Holler's property. Holler agreed to store the hay in his shed beside the road. Kock and Kühl agreed to sound their fire alarms in order to attract an even greater crowd and everyone was to appear with pitchforks and sticks.

On the morning of November 19, Schlüter, a municipal employee (*Gemeind-ediener*) acting as bailiff, walked to Beidenfleth accompanied by two assistants, Afrath and Sötje. Arriving punctually at 11:30, they fetched Kock's ox first, doing so without hindrance and not noticing any signs of the crowd gathered inside Kock's house. At Kühl's farm they chatted uneventfully with Kühl and his father in law as they tied the animals together. Yet barely had they proceeded, out onto the road, when a crowd appeared behind them.

Farmers obstructing the bailiff and his assistants in Beidenfleth, November 19, 1928[64]

Some 150 to 200 people, many of whom were later identified and charged with participating in a civil disturbance, had gathered from more than 20 different villages, testimony to the action's orchestration and the strength of the local com-

64 Photo: Johannes Böge, Itzehoe.

munity network.[65] The first group to appear had waited in Kock's house. A second group from Kühl's farm now joined the others and advanced toward Schlüter, Afrath and Sötje, who continued on. Not far in front of them, however, they saw the hay-bales piled in the road. These were set alight and when Kock and Kühl blew their fire-horns, a third group of farmers emerged from Holler's farm.

Brandishing pitchforks, poles and sticks, the farmers shouted at Schlüter, Afrath and Sötje to return the oxen to their stalls and insisting that they were not to "fall into the hands of the Jews".[66] Some used their forks to carry burning straw from the fire toward the animals, causing them to shy. Others rushed at Afrath, who let the animals loose and fled over the ditch by the side of the road. Sötje was pushed into the ditch, and Schlüter, on receiving a blow, ran off in search of the police, leaving Sötje encircled and alone.

Police, on bicycles, arrive in Beidenfleth, November 19, 1928[67]

The farmers did him no further harm but compelled him to return the oxen to their stalls. After catching the animals and returning Kühl's ox, he too saw a chance to escape, and fled over the fields. Only afterwards did three policemen arrive. On enquiring what had happened, one of the farmers replied that they wanted to teach them not to take their cattle. The farmers then returned home.

65 *Husumer Nachrichten* No. 88, 16.4.1929, p. 7. For the trial, see chapter 4, pp. 140 ff.

66 See the issue of the verdict of the Schöffengericht in Itzehoe from 9-17 April 1929, LAS 309/22569.

67 Photo: Johannes Böge, Itzehoe.

A farmers' deputation later approached Steinburg's county administrator, Göppert, who declined to grant the farmers' tax remission on the grounds that the community was also in need, but he promised that applications for deferment of payment would be widely accepted. He recommended that Kock and Kühl make partial payments and then pay off the remainder in instalments and he agreed to urge the community to wait for the payments until the animals had fattened sufficiently to be fit for sale. Kock and Kühl, however, did not respond to his entreaties and he therefore saw no reason to intercede as proposed. Indeed, Göppert saw to it that 25 policemen were in place to ensure the confiscation be carried out on November 24. They met with no resistance, perhaps as the *Nordischer Kurier* suggested, because the farmers had read in the local *Wilstersche Zeitung* that the confiscation was to proceed two days later.[68] At the last moment, Kühl managed to stop the confiscation of his animal by making a partial payment.[69] Kock, on the other hand, subsequently prevented the auction of his ox in Hamburg with the assistance of Heinrich Hansen (Breitenfelde) and other farmers by threatening the auctioneer that farmers would boycott the market should the sale proceed.[70]

By no means did the events in Beidenfleth blow over with the confiscation of Kock's ox. Not long after, a flyer of unknown source was disseminated in Steinburg and neighbouring counties.[71] Entitled "Forceful expropriation by the government!" (*Gewaltsame Enteignung durch die Regierung*) and signed "the rural people of the County of Steinburg", it advertised a meeting in Itzehoe on November 26.[72]

Some 500 to 600 people attended the meeting which was led by the DNVP Reichstag member, Soth. Among the 13 speakers who addressed the large audience were a number who either bemoaned their own distress, claimed they were condemned to pay taxes from their capital or simply declared that they were no longer able to pay taxes.[73] The prevention of the confiscation of the oxen at Beidenfleth, the police's subsequent seizure of Kock's ox and its transportation to Hamburg were related at the meeting, as was the action in Lunden.

---

68 *Nordischer Kurier* No. 279, 27.11.1928, p. 4; *Wilstersche Zeitung* No. 277, 24.11.1928, p. 2.

69 *Wilstersche Zeitung* No. 278, 26.11.1928, p. 2; Göppert's testimony at the Beidenfleth trial as reported in the *Husumer Nachrichten* No. 87, 15.4.1929, p. 2.

70 Kock and Hansen were subsequently charged with intimidating a public servant (*Beamtennötigung*). Landrat des Kreises Steinburg to the Regierungspräsident in Schleswig concerning "Steuerverweigerungen" Itzehoe, 1.12.1928, in both LAS 301/4695 and 309/22569. For the Beidenfleth trial, see chapter 4.

71 Issue of the verdict of the Schöffengericht in Itzehoe from 9-17 April 1929, LAS 309/22569.

72 *Itzehoer Nachrichten* No. 280, 28.11.1928, p. 3.

73 Unless otherwise noted, the account of the meeting here is based on and the quotations in the text are to be found in the *Itzehoer Nachrichten* No. 280, 28.11.1928, p. 3; *Nordischer Kurier* No. 280, 28.11.1928, p. 5; *Rendsburger Tageblatt* No. 290, 28.11.1928, p. 5.

Anger over the local administration was a further feature of the meeting. In the invitation to the meeting, community chairman Mahlstedt and other community representatives in Beidenfleth were depicted as having betrayed the farmers, for they were branded as servants of the "Jewish expropriation system".[74] Protected by parliamentary immunity, Soth now denounced them once again as "this system's lackeys" (*Büttel dieses Systems*). He declared that the farmers' own colleagues had stabbed them in the back and had helped to deliver them to the Jews, despite being well aware of the farmers' distress.

These local complaints did not dominate the meeting, but yielded instead to more general enumerations of the perceived causes of grievance, which essentially amounted to expressions of hostility to towards the Weimar Republic. The anti-democratic and anti-Semitic elements deserve particular attention because they became distinguishing elements at Landvolkbewegung meetings. Furthermore, as the first meeting of many in which anti-Semitism was to feature so prominently, it would be well to recall that anti-Semitism had a longstanding background in Schleswig-Holstein and that the Rural League and DNVP had played a central role in its development and transmission. As a DNVP member and Rural League leader, Soth – though it should be noted he was only to play a peripheral role in the movement hereon in – conveniently appeared almost as the embodiment of its carry-over into the Landvolkbewegung.

Among other claims, Soth denounced the parliamentary system as "the ruin of the German economy",[75] declared that the government was "consciously" destroying agriculture, and explained that industry and the banks could not be expected to help, because industry was connected to foreign countries via an international ring and the banks were all "in the hands of the Jews". Even the agricultural organisations' failure to work together, Soth ascribed to "Jewry's divisive influence" (*Judentums Spaltpilz*). In a similar vein, Markus Wulf, a farmer from Eisendorf near Rendsburg, declared that "the Jew" was the "chief-culprit" (*der Hauptschuldige*), the "grave-digger of agriculture." To cries of approval from the audience, Wulf explained that social democracy was expropriating the farmers by way of an "ingenious taxation system", and that this was what had led to the crisis.

Poisonous as such words were in ascribing agriculture's difficulties to Jews and democracy, it was the shorter speeches of two other farmers that provoked the greatest reactions. When Heinrich Frauen, a farmer from Hemme in North Dithmarschen, asserted that the farmers were prepared to resort to "illegal measures", he received roaring applause.[76] "In the bible", he continued, "it says that authority comes from God. I declare that this authority does not come from God but from

---

74 *Itzehoer Nachrichten* No. 280, 28.11.1928, p. 3

75 *Nordischer Kurier* No. 280, 28.11.1928, p. 5.

76 See the issue of the verdict of the Schöffengericht in Itzehoe from 9-17 April, LAS 309/22569.

Satan. We have to put the knife to this state's jugular."[77] So strong was the acclaim for Frauen's words, that Soth was stirred to respond, telling the audience that applause was all very well but action was required as well.[78] Next, Hans Kelting (Bahrenfleth) called on young farmers to support their fathers in order to protect their inheritances. If they could not defend their farms by other means, Kelting said, then they would do so with weapons in their hands.[79]

The Itzehoe meeting was testimony to the great excitement the confiscations had occasioned. The farmers again gave voice to their troubles but it was the vehement calls to radical action and the warm reception they received from the large audience which were most striking. As Steinburg's county administrator reported, there was no opposition to the "radical and rabid assertions" made at the meeting.[80]

The radical mood found expression the following night when bombs exploded at the homes of two community chairmen alleged to have betrayed the farmers by authorising confiscations, Thomsen in Hollingstedt and Mahlstedt in Beidenfleth. A third bomb was found, undetonated, in the garage of District Chairman Max Ross in Flederwurth the following day.[81] As attacks on local authorities, the bombings were arguably expressions of local anger. Yet it would be more accurate to say – as the intensity of the anti-system sentiment at the Itzehoe meeting showed – the farmers perceived their misfortune as a consequence of the failings of the local authorities, who had deferred to the greater defects and machinations of the Weimar system. This was underlined by the resolution read out at the meeting's close. The resolution, moreover, endorsed the farmers' direct action.

> "The legitimate demands of the Landvolk have still not been regarded seriously. Now the cold deprivation of the farmers is being carried out even by armed force. Jewish splinter-politics are being used in an attempt to tear the Landvolk's closed front apart. Force is being used to try to break this fighting front against the current system which is supported by dishonest, cowardly and listless colleagues. We however will not allow

---

77 "In der Bibel steht, daß die Obrigkeit von Gott stamme, ich erkläre, daß diese Obrigkeit nicht von Gott, sondern von Satan stammt. Diesem Staat müssen wir das Messer an die Gurgel setzen." Issue of the verdict of the Schöffengericht in Itzehoe from 9-17 April 1929, LAS 309/22569.

78 Issue of the verdict of the Schöffengericht in Itzehoe from 9-17 April 1929, LAS 309/22569.

79 As a consequence of these speeches, Frauen and Kelting were indicted for encouraging disobedience against the laws and tried in April 1929 together with the farmers who prevented the confiscation at Beidenfleth. See the indictment sent from the Oberstaatsanwalt to the Schöffengericht in Itzehoe, Altona, 2.3.1929 and the issue of the verdict of the Schöffengericht in Itzehoe from 9-17 April 1929, both in LAS 309/22569.

80 Copy of a report from the Landrat des Kreises Steinburg to the Regierungspräsident in Schleswig concerning "Steuerverweigerung" Itzehoe, 26.11.1928, sent by the Regierungspräsident to the Oberpräsident in Kiel, Schleswig, 29.11.1928, LAS 301/4695; report to the Regierungspräsident in Schleswig concerning "Steuerverweigerung" Itzehoe, 27.11.1928, LAS 309/22668.

81 See chapter 7.

> ourselves to be driven from our homes and our farms by force, rather we will stand against dead paragraphs and carbines as one. Right is on our side."[82]

To deny the state its dues and disregard the authorities' instructions as the farmers did in Beidenfleth was to threaten the state's means and sovereignty. As a spectacle, Beidenfleth was by implication an expression of hostility, the potency of which was underlined at the Itzehoe meeting and by the bombings. Beidenfleth was not merely a tax-strike but an act of defiance against the communal administration and the state itself. Though this was probably not fully understood by all of the participants at the time of the action, the speeches at the Itzehoe meeting ensured it was interpreted in this way afterwards.

The incidents on the west coast divided rural opinion. This was to be seen in readers' letters in the *Rendsburger Tageblatt* which reflected the immense interest the events had generated, and demonstrated that the public did understand they signified more than simply the farmers' refusal to pay taxes. Indeed, the letters could only reinforce their political significance. The letters revealed that there was consensus the state had failed to assist agriculture, but while some of them criticised the obstruction of the confiscations as violations of the state's authority, others commended it as the only viable method of reprieve until the form of state was radically altered.[83]

The local authorities recognised immediately that the Landvolk's illegal actions and radical criticisms were an "attack" (*Angriff*) on the communal administration.[84] Alarmed at the radicalisation in the countryside, some 30 to 40 community chairmen and farmers met at the county administrator's office and condemned their colleagues' "ill-considered departure from the path of the law" and stood by the Bei-

---

82 "Die berechtigten Forderungen des Landvolkes sind immer noch nicht ernsthaft beachtet worden. Jetzt wird sogar durch bewaffnete Macht die kalte Enteignung der Bauern durchgeführt. Durch jüdische Zersplitterungspolitik versucht man die geschlossene Kampffront des Landvolkes auseinanderzureißen. Man sucht diese Kampffront gegen das heutige System mit Gewalt zu brechen, das unterstützt wird von unehrlichen, feigen und lauen Berufskollegen. Wir sind aber nicht gewillt, uns mit Gewalt von Haus und Hof vertreiben zu lassen, sondern wir werden gegen tote Paragraphen und Karabiner zusammenstehen wie ein Mann. Auf unserer Seite ist das Recht." As cited in the report from the Landrat des Kreises Steinburg to the Regierungspräsident in Schleswig concerning "Steuerverweigerung" Itzehoe, 29.11.1928, in both LAS 301/4695 and 309/22569. NB by "dead paragraphs" and "carbines", the resolution presumably meant "the dead paragraphs of the law" and "police carbines".

83 *Rendsburger Tageblatt* No. 299, 8.12.1928, p. 1, No. 301, 11.12.1928, p. 1 and No. 310, 21.12.1928, p. 12.

84 *Itzehoer Nachrichten* No. 280, 28.11.1928, p. 3; *Rendsburger Tageblatt* No. 290, 28.11.1928, p. 6; *Nordischer Kurier* No. 286, 5.12.1928, p. 3; resolution of the county committee (*Kreisausschuss*), Itzehoe, 3.12.1928, appended to the report from the Landrat des Kreises Steinburg to the Regierungspräsident in Schleswig concerning "Steuerverweigerungen" Itzehoe, 3.12.1929, LAS 309/22569.

denfleth administration's conduct.[85] The chairman of the Chamber of Agriculture, Count Rantzau, also addressed the meeting, warning against the new movement on the basis that its aims were not clear. The following week, Steinburg's Community Association (*Landgemeindeverband*), an assembly of the county's community chairmen, condemned the incidents at Beidenfleth and the Itzehoe meeting, and denounced the bombings. There, County Administrator Göppert acknowledged the crisis, announcing that in particularly critical individual cases he would endeavour to secure a reduction in the national land tax (*Grundsteuer*). But as the head of the association, Community Chairman Augustin (Borsfleth) explained, individual farmers in need were eligible to apply for a deferment of payment, but communal taxes could not be lowered.[86]

Beidenfleth and the related incidents also roused the concern of the provincial authorities. They appeared to have two paths open to them: to suppress insubordination and subdue the radical atmosphere before it got out of control, or to offer relief and influence policy to lessen the economic causes of the farmers' distress and restore, as far as possible, their confidence in the state. It is a severe judgement perhaps, but it seemed that the provincial authorities preferred the first route, simultaneously wishing to be seen as having taken the second, benevolent, route as well.

Among its first steps, the provincial government issued a notice in the press warning against illegal action, maintaining that the state was determined to proceed against it with all its means.[87] It also welcomed the Community Association's response to Beidenfleth, judging it somewhat optimistically as a sign that "cooler heads" had retained the upper hand in the countryside.[88] Nonetheless, its concern about the insurgent atmosphere was demonstrated when it invited representatives of the agricultural organisations, the Chamber of Agriculture, the county assemblies and State Finance Office to a discussion led by the new Regierungspräsident, Waldemar Abegg, on December 7.[89]

A subsequent government press release gave the impression that this long meeting was held to discuss the alleviation of the agricultural crisis and distress in the countryside.[90] The details of the discussion, however, were remarkably brief.

---

85 *Itzehoer Nachrichten* No. 280, 28.11.1928, p. 3.

86 *Nordischer Kurier* No. 286, 5.12.1928, p. 3; report on the events in Beidenfleth by Gemeindevorsteher Augustin, Borsfleth, 3.12.1928, LAS 301/4695.

87 "Die Regierung in Schleswig zu den Vorgängen in der Landwirtschaft" in the *Nordischer Kurier* No. 284, 2.12.1928, p. 2.

88 Regierungspräsident to the Oberpräsident in Kiel, Schleswig, 7.12.1928, LAS 301/4695.

89 Waldemar Abegg (1873-1961), brother of State Secretary Wilhelm Abegg, was a lawyer and career public servant. He was Oberregierungsrat in Schleswig from 1921 and Regierungspräsident of Schleswig-Holstein from October 1928 until October 1932. See E. R. Huber, *Deutsche Verfassungsgeschichte seit 1789 Band VII: Ausbau, Schutz und Untergang der Weimarer Republik* (Stuttgart, Berlin, Cologne and Mainz, 1984), p. 1033.

90 Report of the meeting titled "Verhandlungsniederschrift" sent from the Regierungspräsident to the Oberpräsident in Kiel, Schleswig, 10.12.1928, in both LAS 301/4695 and 309/22921;

Since they were summarised in a single sentence, the proposals for betterment were quite clearly not the Regierungspräsident's priority. Moreover, among the recommendations which had apparently "made a strong impression" on the government was a change in the distribution of school taxes. Not a breakthrough as the press release indicated, this was an older proposal salvaged in order to create the impression that negotiation with agriculture's organisations had been productive.[91] To his credit, Abegg did urge the Ministry of Science, Arts and Education to reduce the burden of school taxes.[92] The press release, however, already pointed to disappointment, referring merely to tax "distribution" and not to imminent action but only to the Regierungspräsident's intention of trying to achieve what was possible.

One might argue that Abegg's approach was at least realistic because the state's capacity to aid the farmers was limited and not likely to meet the farmers' expectations in any case. However, the example of his predecessor showed what he might have done. Adolf Johannsen had initiated tax relief measures, petitioned his superiors and travelled to Berlin to meet with the Reich and Prussian ministers on the farmers' behalf. Abegg, by contrast, could only imitate Johannsen's call for the cooperation of the agricultural organisations. Novel a year earlier, this suggestion now seemed designed merely to shift responsibility to the agricultural associations for the failure of alleviating the crisis.

The press release was the sole tangible outcome of the December 7 meeting. Though it acknowledged the crisis in agriculture and asserted that it was the government's "serious concern" that assistance be provided, its sum and substance was the condemnation of tax-strikes, insubordination and attacks against the authorities.[93] As a report to the Prussian Interior Ministry stated, the discussion at the meeting found expression in the press release.[94] The Regierungspräsident's real objective, therefore, was to suppress insurgency in the countryside, not to relieve the farmers of their distress.

---

press release titled "Korrespondenzbüro 'Nordschleswig' Flensburg, 9.12.1928. 'Die Not der Landwirtschaft. Eine wichtige Besprechung in der Regierung – Schleswig' ", LAS 309/22921.

91 Report from the Landrat to the Regierungspräsident in Schleswig concerning "Steuern der Landwirtschaft" Heide, 16.11.1928, LAS 309/22569; "Verhandlungsniederschrift" sent from the Regierungspräsident to the Oberpräsident in Kiel, Schleswig, 10.12.1928, in both LAS 301/4695 and 309/22921.

92 Regierungspräsident to the Minister für Wissenschaft, Kunst und Volksbildung in Berlin concerning "Aufbringung der Schulabgaben in ländlichen Schulbänden" Schleswig, 7.12.1928, LAS 301/4695.

93 See the press release titled "Korrespondenzbüro 'Nordschleswig' Flensburg, 9.12.1928. 'Die Not der Landwirtschaft. Eine wichtige Besprechung in der Regierung – Schleswig' ", LAS 309/22921.

94 Regierungspräsident to the Preuß. Minister des Innern in Berlin concerning "Besprechung mit den Vertretern des landwirtschaftlichen Organisationen" Schleswig, 8.12.1928, in both LAS 301/4695 and 309/22921.

For their part, some of the agricultural representatives – the Chamber of Agriculture and the Farmers' Association – shared the government's concerns. Nevertheless, they might be criticised for their willingness to engage in talks and cooperate in issuing the press release without using the opportunity to ensure the provincial government undertook more to support their interests.[95]

It is not whether the provincial government should have taken a stand against the Landvolk that is at issue here but the manner in which it did so. Ultimately, it is the suitability of the occasion to discuss relief that should be questioned. Such discussion was imprudent as a response to the use of intimidation by farmers for it was likely to encourage the idea that these tactics were effective. Yet to discuss relief half-heartedly, and one might even say insincerely, was merely to substantiate the farmers' claims that their demands were not being taken seriously.

The Landvolkbewegung's appearance threatened the NSDAP's progress in the province. While the agricultural associations had bickered throughout the year, the NSDAP had continued to expand from South Dithmarschen.[96] Its membership grew dramatically. From approximately 800 on January 1, 1928 to 2,000 by May, it climbed to between 3,000 and 4,000 by the end of the year, the largest increase in any Gau in Germany. Moreover, the party had improved its standing by winning over a number of well-known farmers who then assumed leading positions.[97] Hitler had rewarded these advances, speaking for the first time in the province in Heide on October 14. The subsequent spectacle of an SA march with mounted units and a brass band had drawn a crowd of thousands at the market place.[98]

When the Landvolkbewegung appeared, therefore, tensions quickly surfaced and to such a degree that a combined National Socialist and Landvolk meeting was held in Itzehoe on November 11, 1928. Although the Landvolk and Defence Association had until this time barely given voice to their grievances and plans, Gauleiter Lohse questioned the necessity of the new organisations, claiming they espoused the National Socialists' ideas. The Landvolk's speakers all emphasised that they regretted "bickering among brothers", but they believed the Defence Association promised swifter improvement. As one young Defence Association leader de-

95 See the "Verhandlungsniederschrift" sent from the Regierungspräsident to the Oberpräsident in Kiel, Schleswig, 10.12.1928, in both LAS 301/4695 and 309/22921.

96 See the report from the Landeskriminalpolizeiamt Berlin, Abteilung IA, Heide, 21.3.1929, LAS 309/22668.

97 Rietzler, *"Kampf in der Nordmark"*, pp. 425 and 429-430; Stoltenberg, *Politische Strömungen im schleswig-holsteinischen Landvolk*, pp. 149-150.

98 Rietzler, *"Kampf in der Nordmark"*, pp. 427-429; *Heider Anzeiger* No. 243, 15.10.1928, p. 1. For a transcript of Hitler's speech, see B. Dusik and K. A. Lankheit (eds), *Hitler: Reden, Schriften, Anordnungen: Februar 1925-Januar 1933*, vol. 3 (1) (Munich, New Providence, London and Paris, 1994), pp. 150-152.

clared, it would take much too long before the National Socialists had mustered all "feeble men" (*schlappen Männer*).[99]

It is clear that the National Socialists were troubled by their competitor's surge of activity for they held their very own meeting in response to Beidenfleth. Though smaller than the Landvolk's meeting on November 26, it was nonetheless sizeable for the countryside with some 400 people in attendance. Though he claimed he did not wish to take advantage of Beidenfleth, Lohse had used the excitement to attract the crowd. Very much like the speakers at the Landvolk's meeting in Itzehoe, he attacked the power of "Jewish capital" and the parliamentary system. However, he sought to distance his party from the Landvolk, arguing that actions like Beidenfleth were ineffective, and that they could never change the general state of affairs. He justified his party's parliamentary participation, arguing that it had only entered parliament to sabotage it, and promised nothing less than Germany's liberation.[100]

Lohse's speech was not in the least conciliatory. However, the three Landvolk speakers who responded must have been determined to promote friendly relations at the meeting since they did not return his hostile stance. They conveyed only their unwillingness to wait any longer for the success of the NSDAP's parliamentary endeavours, insisting that action was urgently required or the farmers would be doomed. Knudsen, for instance, asserted that the Landvolk rejected all parties, but also declared that the National Socialists would find great support in their struggle against parliamentarism in the Landvolkbewegung and declared his hope that the NSDAP and Landvolk would avoid conflict in future.[101]

While the county administrator in Rendsburg advised the provincial government against "unnecessary leniency" and called for a "calm and solid course" in regard to tax collection and immediate action against all tax-strikes, Hamkens continued to encourage his fellow farmers against the payment of taxes.[102] At a meeting in Oldenswort at the end of November, he reportedly told his audience simply to refuse to pay taxes and not worry about confiscations, but rather to get in contact with their "trusted men". The government, he said, would weary if police escorts were required at all confiscations.[103] This was one of Hamkens' most forthright calls for tax-strikes and an explicit expression of his hope that Lunden and Beidenfleth-type actions would multiply and spread to such an extent that the authorities become

99 See the *Itzehoer Nachrichten* No. 268, 13.11.1928, p. 3. Also see the report from the Landrat to the Regierungspräsident in Schleswig concerning "Versammlung der Nationalsozialisten" Itzehoe, 27.11.1928, LAS 309/22668.

100 *Nordischer Kurier* No. 283, 1.12.1928, p. 4 and for a more detailed overview of the speeches, the *Wilstersche Zeitung* No. 282, 30.11.1928, p. 2.

101 *Nordischer Kurier* No. 283, 1.12.1928, p. 4; *Wilstersche Zeitung* No. 282, 30.11.1928, p. 2.

102 Report by the Landrat to the Regierungspräsident in Schleswig concerning "Völkische Treibereien" Rendsburg, 26.11.1928, LAS 309/22921.

103 Landjägerei Oldenswort to the Herrn Landrat zu Tönning, Oldenswort, 30.11.1928, LAS 309/22569.

powerless to deal with them. Soon, however, Hamkens was impelled to avoid such direct formulations because the authorities began to keep a closer eye on him.

Legal action was the state's principal weapon against tax-strikes. Before they could be tried, however, Hamkens and the farmers who had participated at Lunden and Beidenfleth remained free to continue their agitation. The authorities, therefore, sought to have legal proceedings accelerated.[104] Additionally, they strove to gather more material against the Landvolk's speakers, arranging for policemen proficient in the local dialects to observe whether they encouraged tax-strikes at meetings.[105]

The provincial government's warnings neither brought calm to the countryside, nor curbed animosity towards the authorities. In December, posters appeared in various counties in the province.[106] One, titled "Jewish treachery against the German people", purported to explain that inflation, taxes and interest on loans amounted to the theft of German property. It complained that successful "self-help" actions had been contorted in the Jewish press and by German traitors who supported this "Jewish expropriation procedure" (*jüdische Enteignungsverfahren*) for fear of losing their offices. Further, it claimed that the rural people wished to protect their children's inheritance but the authorities had responded with warnings, confiscations and twisted reports in the press.[107]

Given the persistence of such censure, it is understandable that an increasing number of community chairmen and county administrators began to grow uneasy. According to the county administrator in Eiderstedt, many community chairmen did not feel the state was offering them sufficient protection and several had given notification of their wish to resign. They had been described in posters and newspapers as "lackeys of the government" and had had their "names dragged through

104 See the report from the Landrat des Kreises Eiderstedt to the Regierungspräsident in Schleswig concerning "Steuerstreik im Kreise Eiderstedt" Tönning, 3.12.1928, Regierungspräsident to the Oberstaatsanwalt in Flensburg, Schleswig, 25.11.1928 and Landrat des Kreises Steinburg to the Regierungspräsident in Schleswig concerning "Steuerverweigerungen" Itzehoe, 1.12.1928, all in LAS 309/22569.

105 Politisches Dezernat der Regierung Schleswig to the Landeskriminal-Polizeistelle Flensburg, Schleswig, 15.12.1928, LAS 309/22921.

106 See the reports from the Landjägereiamt Treia, Kreis Schleswig to the Landjägereiabteilung in Schleswig, Treia, 16.12.1928 and from the Landrat des Kreises Steinburg to the Regierungspräsident in Schleswig concerning "Steuerverweigerung" Itzehoe, 15.12.1928, both in LAS 309/22569; report from the Landjägereiposten Kreis Bordesholm to the Landrat in Bordesholm, Wittorf b. Neumünster, 17.12.1928, the report from the Landrat to the Regierungspräsident in Schleswig concerning "Verbreitung von Druckschriften" Plön, 4.1.1929 and the report from the Landrat des Kreises Eiderstedt to the Regierungspräsident in Schleswig concerning "Unruhen in der Landwirtschaft" Tönning, 21.12.1928, all in LAS 309/22696.

107 See the poster titled "Jüdischer Verrat am deutschen Volk!" and also a further poster titled *Deutsches Weihnachtsfest!*, both appended to the report from the Landjägereiposten Kreis Bordesholm to the Landrat in Bordesholm, Wittorf b. Neumünster, 17.12.1928, LAS 309/22696.

the mud".[108] At the beginning of December 1928 the Regierungspräsident promised Schleswig-Holstein's community chairmen his protection and complete support.[109] But as the bailiff in Garding, Heinrich Frey, was soon to discover, this did not amount to much.

In mid-December Hamkens placed an advertisement in the provincial press, calling for a meeting of the rural people of Eiderstedt County in Garding. The meeting's purpose was, as the advertisement announced, to protest against the Husum finance office and the "... merciless compulsory collection of outstanding taxes".[110] The specific reason for the meeting in this case was the confiscation and impending auction of a young farmer's calf and a widow's furniture in Garding. Neither had heeded the authorities' warnings before the confiscations, presumably as a form of protest. Then, on the evening of December 13, Hamkens visited the community bailiff, Frey, to enquire about the auction. Hamkens questioned Frey's authorising the auction, but when Frey insisted that it was his duty, Hamkens grew angry, calling the bailiffs and community chairmen "arse-lickers" (*Schleim-scheißer*) who clung to their income and office.[111] Frey reported Hamkens' visit to the police and to his county administrator. He requested police protection for himself and his family, noting that he could not be expected to work under such circumstances, and that since the bombings he and his wife had lived in a state of anxiety.[112] Subsequently, Frey also brought charges against Hamkens on the grounds that he had insulted him in public office (*Beamtenbeleidigung*).[113] This charge, alongside those for encouraging tax-strikes, was later brought against Hamkens in Husum on January 3, in a joint trial with the farmers who had attempted to thwart the Lunden confiscation.[114]

---

108 Report from the Landrat des Kreises Eiderstedt to the Regierungspräsident in Schleswig concerning "Unruhen in der Landwirtschaft" Tönning, 21.12.1928, LAS 309/22696.

109 Regierungspräsident to the Amtsvorsteher and Gemeindevorsteher des Bezirks (außer Helgoland), Schleswig, 10.12.1928, LAS 309/22921.

110 *Eiderstedter Nachrichten* No. 8187, 15.12.1928, p. 4.

111 Copy of a statement by Gemeindeerheber Heinrich Frey, Garding, 14.12.1928, LAS 309/22569; indictment, Oberstaatsanwaltschaft, Flensburg, 20.12.1928 and copy of the issue of the verdict of the Schöffengericht, Husum, 3.1.1929, both in GStA PK I. HA Rep. 84a, No. 55359.

112 Copy of a statement made by Gemeindeerheber Heinrich Frey, Garding, 14.12.1928 and a copy of a letter from Gemeindeerheber Heinrich Frey to the Landrat in Tönning, Garding, 15.12.1928, both in LAS 309/22569. For more on the bombings, see chapter 7.

113 Der Landrat des Kreises Eiderstedt to the Regierungspräsident in Schleswig concerning "Beleidigung des Gemeindeerhebers Frey, Kirchsp. Garding" Tönning, 19.12.1928, LAS 309/22569.

114 See chapter 4.

Not enough police were available to provide all bailiffs and community chairmen with adequate protection.[115] The Regierungspräsident assured the county administrator in Eiderstedt that he would make more police available, if more were required.[116] The value of this assurance is dubious. At least, it is hard to imagine Frey mollified by the Regierungspräsident's instructions that he call the nearest police station or his county administrator should he find himself in danger.[117]

The county administrator in Eiderstedt believed that Frey's case showed that swift action against Hamkens was imperative to ensure community chairmen and bailiffs did not resign. He forewarned that the upcoming meeting in Garding did not bode well. Quite rightly, he feared that Beidenfleth could no longer be considered a local incident but a disturbance extending across the entire countryside.[118]

At the meeting in Garding on December 18, Hamkens addressed an audience of around 450 people. His speech revealed a new, more careful method of calling for tax-strikes, one that promised to spare him from future charges. Hamkens was aware of the police spies who observed and documented his speech and he alerted his audience to their presence.[119] They knew, therefore, why he was careful to avoid direct formulations, concealing his summons by denying the farmers' capacity to strike at all. A strike, he said, was impossible, because the farmers simply did not have any money to pay taxes. Recurrent bursts of cheering and applause indicated that his audience clearly understood him.[120]

This style, which Hamkens employed successfully at many meetings thereafter, not only entertained audiences, but also seemed more potent than plain language. This was also illustrated by his observation, "We cannot defend ourselves with guns, grenades and cannons because we do not have any. If we did, then we would

---

115 According to the report from the Landrat des Kreises Eiderstedt to the Regierungspräsident in Schleswig concerning "Unruhen in der Landwirtschaft" Tönning, 16.12.1928, LAS 309/22569.

116 Regierungspräsident to the Landrat in Tönning, Schleswig, 17.12.1928, LAS 309/22569.

117 Regierungspräsident to the Gemeindeerheber Heinrich Frey in Garding, Schleswig, 17.12.1928, LAS 309/22569.

118 Der Landrat des Kreises Eiderstedt to the Regierungspräsident in Schleswig concerning "Unruhen in der Landwirtschaft" Tönning, 16.12.1928, LAS 309/22569.

119 Police report titled "Bericht" Flensburg, 19.12.1928, sent from the Landeskriminalpolizeistelle Flensburg to the Regierungspräsident in Schleswig, Flensburg, 19.12.1928, LAS 309/22569 and appended to the report from the Regierungspräsident to the Preuß. Minister des Innern in Berlin concerning "Unruhe in der Landwirtschaft" Berlin, Schleswig 9.1.1929, 301/4695.

120 "Vernehmungsprotokoll" Garding, 18.12.1928, sent from the Landeskriminalpolizeistelle Flensburg to the Regierungspräsident in Schleswig, Flensburg, 19.12.1928, LAS 309/22569; report from the Landjägerei Abteilung Tönning, Kreis Eiderstedt, sent to the Landrat in Tönning, Tönning, 19.12.1928, LAS 309/22696.

make use of them."[121] It was sure to appeal to those who really wished to hear an exhortation to extreme methods as well as those who understood it merely as a rhetorical gesture, as a colourful call for action in general. Its equivocal nature did not commit Hamkens himself to any particular course of action and also relieved him from responsibility should it have encouraged others to do so.

As with the meeting in Itzehoe, Hamkens and numerous other speakers at Garding went far beyond the immediate concerns of tax-strikes and resistance to the upcoming compulsory auctions, in their complaints about the crisis, expression of anti-system and anti-Semitic sentiment and criticism of the authorities.[122] Among the most interesting speeches was that of Bossen (Neulandshof in South Tondern) who recounted the visit of a farmers' delegation to the county administrator in Niebüll. If we are to believe Bossen, the county administrator's manner gave the impression that he held little respect for the farmers and paid insufficient attention to their concerns. In particular, Bossen's account of the county administrator's suggestions to the farmers, for example that they should plough more extensively, was noted for the way it incited the audience. According to a police report, the audience reaction was to conclude that the authorities simply did not understand agricultural matters at all.[123]

Hamkens' principal purpose at the meeting was to rouse resistance against the upcoming auctions. He avoided calling for action directly, but drew in his audience by asking them to suggest possible strategies. Nevertheless, he implicitly yet unmistakably proffered his plan when he added that he, of course, was going to attend to see if any Jews were there and to see who offered a bid.[124] During the ensuing discussion it was ascertained that all lawful endeavours to have the taxes lowered had failed and among the numerous other speakers, Heinrich Kock related what had happened at Beidenfleth, inspiring cries of indignation and immense excitement.[125] The stage was thus set for a resolution for direct action. Accordingly,

121 Report titled "Vernehmungsprotokoll" Garding, 18.12.1928, LAS 309/22569; report from the Landjägerei Abteilung Tönning, Kreis Eiderstedt, sent to the Landrat in Tönning, Tönning, 19.12.1928, LAS 309/22696.

122 Police report titled "Bericht" Flensburg, 19.12.1928, in both LAS 309/22569 and 301/4695; report titled "Vernehmungsprotokoll" Garding, 18.12.1928, LAS 309/22569; report from the Landjägereiabteilung Tönning, Kreis Eiderstedt, sent to the Landrat in Tönning, Tönning, 19.12.1928, LAS 309/22696.

123 Police report titled "Bericht" Flensburg 19.12.1928, in both LAS 309/22569 and 301/4695.

124 "Vernehmungsprotokoll" Garding, 18.12.1928, LAS 309/22569; report from the Landjägereiabteilung Tönning Kreis Eiderstedt sent to the Landrat in Tönning, Tönning, 19.12.1928, LAS 309/22696.

125 "Vernehmungsprotokoll" Garding, 18.12.1928, LAS 309/22569; report from the Landjägereiabteilung Tönning Kreis Eiderstedt sent to the Landrat in Tönning, Tönning, 19.12.1928, LAS 309/22696; police report titled "Bericht" dated Flensburg 19.12.1928, in both LAS 309/22569 and 301/4695.

when Hamkens repeated that he would be going to the compulsory auction, the audience answered him energetically, calling "We're all coming".[126]

Were the members of the audience fully aware of Hamkens' intentions when they avowed their willingness to accompany him to the auction? A police spy noted that all of the speakers at the meeting advocated "self-help", that they spurred the audience to action without making clear what these actions might be. He had the impression that either the implementation of "self-help" was to be left to the individual or that it was conveyed by innuendo. Containing comments and discussion heard made among the audience, the spy's report reveals broad agreement with the ideas the speakers expressed.[127] Moreover, the tremendous applause and cheering, the cries from the audience and its unified resolve also point to a high level of cohesion at the meeting. Hamkens knew how to consolidate this feeling. At the end of his speech he urged the audience to stand, raise a hand, swear to remain loyal to one another and work together, before crying "One for all, all for one."[128]

Two days later some 200 people crowded to the auction.[129] They did not bid for the widow's furniture and just one paltry bid of RM 30 was made for the young farmer's calf. When this lone voice among the crowd was heard, a sudden spurt of movement was seen. Then there was silence and no further bids followed. As the bid was well under the reserve price, the calf was withdrawn and the crowd jeered at the young man who had made the bid as he departed.[130] Hamkens attempted to identify him, even asking a policeman to name the man, but the officer simply told him that it was none of his business.[131] Hamkens' expectation that the officer should answer his question demonstrated remarkable confidence, and one might say impertinence, in dealing with authority. Undoubtedly, Hamkens' object was to denigrate the bidder. Two days later, at a protest meeting against the German-Polish trade agreement in Rendsburg, the man's name was made public, although it is not clear if or how the farmers did exact revenge.[132]

---

126 "Vernehmungsprotokoll" Garding, 18.12.1928, LAS 309/22569; report from the Landjägereiabteilung Tönning Kreis Eiderstedt sent to the Landrat in Tönning, Tönning, 19.12.1928, LAS 309/22696.

127 See the report titled "Bericht" Flensburg 19.12.1928, in both LAS 309/22569 and 301/4695.

128 Report from the Landjägereiabteilung Tönning Kreis Eiderstedt to the Landrat in Tönning, Tönning, 19.12.1928, LAS 309/22696.

129 Report from the Landjägereiabteilung Tönning Kreis Eiderstedt to the Landrat in Tönning, Tönning, 21.12.1928, LAS 309/22696. Also see the report from the Landrat des Kreises Eiderstedt to the Regierungspräsident in Schleswig concerning "Unruhen in der Landwirtschaft" Tönning, 21.12.1928, LAS 309/22696.

130 *Der Schleswiger* No. 300, 21.12.1928, LAS 309/22569.

131 Report from the Landjägereiabteilung Tönning Kreis Eiderstedt to the Landrat in Tönning, Tönning, 21.12.1928, LAS 309/22696.

132 See report titled "Pressebericht. Landvolkbewegung des Kreises Rendsburg" Rendsburg, 22.12.1928, in both LAS 301/4695 and 309/22569.

This must have had a discouraging effect on anyone thinking of bidding at future compulsory auctions. Bidders did not just undermine the attempt to sabotage auctions; they were a manifestation of disunity. From the beginning, the farmers' movement which had developed into the Landvolkbewegung was a reaction to the divisions the farmers felt even within their own occupation. Their attempt to identify the bidder at the Garding auction underlines their persisting hunger for solidarity.

The auction's outcome was a qualified victory for the farmers. The calf was returned to its owner against a security but was to be sold at a later date.[133] Afterwards, the farmers moved to an inn where they held a meeting, again denouncing the Weimar system and the destruction of their capital (*Substanzvernichtung*). Further, a committee of three members was formed as a "crisis assistance" to serve the interests of the "suffering rural people".[134] There can be little doubt that the committee's innocuous-sounding mission was nothing but a euphemism for organising resistance to confiscations, and it is most likely that such a committee was precisely what Hamkens meant when he urged farmers to contact their "trusted men".[135]

According to one police report, Hamkens' presence alone was probably sufficient to deter bids at the auction.[136] He was clearly an esteemed figure and, indeed, as the Landvolk's principal agitator, a driving force for the entire movement. He may well have had a hand in the Beidenfleth action and his role at Garding was certainly central. However, even he cannot be held accountable for the actions that took place at confiscations and compulsory auctions in other parts of the province over the next few days. Though there are clear signs that networks among the farmers, including the Defence Association, played a role, in essence these appear to have been local initiatives.[137]

At an auction of a pig in Süderlügum near the Danish border on December 20, at the confiscation of three pigs in Oster-Ohrstedt on December 21 and at the auction of a calf in Wittbek on December 22, farmers came together to obstruct proceedings.[138] In each case they employed intimidation. In Süderlügum some 25 peo-

133 Report from the Landrat des Kreises Eiderstedt to the Regierungspräsident in Schleswig concerning "Unruhen in der Landwirtschaft" Tönning, 21.12.1928, LAS 309/22696.

134 See the notice in the *Eiderstedter Nachrichten* No. 8190, 22.12.1928, p. 6 and the report in the *Rendsburger Tageblatt* No. 310, 21.12.1928, pp. 11-12.

135 Oberstaatsanwalt to the Generalstaatsanwalt in Kiel, Flensburg, 27.2.1929, GStA PK I. HA Rep. 84a No. 55363. See also p. 103 above.

136 Report from the Landjägereiabteilung Tönning Kreis Eiderstedt to the Landrat in Tönning, Tönning, 21.12.1928, LAS 309/22696.

137 Report from the Landjägereiabteilung Kreis Husum to the Landrat in Husum, Husum, 23.12.1928, LAS 309/22696.

138 For the auction of Heinsen's pig in Süderlügum, see the report from the Kriminal- und Grenzkommissariat, Niebüll, 20.12.1928, sent by the Landrat in Niebüll to the Regierungspräsident in Schleswig, 22.12.1928, LAS 309/22696. For the bailiff's attempt to confiscate Claus Carstensen's pigs in Oster-Ohrstedt, see the statements made by Vollziehungsbeamter

ple, including three members of the Defence Association who brought along their shotguns, terrified the auctioneer. After reading out the conditions, he barely paused to announce the auction closed as no bids had been made. In Oster-Ohrstedt, 12 neighbours blocked the bailiff's exit from a pigpen. The following day at an inn in Wittbek, when the same bailiff attempted to auction a calf, 60 to 70 assembled farmers threatened him and potential bidders with violence.[139] Furthermore, they swore at the innkeeper, warned that it would be made public that he had aided the finance office and threatened to avoid his inn and stables in future.[140] A subsequent gathering at the finance office in Husum also achieved the desired result. Declaring that their taxes had paid for the "lovely" new building, the farmers refused to depart until the director agreed to release the calf.[141]

The actions in Süderlügum and Oster-Ohrstedt were not unqualified successes. In Süderlügum the community chairman paid the outstanding taxes from his own pocket in order to avoid further trouble and in Oster-Ohrstedt the bailiff returned and confiscated the pigs six weeks later.[142] The Wittbek incident, however, demonstrated the power of intimidation – particularly the threat of boycott and ostracism – on bidders and those branded as collaborating with the authorities. There the innkeeper immediately had the calf removed from the stable and refused to comply with the bailiff's further entreaties.[143] It was not the last time such tactics proved so potent.

These actions showed just how difficult it was for the state to uphold its authority in the countryside when the rural people acted cohesively and adequate police protection was lacking. Only the confiscation in Oster-Ohrstedt resulted in legal action. The owner of the pigs and five of his accomplices were charged with intimidating a public servant (*Beamtennötigung*). Yet even in this case, many of the participants escaped charges. Unable to identify all of the farmers himself, the bail-

---

Degen titled "Finanzamt" Husum, 22.12.1928 and by Claus Carstensen, Oster-Ohrstedt, 22.12.1928, both in LAS 309/22696; the indictment of the Staatsanwalt, Flensburg, 19.2.1929 and issue of the verdict of the Schöffengericht in Husum, 14.3.1929, pp. 1-10, both in GStA PK, I. HA Rep. 84a, No. 55363; *Husumer Nachrichten* No. 63, 15.3.1929, p. 2; *Rendsburger Tageblatt* No. 63, 15.3.1929, p. 2. For the bailiff's attempt to auction the calf in Wittbek, see statement made by Vollziehungsbeamter Degen, Husum, 22.12.1928 and the report by Polizeihauptwachtmeister Tiemann, Husum, 22.12.1928, appended to the report sent from the Landrat in Husum to the Regierungspräsident in Schleswig, Husum, 22.12.1928, LAS 301/4695; *Der Schleswiger* No. 312, 24.12.1928, LAS 309/22569 and the *Husumer Nachrichten* No. 310, 22.12.1928, p. 2.

139 Report by Polizeihauptwachtmeister Tiemann, Husum, 22.12.1928, LAS 301/4695; statement by Vollziehungsbeamter Degen, Husum, 22.12.1928, LAS 301/4695.

140 Report by Polizeihauptwachtmeister Tiemann, Husum, 22.12.1928, LAS 301/4695.

141 *Husumer Nachrichten* No. 301, 22.12.1928, p. 2.

142 Report from the Kriminal- und Grenzkommissariat, Niebüll, 20.12.1928, sent by the Landrat in Niebüll to the Regierungspräsident in Schleswig, 22.12.1928, LAS 309/22696; *Rendsburger Tageblatt* No. 39, 15.2.1929, p. 4.

143 Statement by Vollziehungsbeamter Degen, Husum, 22.12.1928, LAS 301/4695.

iff called on the community chairman to assist him. However, he was reluctant to perform this task, ostensibly because he sympathised with his neighbours. Those who were identified feigned ignorance of the identity of their assistants. Even the local police officers themselves perhaps – or so their failure to ask appropriate follow-up questions would suggest – did not vigorously pursue the investigation, probably because they wished to avoid trouble with the farmers.[144] The six identified faced the *Schöffengericht*, a court presided over by one professional and two lay judges, in Husum in March 1929. Though the farmers were not prepared to acknowledge it, the verdict was mild, fines of RM 100 and RM 75, in place of detention because the court recognised the strain of the agricultural crisis and that none of the accused had prior convictions.[145] The public prosecutor appealed, demanding higher penalties, but the verdict was upheld on account of the convicted's substantial debts.[146]

By no means did all tax-strikes end with confiscations and compulsory auctions. Passive resistance of various kinds was widespread. At a meeting in South Tondern, Max Petersen told his audience that continual submission of applications for tax postponement would eventually persuade the authorities that agriculture was exhausted.[147] Many citizens in the village of Elsdorf near Rendsburg put a similar plan into practice. They had deferred their payments for almost a year and, encouraged by Markus Wulf (Eisendorf), still did not pay, nor reply to the finance office, even after they received warnings.[148] They refused to pay so long as they were sure it was not at their cost, and it was only after the finance office proceeded to confiscate from the majority of the village's taxpayers that they finally began to respond.

Community representatives from Elsdorf met with the county administrator and the sale of the confiscated items was delayed so that taxpayers had further opportunity to submit a detailed application for assessment.[149] Many did so. Evidently,

144 Statement by Vollziehungsbeamter Degen titled "Finanzamt" Husum, 22.12.1928, LAS 309/22696. For the police's failure to question how Carstensen (Oster-Ohrstedt) could claim that he did not know most of the men who had assisted him but, at the same time, declare that they were all married, see the statement made by Claus Carstensen, Oster-Ohrstedt, 22.12.1928 and the report from the Landjägereiabteilung Kreis Husum to the Landrat in Husum, Husum, 23.12.1928, both in LAS 309/22696.

145 Issue of the verdict of the Schöffengericht in Husum, 14.3.1929, GStA PK I. HA Rep. 84a No. 55363; *Der Schleswiger* No. 64, 16.3.1929 and *Westküste* No. 9, 16.3.1929, both in LAS 309/22921; *Husumer Nachrichten* No. 63, 15.3.1929, p. 2; *Rendsburger Tageblatt* No. 63, 15.3.1929, p. 2.

146 Issue of the verdict of the Großer Strafkammer des Landgerichts in Flensburg, 11.6.1929, GStA PK I. HA Rep. 84a No. 55363.

147 Report from the Landeskriminalpolizeistelle Flensburg to the Regierungspräsident in Schleswig, Flensburg, 27.2.1929, LAS 309/22921.

148 *Rendsburger Tageblatt* No. 314, 28.12.1928, p. 4.

149 Report from the Landrat to the Regierungspräsident in Schleswig concerning "Steuerpfändungen in Elsdorf" Rendsburg, 26.1.1929, LAS 309/22696. NB the report did not note the

the farmers were willing to push the state's tolerance as far as they dared. These were not farmers who simply could not pay their taxes. Moreover, that they ignored the finance office's warnings, responding only after the confiscations, reveals that they cared little for the laws, but only feared the repercussions of breaking them. The state's exertions, therefore, were not a sign of its strength, but rather demonstrated that the farmers bore it little respect.

Wulf led the meeting of the "Crisis fellowship of farmers" (*Notgemeinschaft der Landwirte*) in Rendsburg on December 22. If the purpose of this meeting had not been proclaimed as a protest against the trade agreement to be signed with Poland, it would have been difficult to distinguish it from that of any other Landvolk meeting. One of 18 speakers, Wulf complained that agriculture received inadequate protection, that Germany's trade policy had ruined the farmers, and likened the upcoming trade agreement to Damocles' sword hanging over the farmers' heads.[150] But like the Landvolk's meetings in Itzehoe, Garding and elsewhere, slogans of hostility toward investment capital, Jews, the government and the Weimar Republic were common in many of the speeches.[151] With legal proceedings against the Lunden and Beidenfleth demonstrators pending, another common sentiment was that the farmers had been wronged. In this respect, it is important to understand that the Landvolk deemed the state's conduct not only as unfair or heavy-handed, but as completely iniquitous. Convinced of agriculture's central importance for the entire nation, they believed they were entitled to assistance. They thought the state was indebted to them, rather than the other way round, and they were not willing to accept their obligation to pay outstanding taxes, and were outraged that they should face the courts as a consequence. Such was their anger, that a number of speakers declared it was the state and authorities, not the Landvolk, that were criminal. These farmers did not regard their actions as mischievous, nor did they appear to entertain misgivings. They felt perfectly justified in their protests. As Wulf told the audience,

> "They curse honest farmers, portraying them as agitators and whiners, threatening them with the public prosecutor and prison, [farmers] who have proven that they were willing to give everything; four years on the battlefield, passing their lives in the trenches and

---

number of confiscations but only that they were undertaken against "the majority" of Elsdorf's tax-payers.

150 Report titled "Pressebericht. Landvolkbewegung des Kreises Rendsburg" Rendsburg, 22.12.1928, both in LAS 301/4695 as well as LAS 309/22569.

151 See the meeting report from the Polizeiverwaltung Rendsburg to the Oberpräsident in Kiel titled "Versammlungsbericht" Rendsburg, 22.12.1928 and the report titled "Pressebericht. Landvolkbewegung des Kreises Rendsburg" Rendsburg, 22.12.1928, both in LAS 301/4695 as well as LAS 309/22569.

then ten years of hard work to rebuild the economy. Now, when they should in fact be certain of the fatherland's gratitude, they are called "criminals".[152]

Weschke declared that it was the state that belonged in the dock for allowing the Beidenfleth incident, and Heinrich Hansen asked, "Where are the successes of January 28? Who is to blame? .... Berlin must come here. .... The criminals are still steering the ship. .... Incompetent men are in charge. That is why we are being ruined. Farmers help yourselves. German farmers awake!"[153]

Some speakers, such as Pohlmann (Kaltenkirchen), simply called for solidarity and enjoined their colleagues not to take out further loans in order to pay their taxes, but most expressed more extreme views.[154] Frequently, these were anti-Semitic. In fact, two speakers, Guth (Österfeld) and Peters (Tetenbüll), were later tried, although acquitted, for the remarks they made at the meeting.[155] Guth described Weimar as a "Jews' Republic" (*Judenrepublik*), claiming that Jews were using their influence in German politics to enslave the German people, while Peters went further. He called for Germans to band together against Jewish rule, closing his speech, crying "Death and decay to the Jews and the Jews' comrades."[156] Another speaker, Hamkens' brother-in-law, Hönck (Harblek) demanded better press coverage for the farmers' interests, claiming that most newspapers were under Jewish control because they were entirely dependent on the advertisements of large department stores.[157] The *Heider Anzeiger's* reaction illustrated just how serious a slur this was deemed, for it closed its comprehensive report on the meeting stress-

152 "Man beschimpft ehrliche Bauern, stellt sie hin als Hetzer und Schreier, droht ihnen mit dem Staatsanwalt und Gefängnis, die bewiesen haben, daß sie alles zu geben bereit waren, vier Jahre im Felde ihr Leben in die Schanze geschlagen haben, die in 10jähriger schwerer Arbeit die Wirtschaft wieder aufgebaut haben. Nun, wo ihnen der Dank des Vaterlandes gewiß sein sollte, heißt man sie "Verbrecher". Meeting report from the Polizeiverwaltung Rendsburg to the Oberpräsident in Kiel titled "Versammlungsbericht" Rendsburg, 22.12.1928, in both LAS 301/4695 and 309/22569.

153 Report titled "Pressebericht. Landvolkbewegung des Kreises Rendsburg" Rendsburg, 22.12.1928, in both LAS 301/4695 and 309/22569.

154 Report titled "Pressebericht. Landvolkbewegung des Kreises Rendsburg" Rendsburg, 22.12.1928, in both LAS 301/4695 and 309/22569.

155 In September 1929, Guth and Peters were tried and acquitted at the Schöffengericht in Kiel, the former for breaching the "Law for the Protection of the Republic" (*Republikschutzgesetz*), the latter for "inciting class hatred". See the transcript of the proceedings titled "Landvolk-Prozess und Landvolk-Kundgebung am 30.9.1929", in both LAS 301/4696 and 309/22784. Guth, who had joined the NSDAP in 1928, went on to become a party functionary in North Dithmarschen. See S. Lehmann, *Kreisleiter der NSDAP in Schleswig-Holstein: Lebensläufe und Herrschaftspraxis einer regionalen Machtelite* (Bielefeld, 2007), p. 94.

156 Report titled "Versammlungsbericht" Rendsburg, 22.12.1928 from the Polizeiverwaltung Rendsburg to the Oberpräsident in Kiel, in both LAS 301/4695 and 309/22569.

157 Report titled "Pressebericht. Landvolkbewegung des Kreises Rendsburg" Rendsburg, 22.12.1928, LAS 301/4695 and 309/22569.

ing that Schleswig-Holstein's middle-class press (*bürgerliche Presse*) was utterly independent of "Jewish capital and Jewish influences".[158]

At the meeting in Rendsburg, a police report noted that all the speeches, but especially the more radical ones, received strong and universal applause from the large audience.[159] Accounts in the *Heider Anzeiger* and *Nordischer Kurier* support this.[160] Founded on the observations of their agent, two provincial government officials, however, were convinced that the mood of the meeting was not as intense as the press reported.

> "... again and again it was always the same few individuals who made loud exclamations, especially of an anti-Semitic nature, during the speeches, while most of the audience responded passively and only occasionally joined in the applause. Characteristic of the meeting is that there was a considerable consumption of grog, coffee and so on. At this meeting, as in the past, it is apparent that the movement is being driven more by specific individuals and organisations like the Landvolk League and the Young Northern Lands Defence than it is propelling itself."[161]

Since this report was an interpretation based on the observations of someone who was present only for part of the meeting it is less convincing where it conflicts with the police report and the newspapers' accounts. To complicate matters further, a stenographer at the meeting, Dr. Gosch, stated that the meeting had been quite agitated "at times". Gosch later testified in Guth and Peters' trial about the accuracy of his manuscript for it was upon his transcript that the prosecution was based. Incidentally, Dr. Gosch's testimony is of interest not least because it tells us about the reliability of such transcripts. Gosch admitted that he might have omitted individual words. He had had difficulty understanding "...because the excitement in the hall was very great at times. Calls were made and applause was given. In the transcript there are catchwords, not word for word [sic]. It is possible that I made one sentence out of different sentences. I did however use the words that the speakers used wherever possible."[162]

---

158 *Heider Anzeiger* No. 302, 24.12.1928, p. 2.

159 Report from the Polizeiverwaltung Rendsburg to the Oberpräsident in Kiel titled "Versammlungsbericht" Rendsburg, 22.12.1928, in both LAS 301/4695 and 309/22569.

160 *Heider Anzeiger* No. 302, 24.12.1928, pp. 1-2; *Nordischer Kurier* No. 303, 25.12.1928, p. 3.

161 "... es waren immer wieder die gleichen vereinzelten Persönlichkeiten, die laute Zwischenrufe, besonders solche antisemitischen Charakters machten, während sich der größere Teil der Versammlung durchaus passiv verhielt und nur gelegentlich an den Beifallsbezeugungen teilnahm: kennzeichnend für die Versammlung ist, daß während der Reden ein erheblicher Verzehr an Grog, Kaffee und dergl. stattfand. Auch bei dieser Versammlung trat hervor, daß die Bewegung mehr von einzelnen Persönlichkeiten und Organisationen, wie Landvolkbund und Jungnordmarkwehr, vorwärts getrieben wird, als daß die sich selbst trägt." Report from the Regierungspräsident to the Oberpräsident in Kiel, Schleswig, 9.1.1929, in both LAS 301/4695 and 309/22569.

162 "... im allgemeinen konnte ich nur schlecht verstehen, da die Erregung im Saale zeitweilig sehr groß war, es wurden Zwischenrufe gemacht und wurde Beifall kundgegeben. Im Ste-

These various reports are not as discordant as they might appear. Indeed, a cautious synthesis provides a credible assessment of the audience's response. We need only disregard the government officials' mistaken impression that most of the audience only occasionally joined in the applause. To do so is permissible, firstly because the other evidence amasses against it, and secondly because Gosch's statement that the excitement in the hall was very great "at times" explains how the limited attendance of the government officials' agent may have produced such an inconsistency. It is unnecessary, therefore, to resort to speculation over differences in perception.

The considerable consumption of grog and coffee at the Rendsburg meeting lent the occasion the air of a social function. It would appear that the farmers attended the meeting in Rendsburg not only to hear the speeches but also to have a pleasant time. Yet grog and coffee did not make the speeches less radical. All received strong and universal applause, especially the more radical ones. The audience was highly aroused at times and a small number of individuals also made anti-Semitic cries. Difficult though it is to measure satisfactorily the level and extent of support for these views, it is clear that the audience's response was one of approval.

The sudden emergence and swift expansion of the Landvolkbewegung at the end of 1928 were extraordinary. The authorities had received little warning of the coming radicalisation, but by the end of the year, the Landvolk had held meetings and actions along the west coast and beyond: in the counties of North Dithmarschen, Eiderstedt, Steinburg, South Tondern, Husum and Rendsburg. An abrupt shift from the rallies of January 28, 1928, the new movement abandoned hope in government assistance and instead of cooperation with the authorities took a course of defiance and confrontation. Furthermore, whereas the rallies of January 28, 1928 were an attempt to unify the agricultural associations, the Landvolkbewegung aimed to unite the farmers independently.

The level of discontent among the farmers was very high. What is striking is the ease with which the leaders mobilised substantial crowds to actions and meetings, some many hundred strong, and that these were characterised by a spirit of collaboration and solidarity. Strong approval of the radical speeches at Landvolk meetings also indicates a readiness to embrace extremism. Furthermore, the fact that the actions that followed those in Beidenfleth, Lunden and Garding were conducted without the leaders' involvement points to considerable and widespread enthusiasm for the Landvolkbewegung in the countryside.

Numerous factors made the Landvolkbewegung's development and radicalisation possible. First, it should be recalled that disappointed hopes for government

---

nogramm sind Stichworte, nicht wörtlich. Es kann sein, daß ich verschiedene Sätze zu einem Satz zusammengestellt habe. Ich habe aber möglichst die Worte gebraucht, die auch die Redner gebraucht haben." See the transcript of court proceedings titled "Landvolk-Prozess und Landvolk-Kundgebung am 30.9.1929", in both LAS 301/4696 and 309/22784.

assistance fostered resentments of the authorities and Weimar system, while the established agricultural associations' failure to unite and lobby successfully encouraged the province's farmers to seek alternative means of association and allowed new leaders to emerge. Ultimately, however, it was the state's resumption of tax collection in the face of the farmers' expectation of continued leniency, and, specifically, the particular individual local authorities' forceful pursuit of payment which served as the catalyst for conflict. Beidenfleth and Lunden were the turning points which showed farmers the possibilities of collaboration and served as the stimulus for further actions.

The farmers' obstructionist actions, intended to protect their capital, gained greater political significance because the Landvolk viewed those held responsible as servants of the Weimar system and interpreted their local skirmishes as part of a broader struggle against the system itself. In this respect, the farmers' meetings were of particular importance. Far from simply expressing discontent over the crisis, the Landvolk's agitation increasingly centred on fierce criticisms of the authorities and anti-system and anti-Semitic sentiment.

A foundation for the Landvolk's anti-system and anti-Semitic agitation had long been present in rural Schleswig-Holstein in the ideology of political parties, such as the DNVP, and in the veteran and nationalist organisations as well as the Rural League. While reports of anti-Semitic comments and cries are a flimsy gauge of the extent of approval for such sentiments, since it is impossible to establish just who and how many people made them, it is clear that farmer audiences entered an anti-Semitic climate freely, listened to anti-Semitic rhetoric welcomed by the cries of their neighbours, and then showed approval with their applause.

It was, moreover, from the veteran, nationalist and agricultural organisations that many of the Landvolkbewegung's leaders came. Beyond familial and neighbourly ties, the Stahlhelmwestküste, Defence Association and Rural League were essential in establishing the networks upon which the movement formed. Most of the movement's leaders were farmers, but the Defence Association also brought in other seasoned right-wing activists and created connections to further organisations which would later be of significance. The Defence Association melted into the movement. However, it should be noted that the Landvolkbewegung received little active assistance from other groups and certainly not from the NSDAP. After modest beginnings, the party was emerging from its local strongholds in Dithmarschen and had just held its successful rally in Heide when the Landvolkbewegung upstaged it with its actions. For the NSDAP, the Landvolkbewegung was a competitor but one on which it hoped to capitalise as was seen in the party's opportunistic response to Beidenfleth.

From the outset the Landvolkbewegung was propelled by a small number of energetic leaders. Whether they described them as farmers or "right-wing radicals", reports from the police, county administrators and the Regierungspräsident all agreed on this point.[163] These leaders were, moreover, reported to be unified and determined.[164] Indeed, their collaboration was all the more necessary in the absence of an organisational structure. The movement required their impetus and agitation. It cannot be denied, however, that they found fertile soil in Schleswig-Holstein. From the publication of Hamkens and Heim's first calls to resist the payment of taxes, the movement had developed within just two months.

While individual leaders held the Landvolkbewegung together, community solidarity formed the basis of its actions. Growing in their conviction that they had to collaborate and help themselves since the state had failed to do so, the farmers turned to the west coast tradition of independence and defiance. The farmers' consciousness of their history was expressed most clearly in the cry, "Rather dead than a slave". Yet community collaboration also had a more recent precedent in the post-war militias. Cooperation in the face of distress fostered fellowship but there are also indications that the Landvolkbewegung held additional appeal because it presented a social outlet; farmers enjoyed the meetings and also found participating in actions entertaining. The counterpart to these unifying factors was the use of intimidation and ostracism, for they were applied not only against the local authorities acting as bailiffs but also against the few fellow farmers who dared to bid at auctions or thought to oppose the movement.

The authorities' warnings and their threats of legal action incensed the farmers, exacerbated discontent in the countryside and – contrary to their intentions – contributed to the emergence of the Landvolkbewegung. Perhaps greater police protection could have prevented some of the Landvolk's actions. However, the key failure of the Regierungspräsident in particular, but also of some county administrators and community chairmen, was their inability to persuade farmers that the state was showing sufficient understanding of the agricultural crisis. This helped to intensify the anti-Republican character of the movement. It did so not simply because it bolstered the idea that Weimar was failing to give agriculture the priority it warranted. More importantly, it provided fuel for the Landvolk leaders' accusations that the

163 See the report from the Landrat to the Regierungspräsident in Schleswig concerning "Steuern der Landwirtschaft" Heide, 16.11.1928 and the report from the Landrat des Kreises Eiderstedt to the Regierungspräsident in Schleswig concerning "Unruhen in der Landwirtschaft" Tönning, 16.12.1928, both in LAS 309/22569; report from the Landjägereiabteilung Tönning Kreis Eiderstedt to the Landrat in Tönning, Tönning, 21.12.1928, LAS 309/22696; report from the Regierungspräsident to the Preuß. Minister des Innern in Berlin concerning "Unruhe in der Landwirtschaft" Schleswig, 9.1.1929, in both LAS 301/4695 and 309/22569; report from the Regierungspräsident to the Oberpräsident in Kiel, Schleswig, 9.1.1929, in both LAS 301/4695 and 309/22569; report from the Landeskriminalpolizeiamt Berlin, Abteilung IA, Heide, 21.3.1929, LAS 309/22668. Bodo Uhse, editor of the NSDAP's provincial newspaper, also recalled later that it was "a few men" who "drove" the movement. Uhse, *Söldner und Soldat*, p. 172.

164 Report from the Regierungspräsident to the Preuß. Minister des Innern in Berlin concerning "Sprengmittelanschläge in Schleswig-Holstein" Schleswig, 15.12.1928, LAS 301/4704.

authorities were, in fact, not acting in the interests of farmers at all, but carrying out confiscations on behalf of a foreign system.

Chapter Four

The Landvolkbewegung consolidates

After the dramatic obstructive actions at confiscations and compulsory auctions in the winter of 1928/1929, local and provincial authorities predicted that the Landvolkbewegung would soon abate. They supposed the farmers would become busy with the demands of spring work and anticipated that legal action would have a sobering effect.[1] The following months were to show just how misplaced their hopes were. Indeed, the beginning of the new year witnessed a consolidation of the movement. Its actions and meetings continued and spread to other areas of the province, and its conflict with the authorities also entered the courts and was taken up in the movement's own newspaper. Not only did the authorities' pay the Landvolk close attention, so too did other organisations, in particular the agricultural political associations and the NSDAP, which sought to profit from the movement.

On the morning of January 3, 1929, Hamkens and the nine farmers accused of thwarting the confiscation of cattle in Lunden were tried at the Schöffengericht in Husum. Far from contesting the charges, the farmers explained that they had wished to assist their colleague, Haalck, and declared that they would view their punishments as an honour.[2] These punishments were hardly sufficient to make martyrs of them but to the satisfaction of the provincial government, the public prosecutor appealed against the leniency of their fines and the subsequent trial in March in distant Flensburg – where the judges were subject to less psychological and local social pressure than in Husum – resulted in harsher punishments. The ringleader, Mahmens, was fined RM 500, another farmer RM 400 and the remaining six defendants RM 300 each.[3] Charges against Hamkens for affronting Frey were over-

---

1 Report from the Kriminal- und Grenzkommissariat, Niebüll, 20.12.1928, LAS 309/22696; report from the Regierungspräsident to the Oberpräsident in Kiel, Schleswig, 9.1.1929, LAS 301/4695; untitled report, Itzehoe, 28.3.1929, sent from the Landeskriminalpolizeistelle to the Regierungspräsident in Schleswig, Flensburg, 2.4.1929, LAS 309/22921.

2 *Husumer Nachrichten* No. 3, 4.1.1929, pp. 1-2.

3 Issues of the verdicts of the Schöffengericht Husum, 3.1.1929, pp. 1-8 and of the Landgericht Flensburg, 19.3.1929, pp. 1-8, both in GStA PK I. HA Rep. 84a No. 55359; note in the Regierungspräsident's files titled "Vermerk über die Gerichtsverhandlung gegen Mahmens und Gen. und Wilhelm Hamkens am 3.d.Mts in Husum" Schleswig, 5.1.1929, LAS 309/22569. For the trial in Flensburg, see the report from the Regierungspräsident to the Preuß. Minister des Innern in Berlin concerning "Gerichtsverhandlung gegen Hamkens - Tetenbüll wegen Aufforderung zum Steuerstreik und Beamtenbeleidigung und gegen Mahmens und Genossen wegen Pfandbruchs" Schleswig, 20.3.1929, LAS 301/4695. Also see the *Rendsburger Tageblatt* No. 67, 20.3.1929, p. 9. For Claus Wallichs's brief imprisonment on failing to pay the fine, see the Oberstaatsanwalt to the Regierungspräsident in Schleswig, Flensburg, 29.12.1931, LAS 309/22924.

turned, but whereas he had only received a fine in Husum, he received a one-month prison term and a fine of RM 50 for encouraging tax-strikes.[4]

As many as 300 farmers rallied in solidarity in Husum on January 3. Most had gathered at the livestock market where a farmer had invited them to go to greet their "brothers sitting in the docks".[5] In front of the courthouse, Heinrich Hansen addressed the crowd, complaining that the rural people could no longer tolerate such treatment and subjugation. He shouted, "Rather dead than a slave" and called for three cheers for the accused.

The noise interrupted proceedings in the court.[6] Four policemen soon arrived and instructed the farmers to disperse but the farmers answered this and each subsequent warning with derisive laughter. Realising that they were too few to remove the crowd by force, the officers began to identify individuals. When one farmer refused to name himself, they set about escorting him to the courthouse prison to establish his identity there but the crowd grew unruly and a rock was thrown, striking one of the officers.[7] Not until a contingent of country policemen arrived, were the police finally able to drive the farmers back. However, it was only after a farmer standing at the rear gave a signal that the crowd actually moved on.[8] That they followed this hand-signal and departed "almost in formation" suggests that what on first impression appeared a spontaneous march from the market was, in fact, a premeditated and planned course of action.

Further displays of solidarity made a mockery of the state's attempt to assert its authority. By the time the trial had finished some 500 to 600 farmers had gathered at a hall nearby to welcome the convicted and listen to Hamkens who, unperturbed by his conviction, again brazenly encouraged the farmers to resort to "self-help".[9]

---

4 Issues of the verdict of the Schöffengericht Husum, 3.1.1929, pp. 1-8 and of the verdicts of the Landgericht Flensburg, 19.3.1929, pp. 1-6 and pp. 1-10, all in GStA PK I. HA Rep. 84a No. 55359. Also see the note in the Regierungspräsident's files titled "Vermerk über die Gerichtsverhandlung gegen Mahmens und Gen. und Wilhelm Hamkens am 3.d.Mts in Husum" Schleswig, 5.1.1929, LAS 309/22569.

5 Unless otherwise noted, the narrative here is based on the report titled "Auflauf", Husum, 3.1.1929, in both LAS 301/4695 and 309/22569.

6 Indictment issued by the Oberstaatsanwaltschaft Flensburg, 28.3.1929, the addendum to the indictment, Flensburg, 12.4.1929 and the report from the Landgerichtspräsident and Oberstaatsanwalt in Flensburg to the Oberlandesgerichtspräsidenten in Kiel, Flensburg, 12.1.1929, all in GStA PK I. HA Rep. 84a No. 55397; *Husumer Nachrichten* No. 2, 3.1.1929, p. 3.

7 Only two projectiles, not several, were thrown at the police, and the farmers certainly did not "stone" the officers as Stoltenberg and Farquharson contended. See Stoltenberg, *Politische Strömungen im schleswig-holsteinischen Landvolk* p. 133; Farquharson, *The Plough and the Swastika*, p. 8.

8 Note in the Regierungspräsident's files titled "Vermerk über die Gerichtsverhandlung gegen Mahmens und Gen. und Wilhelm Hamkens am 3.d.Mts in Husum" Schleswig, 5.1.1929, LAS 309/22569.

9 *Husumer Nachrichten* No. 3, 4.1.1929, pp. 2-3.

The harsher penalty he later received in Flensburg roused anger in farming circles. The *Rendsburger Tageblatt* commented that he had acted not for his own benefit but for his colleagues and the preservation of agriculture, and that many farmers would therefore feel that his punishment was directed against themselves.[10]

An additional exhibition of solidarity followed with the prosecution of those who had demonstrated outside the Husum courthouse on January 3.[11] Farmers compiled a list, reportedly containing 420 names, and sent it to the county court, apparently to save it the trouble of determining who was present.[12] Most of the 57 who were charged with disobeying police instructions and who faced proceedings in Husum at the end of May showed no remorse, but rather displayed their antagonistic intent. They explained they had been on the verge of leaving but had remained simply because the police had ordered them to disperse.[13] Not surprisingly, just 11 of the accused were acquitted. The remainder received fines of RM 100, later reduced to RM 50 on appeal on the grounds that the demonstrators had not intended to influence the course of justice.[14] The public prosecutor considered the initial penalty lenient, while the district administrator deemed it severe given the economic circumstances.[15] There was no significant disruption of the proceedings on this occasion, but again farmers supported their colleagues. Paul Adam Ross had to be restrained from calling three cheers for the rural people and once more a meeting was held during the trials.[16]

Indicative of the interest the Landvolkbewegung had begun to generate beyond the borders of Schleswig-Holstein, the *Frankfurter Zeitung* reported on the trial. It judged the state's authority so badly shaken that many years' work would be required to restore it. This was a belated appraisal four months after Beidenfleth, but

---

10 *Rendsburger Tageblatt* No. 67, 20.3.1929, p. 9.

11 Regierungspräsident to the Preuß. Minister des Innern in Berlin, Schleswig, 12.1.1929, in both LAS 301/4695 and LAS 309/22569.

12 Landrat to the Regierungspräsident in Schleswig concerning "Unruhe in der Landwirtschaft" Husum, 9.1.1929, LAS 309/22569.

13 *Husumer Nachrichten* No. 118, 23.5.1929, pp. 2 and 7, No. 119, 24.4.1929, No. 120, 25.5.1929, p. 2 and No. 121, 27.5.1929, p. 8.

14 Issues of the verdicts of the Amtsgericht Husum, 22.5.1929, 23.5.1929, 24.5.1929 and 5.5.1929 and issue of the verdict of the Landgericht Flensburg, 3.10.1929, all in GStA PK I. HA Rep. 84a No. 55397; report from the Landrat to the Regierungspräsident in Schleswig concerning "Verhandlung gegen Landleute vor dem Husumer Gericht" Husum, 30.5.1929, in both LAS 301/4695 and 309/22569; *Husumer Nachrichten* No. 124, 30.5.1929, p. 2; *Rendsburger Tageblatt*, No. 124, 30.5.1929, p. 4 and No. 233, 4.10.1929, p. 4.

15 Oberstaatsanwalt to the Preuß. Justizminister in Berlin concerning "Ausschreitungen von Landleuten in Schleswig-Holstein" Flensburg, 13.7.1929, GStA PK I. HA Rep. 84a No. 55397.

16 Oberstaatsanwalt to the Regierungspräsident in Schleswig, Flensburg, 21.6.1929, LAS 309/22668. For the meeting held during the trial, see *Das Landvolk* No. 66, 26.5.1929, p. 2.

it was correct nonetheless. Not so accurate was its condemnation of the Rural League for the radicalisation in the countryside.[17]

The Rural League was one of many forces contributing to the radicalisation in Schleswig-Holstein's countryside. However, it was not the driving force. Like the NSDAP and Ludendorff's *Tannenbergbund*, a völkisch organisation of military and veteran associations formed in 1925, it was trying to use the Landvolkbewegung for its own ends. Various Tannenbergbund speakers had soon followed Ludendorff who visited the province in autumn 1928, and NSDAP agitation rapidly intensified in January and February 1929.[18] But as the results of a Berlin police investigation noted, "Only after the right-wing political parties and organisations had recognised that the dissatisfied and aroused peasantry offered them an extraordinarily propitious area for their propaganda did they send speakers to all of the farmers' meetings to promote their interests."[19]

In the winter of 1928/1929 the Landvolkbewegung was the dynamic force in Schleswig-Holstein's countryside. It did not aspire to become a professional agricultural organisation, but developed as an autonomous political alternative to them.[20] The movement offered a rallying point for all right-wing rural discontents, yet remained essentially independent. The school teacher, Rural League functionary and passionate Landvolk leader, Peter Petersen (Ahrenholz) boasted that the Landvolkbewegung was formed to propel all national associations and organisations, and that it welcomed their cooperation without ever binding itself to them.[21] The movement was never a by-organisation of the Rural League, even if it attracted various Rural League leaders to its ranks such as Guido Weschke, its secretary in Steinburg County, Ernst Bossen and Max Petersen in South Tondern, and later Jo-

17 *Frankfurter Zeitung* No. 220, 23.3.1929, LAS 309/22921.

18 Report from the Landeskriminalpolizeistelle Flensburg concerning "Bauernbewegung in der Provinz Schleswig-Holstein" Flensburg, 24.1.1929, in both LAS 301/4589 and 309/22696; report from the Regierungspräsident to the Preuß. Minister des Innern concerning "Unruhe in der Landwirtschaft" Schleswig, 18.2.1929, in both LAS 301/4695 and LAS 309/22696. For Wilhelm Koops' similar assessment of the Tannenbergbund's response to the Landvolkbewegung in South Tondern, see Koops, *Südtondern in der Zeit der Weimarer Republik*, pp. 259-261.

19 "Erst nachdem von den politischen Rechtsparteien und Organisationen erkannt worden war, daß sich ihnen in der unzufriedenen und erregten Bauernschaft ein außerordentlich günstiges Agitationsfeld böte, wurden von ihnen in alle Bauernversammlungen Redner entsandt .... um für ihre Belange zu werben." Report from the Abteilung IA, Außendienst II. G. St., Berlin, 27.12.1928, appended to the report from Regierungspräsident to the Oberpräsident in Kiel concerning "Ergebnis der Ermittelungen der Beamten des Polizeipräsidiums Berlin" Schleswig, 9.1.1929, LAS 301/4695.

20 Report from the Regierungspräsident to the Oberpräsident in Kiel concerning "Entwicklung des landwirtschaftlichen Berufsorganisationswesens in Schleswig-Holstein" Schleswig, 22.1.1930, LAS 309/22765.

21 Schimmelreiter, *Unter der schwarzen Bauernfahne*, pp. 23-24.

hannes Kühl, who wrote many of the leading articles in the Rural League's *Landbote*.[22]

Concerned that its members were deserting it for the Farmers' League, the Rural League sought to retain its membership by radicalising, and appointed Peter Petersen as the business administrator of its county group in Plön. Petersen used the position to foster the Landvolkbewegung in the county. Although the Landvolkbewegung did not gain much support from estate owners, and consequently never grew to be as popular in Schleswig-Holstein's eastern counties as it was in the west, Plön's administrator regarded Petersen as a danger. He predicted that Petersen would radicalise the farmers, who would in time desert the Rural League and join the National Socialist camp. Consequently, he pushed to have Petersen removed from his teaching post.[23]

In the Landvolkbewegung's shadow, the Rural League sought to capitalise on, and thereby contributed to, ill feeling toward the authorities and the atmosphere of crisis in the province. The *Landbote*, for instance, joined the Landvolkbewegung in its criticisms, helping to spread the opinion that taxes were being collected mercilessly.[24] Moreover, its provincial chairman, Count Kielmansegg, more or less sided with the Landvolk when he publicly objected to the auction of the widow's furniture in Garding in December, asking, "Is this the care and understanding that the state promised to give the rural population? On one side the rural population can only feel the state's strangling hand and on the other side they hear promises which stand in crass contradiction to the facts."[25]

The Farmers' League, by contrast, condemned the Landvolkbewegung's illegal activities but used them as leverage in its endeavours to secure relief for agriculture. It stressed that it had nothing to do with what it called the "acts of terror on the west coast", but warned the authorities that the "desperate situation" could only be overcome by tax cuts, not by threats of legal action.[26] Objecting to the looming German-Polish trade agreement, which would add further strain on the livestock sector, the Farmers' League leaders, Tönnsen and Köhler, used the Landvolk's pre-

22 For examples of the *Landbote's* sympathetic accounts of the Landvolkbewegung's meetings and actions, see the *Landbote für Schleswig-Holstein* No. 98, 9.12.1928, p. 1 and No. 6, 17.1.1929, p. 1.

23 Report from the Landrat to the Regierungspräsident in Schleswig, Plön, 12.3.1929, LAS 309/22696; report from the Landrat to the Regierungspräsident in Schleswig concerning "Landvolkbewegung" Plön, 23.7.1929, LAS 309/22668; report from the Regierungspräsident to the Oberpräsident in Kiel concerning "Landvolkbewegung" Schleswig, 5.8.1929, LAS 301/4700; Petersen, *Fliegender Sand*, p. 77.

24 *Landbote für Schleswig-Holstein* No. 3, 13.1.1929, LAS 309/22569; Regierungspräsident to the Oberstaatsanwalt in Flensburg, Schleswig, 23.1.1929, LAS 309/22696.

25 *Landbote für Schleswig-Holstein* No. 104, 30.12.1928, p. 1.

26 Letter from the Schleswig-Holsteinischer Bauernbund e.V. to the Regierungspräsident in Schleswig, Bad Segeberg, 4.12.1928, LAS 309/22921. This letter was also published in the local press, see the *Heider Anzeiger* No. 287, 6.12.1928, p. 1.

vention of compulsory auctions and attacks on those held responsible for the compulsory collection of taxes as their principal illustration of the severity of the agricultural crisis.[27] The Farmers' League in South Dithmarschen went further, demanding the re-establishment of profitability, and claiming that Beidenfleth and Husum indicated what would happen if the farmers' organisations and their leaders continued to return from Berlin "empty handed". The farmers, it warned, could not be held responsible for what might happen if their demands were not met.[28]

While the Farmers' League did not enter into or openly support illegal actions, it largely cooperated with the Landvolkbewegung. Of course, the boundaries between them were at times obscure. Under Hamkens' leadership, the Unity Organisation in Eiderstedt was almost indistinguishable from the Landvolkbewegung. Significantly, the authorities recognised that the bond between the two groups was having a radicalising effect on the Farmers' League.[29]

This development did not please all farmers, particularly those who remained loyal to the old Farmers' Association. In the County of Eckernförde, the Farmers' Association was not prepared to join the Farmers' League until the Farmers' League formulated an intelligible programme.[30] Within the Farmers' League too, the Landvolkbewegung had its critics. At its Rendsburg County annual general meeting, some members disapproved of the Landvolkbewegung for moving into revolutionary waters. Yet here, as on other occasions, Landvolk and Farmers' League leaders pushed for continued cooperation.[31]

Tax-strikes and protests at confiscations were advantageous for farmers, even in cases where they only delayed further attempts to remove their capital. Debtors benefited from the respite the actions effected by using the interval to apply for

---

27 See the letter to the president of the Vereinigung der Deutschen Bauernvereine, Andreas Hermes, who was to lead the German commission to settle the German-Polish trade agreement. *Rendsburger Tageblatt* No. 301, 21.12.1928, p. 1; *Nordischer Kurier* No. 302, 23.12.1928, p. 6.

28 See the resolution from the Kreisbauernbund Süderdithmarschen's general meeting on 8.1.1929 titled "Entschließung" sent to the Regierungspräsident in Schleswig, Meldorf, 11.1.1929, LAS 309/22765.

29 Report from the Regierungspräsident to the Oberpräsident in Kiel concerning "Entwicklung des landwirtschaftlichen Berufsorganisationswesens in Schleswig-Holstein" Schleswig, 22.1.1930, LAS 309/22765; untitled report from the Landeskriminalpolizeiamt, Abteilung IA, Berlin, 2.4.1929, LAS 309/22668.

30 Schleswig-Holsteinischer Bauernverein, Kreisgruppe Eckernförde, to the "Vertrauensmänner", Eckernförde, 23.11.1928, LAS 309/22696.

31 See the untitled report from the Polizeiverwaltung Rendsburg to the Oberpräsident in Kiel, Rendsburg, 14.1.1929, in both LAS 301/4695 and 309/22765. Also see the *Kieler Zeitung* No. 16, 15.1.1929, p. 4 and for a further example, see Hamkens' speech at the annual general meeting of the Eiderstedt Landvolk-Vereinigung: *Flensburger Nachrichten* No. 36, 12.2.1929, p. 5.

postponements and reductions.[32] The code of civil procedure prevented the Regierungspräsident from relocating compulsory auctions en masse but he arranged for greater police protection and he counselled the county administrators to move auctions to larger, more distant locations if they suspected that interference might occur.[33] This became increasingly difficult as the Landvolk threatened to boycott markets which sold confiscated animals. The pressure was effective. After Beidenfleth and the attempt to auction Kock's ox in Hamburg, the Chamber of Agriculture and the Hamburg market declined to sell confiscated animals.[34] Here the Rural League also played a part, publicly demanding that the Chamber of Agriculture refrain from acting as a "lackey for the system" by assisting in confiscations and compulsory auctions again.[35] The pressure also mounted on community chairmen who were compelled to act as bailiffs. Some simply resigned rather than proceed with confiscations.[36] Arguably, the Landvolk's actions were also a success in that the authorities increasingly confiscated furniture instead of livestock, reluctantly recognising that auctioning animals at the markets in Husum or Hamburg was unlikely to succeed.[37] In this way, the farmers at least encouraged the state to refrain from taking their capital.

For all the attention tax obstruction received, its material effect on the state's revenue appears to have been negligible. In January 1929, the provincial gov-

---

32 For the success of this tactic in South Tondern, see the reports from the Kriminal- und Grenzkommissariat, Niebüll, 10.6.1929 and 12.6.1929 sent by the Landeskriminalpolizeistelle Flensburg to the Regierungspräsident in Schleswig, Flensburg, 17.6.1929, LAS 309/22668.

33 Regierungspräsident to the Landräte des Bezirks, Schleswig, 3.1.1929, LAS 309/22696; Regierungspräsident to the Landräte, Magistrate, Bürgermeisterämter des Bezirks, Polizeipräsidenten in Kiel and Altona, Polizeiverwaltung in Flensburg and Neumünster, Katasterämter and Kreiskassen des Bezirks, Schleswig, 28.6.1929, LAS 309/22668; Oberlandesgerichtspräsident to the Regierungspräsident in Schleswig concerning "Ort der Versteigerung" Kiel, 6.3.1930, LAS 309/22590.

34 Note in the Regierungspräsident's files titled "Vermerk: Ferngespräch des Sachbearbeiters mit Landrat Kracht in Heide" Schleswig, 13.1.1929, LAS 309/22696; *Hamburger Echo* No. 359, 28.12.1928, LAS 309/22569; *Das Landvolk* No. 1, 3.1.1929, p. 3.

35 *Der Schleswiger* No. 293, 13.12.1928, LAS 309/22569.

36 *Der Schleswiger* No. 293, 13.12.1928, LAS 309/22569; report from the Landrat to the Regierungspräsident in Schleswig concerning "Kommunalpolitische Verhältnisse im Kreise Segeberg" Bad Segeberg, 23.6.1929, in both LAS 301/4695 and 309/22668.

37 See the report from the Gemeindevorsteher to the Landrat in Niebüll concerning "Zwangsversteigerung von 2 Ochsen des Landmannes Ludolf Jessen, hier" Niebüll, 23.4.1929, LAS 309/22921; report from the Landrat to the Regierungspräsident in Schleswig concerning "Steuerbeitreibungen auf der Insel Fehmarn" Oldenburg in Holstein, 1.6.1929, LAS 309/22569 and the list farmers disseminated throughout the countryside listing household items and farming inventory not permitted to be confiscated titled "Gesetzliche Schutzvorschriften bei Pfändungen" sent by the Kriminal- und Grenzkommissariat to the Landeskriminalpolizeistelle in Flensburg, Niebüll, 30.7.1929, LAS 309/22668.

ernment reported that it had not yet detected a significant discrepancy.[38] Indeed, tax-strikes may not even have been as prevalent as feared. Most farmers remained unwilling to go beyond delaying their payments. In North Dithmarschen, for example, taxes came in slowly but farmers were nevertheless still paying as prescribed.[39] At a meeting in South Dithmarschen, farmers drafted a resolution not to pay taxes before the profitability of their farms had been restored. However, 16 of the 60 farmers who signed promptly continued to pay.[40] It seems, then, that some farmers joined the Landvolkbewegung in protest, or succumbed to peer pressure at meetings, but that their sense of obligation to the movement was limited. The Landvolkbewegung's open structure, that is, the fact that the farmers were not actually members of the movement, doubtlessly reduced its influence in this respect. Then again, the farmers' readiness to participate in tax-strikes may simply have been outweighed by their fear of possible legal action and mounting arrears.

The authorities, nevertheless, feared the potential financial repercussions should tax-strikes and resistance to confiscations and compulsory auctions expand. The Regierungspräsident warned that if these actions were to multiply, they would not only seriously harm the state's authority, but the essential finances of the Reich, Prussia and communities as well.[41] Ultimately, they did not become so frequent as to damage the state's finances. However, they did remain the cause of considerable vexation for the provincial authorities.

At the end of November 1928, the state finance office in Kiel assured a Farmers' League commission that it would avoid all severity in collecting taxes and promised to examine every application for remission benevolently.[42] At the local level, the Farmers' League initiated the formation of commissions to decide on the remission of community taxes.[43] One of the first of these commissions was set up at a meeting of community representatives in Lunden and Paul Adam Ross was elected one of its six members.[44] Within two months, however, he had resigned.[45] At a Landvolk meeting on February 8, the committees were branded a failure, hav-

---

38 Report from the Regierungspräsident to the Preuß. Minister des Innern in Berlin concerning "Unruhe in der Landwirtschaft" Schleswig, 9.1.1929, in both LAS 301/4695 and 309/22569.

39 Regierungspräsident to the Oberpräsident in Kiel, Schleswig, 27.11.1928, in both LAS 301/4695 and 309/22569; note in the Regierungspräsident's files titled "Vermerk: Ferngespräch des Sachbearbeiters mit Landrat Kracht in Heide" Schleswig, 13.1.1929, LAS 309/22696.

40 Landrat to the Regierungspräsident in Schleswig concerning "Steuerverweigerungen der Landwirte" Meldorf, 27.12.1929, LAS 309/22569.

41 Regierungspräsident to the Landräte des Bezirks, Schleswig, 3.1.1929, LAS 309/22696.

42 *Heider Anzeiger* No. 287, 6.12.1928, p. 1; *Rendsburger Tageblatt* No. 297, 6.12.1928, p. 5.

43 *Heider Anzeiger* No. 283, 1.12.1928, p. 2.

44 *Heider Anzeiger* No. 283, 1.12.1928, p. 2.

45 *Dithmarscher Zeitung* No. 5515, 29.1.1929, LAS 309/22569.

ing not sufficiently represented the farmers' interests.[46] By the beginning of March, the Landvolk rejected them outright.[47]

Gerhard Stoltenberg contended that the Landvolk did not recognise the commissions but this is not quite correct.[48] Ross's resignation and the commissions' failure appear to have occurred only after the provincial government began to regulate them. Ross was the community chairman of his village but by no means compliant with the provincial government under its Republican-friendly Regierungspräsident, Abegg. A staunch Landvolk leader, Ross had undoubtedly joined the tax commission to hinder compulsory tax collection.[49] Under the provincial government's guidance, the commissions' powers were diluted. Divested of any genuine influence, they were applied only to assist the authorities in assessing the taxation of agriculture by preparing detailed reports to determine a farm's viability if it paid dues levied on it and to check if payment by instalments was feasible.[50] In North Dithmarschen the commissions dealt with 403 applications for a reduction of community taxes of which 125 were granted, but in many cases it was not farmers but pensioners who applied.[51]

If the rural people rejected the tax commissions because they were taken over by the provincial government, then they did so with good reason. While the Regierungspräsident informed the county administrators that the commissions were to preclude "unwarranted severity" in the collection of taxes, a dispatch to the president of the Chamber of Agriculture created a different impression.[52] Here, Abegg disclosed that the tax commissions had been established in order to combat the "mistaken opinion" that taxes were being collected with "unwarranted severity".[53] Seen from this perspective, the tax commissions were nothing more than tokenism. Perhaps Abegg's intervention was necessary to thwart those who wished to obstruct tax collection. Once again, however, it appears that the provincial government's priority was not to restore the province's farms to profitability, not to address the actual causes of discontent, but merely to silence its critics in the countryside.

---

46 *Nordischer Kurier* No. 33, 8.2.1929, p. 7.

47 In March the Landvolkbewegung established the Crisis Assistance committees in their stead. See chapter 5.

48 Stoltenberg, *Politische Strömungen im schleswig-holsteinischen Landvolk*, p. 126.

49 For Ross's criticisms of tax collection, see his letter in the *Heider Anzeiger* No. 286, 5.12.1928, p. 2.

50 See the drafts of the constitution for the assessment committees titled "Satzung des Gutachter-Ausschusses" undated, LAS 309/22696. For a description of their function, see the *Nordischer Kurier* No. 13, 16.1.1929, p. 6.

51 Landrat to the Regierungspräsident in Schleswig concerning "Tätigkeit der Gemeindesteuerkommissionen" Heide, 4.2.1929, LAS 309/22696.

52 Regierungspräsident to the Landräte des Bezirks, Schleswig, 3.1.1929, LAS 309/22696.

53 Regierungspräsident to the Präsident der Landwirtschaftskammer in Kiel, Schleswig, 3.1.1929, LAS 309/22696.

In his endeavours to check the increasing radicalisation in the countryside, Abegg would have liked to have monitored Landvolk leaders more closely. For instance, he sought to tap Weschke's telephone calls and telegrams but failed to persuade the public prosecutor whose permission was required.[54] The public prosecutor believed that such action would be fruitless, since those connected with Beidenfleth had become extremely cautious. A house-search of Hamkens' property failed to uncover any incriminating material.[55]

The Landvolkbewegung's actions continued and a pattern of obstruction leading to conflict with the authorities, accompanied by meetings and followed by court hearings, was now firmly established. Yet increasingly, the Landvolkbewegung's attention revolved just as much around the conflict with the authorities as the actual prevention of confiscations and auctions themselves. When police banned a rally intended to accompany Hamkens to the finance office in Husum on February 7, following the confiscation of his dairy earnings (*Milchgelder*), up to 100 farmers defiantly gathered inside and outside the building.[56] Only after repeated police orders did the crowd reluctantly leave and proceed to a nearby inn to wait for Hamkens there.[57] A consequence of their recurring clashes was that farmers in the 300 to 400 strong audience at the meeting which followed expressed their growing resentment toward the police, protesting that they never banned communist rallies but only those of "upright citizens" (*ehrliche Bürger*).[58] They would soon have further cause for complaint. Ten farmers, who had refused to comply with the police's orders, were subsequently convicted of trespass (*Hausfriedensbruch*) and received fines of up to RM 80.[59]

---

54 Note in the Regierungspräsident's files titled "Vermerk" Schleswig, 28.1.1929, LAS 309/22569.

55 Note in the Regierungspräsident's files titled "Vermerk" Schleswig, 3.2.1929, LAS 309/22569.

56 Note in the Regierungspräsident's files titled "Vermerk über einen Anruf des Bürgermeisters in Husum" Schleswig, 6.2.1929 and Landrat to the Regierungspräsident concerning "Landvolkbewegung" Husum, 8.2.1929, both in LAS 309/22696.

57 Report titled "Gemeinschaftlicher Hausfriedensbruch" Husum, 7.2.1929, LAS 309/22696; *Der Schleswiger* No. 35, 8.2.1929, in both LAS 309/22696 and 301/4695; *Rendsburger Tageblatt* No. 33, 8.2.1929, p. 13.

58 Untitled meeting report, Husum, 7.2.1929, LAS 309/22696; *Nordischer Kurier* No. 33, 8.2.1929, p. 7.

59 In May, eight of ten prosecuted were found guilty and fined RM 80, later reduced to RM 25 on appeal, while in a further trial in February 1930, two farmers were fined RM 10. See the indictment issued by the Oberstaatsanwalt in Flensburg, 6.5.1929, issue of the verdict of the Schöffengericht in Husum, 30.5.1929 and issue of the verdict of the Landgericht in Flensburg, 8.10.1929 and the issue of the verdict of the Schöffengericht in Husum, 27.2.1930, all in GStA PK I. HA Rep. 84a No. 55398; report from the Landrat to the Regierungspräsident in Schleswig concerning "Prozeß gegen Landleute wegen Hausfriedensbruchs" Husum, 29.6.1929, LAS 309/22668; *Husumer Nachrichten* No. 125, 31.5.1929, p. 2; *Rendsburger Tageblatt* No. 237, 9.10.1929, p. 5; *Das Landvolk* No. 166, 10.10.1929, p. 2.

Throughout the spring and thereafter, farmers blocked a number of compulsory enforcement proceedings, mostly auctions, by intimidating bidders, much as they had done in Garding and Wittbek in December. The movement always remained strongest on the west coast but the fact that two of these actions took place on the island of Fehmarn in Oldenburg County were symptoms of the movement's growth across the province.[60]

The movement's expansion was by no means uniform. In the County of Segeberg, pockets of support were to be found relatively early but it was not until the summer of 1929 that it held large, regular meetings. The county administrator, who even went as far as to acknowledge the overzealousness of one community chairman, in Wiemersdorf, ensured a more lenient approach to tax collection and successfully averted insurgency in the county.[61] The spread of tax-strikes, however, was merely delayed. By April Wiemersdorf and some 20 other communities were behind in their payments. Community chairmen had resigned and replacements

---

60 For a failed compulsory auction on the island of Fehmarn in April 1929, see the *Hamburger Nachrichten* No. 166, 10.4.1929, LAS 309/22668 and *Rendsburger Tageblatt* No. 83, 10.4.1929, p. 11. For a compulsory auction of oxen in Lunden under police protection in April at which the farmers purchased the animals to return them to their owner, see the report from the Vorsitzender des Kreisausschusses, Heide, 18.4.1929, the report from the Landrat to the Regierungspräsident in Schleswig concerning "Steuerpfändung" Heide, 18.4.1929, the police report titled "Bericht über die Tätigkeit der anläßlich der Zwangsversteigerung in Lunden nach dort entsandten Streife am 22. April 1929" Schleswig, 22.4.1929 and the report from the Amts- u. Kirchspielsgemeindevorsteher to the Landrat in Heide, Lunden, 3.5.1929, all in LAS 309/22668. For the prevention of a compulsory auction in South Tondern in April, see the report from the Gemeindevorsteher to the Landrat in Niebüll concerning "Zwangsversteigerung von 2 Ochsen des Landmannes Ludolf Jessen, hier" Niebüll, 23.4.1929, LAS 309/22921. For a failed auction in Tönning in May, see the reports from the Landrat to the Regierungspräsident in Schleswig concerning "Die in Tönning am 4. d. Mts. stattfindende Versteigerung" Tönning, 2.5.1929 and "Versteigerung in Tönning" Tönning, 4.5.1929, both in LAS 309/22921; report from the Regierungspräsident to the Oberpräsident in Kiel, Schleswig, 7.5.1929, LAS 301/4695; *Das Landvolk* No. 49, 4.5.1929, p. 6. For further confiscations carried out under police protection on the island of Fehmarn in May, see the report from the Landjägerabteilung Oldenburg to the Landrat in Oldenburg, Oldenburg i. Holst., 29.5.1929 and the Landrat to the Regierungspräsident in Schleswig concerning "Steuerbeitreibungen auf der Insel Fehmarn" Oldenburg in Holstein, 1.6.1929, both in LAS 309/22569. For the prevention of a further auction in South Tondern in June, see the reports from the Kriminal- und Grenzkommissariat, Niebüll, 10.6.1929 and 12.6.1929, sent by the Landeskriminalpolizeistelle Flensburg to the Regierungspräsident in Schleswig, Flensburg, 17.6.1929, both in LAS 309/22668. For Markus Wulf's attempt to prevent a confiscation at his farm in Eisendorf near Rendsburg in September 1929 and the subsequent trial, see *Das Landvolk* No. 41, 18.2.1930, p. 5.

61 Landrat to the Regierungspräsident in Schleswig concerning "Kundgebung in der Landwirtschaft" Bad Segeberg, 28.11.1928, LAS 309/22921; Landrat to the Regierungspräsident in Schleswig concerning "Steuerverweigerungsbewegung" Bad Segeberg, 5.1.1929 and the appended meeting report, both in LAS 309/22696.

were becoming difficult to find. The county administrator warned that compulsory auctions appeared futile, that all confiscated items would have to be sold elsewhere, and that the increase of such difficulties threatened the operation of the county's administration.[62] The county administrator knew the Landvolkbewegung was working vigorously underground but found its activities almost impossible to monitor. He also reported increasing activity among other groups of the radical right. The county's small Tannenbergbund group was busier, the National Socialists had held their first meetings and there was a noticeable radicalisation among local Stahlhelm groups.[63]

The foundation of its own newspaper further aided the Landvolkbewegung's growth. From its first meetings, the Landvolkbewegung had criticised the established regional press.[64] To counter its alleged failings, to represent rural interests independent of the political parties, associations and professional organisations *Das Landvolk* in Itzehoe was born. It appeared weekly from the beginning of January 1929 and daily as of March.[65] It provided a centre of coordination and means of communication as the movement expanded, and this was all the more important because of the movement's rejection of other forms of structured organisation.

Established as a limited company, *Das Landvolk* started with RM 80,000 capital, Hamkens and Heim being the main investors.[66] Given that the farmers were in crisis, it seems paradoxical that they could afford such an undertaking. There is evidence, however, that they chose to run up debts in order to do so.[67] By mid-1930 *Das Landvolk* had some 36 shareholders with voting rights, most with RM 1,000

62 Report from the Landrat to the Regierungspräsident in Schleswig concerning "Kommunalpolitische Verhältnisse im Kreise Segeberg" Bad Segeberg, 23.6.1929, in both LAS 301/4695 and 309/22668.

63 See the reports from the Landrat to the Regierungspräsident in Schleswig concerning "Stand der politischen Bewegung in der Landwirtschaft" Bad Segeberg, 29.4.1929, LAS 309/22569 and concerning "Kommunalpolitische Verhältnisse im Kreise Segeberg" Bad Segeberg, 23.6.1929, in both LAS 301/4695 and 309/22668. Also see the copies of the report on the Landvolk's first large public meeting in the county from the Landrat to the Oberpräsident in Kiel concerning "Landvolkversammlung" Bad Segeberg, 10.6.1929, in both LAS 309/22668 and 301/4695.

64 See, for example, chapter 3, pp. 108 and 117.

65 See the leading article in *Das Landvolk*'s first issue, No. 1, 3.1.1929, p. 1.

66 Copy of a letter from Wilhelm Hamkens, Tetenbüll, to the Oberstaatsanwalt in Kiel, Altona, 14.5.1930, in both LAS 301/4697 and 309/22590; undated extracts from the examination transcripts of Claus Heim titled "Klaus Heim, 24.3.1884 in Oesterfeld" and of Bruno von Salomon titled "Bruno von Salomon, 7.2.1900 in Stettin, Hauptschriftleiter 'Landvolk' in Itzehoe ...", both in LAS 309/22784.

67 Hamkens borrowed as much as RM 50,000 to support the newspaper. Report from the Regierungspräsident to the Oberpräsident in Kiel concerning "Landvolkbewegung" Schleswig, 20.6.1929, LAS 301/4695. Also see the letter from Wilhelm Hamkens to the Oberstaatsanwalt in Kiel, Altona, 14.5.1930, LAS 301/4697.

investments.[68] Heim, Nickels, Hönck and many others travelled the countryside, selling shares and subscriptions and taking donations.[69] They sought out "trusted men" to help them promote the newspaper and also to collect contributions to so called "fighting funds" (*Kampffonds*) in order to pay for followers' court fees and fines as well as assist colleagues in financial difficulties.[70]

In March, when production of the first daily issue commenced, *Das Landvolk* had a circulation of 4,500.[71] By the beginning of April the circulation had reached somewhere between six and seven thousand. This compared favourably with the National Socialist *Schleswig-Holsteinische Tageszeitung* which, at that time, appeared to have a greater circulation, estimated between eight and twelve thousand copies, but fewer subscriptions, 4,500 in August 1929, climbing to 5,000 in January 1930.[72] When it peaked in October 1929, *Das Landvolk* printed twelve thousand copies daily, of which 6,500 were subscriptions delivered by mail and 2,000 distributed, mainly at meetings, for promotional purposes.[73] This made it the one of the most widely read newspapers in the province.[74]

Despite this swift growth, *Das Landvolk* experienced financial difficulties on several occasions throughout 1929, often having to operate on substantial deficits.[75] In November 1929 however, a registered association, the *Volksmacht* (People's Power), appeared in the province with the self-proclaimed purpose of supporting

---

68 Report from the Polizeiverwaltung, Itzehoe, 18.8.1930, LAS 301/4697.

69 See, for example, *Das Landvolk* No. 3, 17.1.1929, p. 1 and No. 7, 16.2.1929, p. 4 and for Heim and Nickels' promotional activities, see the undated extracts from examination transcripts of Claus Heim titled "Klaus Heim, 24.3.1884 in Oesterfeld" and of Hans Nickels titled "Kurzer Auszug aus der Vernehmungsniederschrift des Polizeihauptmanns a.D. Hans Nickels ...", both in LAS 309/22784.

70 Report from the Landeskriminalpolizeistelle Flensburg concerning "Bauernbewegung in der Provinz Schleswig-Holstein" Flensburg, 24.1.1929, in both LAS 301/4589 and 309/22696.

71 This figure, given by Peter Petersen (Ahrenholz), must be treated with caution but is, nevertheless, plausible as it conforms with the later circulation. See Schimmelreiter, *Unter der schwarzen Bauernfahne*, p. 41.

72 Report from the Landeskriminalpolizeiamt, Berlin, 2.4.1929, sent to the Oberpräsident by the Polizeipräsident, Landeskriminalpolizeiamt, Abteilung IA, Berlin, 24.4.1929, LAS 301/4695; M. Oddey, *Unter Druck gesetzt. Presse und Nationalsozialismus in Schleswig-Holstein. Struktur – Wahrnehmung – Herrschaftsakzeptanz* (Eutin, 2006), pp. 64-65.

73 Report from the Landeskriminalpolizeistelle Altona, Abteilung IA, titled "Bericht über die Landvolkbewegung" Altona, 5.10.1929, in both LAS 301/4696 and 309/22784.

74 Two thirds of the 80 newspapers published in the province in 1926 had a circulation under 3,000 and just seven higher than 10,000. See T. Steensen, ' "In Land und Stadt das Heimatblatt" Zur Zeitungslandschaft in Schleswig-Holstein – insbesondere Nordfriesland – während der Weimarer Republik' in W. Paravicini (ed.), *Mare Balticum: Beiträge zur Geschichte des Ostseeraums in Mittelalter und Neuzeit* (Sigmaringen, 1992), pp. 391-412.

75 Undated extract from the examination transcript of Claus Heim titled "Klaus Heim, 24.3.1884 in Oesterfeld", LAS 309/22784.

the independent nationalist press.[76] In reality, it was formed by a number of Landvolk followers to support *Das Landvolk*, and by the beginning of the new year the financial position of the paper had improved considerably. The Volksmacht, which had doubled its membership to approximately 300 within those two months, had reportedly collected some RM 40,000, all of which it donated to *Das Landvolk*.[77]

The most important figures at *Das Landvolk* were the editors, Bruno von Salomon, Johannes Kühl and, later, Otto Winter. The printer and publisher was Ferdinand Pramor, and the business directors were Hamkens and Johnsen, later replaced by Hönck.[78] Weschke signed as the editor for the *Kämpfendes Landvolk*, an edition distributed outside Schleswig-Holstein.[79]

Kühl came to *Das Landvolk* having previously worked for the DNVP and the Rural League's *Landbote*. Winter came from Thuringia, from the *Artam* movement, a nationalist-völkisch organisation which aimed to halt urbanisation and foreign farm labour by promoting rural employment and settlement.[80] Pramor, who had been a member of the NSDAP from 1925 to 1928 and printed material for the party, then published the Stahlhelmwestküste's *Westküste* and had already begun to

---

76 Reports from the Landeskriminalpolizeistelle Altona, Abteilung IA, concerning the "Landvolkbewegung" Itzehoe, 6.11.1929 and from the Preuß. Polizeipräsident Altona-Wandsbek in Altona to the Regierungspräsident in Schleswig concerning "Landvolkbewegung" Altona, 13.11.1929, both in LAS 301/4697 as well as 309/22783; report from the Landrat in Heide to the Oberpräsident in Kiel concerning "Volksmacht" Heide, 19.11.1929, LAS 301/4697; copy of the Volksmacht's statutes, Heide, 29.1.1930, LAS 309/22590.

77 Reports from the Landeskriminalpolizeistelle Altona, Abteilung IA, concerning "Landvolkbewegung" Altona, 8.1.1930 and Altona, 5.2.1930, both in LAS 301/4697 as well as 309/22783; reports from the Landeskriminalpolizeistelle Altona, Abteilung IA, concerning "Landvolkbewegung" Altona, 8.3.1930 and from the Polizeiverwaltung, Itzehoe, 18.8.1930, both in LAS 301/4697.

78 Report from the Polizeiverwaltung, Itzehoe, 18.8.1930, LAS 301/4697; undated extract from the examination transcript of Johann Kühl titled "Redakteur bei der Landvolkzeitung G.m.b.H. Johann Kühl, ...", LAS 309/22784.

79 Undated extract from the examination transcript of Guido Weschke titled "Syndikus Guido Hermann, Hans Weschke, 6.9.1893 in Pichelsdorf …", LAS 309/22784

80 Report from the Landeskriminalpolizeistelle Altona, Abteilung IA, concerning "Landvolkbewegung" Altona, 8.1.1930, in both LAS 301/4697 and 309/22783. The Artam movement was anti-urban, anti-Slavic and idealised healthy rural living, swiftly embraced anti-Semitism, aspirations of racial renewal and ambitions to colonise in Germany's East. Many members of the Artam, known as *Artamanen*, were also members of the NSDAP, but thanks to August Georg Kenstler, who headed the first contingent of young urban Germans to work in the countryside in 1924, the Artam maintained close contact with the Landvolkbewegung. See Bergmann, *Agrarromantik und Großstadtfeindschaft*, pp. 247-297; G. Hoch, 'Artamanen in Schleswig-Holstein' in E. Hoffmann and P. Wulf (eds), *"Wir bauen das Reich" Aufstieg und erste Herrschaftsjahre des Nationalsozialismus in Schleswig-Holstein* (Neumünster, 1983), pp. 137-148; M. H. Kater, 'Die Artamanen – Völkische Jugend in der Weimarer Republik', *Historische Zeitschrift*, 213 (1971), pp. 577-638.

support the Landvolkbewegung with the short-lived *Grüne Front*, a supplement edited by Peter Petersen (Ahrenholz) on behalf of the Defence Association.[81]

Bruno von Salomon was a stranger both to the province and agriculture. An erstwhile member of the NSDAP – for short periods its Gauleiter in East Pomerania and then secretary in Hamburg – he had lost interest in the party because of its parliamentary intentions after the ban upon it was lifted.[82] According to both the fictional and autobiographical writings of his brother, Ernst von Salomon, Heim recruited Bruno from the nationalist *Deutsche Front* in Hamburg.[83] It was the insurgent atmosphere in Schleswig-Holstein which appealed to Bruno, and it was with this very enticement that he lured Ernst to the province. He wrote, "In Berlin you still do not seem to have noticed at all that it is starting to become serious here. Have you perhaps completely lost your nose for it?"[84]

Ernst von Salomon, accomplice in the murder of Rathenau, turned writer, soon visited Itzehoe to report on the Beidenfleth trial but he also sent numerous contributions from Berlin.[85] Ehrhardt's circle, with which Ernst was associated, also sent the paper an abundance of material.[86] It was Ernst von Salomon and Ehrhardt's secretary, Hartmut Plaas, who facilitated the employment of eminent lawyer Wal-

81 Report from the Landeskriminalpolizeistelle Flensburg concerning "Bauernbewegung in der Provinz Schleswig-Holstein" Flensburg, 24.1.1929, in both LAS 301/4589 and 309/22696; report from the Regierungspräsident to the Oberpräsident in Kiel, Schleswig, 24.1.1929, LAS 301/4695. Also see the issues of the *Grüne Front* No. 4, 18.12.1928 and the *Westküste* No. 24 (undated) 1928, No. 1, 10.1.1929 and No. 2, 17.1.1929, all in LAS 301/4695 and Oddey, *Unter Druck gesetzt*, pp. 87-88.

82 Report from the Landeskriminalpolizeiamt, Abteilung IA, Außendienst VII, Berlin, 5.6.1930, LAS 309/22701.

83 See von Salomon, *Die Stadt*, pp. 20-21. In this fictional account of the Landvolkbewegung, Bruno is represented by Iversen. Klein also attributes Bruno's recruitment to Heim. See M. J. Klein, *Ernst von Salomon: Eine politische Biographie mit einer vollständigen Bibliographie* (Limburg /Lahn, 1994), p. 181.

84 "Ihr scheint in Berlin noch garnicht zu bemerken, daß es hier so langsam ernst wird. Habt ihr denn schon gänzlich die diesbezügliche Nase verloren?", cited in Klein, *Ernst von Salomon*, p. 176.

85 Ernst von Salomon (1902-1972) was a Freikorps soldier in the Baltic in 1919 and Silesia in 1921 and member of Organisation Consul. He was sentenced to five years' imprisonment for accessory to the murder of Walter Rathenau and to a further five years' term for an attempted "Feme" murder but granted an amnesty and released in 1927. He worked as a script writer for film from 1933 and was later active in the anti-nuclear movement but was best known for his autobiographic novels, *Die Geächteten* (1930) and *Der Fragebogen* (1961). See Klein, *Ernst von Salomon*.

86 See the undated report titled "Ehrhardt Kreis" appended to the undated list of arrested bombings suspects titled "Festgenommene Personen anläßlich der Sprengstoffanschläge", LAS 309/22784.

ther Luetgebrune to take over the defence of the farmers at the pending Beidenfleth trial.[87]

Bruno von Salomon played a significant role in the Landvolkbewegung as editor in chief at *Das Landvolk* and also in providing useful connections. Hamkens and Heim, however, ensured that the Landvolkbewegung remained a farmers' movement, discounting von Salomon as a potential speaker and turning down his offer to organise more help from "qualified putschists".[88] Hamkens, Heim and their fellow farming patrons retained an active influence over the operation and content of *Das Landvolk*. Heim and Hamkens always directed it in their interests, and it is clear from Ernst von Salomon's writings, that he and Bruno deferred to them.[89] This is best illustrated with his description of Heim putting an end to discussion about what to call the paper by "… bringing his fist down on the table with a crash and announcing: It's called *Das Landvolk* and that's that."[90]

In its first issues *Das Landvolk* summarised the movement's progress, telling its readers of the movement's actions at confiscations and compulsory auctions, and reprinting Heim's letter to his community chairman as well as excerpts from Hamkens' speech in Garding.[91] Reporting the movement's achievements and meetings, and transcribing or summarising the speeches for all to read continued to be one of the paper's most important functions. *Das Landvolk* also operated as a herald for the movement, advertising meetings and publicising confiscations and compulsory auctions.[92] In addition, it blacklisted those who dared to bid at compulsory auctions.[93] *Das Landvolk* reported on professional agricultural matters, but it was

---

87 Walter Luetgebrune (1879-1949), based in Göttingen, specialised in political trials after the First World War and rose to prominence defending various high-profile officers and nationalists, including von Jagow (Kapp Putsch) in 1921, Ehrhardt in 1922 and Ludendorff at the "Hitler Trial" in 1924. Luetgebrune conducted the defence at a number of Landvolk trials and in 1932 was appointed chief law advisor to the SA. He was arrested and subsequently expelled from the party in 1934 after representing expropriated Jewish industrialists and was unable to practise again until 1937. See R. Heydeloff, 'Staranwalt der Rechtsextremisten: Walter Luetgebrune in der Weimarer Republik', *Vierteljahrshefte für Zeitgeschichte*, 32 (1984), pp. 373-421. Also see Klein, *Ernst von Salomon*, pp. 128-132, 134-135, 144 and 146.

88 Examination transcript of Bruno von Salomon, Berlin, 4.11.1929, LAS 309/22701.

89 Report from the Landeskriminalpolizeistelle, Abteilung IA, concerning "Landvolkbewegung" Altona, 8.3.1930, LAS 301/4697; report from the Landeskriminalpolizeiamt, Abteilung IA, Außendienst VII, Berlin, 5.6.1930, LAS 309/22701.

90 E. v. Salomon, *Der Fragebogen*, p. 226 and for the translation here, see E. v. Salomon, *Der Fragebogen (The Questionnaire)* (New York, 1955), p. 130. In *Die Stadt* Heim is introduced as "always the first among equals": v. Salomon, *Die Stadt*, p. 12.

91 See *Das Landvolk* No. 1, 3.1.1929, pp. 3-5, No. 2, 9.1.1929, p. 1 and No. 5, 1.2.1929, p. 1.

92 For example, *Das Landvolk* No. 118, 15.8.1929, p. 4, No. 119, 16.8.1929, p. 4 and No. 120, 17.8.1929, p. 4.

93 For example, *Das Landvolk* No. 1, 3.1.1929, p. 3 and No. 40, 23.4.1929, p. 2.

always much less an agricultural than a political medium and articulated an extreme nationalist and anti-system perspective.

Das Landvolk

»Lewwer duad üs Slaav!«

Ueberparteiliche, unabhängige Tageszeitung für das deutsche Volk in Land und Stadt

2. Jahrgang
Nummer 249

Verlagsort Itzehoe

Preis monatlich 1.80 RM. außer Zustellgeld Einzelpreis 15 Pfg. Fernruf: 2513 u. 2500 Drahtanschrift: Landvolk, Itzehoe

Mittwoch
22. Oktober 1930

Dr. Luetgebrune spricht weiter

„Und ist das Dunkel noch so groß Ein Weg zum Licht ist immer frei!"

Knalldemonstrationen

Verschiedene Wege Heims und Hamkens

Dr. Luetgebrune gab nunmehr eine Gliederung seiner weiteren Darlegungen; er führte dazu aus, daß er, wenn er auf die einzelnen Taten eingehe, damit keineswegs eine Entschuldigung der Angeklagten beabsichtige.

Er fuhr fort: Die Besprechung Heim—Hamkens hat nicht nur eine Bedeutung für die Anklage gegen Hamkens, sondern vor allem für die Frage,

ob das Handeln der Angeklagten gerechtfertigt ist durch den übergesetzlichen Notstand, der eine ausschlaggebende Rolle spielt für die Untersuchung,

ob nun derjenige, der sich auf diesen Rechtfertigungsgrund

werden, um den Gemeindevorstehern einen Wink zu geben! Alle Entschlüsse dieser Art hat Heim aber erst nach Hamkens Reise gefaßt.

Rathjens Komperit — Volcks Adjutantentätigkeit

Es kommt dann der Auftrag Volck; Volck setzt sich mit Rathjen in Verbindung, der ihm von Heim genannt war. Volck hat immer nur von Feuerwerkskörpern gehört, denn Rathjen hat den Ausdruck Komperit nicht gebraucht und auch von der Natur als Sprengstoff nichts mitgeteilt. Wenn Volck [illegible] gesagt hat, [illegible] einen Kanonenschlag abbrennen, so deckte sich das mit den Angaben Weschkes, der glaubte, es solle sich um einen Knallkörper handeln,

Man muß sich aber auch überlegen, daß Sprengstoff nicht nur ein Qualitäts-Begriff, sondern auch ein Quantitäts-Begriff ist, und wenn Volck hingeht und sich von Rathjen genau beschreiben läßt, welche Dosierung wirkt, so spricht das doch dafür, daß er den Sprengstoff der Quantität nach haben wollte. Wenn ich den Sprengstoff in solcher Quantität nehme, daß er überhaupt nicht als Sprengstoff wirken kann, so will ich kein Sprengstoffattentat, sondern nur eine Knalldemonstration. Damit ist zu vergleichen jene schon angeführte Reichsgerichtsentscheidung, nach der Schrecken nicht als unerlaubter Zweck im Sinne des Sprengstoffgesetzes, nicht als die Absicht Gesundheit, Leben oder Eigentum zu gefährden gilt. Wenn nun die Angeklagten sich auch vorgestellt hätten, daß möglicherweise ein Erfolg einträte, den sie nicht wollten, dann sei nach der Ansicht des Reichsgerichts nur bewußte Fahrlässigkeit anzunehmen, die allerdings zur Anwendung des § 5 des Sprengstoffgesetzes nicht ausreichen würde.

Front page of Das Landvolk, 22.10.1930[94]

As a political instrument first and foremost, *Das Landvolk* attacked parliamentary democracy, the pro-Republican parties and the communists, finance capital and war reparations. *Das Landvolk* deliberately and candidly set out to indoctrinate farmers, with contributions from nationalist intellectuals, including Ernst Jünger, and it introduced its "school of politics" columns, for example, as providing the "intellectual foundation" for the movement.[95] It also sought to promote regional patriotism, for instance running an extensive special daily series on the history of the free northern German farmers.[96] The first issue alone included articles on the lies about Germany's war guilt, the influence of Jewish representatives in the Reichstag and the importance of national pride.[97] The leading article in the third

94 *Das Landvolk* No. 249, 22.10.1930, p. 1, Schleswig-Holsteinische Landesbibliothek.

95 These were written by Gerhard and Albrecht Erich Günther, the latter known as the co-editor of the journal *Deutsches Volkstum*. See *Das Landvolk* No. 16, 9.3.1929, p. 8 and for examples, No. 21, 30.3.1929, p. 10, No. 26, 6.4.1929, p. 9 and No. 61, 19.5.1929, p. 12. For an example of an article by Jünger, see *Das Landvolk* No. 61, 19.5.1929, p. 9.

96 See *Das Landvolk* No. 38, 14.2.1930, p. 9 – No. 81, 5.4.1930, p. 7 [passim].

97 *Das Landvolk* No. 1, 3.1.1929, p. 1.

issue listed examples of public spending follies under the headline "What you are paying taxes from [your] capital for".[98]

Anti-Semitism was always a prominent element. Sometimes objections to "Jewish influences" were expressed in individual articles and, on other occasions, they were made in combination with other grievances. Not only did *Das Landvolk* promote anti-Semitism with stereotypical resentments such as the connection between communism and Jews, by attacking "Jewish department stores" or drawing on the Barmat and Sklarek scandals as symptomatic of wider Jewish and Republican corruption, it also did so by discussing anti-Semitism as a subject in itself.[99] Perhaps borrowing from Schleswig-Holstein chronicler and literary historian, Adolf Bartels, one article opened, "For reasons which are yet to be investigated, every healthy German who thinks and feels naturally is an 'anti-Semite'."[100]

Anti-Semitic, anti-democratic, anti-capitalist and anti-socialist sentiment had all been strongly and frequently voiced in the Landvolkbewegung before. The movement was already radicalising prior to the foundation of *Das Landvolk*. Yet there can be no doubt that *Das Landvolk* incited the rural population further. In his novel, *Die Stadt*, Ernst von Salomon suggested that his brother's aim was to make the movement the "shock troops of a new reality, ... [the] germ cell of a new state".[101] But however far Bruno von Salomon's hopes may have stretched, the mere employment of this non-farmer at *Das Landvolk* was one more indication that the movement was no longer primarily concerned with combating the agricultural crisis. Combating "the system" had become its raison d'être. In September 1929 one farmer ventured in the *Rendsburger Tageblatt* that since the foundation of *Das Landvolk*, the movement had become so radical its followers simply saw the "administrative apparatus" as their enemy. By no means did this farmer champion the Republic because he too demanded a new state. He merely wished to protest that the Landvolkbewegung had overstepped the mark.[102]

Unsurprisingly, it was Schleswig-Holstein's authorities who were most troubled by *Das Landvolk*. Schleswig-Holstein's Oberpräsident judged that it was simply founded for the purpose of producing propaganda "totally hostile" to the Republic and the state's institutions, stating that it "agitated against the government in an

---

98 *Das Landvolk* No. 3, 17.1.1929, pp. 1-2.

99 See, for example, the list titled "The Jews who govern Russia" in *Das Landvolk* No. 4, 25.1.1929, p. 6 and the articles "The German department-store", "The role of the Jews" as well as the list of the number of department-store branches in Germany in No. 6, 11.2.1929, pp. 1-3. For an example of the Barmat and Sklarek scandals used to reinforce the idea that corruption and discrimination were rife, see *Das Landvolk* No. 173, 18.10.1929, p. 1.

100 *Das Landvolk* No. 14, 7.3.1929, p. 5. According to Bartels in 1913, "Whoever is not an Anti-Semite in our times, is not a good German" cited in Bergmann, *Agrarromantik und Großstadtfeindschaft*, p. 211.

101 v. Salomon, *Die Stadt*, p. 37.

102 *Rendsburger Tageblatt* No. 222, 21.9.1929, pp. 1-2.

almost unsurpassable manner on a daily basis".[103] Even the puzzles in the entertainment segment required solutions deriding the authorities.[104]

The government used the threat of legal action to compel *Das Landvolk* to correct misleading and false reporting. *Das Landvolk* only did so grudgingly, questioning the corrections wherever possible and mocking the authorities in the process, even when it risked litigation.[105] Given that the authorities scrutinised the newspaper hoping to find material that would justify litigation, it is perhaps surprising that *Das Landvolk* was not banned more often. For the most part, however, its editors simply proved too shrewd. The public prosecutor suspected they had received expert legal advice enabling the paper to indulge in "provocative language" yet avoid entanglements with the law.[106]

Nevertheless, *Das Landvolk*, its auxiliaries, substitutes, successors and respective editors faced legal action on numerous occasions, either for libel or for breaching the "Law for the protection of the Republic" (*Republikschutzgesetz*). This resulted in fines for the editors and temporary bans of the paper.[107] *Das Landvolk* was banned on four occasions.[108] The first time, it was banned for four weeks as a consequence of two items, one a letter alleging the Republic was "born in lies and deception" and "selling the people into slavery", the other accusing "social democracy" of treason.[109] The editors appealed, which resulted in the Reich court reducing the ban to two weeks on the grounds that the second article pertained only to

103 Oberpräsident's decree banning *Das Landvolk* printed in *Das Landvolk* No. 20, 14.3.1929, p. 1.

104 See, for example, *Das Landvolk* No. 121, 18.9.1929, p. 9.

105 For possibly the most complicated case in point, see the articles titled "Achtung! Neumünster!" in *Das Landvolk* No. 262, 6.11.1930, p. 7, "Nun erst Recht" and the "Offener Brief an Herrn Abegg" in No. 264, 8.11.1930, p. 1, the corrections to the article "Achtung! Neumünster!" from No. 262 in "Abegg berichtigt und klagt" No. 266, 11.11.1930, p. 3, further corrections in the article "Eine Flut von Abeggschen 'Berichtigungen' " and also see the article "Für sozialdemokratische Denunziationen ist Herr Abegg zu haben" in No. 268, 13.11.1930, p. 3, "Fragen an Regierungspräsident Abegg" and "Abegg berichtigt weiter" in No. 269, 14.11.1930, p. 1, "Abegg berichtigt immer weiter" in No. 270, 15.11.1930, p. 3 and further corrections in "Herzliches Beileid, Herr Abegg" No. 271, 17.11.1930, p. 3. For the legal action which resulted, see chapter 6, pp. 225-226.

106 See the note in the Regierungspräsident's files titled "Vermerk: Über eine Besprechung des Sachbearbeiters mit Staatsanwaltschaftsrat Dr. Junker" Schleswig, 3.2.1929, LAS 309/22569.

107 See, for example, the report from the Regierungspräsident to the Oberpräsident in Kiel concerning "Strafverfahren gegen den Schriftleiter Kühl, Itzehoe" Schleswig, 16.8.1930, LAS 301/4697.

108 It was banned twice in 1929 and twice in 1931. For the later bans, see chapter 8.

109 See the Oberpräsident's decree, Kiel, 13.3.1929, LAS 301/4705; *Das Landvolk* No. 15, 8.3.1929, p. 6 and No. 16, 9.3.1929, p. 1.

the SPD and not the Republic.[110] Nonetheless, *Das Landvolk* sought to circumvent the ban by publishing issues of the *Westküste* and the *Grüne Front* instead, but since the Oberpräsident had anticipated and expressly prohibited this, they too were swiftly banned and confiscated by the police.[111]

In June *Das Landvolk* was banned for the second time under similar circumstances. Renewed attempts to publish replacements, which included a supplement named *Der Kürbis* (*The Pumpkin*) ridiculing the Oberpräsident, resulted in a round-the-clock police watch at the newspaper's offices as well as legal proceedings against those responsible.[112]

The foundation of *Das Landvolk* not only intensified conflict with the authorities, but also the rivalry between the Landvolkbewegung and the NSDAP. Like *Das Landvolk,* the NSDAP's own provincial newspaper, the *Schleswig-Holsteinische Tageszeitung*, started publication at the beginning of January 1929, was also based in Itzehoe and, at RM 1.80, cost exactly the same for a monthly subscription.[113] From the very first issue, the *Schleswig-Holsteinische Tageszeitung* sought to woo the rural people. According to Friedrich Wilhelm Heinz, editor of

---

110 See the appeal from Gries – Danican, Rechtsanwalt u. Notar, "Beschwerde der Tageszeitung 'Das Landvolk' gegen das Verbot vom 13. d. Mts. O.P. 673" to the Oberpräsident in Kiel, Itzehoe, 16.3.1929, LAS 301/4705 and the decision of the Reichsgericht, 4. Strafsenat from 26.3.1929, both in LAS 301/4705. Also see *Das Landvolk* No. 20, 14.3.1929, p. 1; *Husumer Nachrichten* No. 62, 14.3.1929, p. 1.

111 Copies of the *Grüne Front* Folge 8, March 1929 and the "Telegramm" titled "Auch die Westküste verboten!", LAS 301/4705; the report from the Landeskriminalpolizeistelle to the Regierungspräsident in Schleswig, Flensburg, 2.4.1929, LAS 309/22921; the Preuß. Minister des Innern to the Oberpräsident, Berlin, 27.3.1929, LAS 301/4705; issue of the verdict of the Schöffengericht in Itzehoe, 9.8.1930, LAS 301/4700. Also see Ernst von Salomon's humorous account of these bans: v. Salomon, *Der Fragebogen*, pp. 226-227.

112 See the Oberpräsident's decrees, Kiel, 22.6.1929 and Kiel, 26.6.1929 (the second extending the ban across the entire Reich) and also see the offending article, a report on a speech made by Guido Weschke, *Das Landvolk* No. 85, 18.6.1929, p. 5. Also see the report from the Oberpräsident to the Preuß. Minister des Innern in Berlin concerning "Das Verbot der Tageszeitung 'Das Landvolk' " Kiel, 28.6.1929 and for the police's measures and confiscations of the *Grüne Front* and *Kämpfendes Landvolk*, and subsequent legal proceedings against the editors and publishers, see the untitled report from the Polizeiverwaltung, 29.6.1929 and the report from the Landrat des Kreises Steinburg to the Oberpräsident in Kiel concerning "Das Verbot der Zeitung Landvolk" Itzehoe, 2.7.1929, Generalstaatsanwalt to the Oberpräsident in Kiel concerning "Strafsache gegen den Schriftleiter Bruno von Salomon in Itzehoe" Kiel, 27.8.1929, the report from the Regierungspräsident to the Oberpräsident in Kiel concerning "Die 'Westküste' " Schleswig, 11.7.1929 and Generalstaatsanwalt to the Oberpräsident in Kiel, Kiel, 2.11.1929, all in LAS 301/4705. In contrast to his brother's account, Bruno von Salomon stated that *Der Kürbis* only appeared as a supplement during the second ban on *Das Landvolk.* See the examination transcript of Bruno von Salomon, 1.10.1929, GStA PK I. HA Rep. 84a No. 55381. Also see E. v. Salomon, *Der Fragebogen*, p. 227.

113 Oddey, *Unter Druck gesetzt*, pp. 59-66.

the party's newspaper in Hanover, the difference in political stance between the two newspapers was negligible.[114] Palpably similar to the Landvolkbewegung, the *Schleswig-Holsteinische Tageszeitung*, for example, attributed the agricultural crisis and destruction of capital to the machinations of high finance which contrived to turn Germany into a "slave colony" by way of credits at excessive rates of interest and taxes to pay war reparations.[115]

Though Bruno von Salomon soon developed a close friendship with his National Socialist counterpart, Bodo Uhse, their hope of forging closer ties between the movements was firmly prevented by Gauleiter Lohse.[116] The reason for this was that the NSDAP demanded sole representative status. In fact, its claim to be the only group to represent agricultural interests was to become one of the simple but essential elements of its rural propaganda. The party insisted in a flyer, for example, that only it could release Germany from reparations, and that because members of all other parties had supported the Dawes Plan, "German Farmers, ... only we National Socialists .... have the moral right to appeal to you."[117]

The *Schleswig-Holsteinische Tageszeitung* strove to win over the Landvolk's followers by printing special agricultural issues. One of the first of these was titled "Landvolk in crisis" (*Landvolk in Not*), a slogan that echoed the Landvolkbewegung's own vocabulary. Timed to appear just as *Das Landvolk* began to be published daily, the special issue was promoted in advance and 30,000 copies were distributed.[118] The content was not novel but interesting in so far as it confirmed the

---

114 Letter from F. W. Heinz to Bruno von Salomon, Hanover, 16.5.1929, cited in the list of confiscated material titled, "V. Auslese wichtigen Materials aus den von der Redaktion der Landvolk – Zeitung beschlagnahmten Ordnermappen", LAS 309/22784.

115 See the article titled "Der Nationalsozialismus und die deutsche Landwirtschaft" in the *Schleswig-Holsteinische Tageszeitung* No. 1, 3.1.1929, p. 3.

116 Undated extract from the examination transcript of Bruno von Salomon titled "Bruno von Salomon, 7.2.1900 in Stettin, Hauptschriftleiter 'Landvolk' in Itzehoe ...", LAS 309/22784; letters from Bruno von Salomon to F. W. Heinz, 17.5.1929 and 9.7.1929, cited in the list of confiscated material titled, "V. Auslese wichtigen Materials aus den von der Redaktion der Landvolk – Zeitung beschlagnahmten Ordnermappen" and the letter from Bodo Uhse to Bruno von Salomon concerning "Gegenseitige persönliche Reibungen" Itzehoe, 29.7.1929" cited in the list of confiscated material titled "IV. In der Redaktion der Landvolk-Zeitung am 9.9.1929 beschlagnahmten Materials", all in LAS 309/22784. Also see Uhse, *Söldner und Soldat*, pp. 171 and 203. For the friendship between von Salomon and Uhse, see the report from the Regierungspräsident to the Preuß. Minister des Innern in Berlin concerning "Landvolkbewegung und die politischen Sprengstoffanschläge" Schleswig, 30.9.1929, in LAS 301/4696 and 28.9.1929 in 309/22784 and Klein, *Ernst von Salomon*, pp. 263-264.

117 Undated leaflet of the NSDAP titled "Bauern, Ihr werdet enteignet!", LAS 301/4695.

118 *Schleswig-Holsteinische Tageszeitung* No. 46, 24.2.1929, pp. 9 and 11, No. 47, 26.2.1929, p. 8, No. 48, 27.2.1929, pp. 2 and 10, No. 49, 28.2.1929, pp. 3 and 8; R. Rietzler, 'Gegründet 1928/29: Die Schleswig-Holsteinische Tageszeitung: Erste Gau-Tageszeitung der NSDAP' in E. Hoffmann and P. Wulf "*Wir bauen das Reich" Aufstieg und erste Herrschaftsjahre des Nationalsozialismus in Schleswig-Holstein* (Neumünster, 1983), p. 123.

party's strategy in dealing with the Landvolkbewegung, criticising it on the one hand but sympathising with the farmers on the other. The leading article was an appeal from Hitler that had already appeared in the *Völkischer Beobachter* at the end of December 1928.[119] Hitler attacked the Landvolkbewegung as "farmers' associations acting radically", which posed no threat to the "beneficiaries of the current system". He asked, "What do the current holders of power care if some small tax-collectors ... are struck down, if a bomb is thrown at a minor official ... so long as they themselves and their system are left untouched?" At the same time, however, he also flattered the Landvolk activists, suggesting that long prison terms awaited the "best and most resolute" among them. Not for the last time, he warned of attempts to associate the NSDAP with the Landvolk's actions as a pretext for taking action against his movement, and ordered members not to take part in any illegal actions.[120]

At the "Beidenfleth Trial", which commenced in Itzehoe on April 9, 1929, 57 farmers stood trial for their participation and organisation of the demonstration of November 19, 1928 and other related incidents. Security was tight as police cordoned off the town hall hosting the Schöffengericht to prevent disturbances. With the exception of a substantial number of journalists, only distinguished guests (*Prominente*) were admitted to the courtroom, and police road blocks thwarted protests planned for the trial's final day by stopping trucks loaded with farmers headed to the town.[121]

Although some 150 to 200 people had demonstrated at Beidenfleth, only those who could be identified were indicted.[122] On Kock's bidding, Weschke had arranged for a photographer to take pictures on November 19, and although unable to capture the critical moment when the oxen shied, three clear photos were taken. It was this evidence that incriminated many of the farmers.[123]The trial also dealt with Kelting and Frauen's speeches at the Landvolk meeting in Itzehoe on November 25. They were tried for publicly encouraging "disobedience against the laws". Kock and Hansen were tried for intimidating a public servant in the course of the sale of Kock's ox in Hamburg.[124] On a separate charge, Hansen was also tried for

---

119 "Bauern, hütet Euch vor Provokateuren!" *Völkischer Beobachter* 23./24.12.1928 in K. A. Lankheit (ed.), *Hitler: Reden Schriften Anordnungen: Februar 1925 bis Januar 1933*, vol. 3 (2) (Munich, 1994), pp. 369-371.

120 See the article retitled as "Bauern, seid auf der Hut" in *Schleswig-Holsteinische Tageszeitung* No. 52, 3.3.1929, p. 1.

121 *Itzehoer Nachrichten* No. 83, 10.4.1929, p. 4 and No. 91, 19.4.1929, p. 4; *Husumer Nachrichten* No. 82, 9.4.1929, p. 2 and No. 90, 18.4.1929, pp. 1 and 5.

122 Indictment issued by the Staatsanwalt Altona, 4.2.1929, and Vermerk IV f301 G. 3/29, both in GStA PK I. HA Rep. 84a No. 55362.

123 *Husumer Nachrichten* No. 83, 10.4.1929, p. 1.

124 Indictment issued by the Staatsanwalt Altona, 13.12.1928, GStA PK I. HA Rep. 84a No. 55365.

threatening his community chairman after he was informed that the bailiff might confiscate one of his animals for outstanding taxes.[125]

The principal question concerning the court was the extent to which the action in Beidenfleth had been planned. Witnesses could neither remember who had called the meeting at which the action was arranged, nor specifically what had been discussed there.[126] Many of the farmers, too, who went to Beidenfleth on November 19, claimed to have known nothing of Kock and Kühl's plans, stating they believed they were merely attending a demonstration of the rural people's difficulties or thought that there was a real fire when the horns sounded.[127]

The farmers probably feigned ignorance and forgetfulness to escape punishment and to protect their fellows. The public prosecutor maintained that the defendants had orchestrated in advance the evidence they were going to give. He explained that they must have known what was going to happen on the morning of November 19. His most convincing argument was that the failure of Municipal Employee Schlüter and his assistants to notice the presence of such a large gathering of people could not be explained other than that the farmers must have somehow hidden in Kock, Kühl and Holler's houses.[128] The court, however, acquitted those whose intentions to take part in the "disturbance" could not be determined.[129]

Whatever their intentions, all of the defendants disagreed that violence had been directed or was ever intended against Schlüter and his assistants.[130] In this regard, Sötje and Schlüter's testimonies aided the defendants, for they both said that it was possible they had been struck in the turmoil accidentally.[131] Once again, the court was obliged to give the farmers the benefit of doubt, unable to disprove the rather dubious claims that they had brought the pitchforks to put out the fires or simply for "effect".[132]

Though some defendants denied that they had intended to prevent the removal of the cattle, they nevertheless made it quite clear that they condemned the confiscations. Indeed, many of the farmers exploited the opportunity to speak in court in

---

125 Indictment issued by the Staatsanwalt Altona, 8.12.1928, GStA PK I. HA Rep. 84a No. 55364; issue of the verdict of the Schöffengericht in Itzehoe from 9-17 April 1929, LAS 309/22569.

126 *Itzehoer Nachrichten* No. 83, 10.4.1929, p. 4; *Husumer Nachrichten* No. 83, 10.4.1929, p. 5 and No. 86, 13.4.1929, p. 9.

127 Issue of the verdict of the Schöffengericht in Itzehoe from 9-17 April 1929, LAS 309/22569; *Itzehoer Nachrichten* No. 85, 12.4.1929, p. 3.

128 *Itzehoer Nachrichten* No. 89, 17.4.1929, p. 4.

129 *Itzehoer Nachrichten* No. 91, 19.4.1929, p. 3.

130 See the reports of the various testimonies in the *Itzehoer Nachrichten* No. 84, 11.4.1929, pp. 3-4 and No. 85, 12.4.1929, p. 3.

131 *Husumer Nachrichten* No. 86, 13.4.1929, p. 2; *Itzehoer Nachrichten* No. 87, 14.4.1929, p. 3.

132 *Husumer Nachrichten* No. 83, 10.4.1929, p. 5; *Itzehoer Nachrichten* No. 85, 12.4.1929, p. 3.

order to draw public attention to the crisis in agriculture. Most told of their own hardships and the authorities' insufficient understanding for their plight.[133]

Significantly, many farmers expressed their anger that Kock and Kühl's oxen had been confiscated before they had reached their prime weight and at a time prices were disadvantageous. Ignoring Koch and Kuhl's provocative actions, and forgetting that the county administrator had subsequently made concessions, they viewed the authorities' failure to postpone the confiscations until the animals were ready for slaughter as evidence, not just of the authorities' deficient agricultural knowledge, but also of indifference and even cold-heartedness.[134]

The presiding judge could not be accused of any such lack of understanding. He allowed the defendants to tell of their difficulties at length and even spoke to them in the local dialect during proceedings.[135] He was incensed, however, at the contemptuous behaviour of some of the defendants, who in protest against Weimar justice, read newspapers in the courtroom.[136] Two moments underlined that the farmers were united in their anger and convinced of the rectitude of their actions. The defendants and their sympathisers among the audience cheered when in the course of giving evidence Johannes Schade (Kathen) declared in a raised voice, "Not we, but the state belongs here in the dock!"[137] Later, they again burst into applause when a witness, asked whether Frauen, at the Itzehoe meeting, had said that the farmers in Lunden approved of the action at Beidenfleth, answered "The joy was universal."[138]

The public prosecutor ultimately helped the defence by acknowledging that the farmers' actions were a response to the crisis. However, he then argued that Kock and Kühl's actions could not solely be attributed to the pressure of excessively high taxes, and rebuked them for their failure to apply for deferment of payment, maintaining that they would have been granted alleviation had they done so.[139] Chief defence lawyer, Luetgebrune, Germany's best according to *Das Landvolk*,[140] in-

133 See the reports of the testimonies in the *Husumer Nachrichten* No. 84, 11.4.1929, p. 1 and p. 9, No. 85, 12.4.1929, pp. 1 and 5 and in the *Itzehoer Nachrichten* No. 84, 11.4.1929, p. 4 and No. 85, 12.4.1929, p. 3.

134 *Husumer Nachrichten* No. 84, 11.4.1929, pp. 1-2; *Itzehoer Nachrichten* No. 85, 12.4.1929, p. 3.

135 Vermerk IV f301 G. 6/29, GStA PK I. HA Rep. 84a No. 55362.

136 Landgerichtspräsident and Oberstaatsanwalt to the Oberlandesgerichtspräsidenten and Generalstaatsanwalt in Kiel concerning "Beidenflether Bauernprozeß" Altona, 7.5.1929 and report from the Landgerichtspräsident to the Preuß. Justizminister, 7.5.1929, both in GStA PK I. HA Rep. 84a No. 55362.

137 *Itzehoer Nachrichten* No. 85, 12.4.1929, p. 3; *Husumer Nachrichten* No. 86, 13.4.1929, p. 2.

138 *Husumer Nachrichten* No. 86, 13.4.1929, p. 2; *Itzehoer Nachrichten* No. 87, 14.4.1929, p. 4.

139 *Husumer Nachrichten* No. 88, 16.4.1929, p. 7; *Itzehoer Nachrichten* No. 89, 17.4.1929, p. 4.

140 *Das Landvolk* No. 35, 17.4.1929, p. 1.

sisted that the agricultural crisis, or at least the farmers' belief in the agricultural crisis, constituted a mitigating circumstance. Excessively high taxes, he argued, were the only point of contact between the farmers and the authorities. He attacked the prosecution on points of law and with patriotic sentiment, maintaining that the confiscations had not been carried out in accordance with procedure, and that the farmers had acted not to enrich themselves but to preserve their vocation and their "tribe".[141]

Albert Kühl (left) and Heinrich Kock pose with Walter Luetgebrune (middle) during the Beidenfleth trial in Itzehoe, April 1929[142]

Principally because they were deemed the "ring-leaders", Kock and Kühl were both sentenced to eight months' imprisonment. A further 23 farmers received six-month prison terms but were granted three years' probation on bail. The remainder were acquitted. Kelting and Frauen were acquitted because their calls to violent methods had been general and were not made in connection with any specific

141 *Husumer Nachrichten* No. 89, 17.4.1929, p. 5.

142 Photo: Stadtarchiv Neumünster.

laws.[143] On further charges concerning the sale of Kock's ox in Hamburg, Kock and Hansen were acquitted, but Hansen was fined RM 150 for threatening his community chairman after being informed that the bailiff might confiscate one of his animals for outstanding taxes.[144]

The 57 farmers on trial in Itzehoe for their participation in the Beidenfleth protest pose with Walter Luetgebrune (left of centre), April 1929[145]

The outcome was a victory for the Landvolkbewegung since the court recognised, and drew the public's attention to, Schleswig-Holstein's agricultural crisis. The court accepted that many farmers feared the loss of their farms should the crisis continue, and even that the farmers believed that no measures had been instituted to alleviate their difficulties.[146] Of course, the farmers were punished for breaking the law. In this respect, the provincial government welcomed the verdict, satisfied that the "ring-leaders", Kock and Kühl, were to be punished without the other defendants acquiring "martyr" status. Moreover, the provincial government

143 According to §110 of the *Reichsstrafgesetzbuch*, an infringement required that a specific law be breached. See L. Ebermayer, A. Lobe and W. Rosenberg, *Kommentar zum Reichs-Strafgesetzbuch* (Berlin and Leipzig, 1925), p. 384.

144 Issue of the verdict of the Schöffengericht in Itzehoe from 9-17 April 1929, sent from the Oberstaatsanwalt to the Regierungspräsident in Schleswig, Altona, 28.5.1929, LAS 309/22569. Also see the *Husumer Nachrichten* No. 90, 18.4.1929, p. 5.

145 Photo: Stadtarchiv Neumünster.

146 Issue of the verdict of the Itzehoe Schöffengericht from 9-17 April 1929, LAS 309/22569.

appeared relieved that Weschke, whom it deemed the architect of the disturbance, was among those who received a six month suspended term.[147]

Yet farmers throughout the province sympathised with the convicted.[148] As they left the town hall, girls presented them with bouquets tied with ribbons in the old imperial colours, and *Das Landvolk* objected bitterly to the "draconian" and "false" verdict, and claimed that in the "people's hearts" the convicted were not guilty.[149] From the beginning of the trial, *Das Landvolk* inverted prosecution and defence, presenting the "accused as accusers" against "the system" with its "dry legal paragraphs".[150] It printed numerous telegrams of support for the defendants from all over Germany and portrayed a friendless Republic which protected itself with the constabulary which "harassed" farmers "on pilgrimage" to Itzehoe; that is, farmers who wished to rally in the town in solidarity with their convicted colleagues.[151]

The Rural League supported the defendants too, explained their actions as driven by their despair and printed a letter praising the Landvolkbewegung for furthering the rural people's cause. However, the Rural League also dissociated itself from the movement, taking exception to claims that it was responsible for Beidenfleth.[152] The reports in its *Landbote* were consistent with previous sympathetic treatment of the Landvolkbewegung, but lacked the passion and verve necessary to benefit from the affair.

By contrast, the National Socialists were determined to use the trial for their own ends. The *Schleswig-Holsteinische Tageszeitung's* coverage of the trial was only slightly less extensive than that of *Das Landvolk* and it too reckoned the "verdict of the system" "draconian".[153] Like *Das Landvolk*, it attacked the authorities' lack of understanding of agriculture, although particular to its coverage was its admiration for the defendants as "splendid Nordic people" and "blond giants".[154] The NSDAP's sympathy for the defendants was unquestionable. It supported them as if their cause were its own, but where *Das Landvolk* wrote of a "Landvolk trial", the

147 Regierungspräsident to the Preuß. Minister des Innern in Berlin concerning "Aburteilung der Beidenflether Bauernunruhen" Schleswig, 22.4.1929, LAS 301/4695. Following further convictions, Weschke ultimately served part of his sentence. Decrees of the Landgericht, Altona 15.9.1931 and Altona 23.11.1931, and of the Schöffengericht, Husum 8.6.1932, all in GStA PK I. HA Rep. 84a No. 55362.

148 According to a report from the Landeskriminalpolizeistelle Flensburg to the Regierungspräsident in Schleswig concerning "Nothilfe" Flensburg, 22.4.1929, LAS 309/22921.

149 *Das Landvolk* No. 36, 18.4.1929, pp. 1-2.

150 *Das Landvolk* No. 27, 8.4.1929, p. 1 and No. 29, 10.4.1929, p. 1.

151 *Das Landvolk* No. 29, 10.4.1929, p. 1, No. 31, 12.4.1929, pp. 1-2, No. 35, 17.4.1929, p. 1 and No. 36, 18.4.1929, p. 2.

152 *Landbote für Schleswig-Holstein* No. 30, 18.4.1929, p. 1 and No. 31, 21.4.1929, p. 1.

153 *Schleswig-Holsteinische Tageszeitung* No. 90, 17.4.1929, p. 1 and No. 92, 19.4.1929, p. 1.

154 See, for example, the *Schleswig-Holsteinische Tageszeitung* No. 86, 12.4.1929, p. 3.

National Socialists did not mention the Landvolkbewegung at all.[155] The *Schleswig-Holsteinische Tageszeitung* did not report that the substantial police presence in Itzehoe was to prevent Landvolk demonstrations but associated it with the unjustifiable suppression of National Socialism. Once more, the NSDAP appealed to the rural people as if it were the only body representing them. Characteristically, the *Schleswig-Holsteinische Tageszeitung* concluded its coverage of the trial insisting that the NSDAP was the solution to the agricultural crisis and Germany's distress.[156]

The Beidenfleth trial generated one further notable offshoot. During the trial, Luetgebrune had requested that a letter from Regierungspräsident Abegg to the public prosecutor be read to the court. In it, Abegg had sought the acceleration of preliminary proceedings against Frauen and asked that they be so conducted that they would produce a "lasting effect against the spreading tax resistance movement" on the public.[157] As he continued to argue in his subsequent writings, Luetgebrune had hoped to show that Abegg wanted the trial conducted "in a particular manner" (*in besonderer Weise*).[158] However, the court rejected Luetgebrune's request on the grounds that the letter did not constitute evidence and was irrelevant to the trial's outcome.[159] It found Abegg had not intended to influence the verdict and that therefore his conduct did not constitute impermissible intervention in the judicial process.[160]

Nevertheless, one might be excused, as Bruno von Salomon ultimately was, in a subsequent defamation case, for believing it. Responding in the first place to a Prussian Press Service (*Amtlicher Preußischer Pressedienst*) report claiming *Das Landvolk's* treatment of the issue had been entirely tendentious, von Salomon accused Abegg of a cover up and insulted him.[161]

---

155 For an example in *Das Landvolk*, see No. 36, 18.4.1929, p. 2.

156 *Schleswig-Holsteinische Tageszeitung* No. 92, 19.4.1929, p. 2.

157 Regierungspräsident to the Oberstaatsanwalt in Altona, Schleswig, 4.12.1928, LAS 309/22569; Oberpräsident to the Preuß. Minister des Innern in Berlin, Kiel, 12.12.1928, GStA PK I. HA Rep. 84a No. 55362.

158 Luetgebrune, *Neu-Preußens Bauernkrieg*, pp. 46-55.

159 *Husumer Nachrichten* No. 87, 15.4.1929, p. 2 and No. 88, 16.4.1929, p. 7; report from the Regierungspräsident to the Preuß. Minister des Innern in Berlin concerning "Aburteilung der Beidenflether Bauernunruhen" Schleswig, 22.4.1929, LAS 301/4695.

160 Werner, *Die Prozesse gegen die Landvolkbewegung*, pp. 150-160.

161 Regierungspräsident to the Oberpräsident in Kiel concerning "Presseangriffe gegen mich anläßlich des Beidenflether Bauernprozesses" Schleswig, 27.4.1929, LAS 301/4695. In April 1930, the Schöffengericht in Itzehoe found von Salomon guilty of defamation and a breach of the press laws and fined RM 110. See the issue of the verdict of the Schöffengericht in Itzehoe from 4.4.1930, in both LAS 301/4697 and 309/22759. For the Amtlicher Preußischer Pressedienst's report from 16.4.1929 and Abegg's press release from 19.4.1929, see LAS 309/22579. The Pressedienst report dated Abegg's letter December 28, rather than December 4, and claimed the letter did not concern the Beidenfleth trial. For von Salomon's attacks on Abegg, see the articles "Wer lügt?" in *Das Landvolk* No. 36 from

If Abegg's letter to the public prosecutor be considered imprudent for the interfering impression it made on the farmers, then his response to von Salomon's accusations was a dreadful blunder. The Press Service's notice had contained several inaccuracies. In particular, its elucidation of Abegg's letter omitted his stated purpose, to produce a "lasting effect against the spreading tax resistance movement". A subsequent clarification from Abegg himself also neglected this critical sentence before the text was finally published in its entirety. Once the farmers became aware of the blatant omission of this sentence, they were convinced that their Regierungspräsident was dealing with them in an underhanded fashion and that he had indeed wished to influence proceedings in the "particular manner" Luetgebrune had supposed.

Following the Beidenfleth incident, the *Rendsburger Tageblatt* interpreted the unrest among the province's habitually peaceful farmers as evidence that the pressure of taxation had reached unbearable levels.[162] The local and county authorities, by contrast, suggested that Kock and Kühl had acted provocatively and maintained they were able but simply unwilling to pay their taxes.[163] They believed the theatrical mechanisms the farmers had employed, the forks, fire and fire horns as well as the engagement of a photographer, showed they had wished to create a sensation. Kock, moreover, had used the news that he had managed to recover and sell his ox himself – in the words of the County Administrator Göppert – to "make propaganda".[164] Presumably Göppert meant that Kock had boasted publicly that the action had been a success and that he had outfoxed the authorities.

The authorities, however, had been too hasty to dismiss their distress. Kühl's severe financial hardship, for example, was so evident that Göppert later sought to help him, asking the senior public prosecutor to postpone his prison sentence because of the pressing seasonal obligations of his farm.[165] The spectacle Kock and Kühl created at Beidenfleth was directed against the authorities but at its core were genuine economic grievances. Their actions stemmed from both financial distress

---

18.4.1929, p. 2, "Abegg macht einen Fluchtversuch" in *Das Landvolk* No. 39, 22.4.1929, p. 2, and, after Abegg's request was printed in the press in its entirety, "Herr Abegg gibt endlich zu" in *Das Landvolk* No. 40, 23.4.1929, p. 2.

162 *Rendsburger Tageblatt* No. 297, 6.12.1928, p. 5.

163 For the views of Community Chairman Mahlstedt, Community Association Chairman Augustin and County Administrator Göppert, see the *Nordischer Kurier* No. 286, 5.12.1928, p. 3 and the report on the events in Beidenfleth by Gemeindevorsteher Augustin from Borsfleth, 3.12.1928, LAS 301/4695.

164 Landrat des Kreises Steinburg to the Regierungspräsident in Schleswig concerning "Steuerverweigerungen" Itzehoe, 1.12.1928, in both LAS 301/4695 and 309/22569.

165 Letter from the Landrat Kreis Steinburg to the Oberstaatsanwalt in Altona, 1.9.1930, LAS 309/22590.

and political motivations. Indeed, this was also the way the director of the Chamber of Agriculture assessed refusals to pay the chamber levy.[166]

There can be no question that the Landvolkbewegung increasingly turned the rural people's distress over the agricultural crisis into a political struggle. Nevertheless, it is important not to overlook the farmers' deteriorating economic circumstances. In some cases, of course, their impoverishment was relative. According to one police report, the farmers' difficulties could not have been so severe if they were able to pay for motorcars.[167] Yet even in such cases, the fear of ruin remained while debts mounted and prices remained unstable. They did not have to be on the brink of ruin to fear the possible consequences of the crisis. As Luetgebrune argued at the Beidenfleth trial, and Jeremy Noakes underlined in his study of Lower Saxony, what is important is that the farmers considered themselves to be in distress.[168] Furthermore, the farmers' anxieties were heightened by their emotional attachment to their properties, a factor even the courts acknowledged, that is, that their "weal and woe" depended upon their farms, that their fields were not simply a livelihood, but "occupied all [their] thoughts and feelings".[169]

Farmers required little imagination to picture their ruin. On March 27, 1929, a farmer from Steinburg County committed suicide in a courtroom in Altona. His failure to pay outstanding taxes of RM 56 and unpaid interest on a loan of RM 5,000 had led to the foreclosure of the loan and subsequently to the compulsory auction of his farm. On March 27 he was to attend an appointment at the Altona court for the distribution of the outcome of the auction. Before proceedings began, however, he drew a revolver and shot himself in the back of the head.[170] Whatever other causes may have driven him to take his life, it is improbable that the province's farmers considered it as anything other than the corollary of his ruin. Certainly this is how it was portrayed in the press, the *Deutsche Zeitung* writing that he had preferred death than face the loss of the family farm.[171]

The farmers' distress, then, was partially a matter of perception. Regierungspräsident Abegg's response to the Landvolkbewegung, however, reflected his conviction that the farmers in the movement had greatly exaggerated their difficulties. He firmly believed the state was showing sufficient consideration for the farmers and focussed so strongly on the radical motivations of their obstruction that he made too little of their hardship.

---

166 Thyssen, *Bauer und Standesvertretung*, p. 250.

167 Report from the Landeskriminalpolizeiamt Berlin, Abteilung IA, Heide, 21.3.1929, LAS 309/22668.

168 *Husumer Nachrichten* No. 89, 17.4.1929, p. 5; Noakes, *The Nazi Party in Lower Saxony*, p. 111.

169 Issue of the verdict of the Landgericht in Kiel from June 27, 1929, p. 8, LAS 309/22668.

170 Copy of a report from the Polizeipräsident Altona-Wandsbek, Abteilung IA, to the Regierungspräsident in Schleswig concerning "Selbstmord des Landwirts Carl Gottlieb Fiebig aus Niendorf" Altona, 28.3.1929, LAS 309/22921.

171 *Deutsche Zeitung* No. 75, 29.3.1929, LAS 309/22921.

In April 1929 Abegg charged the province's county chairmen to report to him on the extent of compulsory auctions for outstanding taxes for the period April 1928 to March 1929. His explicit intention was to provide "numerical material" to combat the recurrent "false claim" that farmers had been subject to "particularly rigorous" collection in "very many cases".[172] A subsequent press release claimed that investigations had shown that taxes were not collected too "rigorously" and that only "malicious" tax debtors had faced confiscations.[173] However, the press release did not contain the numerical evidence which Abegg had so urgently demanded.

In fact, most of the county administrators' reports neither confirmed that taxes were collected too rigorously, nor did they provide proof that only "malicious" tax debtors had faced confiscations. In this respect, there were only isolated implications to give any support to Abegg's supposition. One community chairman in Bordesholm County had responded to the enquiries of his community administrator, informing him forthrightly that he did not carry out compulsory auctions because the farmers had contrived to boycott anyone who bid, and South Dithmarschen's administrator noted that he had heard complaints that farmers received too much protection. However, only the county administrator in Stormarn actually remarked that the communities did not pursue taxes rigorously.[174]

If the number of compulsory auctions is to be considered indicative of whether taxes were collected rigorously, as Abegg's request to his county administrators would have it, they did not warrant his claims in the press release. It could only have done so if very small. Yet the data, though incomplete, revealed a considerable number of compulsory auctions on account of public obligations for the period.[175] A hastily compiled table from the administrators' reports in the Regierungspräsident's files listed 92 compulsory auctions in the province for outstanding communal, state and Reich taxes and other public duties, of which eight had been moved to other locations while a further 24 had failed completely.[176] Moreover, the

172 Regierungspräsident to the Landräte (except for Helgoland), Schleswig, 18.4.1929, LAS 309/22590.

173 Press release from the Amtlicher Preußischer Pressedienst, 12.7.1929, LAS 309/22590 and see, for example, the *Itzehoer Nachrichten* No. 164, 16.7.1929, p. 5.

174 Reports to the Regierungspräsident in Schleswig from the Landrat concerning "Zwangsversteigerungen bei Landwirten" Bordesholm, 25.4.1929, from the Landrat concerning "Zwangsversteigerungen" Heide, 24.4.1929 and the Landrat concerning "Zwangsversteigerung von Gegenständen bei Landwirten wegen rückständiger Steuern" Wandsbek, 27.4.1929, all in LAS 309/22590.

175 For the limitations and difficulties in completing the enquiries within the time-period the Regierungspräsident had demanded, see the report from the Landrat to the Regierungspräsident in Schleswig concerning "Zwangsweise Beitreibung öffentlicher Abgaben" Ratzeburg i/Lbg, 23.4.1929 and a record simply titled "Niederschrift", both in LAS 309/22590.

176 See the undated and untitled hand-written table that follows the county administrators' reports in LAS 309/22590.

county administrators' reports actually reveal that circumstances were much worse. Abegg had only enquired into the number of compulsory auctions. He had not taken into account the confiscation of milk, garnishment from dairy earnings and the sale of other produce as well as confiscations which had impelled farmers to pay their taxes subsequently in order to avoid yet more costly arrears. He had also overlooked confiscations which had not yet proceeded to compulsory auction, delayed as reckoned likely to fail in the face of the farmers' opposition.[177]

According to a note in the files, Abegg justified not revealing the data for fear of endangering informants.[178] This appears unlikely since it had not concerned him when he requested the information. He almost certainly held the data back because it did not further his purpose. Though the investigations did not prove that taxes were collected too rigorously, the findings should have encouraged Abegg at least to reconsider the farmers' complaints. Abegg, however, was obstinate in his reluctance to take their difficulties seriously. Instead, he was much more concerned with promoting the appearance that the state was dealing with them appropriately.[179] Intent upon suppressing the Landvolkbewegung, he deceived the public. But without evidence, his claim that only malicious tax debtors had faced confiscations was unlikely to pacify the farmers. It could not inspire their confidence that the state would treat them justly should they too find themselves unable to pay their taxes.

Of course the Landvolk leaders' radicalism was politically driven, but Abegg seemed unwilling to accept that their distress, too, was genuine. Unbending, he instigated a further investigation in the summer of 1929, this time into the financial circumstances of the Landvolk leaders. Again he was looking for evidence to use against the rebellious farmers, perhaps in the form of another press release, to show that their dissent was unwarranted and merely political. But again the reports he received disappointed his expectations. Though some of the reports mentioned poor management of their farms, and in one case even the expense of journeys undertaken for the Landvolkbewegung, they revealed that many of the Landvolk's lead-

177 See the reports to the Regierungspräsident in Schleswig from the Landrat concerning "Zwangsversteigerungen" Tönning, 24.4.1929, from the Landrat concerning "Zwangsversteigerungen" Heide, 24.4.1929, from the Landrat concerning "Zwangsversteigerungen bei Landwirten wegen Steuerrückstände" Husum, 25.4.1929, from the Landrat concerning "Beitreibung von Steuern" Rendsburg, 24.4.1929, from the Landrat concerning "Zwangsversteigerung wegen Steuerrückstände" Itzehoe, 20.4.1929, from the Landrat concerning "Zwangsversteigerung von Gegenständen bei Landwirten" Pinneberg, 24.4.1929 and from the Landrat to the Regierungspräsident in Schleswig concerning "Zwangsversteigerung von Gegenständen die bei Landwirten wegen rückständiger Steuern gepfändet waren" Oldenburg in Holstein, 25.4.1929, all in LAS 309/22590.

178 See the note in the Regierungspräsident's files titled "Niederschrift", LAS 309/22590.

179 Also see here Abegg's instructions in July 1929 to the county administrators to ensure that the press in general was informed of the "facts" about compulsory enforcement proceedings before publishing reports: Regierungspräsident to the Landräte and Polizeipräsidenten in Kiel and Altona, Schleswig, 14.7.1929, LAS 309/22784.

ers were truly in considerable difficulty or had at least suffered significant financial losses.[180]

Contrary to the authorities' expectations, the Landvolkbewegung continued to attract large numbers of farmers to its actions and meetings in 1929. By the end of March two dozen compulsory auctions had been prevented and numerous others relocated. The movement spread into other counties, Segeberg, Plön and Oldenburg, and the readership of its newspaper, *Das Landvolk*, increased rapidly. Just a month after it began to appear daily, its circulation was already set to climb over seven thousand copies.

Rather than calm the rural population, the first trials involving the Landvolkbewegung simply furthered the radicalisation in the province. Inside the courtrooms the accused gave voice to their grievances and attacked the authorities and the system. Outside, they incited additional meetings and marches of solidarity. The trials and these further incidents, which increasingly involved the police, compounded the division between the farmers and the Prussian authorities, but at the same time strengthened the identity and cohesion of the movement in the absence of organisational structure.

The founding of *Das Landvolk* had similar consequences, giving the movement a centre and amplifying the farmers' attacks against finance capital, Jews, parliamentary democracy, the pro-Republican parties and the authorities considered adherent to "the system". It is, nevertheless, important to note that the Landvolkbewegung was already radicalising before *Das Landvolk* began its vicious propaganda. Trade and agricultural policy had increasingly fallen into the background toward the end of 1928, as hostility to the Republic grew. *Das Landvolk* intensified this shift but did not induce it. This further radicalisation was facilitated by the small number of individuals who joined the movement from outside agriculture. The Landvolkbewegung, however, did not become their instrument as Hans Beyer

180 Regierungspräsident to the Landräte and Polizeipräsidenten in Kiel and Altona, Schleswig, 14.7.1929 and the resultant reports from the Landrat to the Regierungspräsident in Schleswig concerning "Wirtschaftliche Lage von Führern der Landvolkbewegung" Bordesholm, 17.8.1929, from the Landrat des Kreises Eckernförde to the Regierungspräsident in Schleswig concerning "Wirtschaftliche Lage von Führern der Landvolkbewegung" Eckernförde, 24.7.1929, the Amtsvorsteher to the Landrat in Heide concerning "Wirtschaftliche Lage von Führern der Landvolkbewegung" Lunden, 23.7.1929, from the Landrat to the Regierungspräsident in Schleswig concerning "Landvolkbewegung" Plön, 25.7.1929, from the Landrat to the Regierungspräsident in Schleswig concerning "Landvolkführer" Rendsburg, 6.8.1929, from the Landrat to the Regierungspräsident in Schleswig concerning "Führer in der Landvolkbewegung" Bad Segeberg, 1.8.1929, from the Landrat des Kreises Eiderstedt to the Regierungspräsident in Schleswig concerning "Wirtschaftliche Schwierigkeiten der Landvolkführer" Tönning, 19.7.1929, all in LAS 309/22784.

later claimed.[181] It remained a rural people's movement. It was farmers who ran and participated in its meetings and actions, and Hamkens and Heim and other farming figures who continued to provide leadership.[182]

The agricultural associations and their tortuous endeavours to merge in the "unity organisation" appeared languid in contrast to the Landvolk's actions and boisterous meetings. In the Landvolk's wake, they attempted to use its agitation and activism, pointing to the unrest in the countryside to give weight to their calls for assistance. Though the Landvolk also sought to unite the rural people, it did not seek to draw them away from the associations. The associations scarcely affected the considerably more dynamic movement's course.

In much the same way, other right-wing groups, principally the Tannenbergbund and the NSDAP, were responding to the Landvolk's radicalisation of the countryside and seeking to benefit from it. The NSDAP operated with greater reference to the Landvolkbewegung than vice versa. As a force for all nationalist groups, the Landvolkbewegung sought friendly relations with the NSDAP. For the party which had already viewed the Landvolkbewegung as a rival, however, the founding of *Das Landvolk* posed further competition. Much more than the agricultural associations, the NSDAP vigorously schemed to profit from the Landvolkbewegung and the radical atmosphere it had largely created, criticising the movement, while all the time sympathising with the farmers and expanding its rural propaganda where the Landvolk agitated.

Police intervention and legal action were the authorities' principal means of dealing with the Landvolk's insubordination. The provincial government's priority was to maintain law and order. Abegg rightly recognised the Landvolkbewegung's provocative nature but, seeing only its maliciousness, neglected the farmers' distress. His undertakings – the regulation of the tax commissions as well as the investigations of the extent of compulsory auctions and the Landvolk leaders' financial circumstances – were intended to demonstrate that the province was dealing with agriculture and the movement appropriately. Abegg knew very well that his claims were not substantiated but simply ignored the results of the investigations hoping to bolster the state's authority. Yet his neglect of the farmers' distress was part of the problem. In fact, his conduct makes it difficult to dismiss the Landvolkbewegung's criticism that the authorities lacked understanding and sympathy for agriculture.

It appears that the provincial government also failed to comprehend that the farmers' despair and their turn to "self-help" were exacerbated by their feeling that circumstances were beyond their control. Abegg might have encouraged their cooperation by involving them in the process of decision-making, and the tax com-

---

181 Beyer, 'Die Landvolkbewegung Schleswig-Holsteins und Niedersachsens 1928-1932', pp. 189-109; similarly, Coppin, *Landvolkbewegung und Nationalsozialismus im schleswig-holsteinischen Marschgebiet*, p. 125.

182 See, for example, the report from the Landeskriminalpolizeiamt titled "Bericht" Altona, 7.10.1929, LAS 309/22784.

missions presented such an opportunity. This is not to suggest that Abegg could have allowed the farmers free rein. Preventing their influence, however, could only heighten their feeling of helplessness and alienation, and widen the perceived gap between the farmers and authorities. It was clear within the first months of 1929 that one-sided endeavours to suppress the Landvolkbewegung had had little effect. Moreover, there was nothing to suggest that the continued pursuit of this strategy would prevent the future intensification of the conflict between the farmers and authorities.

Chapter Five

Further growth and growing extremism

Continued growth and increasing militancy characterised the Landvolkbewegung's development from the spring of 1929. In their endeavours to increase the movement's appeal, Landvolk leaders held countless meetings throughout the province. Additionally, they founded "Crisis Assistance" (*Nothilfe*) committees in many counties. These both promoted tax obstruction and challenged the authority of the provincial administration. Despite its expansion, however, the Landvolk refused to adopt customary organisational structures. Moreover, further militant actions were not always directly related to the agricultural crisis, rather defiance itself appeared to be ever more their essential purpose. Noteworthy for its carnival aspect and its increasing belligerence, the Landvolk's militancy produced fresh sources of conflict with the authorities as well as heightened tensions with the NSDAP.

The most important single factor driving the Landvolkbewegung's growth was the many meetings it held throughout the province and which typically attracted audiences of 300 to 400 people.[1] The Landvolk's first meetings in the winter of 1928/1929 had been more or less spontaneous affairs, usually held before or after confiscations and compulsory auctions. From the spring of 1929, however, more and more meetings were planned events to promote the movement. Chiefly for this reason, but also, perhaps, because Landvolk leaders had gradually grown familiar with their roles, the meetings and the speeches settled into a routine, acquiring greater method both in structure and content. At least three speakers spoke at Landvolk meetings. The meetings' convener briefly welcomed the audience and introduced the main speaker. This was usually Hamkens whose speeches lasted up to two hours.[2] After Hamkens, shorter speeches from others such as Kühl, Weschke, Wulf or Ross followed.

The speeches Hamkens and other main speakers delivered from the spring of 1929 and into 1930 were by no means the same but they invariably contained the same key elements and they soon followed a routine which changed little thereafter.[3] The centrepiece of Hamkens' speeches was an account of developments up

1 See, for example, the reports on meetings in Heide on May 18, 1929, in Preetz on May 25, 1929, in Kaltenkirchen on June 1, in Stenderupau on August 21, 1929 and in Gelting on August 24, 1929. See fn. 3 below.

2 See, for example, *Schleswiger Nachrichten* No. 290, 11.12.1929, p. 2; *Das Landvolk* No. 230, 25.12.1929, p. 9.

3 For Hamkens' speeches and those of his colleagues, see the following: report on a meeting in Kellinghusen in the *Nordischer Kurier* No. 30, 5.2.1929, p. 7; reports on a unity organisation meeting in Marne, in the *Marner Zeitung* No. 73, 27.3.1929 and No. 74, 28.3.1929, LAS 301/4695; the report titled "Bericht über die im Heider Anzeiger angekündigte 'Versammlung des notleidenden Landvolks in Stadt und Land' am Sonnabend, den 18. Mai

to the agricultural crisis, or as he put it, of the enslavement of the farmers and the German people. While he sometimes referred to the fighting "spirit of 1914" as a model for unity and camaraderie, he did not blame England, France or Russia for

1929 um 9 ½ Uhr im Stadttheater in Heide" 19.5.1929, sent from the Landrat to the Regierungspräsident in Schleswig, Heide, 23.5.1929, LAS 301/4695; report on the meeting in Neukrug on 7.6.1929 in the *Flensburger Nachrichten* No. 132, 8.6.1929, p. 7; report on a meeting in Kiel from the Polizeipräsident to the Oberpräsident in Kiel concerning "Landvolkbewegung, Versammlung am 27. Juni d. Js. in der 'Deutschen Wacht' " Kiel, 1.7.1929, in both LAS 301/4695 and 309/22668; report on a meeting in Satrup in the *Schleswiger Nachrichten* No. 154, 4.7.1929, p. 8; report on a meeting in Husby on 4.7.1929 in *Der Schleswiger* No. 156, 5.7.1929, LAS 301/4695; report on a meeting in Wesermünde (Hanover) from the Landeskriminalpolizeistelle, Abteilung IA, concerning "Landvolk-Kundgebung" Wesermünde, 5.8.1929, LAS 309/22668; report on a meeting in Stenderupau on August 22, 1929, sent from the Landeskriminalpolizeistelle Flensburg to the Regierungspräsident in Schleswig, Flensburg, 22.8.1929, LAS 309/22668; report on a meeting in Gelting on August 24, 1929 titled "Bericht" Flensburg, 25.8.1929, in both LAS 301/4696 and 309/22668; report on a meeting in Bad Segeberg in the *Segeberger Kreis- und Tageblatt*, No. 199, 26.8.1929, in both LAS 301/4696 and 309/22668; report titled "Bericht über die am 2. Okt. 1929, 11 Uhr vorm. in Burg i/Dith. stattgefundene Versammlung der Landvolkbewegung" Altona, 3.10.1929, in both LAS 301/4696 and 309/22784; report from the Landeskriminalpolizeistelle Altona, Abteilung IA, titled "Bericht über eine Landvolkversammlung in Wrist" Itzehoe, 18.10.1929, in both LAS 301/4696 and 309/22784; report on a meeting in Schafflund on 19.10.1929 from the Polizeipräsident to the Regierungspräsident concerning "Landvolkbewegung" Flensburg, 20.10.1929, LAS 309/22784; report from the Landeskriminalpolizeistelle, Abteilung IA, titled "Bericht" concerning "Landvolkkundgebung am 30.10.29 in Leezen" Kiel, 31.10.1929, LAS 309/22783; report on a meeting in Wesselburen from the Landrat to the Regierungspräsident in Schleswig concerning "Landvolkbewegung" Heide, 4.11.1929, LAS 309/22590; transcript from the Landesthing from the Polizeiverwaltung in Rendsburg titled "Landesthing des Schleswig-Holsteinischen Landvolkes, am 10. November 1929 in Rendsburg im 'Schützenhof' nachmittags 3 Uhr", in both LAS 301/4696 and 309/22590; for meetings in Burg on Fehmarn and in Oldenburg in Holstein, see the report from the Polizeiverwaltung, Burg a/F, 7.12.1929 and from the Polizeiverwaltung to the Landrat concerning "Landvolkversammlung in Oldenburg am 5. ds. Mts" Oldenburg in Holstein, 6.12.1929, both in LAS 309/22783; for meetings in Glücksburg and Süderbrarup, see the report from the Polizeipräsident, Landeskriminalpolizeistelle, concerning "Landvolkversammlungen" Flensburg, 11.12.1929, in both LAS 301/4697 and 309/22783; for a meeting in Glückstadt, see the reports titled "Bericht" concerning "Versammlung des Landvolkbundes im hiesigen Lokal Franscher Garten" Glückstadt, 14.12.1929 and from the Landjägerei Kreis Steinburg titled "Bericht über die Landvolkversammlung in Glückstadt am 13.12.1929 im Lokal Englischer Garten" Itzehoe, 15.12.1929, both in LAS 301/4697 as well as 309/22783; reports on the meeting in Niebüll on 4.1.1930 from the Kreisleiter der Landjägerei des Kreises Südtondern sent to the Landrat in Niebüll, Niebüll, 6.1.1930 and on the meeting in Leck from the Landjägereiposten Klixbüll, Kreis Südtondern, sent to the Landrat in Niebüll, Klixbüll, 6.1.1930, both in LAS 301/4697 as well as 309/22783; copies of a report from the Kriminal- und Grenzkommissariat titled "Bericht" Niebüll, 6.1.1930 and of a report titled "Bericht über die am 4.1.1930 in Niebüll stattgefundene Landvolkversammlung", both in LAS 309/22590.

the war and Germany's defeat, nor did he at all consider the complex relationship between Germany's need for capital, the international capital market and allied debts for Germany's financial difficulties. Instead, he constructed a story of the crisis, building on earlier anti-Semitic conceptions and fears which had been developed and passed on by the radical right, and especially the Rural League and DNVP.[4] For Hamkens, "international powers" or "international capital" – the latter was interchangeable with the term "international Jewish capital" – were the enemies of agriculture and the nation. In 1918 their puppets, the revolutionary forces, had given Germany a "stab in the back" and caused defeat. Then, Jewish money and the Jewish press had given rise to the parliamentary system. The many parties did not work for the people but simply squabbled among themselves and were played off against one another by international forces. In the meantime, they robbed the people of their property. They accomplished this by way of inflation, international treaties and trade agreements, Germany's dependence on "Jewish capital" and Erzberger's taxation system. As a result, the farmers' capital was under attack with the intention of destroying agriculture and, therefore, the German people. After January 28, 1928 the farmers had expected Berlin to help, but "the system", Hamkens maintained, could not rescue Germany. Those who held office, in particular the Regierungspräsident, the Oberpräsident and "foreign county administrators", were the instruments of international capital. Organisations were tied to the system and so they bickered among themselves as did the right-wing organisations. Having thus reduced the farmers' predicament to a struggle between the Landvolk and the international powers, the "Jewish parliamentary system" and its "entourage", Hamkens then called for unity, for the audience to support the Landvolkbewegung, to protect their capital, reject all "positive cooperation" with the system and combat it with all legal means.

Hamkens and others' speeches continued a trend that had already begun to develop within the first months of the movement's emergence. Discussion of the crisis itself, of the farmers' hardships and desired changes to government policy, increasingly made way for anti-system and anti-Semitic agitation.[5] Wulf's opening words at a meeting in November in 1929 were illustrative in this respect, "To tell you here about the great crisis, I think, is superfluous. What is important for us is to recognise the causes of this system and, after these have been identified, to eliminate them."[6] This is not to suggest that calls for tax obstruction decreased. Indeed speakers continued to harness the tax issue to incite ill-feeling against the authorities and the Weimar system.

---

4 For the development of these conceptions in rural circles, including their roots before the First World War, see chapter 1, especially pp. 34-36.

5 See chapter 3.

6 Wulf quoted in the transcript of the Landesthing from the Polizeiverwaltung in Rendsburg titled "Landesthing des Schleswig-Holsteinischen Landvolkes, am 10. November 1929 in Rendsburg im 'Schützenhof' nachmittags 3 Uhr", in both LAS 301/4696 and 309/22590.

During the course of his speeches, Hamkens encouraged his fellow farmers not to pay taxes from their capital. He did so using irony, a rhetorical device which he also used to great effect in disparaging the authorities.[7] For example, he mocked Prussian Agricultural Minister Steiger's career advancement from a teacher in the province at the agricultural school in Hohenwestedt, declaring, "I do not wish to insult Mr. Steiger. I will just say, Schleswig-Holstein's farmers were too dumb and he was too clever."[8] After his initial overt calls for tax obstruction had led to trouble with the law, in December 1928, Hamkens typically told audiences to pay taxes and to keep paying them until they bled white, so that the ministers in Berlin would continue to receive their salaries. The farmers knew very well that Hamkens was being ironic and they showed their appreciation and approval with laughter, applause and by stamping their feet on the ground.[9] All the same, to ensure that he had made himself clear, Hamkens sometimes reiterated quite candidly, albeit more gingerly in order to avoid prosecution for inciting tax-strikes, that he had always argued that taxes should be paid, but only for "our people" and not for "international Jewry".[10]

After Hamkens, further speakers repeated the denunciations of international capital and the system as well as calls to protect the farmers' assets. Usually they also reported on the movement's recent developments, such as actions at confiscations or trials against Landvolk followers, and they frequently complained about their unjust treatment at the hands of the law.[11] Toward the close of the meeting either they or Hamkens also condemned the "Jewish influence" on the press, called on the audience to read "their" newspaper, *Das Landvolk,* and appealed to them to lend the movement financial support.

The near standardisation of Landvolk speeches as outlined here was evident from the spring of 1929 and one could almost say they were regimented by October, when guidelines were provided by *Das Landvolk* for speakers at gatherings in

---

7 For Hamkens' first calls to tax-strikes, see chapter 3.

8 Report from the Polizeipräsident, Landeskriminalpolizeistelle, to the Regierungspräsident in Schleswig concerning "Landvolkversammlung in Flensburg" Flensburg, 24.2.1930 and *Der Schleswiger* No. 46, 24.2.1930, both in LAS 309/22783.

9 Issue of the verdict of the Schöffengericht in Itzehoe from 9.8.1930, p. 58; report from the Preuß. Minister des Innern to the Oberpräsident in Kiel "Auf die Berichte vom 10./11. Juni 1929" Berlin, 22.6.1929, both in LAS 301/4700.

10 See, for instance, the *Marner Zeitung* No. 74, 28.3.1929, LAS 301/4695.

11 For further examples, see the report sent to the Landrat in Bad Segeberg from the Landjägeramt Nahe, Kreis Segeberg, Nahe, 3.6.1929, in both LAS 301/4695 and 309/22668 and the *Kaltenkirchener Zeitung* No. 127, 3.6.1929, p. 3; for a meeting in Krempe, see *Das Landvolk* No. 222, 15.12.1929, p. 6; for a meeting in Leck on 5.1.1930, see the report from the Landjägereiposten Klixbüll, Kreis Südtondern, sent to the Landrat in Niebüll, Klixbüll, 6.1.1930, in both LAS 301/4697 and 309/22783.

smaller areas.[12] Nor did their repetitive nature escape the police who monitored them. By the beginning of 1930, a police official in Flensburg saved himself the trouble of providing a comprehensive report by summarising the speeches as the "known" and "usual" talk of the Landvolk's struggle against the "Jewish-democratic-parliamentary system"[13]

Issues of the day, such as the rallying of support for Hugenberg's referendum against the Young Plan and criticism of the Chamber of Agriculture's poor financial management of a major agricultural show, did provide additional thematic variation.[14] Nevertheless, the Landvolkbewegung's core ideas proved steadfast. The only really noteworthy development after the spring of 1929 was an increase in völkisch thinking, that is, racially exclusive nationalism. Although present from the beginning, the expression of völkisch thought grew considerably from the autumn of 1929 and this was clearly marked by a major Landvolk rally, the "Landesthing", in Rendsburg in November. Several thousand farmers attended this symbolic gathering of the tribes as a court of "free Germans", as opposed to the ministers and public servants who did not all have German interests at heart and who cooperated with a corrupt system thrust upon Germany by the rule of Jewish finance capital.[15]

---

12 See the instructions for leaders of Landvolk meetings titled "Anweisungen für die Versammlungsleiter der Landvolkbewegung" sent to the Oberpräsident in Kiel from the Polizeipräsident Altona-Wandsbek, Altona, 7.10.1929, LAS 301/4696.

13 See the reports from the Polizeipräsident, Landeskriminalpolizeistelle, to the Regierungspräsident in Schleswig concerning "Landvolkversammlung in Niebüll" Flensburg, 10.1.1930, concerning "Landvolkversammlung in Flensburg" Flensburg, 24.2.1930 and concerning "Landvolkversammlung" Flensburg, 3.3.1930, all in LAS 309/22783. For a similar assessment, also see the report from the Polizeipräsident to the Polizeipräsident, Landeskriminalpolizeiamt in Berlin, concerning "Landvolkbewegung" Kiel, 16.1.1930, LAS 309/22783.

14 For Hamkens and Kühl's criticisms of the agricultural show, the "GROLA", see their speeches at meetings in Bad Segeberg in August 1929 and in Satrup and Neukrug in May 1930 and in Bordesholm in July 1930 as reported in the *Segeberger Kreis- und Tageblatt* No. 199, 26.8.1929, LAS 301/4696; the reports from the Polizeipräsident, Landeskriminalpolizeistelle, to the Regierungspräsident in Schleswig concerning "Bauernversammlung in Satrup" Flensburg, 16.5.1930, from the Polizeipräsident, Landeskriminalpolizeistelle, to the Regierungspräsident in Schleswig concerning "Landvolkversammlung in Neukrug" Flensburg, 24.5.1930 and from the Landjägereiamt Bordesholm to the Landjägereiabteilung Bordesholm, Bordesholm, 27.7.1930 as well as the pamphlet distributed at these meetings titled "Der Grolaschwindel", all in LAS 309/22590. For Kühl's criticisms of the Young Plan and his promotion of Hugenberg's referendum, see, for example, the report on a meeting in Gelting on August 24, 1929 titled "Bericht" Flensburg, 25.8.1929, in both LAS 301/4696 and 309/22668; the report on a meeting in Bad Segeberg in the *Segeberger Kreis- und Tageblatt*, No. 199, 26.8.1929, in both LAS 301/4696 and 309/22668 and the report from the Landeskriminalpolizeistelle Altona, Abteilung IA, titled "Bericht über eine Landvolkversammlung in Wrist" Itzehoe, 18.10.1929, in both LAS 301/4696 and 309/22784.

15 For the considerable effort made to promote this meeting, see, for example, *Das Landvolk* No. 175, 20.10.1929, p. 9, No. 176, 22.10.1929, p. 1, No. 188, 5.11.1929, p. 1, No. 190,

Provincial writers had accentuated the special importance of farmers and rural life since the 1890s and the ideal of a *Volksgemeinschaft*, a "national community", had made its way into the programme of the Landespartei in the early 1920s.[16] Völkisch ideas, therefore, had been circulating in the province for some time. Nonetheless, it was two individuals who took prominent roles in the Landvolkbewegung from the autumn of 1929 who were very much responsible for introducing their more extreme variants from this time. The first was Hugo Freiherr von Lamezan, a Stahlhelm leader from County Segeberg. His idea of a Volksgemeinschaft reflected Chamberlain's understanding of history as racial struggle, and he extolled the sanctity of marriage, condemned abortion and called for a higher birthrate in the interests of national prosperity.[17]

> "[The] Volksgemeinschaft is nothing other than a fighting community with the ultimate aim of collectively fighting for our individual families' existence against other peoples. .... The current system negates the struggle for the people's existence against other peoples. It negates the German military spirit, the spirits of a Bismarck, of a Frederick the Great, of a Luther, ... The system negates the spirit which has led the German people through the millennia ..."[18]

The second was August Georg Kenstler, one of the principal leaders of the Artamanen.[19] From Thuringia, Kenstler expounded on völkisch ideas for a special section of *Das Landvolk* titled "Blood and Soil" (*Blut und Boden*) in 1930.[20] He also put his own journal of the same name at the Landvolkbewegung's service, and

---

7.11.1929, pp. 7 and 9, No. 191, 8.11.1929, pp. 1 and 3. For attention in other regional newspapers, see, for example, *Der Schleswiger* No. 240, 23.10.1929 in LAS 301/4697. For the meeting, see the report titled "Landvolkkundgebung am 10.11.29 in Rendsburg" from the Polizeiverwaltung, Rendsburg, 12.11.1929, in both LAS 301/4696 and 309/22590; *Das Landvolk* No. 194, 12.11.1929, pp. 1-2; *Nordischer Kurier* No. 266, 12.11.1929, p. 7 and Schimmelreiter, *Unter der schwarzen Bauernfahne*, pp. 44-46.

16 See chapter 1, pp. 23 and 28-29.

17 See, for example, the report from the Landjägerei-Kreisleiter Kreis Steinburg titled "Bericht über die Landvolkversammlung in Glückstadt am 13.12.1929 im Lokal Englischer Garten" Itzehoe, 15.12.1929, in both LAS 301/4697 and 309/22783. For a similar warning, see the flyer titled "An die Front" authorised by Johannes Kühl and distributed in various areas of the province at the beginning of 1930. See the flyer and the report from the Regierungspräsident to the Preuß. Minister des Innern concerning "Flugblatt der Landvolkbewegung" Schleswig, 9.1.1930, in both LAS 301/4697 and 309/22783.

18 Von Lamezan quoted in the transcript of the Landesthing from the Polizeiverwaltung in Rendsburg titled "Landesthing des Schleswig-Holsteinischen Landvolkes, am 10. November 1929 in Rendsburg im 'Schützenhof' nachmittags 3 Uhr", in both LAS 301/4696 and 309/22590.

19 For the Artam movement, see chapter 4, p. 132, fn. 80.

20 For examples of the "Blut und Boden" column, see *Das Landvolk* No. 82, 6.4.1930, p. 11, No. 88, 13.4.1930, p. 9, No. 93, 20.4.1930, p. 9, No. 104, 5.5.1930, p. 11.

Landvolk activists subsequently contributed many articles to it.[21] Though not the only Landvolk leader to employ racism[22], others rarely argued like Kenstler along explicitly biological, pseudo-scientific lines. Kenstler maintained, for example, that cultural decline resulted from the weakening of the blood of the culture's creators and bearers. He called for "natural" racial divisions of society rather than class divisions and he lauded Hans F. K. Günther, whose differentiation between Aryans and Jews later helped provide the theoretical framework for National Socialist racial policy.[23]

While Kenstler and von Lamezan were prepared to take völkisch thought further, its actual appeal and function were generally much simpler. Völkisch thought allowed Landvolk leaders to sidestep debate about agriculture's financial sustainability and international competitiveness and furnished them with apparently selfless grounds for pursuing the preservation of farming property. Sublimating their resistance to tax and interest payments to the defence of their farms, to the defence of their ancestral fields, to the defence of the nation against international capital and the system, they presented their struggle as serving the national interest. They argued that agriculture was the basis of the economy, that German property required German ownership if German interests were to be served, that Germany's survival depended on their survival.[24] At their most fervent, they described this struggle as "a holy war for blood and soil".[25]

---

21 Kenstler corresponded with Bruno von Salomon from the summer of 1929, organised for Hamkens to speak in Silesia and soon began to contribute to *Das Landvolk*. An editorial position was kept open for him for several months but the authorities prevented him from assuming it, deporting him from Prussia as a "troublesome foreigner" because he was a Hungarian citizen. Letters from Kenstler to B. v. Salomon in June and July 1929 were listed in the inventory of confiscated material titled "III. Bei Bruno von Salomon (Wohnung) am 10.9.1929 beschlagnahmtes Material" in both LAS 301/4704 and 309/22784; report from the Landeskriminalpolizeistelle Altona, Abteilung IA, titled "Bericht über die Landvolkbewegung" Altona, 5.10.1929 in both LAS 301/4696 and 309/22784; Regierungspräsident to the Landräte, Polizeipräsidenten in Kiel, Altona, Flensburg and the Polizeiverwaltung in Neumünster, Schleswig, 24.10.1929 in LAS 309/22784; reports from the Landeskriminalpolizeistelle Altona, Abteilung IA, concerning "Landvolkbewegung" Itzehoe, 6.11.1929 and Itzehoe, 14.11.1929, both in LAS 301/4697 as well as 309/22783 and concerning "Landvolkbewegung" Altona, 8.1.1930, LAS 301/4697.

22 For complaints about marriages as a "racial disgrace", see *Das Landvolk* No. 4, 5.1.1929, p. 9 and Reimers (Bleckede) quoted at a meeting in Soltau in the *Soltauer Nachrichten* No. 279, 27.11.1929, LAS 301/4696.

23 *Das Landvolk* No. 3, 4.1.1930, p. 7.

24 See Hamkens' speech in Niebüll on 6.1.1930 quoted in *Das Landvolk* No. 15, 7.1.1930, p. 8; *Das Landvolk* No. 40, 16.2.1930, p. 1; Hartmut Plaas' flyer "Erweitet den Kampf! Ein Vorschlag" from mid-1930, appended to the report from the Landeskriminalpolizeistelle, Abteilung IA, titled "Bericht" Altona, 8.6.1930, in both LAS 301/4697 and 309/22590.

25 Schimmelreiter, *Unter der schwarzen Bauernfahne*, p. 42.

Völkisch thought also helped to succour a feeling of common fate, beyond misfortune, and to provide a sense of belonging and unity that had been absent among the agricultural associations. Defining German identity by blood and the bond to the soil, the Landvolk argued that its followers shared the inheritance of thousands of years of common ancestry.[26] Though this excluded all others as foreigners, the racial definition itself was used to extend invitations to join the movement beyond rural circles.[27] Combined with regional patriotism and occasionally with anti-urban sentiment, or at least an estimation of healthy rural above unwholesome and "rootless" urban life, völkisch thought also gave Schleswig-Holstein's farmers a special position on account of the purity of the "Nordic blood" of its old tribes.[28] Völkisch thought perceived farmers as occupying an almost sacred place in German society and accorded their occupation a right to special entitlement.

Yet the Landvolk did not simply conduct meetings and encourage a continuation of earlier strategies. In the spring of 1929 it devised a new institution, the "Crisis Assistance" committees. Conceived as a practical response to the crisis and, in particular, the compulsory payment of outstanding taxes, their purpose was to represent citizens in their dealings with the authorities and financial institutions.[29] In effect, these supra-party bodies sought to institutionalise resistance to confiscations and compulsory auctions. Further, because they assumed an official character, they threatened to undermine the structures and operation of the province's administration.

There is no evidence to indicate where the Landvolk found inspiration for the Crisis Assistance. Twice Hamkens had already initiated the creation of committees of "trusted men", in Eiderstedt in December and in Husum in February.[30] To put these into practice and establish others elsewhere, meetings were called for March 4, 1929. Following careful planning[31] and extensive advertising coordinated by

26 For examples, see *Das Landvolk* No. 153, 25.9.1929, p. 2 and No. 160, 11.7.1930, p. 1.

27 See, for example, Johannes Kühl's speech in Glückstadt in December 1929: report titled "Bericht" concerning "Versammlung des Landvolkbundes im hiesigen Lokal Franscher Garten" Glückstadt, 14.12.1929, in both LAS 301/4697 and 309/22783. Also see *Das Landvolk* which explained that only the "pure blood, racially proud Nordic" could be a "bearer" of the Landvolkbewegung: *Das Landvolk* No. 151, 22.9.1929, p. 1.

28 See, for example, *Das Landvolk* No. 124, 22.8.1929, p. 1, No. 153, 25.9.1929, p. 2, No. 215, 7.12.1929, p. 1 and No. 233, 31.12.1929, p. 1.

29 Issue of the verdict of the Schöffengericht in Itzehoe, 9.8.1930, pp. 10-11, LAS 301/4700.

30 Committees had been appointed at the meeting after the obstruction of the auction in Garding on December 20 and at the meeting following the gathering at the Husum finance office on February 7. For the former, see chapter 3. For the latter, see chapter 4; *Nordischer Kurier* No. 33, 8.2.1929, p. 7; *Der Schleswiger* No. 33, 8.2.1929, LAS 301/4695.

31 For the careful preparations for the meetings, see the issue of the verdict of the Schöffengericht in Itzehoe, 9.8.1930, pp. 10-11 and 24-25, LAS 301/4700; meeting transcript from the Kriminal- u. Grenzkommissariat to the Landrat in Niebüll, Niebüll, 4.3.1929, LAS 301/4695; reports from the Landrat to the Regierungspräsident in Schleswig concerning

Landvolk leaders across the province[32], as many as five thousand people attended the meetings in Itzehoe, Tönning, Leck, Niebüll, Eckernförde, Rendsburg, Husum, Bordesholm and on the island of Fehmarn.[33] The fact that women did not attend, although they attended other Landvolk meetings, was an indication of the gravity the movement's leaders accorded this project.[34] That they considered the Crisis Assistance committees men's business highlighted their rebellious intentions, and suggested that some form of confrontation with the authorities was anticipated.

"Kundgebung der Notbewegung" Husum, 5.3.1929, from the Landeskriminalpolizeistelle Flensburg to the Regierungspräsident in Schleswig concerning "Landvolkbewegung" Flensburg, 6.3.1929 and from the Landrat to the Regierungspräsident in Schleswig concerning "Aufruf der Interessenvertretung des notleidenden Volkes des Kreises Steinburg" Itzehoe, 16.3.1929, all in LAS 309/22921.

32 See the advertisements of the "German crisis movement" in *Das Landvolk* No. 9, 1.3.1929, p. 3 and No. 10, 2.3.1929, p. 5; *Itzehoer Nachrichten* No. 52, 2.3.1929, p. 8; *Lecker Anzeiger* No. 51, 1.3.1929, p. 4 and No. 52, 2.3.1929, p. 4; *Nordischer Kurier* No. 52, 2.3.1929, p. 12; *Rendsburger Tageblatt* No. 51, 1.3.1929, p. 9; *Wilstersche Zeitung* No. 51, 1.3.1929, p. 3.

33 Attendances ranged from 100 in Eckernförde to 2,000 in Itzehoe: report from the Regierungspräsident to the Oberpräsident in Kiel concerning "Landvolkversammlungen am 4.3.1929" Schleswig, 20.3.1929, LAS 301/4695; *Eckernförder Zeitung* No. 54, 5.3.1929, p. 2. The total here is an estimate, since no figure was given for the meeting on Fehmarn. Stoltenberg noted 12 meetings but this figure was too high. See Stoltenberg, *Politische Strömungen im schleswig-holsteinischen Landvolk*, p. 130. An undated report titled "Zusammenfassung der Berichte der Landräte Niebüll, Husum, Tönning, Eckernförde, Rendsburg und Bordesholm über die Landvolkkundgebungen am 4.3.1929", LAS 309/22921 and the *Schleswig-Holsteinische Tageszeitung* No. 55, 7.3.1929, p. 5 both recorded ten meetings. In sum, nine locations were actually named in the provincial authorities' records, see LAS 301/4695, LAS 309/22668 and 309/22921. The same number were named in the reports in *Das Landvolk* No. 12, 5.3.1929, p. 5, No. 13, 6.3.1929, p. 6, No. 14, 7.3.1929, p. 6 and No. 16, 9.3.1929, p. 6.

34 The absence of women at the Crisis Assistance meeting in Husum was significant enough for it to be noted in the police report titled "Bericht über die öffentliche Versammlung der Notbewegung am 4. März 1929" and dated Husum 4.3.1929, LAS 309/22921. Wulf opened his speech in Rendsburg with "Meine Herren" rather than "Meine Damen und Herren" suggesting that only men were in attendance in Rendsburg also. See the report titled "Willenskundgebung zur Selbsthilfe im Schützenhof zu Rendsburg am 4.3.1929" Rendsburg, 4.3.1929, in both LAS 309/22921 and 301/4695. For the attendance of women at other Landvolk meetings, see, for example, the report from the Landrat to the Regierungspräsident in Schleswig concerning "Auftreten Hamkens – Tetenbüll" Plön, 12.11.1929, in both LAS 301/4696 and 309/22783; the report from the Landjägereikreisleiter, Kreis Steinburg, titled "Bericht über die Landvolkversammlung in Krempe am 12.d.Mts" Itzehoe, 14.12.1929, in both LAS 301/4697 and 309/22783; the copy of the report titled "Bericht über die am 4.1.1930 in Niebüll stattgefundene Landvolkversammlung", LAS 309/22590; Landrat to the Regierungspräsident in Schleswig concerning "Auflösung einer Versammlung" Heide, 24.6.1930, LAS 301/4698.

Held concurrently in nine different locations, the meetings required an unusually large number of speakers. For this reason, they provided a unique insight, showing that it was not just the Landvolkbewegung's principal speakers, like Hamkens, but leaders everywhere who adhered to a common body of ideas. All Crisis Assistance meetings included a speech or series of speeches, in which the speakers combined immediate concerns reflecting common rural grievances – claims of the tax commissions' failure to represent agriculture, criticisms of excessive rateable values and "extortionate" (*Wucher*) rates of interest – with the customary criticisms of the parliamentary system and denunciations of "international capital", "international powers" or "dark powers" as accountable for the farmers' plight.[35] Anti-Semitism featured in both the expressions of immediate grievances and in the accounts of the causes of Germany's misfortune, for example, in a passing reference to those who sought to take advantage of the crisis by purchasing farming properties below value, and in more substantial charges of sinister Jewish schemes to exploit Germany.[36] Some speakers also harnessed provincial patriotism, that is they pointed to the region's history of freedom and independence and its traditions, language and "blood" both to explain and promote the movement's exemplary unity as well as the strength of its struggle against "the system".[37]

The speeches were followed by the reading of resolutions and the establishment of the committees. The six resolutions presented at all of the meetings were: 1) the establishment of Crisis Assistance committees, 2) the publishing of names in *Das Landvolk* of those who assisted the authorities in confiscations or purchased confiscated items, 3) the advance payment and transfer of dairy earnings so the

35 Meeting transcript from the Kriminal- u. Grenzkommissariat to the Landrat in Niebüll, Niebüll, 4.3.1929 and the report titled "Willenskundgebung zur Selbsthilfe im Schützenhof zu Rendsburg am 4.3.1929" Rendsburg, 4.3.1929, both in LAS 301/4695 as well as 309/22921; *Das Landvolk* No. 12, 5.3.1929, p. 5, No. 13, 6.3.1929, p. 6, No. 14, 7.3.1929, p. 6 and No. 16, 9.3.1929, p. 6; *Eckernförder Zeitung* No. 54, 5.3.1929, p. 2; *Eiderstedter Nachrichten* No. 8222, 7.3.1929, p. 2; *Husumer Nachrichten* No. 54, 5.3.1929, pp. 7-8; *Itzehoer Nachrichten* No. 55, 6.3.1929, pp. 3-4; *Kieler Neueste Nachrichten* No. 55, 6.3.1929, p. 10; *Lecker Anzeiger* No. 54, 5.3.1929, p. 3; *Nordischer Kurier* No. 54, 5.3.1929, p. 5 and No. 55, 6.3.1929, pp. 5-6; *Rendsburger Tageblatt* No. 54, 5.3.1929, p. 8 and No. 55, 6.3.1929, p. 5; *Schleswiger Nachrichten* No. 54, 5.3.1929, p. 3; *Schleswig-Holsteinische Tageszeitung* No. 54, 6.3.1929, p. 4; *Wilstersche Zeitung* No. 54, 5.3.1929, pp. 2-3.

36 See, for example, the speeches at meetings in Itzehoe, Eckernförde, Niebüll as well as anti-Semitic cries from the audience in Rendsburg: *Das Landvolk* No. 12, 5.3.1929, p. 5; *Eckernförder Zeitung* No. 54, 5.3.1929, p. 2; meeting transcript from the Kriminal- u. Grenzkommissariat to the Landrat in Niebüll, Niebüll 4.3.1929, in both LAS 301/4695 and 309/22921; report titled "Willenskundgebung zur Selbsthilfe im Schützenhof zu Rendsburg am 4.3.1929" Rendsburg, 4.3.1929, in both LAS 301/4695 and 309/22921.

37 See the speeches from Muthmann in Eckernförde in the *Eckernförder Zeitung* No. 54, 5.3.1929, p. 2 and Stadtler, the Stahlhelm functionary and guest speaker in Itzehoe, *Das Landvolk* No. 12, 5.3.1929, p. 5.

authorities would not be able to confiscate them, 4) the prosecution of authorities who breached Article 164 of the Reich Constitution by the deprivation of the middle class, 5) the collection of tax and confiscation notifications in sacks for their return to the finance office, and 6) the rejection of all communal budgets and for the Crisis Assistance committees to superintend these.[38]

Whether read out in the form of catechisms, as was the case in Niebüll, or distributed among members of the audience and then subsequently submitted to create the appearance that the audience had devised the proposals itself, as in Rendsburg, the presentation of the resolutions was contrived in order to maximise consensus.[39] In Rendsburg, even a proviso prohibiting Jews and Freemasons from joining the committees came in the form of a cry from the audience, though at other meetings, it was speakers who told the audience that Jews and Freemasons could not join.[40] Votes on the resolutions were not held. Determined to avoid procedure resembling democratic-parliamentary deliberation and compromise, the Landvolk leaders sought to secure a unified resolve by simply asking audiences whether there were any objections to the resolutions or to raise their right hands to sanction them. As no objections were voiced, they were passed unanimously everywhere.[41]

Although it is clear that there was enthusiasm for the speeches, it is difficult to assess the audiences' support for the committees when no active commitment, such as membership or signatures, was required from them.[42] The collection of taxation documentation in sacks might have offered some indication, but the authorities

---

38 Report from the Regierungspräsident to the Oberpräsident in Kiel concerning "Landvolkversammlungen am 4.3.1929" Schleswig, 20.3.1929, LAS 301/4695; undated report titled "Zusammenfassung der Berichte der Landräte Niebüll, Husum, Tönning, Eckernförde, Rendsburg und Bordesholm über die Landvolkkundgebungen am 4.3.1929", LAS 309/22921.

39 See the meeting transcript from the Kriminal- u. Grenzkommissariat to the Landrat in Niebüll, Niebüll 4.3.1929, in both LAS 301/4695 and 309/22921; report from the Polizeiverwaltung to the Regierungspräsident in Schleswig, Rendsburg, 5.3.1929, LAS 309/22921.

40 This proviso was recorded at four of the meetings and possibly made at others. See the meeting transcript from the Polizeiverwaltung Rendsburg titled "Willenskundgebung zur Selbsthilfe im Schützenhof zu Rendsburg am 4.3.1929" Rendsburg, 4.3.1929; meeting transcript from the Kriminal- u. Grenzkommissariat to the Landrat in Niebüll, Niebüll, 4.3.1929, both in LAS 301/4695 as well as 309/22921; *Eckernförder Zeitung* No. 54, 5.3.1929, p. 2.

41 Report from the Regierungspräsident to the Oberpräsident in Kiel concerning "Landvolkversammlungen am 4.3.1929" Schleswig, 20.3.1929, LAS 301/4695; undated report titled "Zusammenfassung der Berichte der Landräte Niebüll, Husum, Tönning, Eckernförde, Rendsburg und Bordesholm über die Landvolkkundgebungen am 4.3.1929", LAS 309/22921.

42 For the strong approval, for example, of Muthmann's speech in Eckernförde, see the *Eckernförder Zeitung* No. 54, 5.3.1929, p. 2.

could not determine the extent to which audiences took part in this.[43] The way the proposals were presented did not encourage dissent, nor did the risk of condemnation from the meetings' leaders and fellow farmers. This did not mean that audiences were completely unwilling to voice criticism. Interrupting proceedings in Itzehoe, one member of the audience complained that the RM 1 admission fee was unreasonable in the prevailing economic circumstances and that hundreds of farmers who had wished to attend the meeting had been forced to remain outside.[44]

Only in Leck, however, did a member of the audience actually voice objections to the resolutions. Though only speaking out after they had been passed, Hogrefe, a leader of the nationalist paramilitary organisation the *Jungdeutscher Orden*, argued that the only solution to the crisis was to replace the state with a "people's state" (*Volksstaat*). This was the Jungdeutscher Orden's comparatively clearly-defined concept of a corporative state, free from party politics.[45] The suggestion met with a cry of disapproval from the audience and boisterous disorder ensued with cries for and against Hogrefe continuing his speech.[46] Yet Hogrefe's objections did not so much undermine support for the Crisis Assistance committee in Leck, as simply destroy the high level of solidarity at the meeting. The meeting's leaders had carefully crafted an atmosphere of unity, rallying the audience in their common concerns about the crisis and in anger toward the system.[47] Though evidently in agreement that Weimar was incapable of providing answers to their grievances, Hogrefe's objections showed that the rural people were divided when it came to finding a solution to their problems.

The Crisis Assistance committees not only posed a more prolific organisation of resistance to confiscations and compulsory auctions, they also constituted an attempt to gain greater control over local and provincial affairs. The resolutions concerned taxation, confiscations and communal budgets, but prior to presenting these, some speakers voiced insurrectionary aspirations. Though equivocal, they indicated that the committees might supplement or supplant the local and provincial authorities. Claiming that Rendsburg's county administrator had said that there was no longer a state, merely an "administrative apparatus" (*Verwaltungsapparat*), they

43 See the police report titled "Bericht über die öffentliche Versammlung der Notbewegung am 4. März 1929" Husum, 4.3.1929; report from the Landrat des Kreises Eiderstedt to the Regierungspräsident in Schleswig concerning "Notbewegung" Tönning, 6.3.1929, both in LAS 309/22921; *Das Landvolk* No. 12, 5.3.1929, p. 5; *Eckernförder Zeitung* No. 54, 5.3.1929, p. 2; *Rendsburger Tageblatt* No. 55, 6.3.1929, p. 5.

44 *Schleswig-Holsteinische Tageszeitung* No. 54, 6.3.1929, p. 4.

45 For the Jungdeutscher Orden's concept of "Volksstaat", see Diehl, *Paramilitary Politics in Weimar Germany*, pp. 225-227.

46 *Lecker Anzeiger* No. 54, 5.3.1929, p. 2.

47 Meeting transcript from the Kriminal- u. Grenzkommissariat to the Landrat in Niebüll, Niebüll, 4.3.1929, in both LAS 301/4695 and 309/22921.

said that the rural people, therefore, had to govern themselves.[48] In Itzehoe the newly formed committee went so far as to announce that it, and not the authorities, was now to be followed.[49] The Landvolk's leaders, it seemed, had grander ambitions for the Crisis Assistance committees than the resolutions allowed.

At the same time, or so Hamkens insisted, the movement was to proceed using "legal means".[50] Purportedly, the Crisis Assistance was legal, because it was based on the first article of the constitution, that "All power comes from the people."[51] However, this attempt to give the Crisis Assistance a semblance of constitutional legitimacy was questionable when the Landvolkbewegung was so expressly anti-Republican. Since its claim to represent "the people" was tenuous, the reference to the constitution, alongside the criticisms of the "administrative apparatus", amounted to little more than a provocative criticism of the authorities' failure to represent the people.

The resolutions of the Crisis Assistance meetings, the aspirations to self-government, the constitutional pretensions and illegal strategies were a bewildering mix. It is debatable whether the Landvolk's adoption of article 164 of the constitution, which safeguarded the "independent middle classes" from "overburdening and deprivation"[52] gave any validation to the return of tax notifications to the finance office, or amounted to an inconsistency, that is, arbitrary adherence to the laws. Yet in the absence of any explanation, say, that the use of legal means was to fight the Republic with its own weapons, or that "the system" might be transformed from within, audiences had good cause to be baffled by speakers who attacked "the system" so viciously and then drew on the very constitution upon which "the system" was based. It is, therefore, quite appropriate that Eiderstedt's administrator reckoned the resolutions "completely contradictory" (*völlig widersinnig*).[53] Less tactfully, the socialist *Schleswig-Holsteinische Volkszeitung* derided what it called "the new government of the völkisch Rural Leaguers" with its premier (*Ministerpräsi-*

48 Meeting transcript from the Polizeiverwaltung Rendsburg titled "Willenskundgebung zur Selbsthilfe im Schützenhof zu Rendsburg am 4.3.1929" Rendsburg, 4.3.1929; meeting transcript from the Kriminal- u. Grenzkommissariat to the Landrat in Niebüll, Niebüll, 4.3.1929, both in LAS 301/4695 as well as 309/22921; *Das Landvolk* No. 12, 5.3.1929, pp. 5-6; *Rendsburger Tageblatt* No. 55, 6.3.1929, p. 5; *Wilstersche Zeitung* No. 54, 5.3.1929, p. 3.

49 Issue of the verdict of the Schöffengericht in Itzehoe from 9.8.1930, p. 12, LAS 301/4700.

50 Report from the Landrat des Kreises Eiderstedt to the Regierungspräsident in Schleswig concerning "Notbewegung" Tönning, 6.3.1929, LAS 309/22921.

51 See Petersen's speech in Niebüll and the Steinburg committee's public declaration: meeting transcript from the Kriminal- u. Grenzkommissariat to the Landrat in Niebüll, Niebüll, 4.3.1929, in both LAS 301/4695 and 309/22921; *Das Landvolk* No. 17, 11.3.1929, p. 5.

52 'Die Verfassung des Deutschen Reichs vom 11. August 1919' in E. R. Huber (ed.), *Dokumente zur deutschen Verfassungsgeschichte,* vol. 4 (Stuttgart, Berlin and Cologne, 1992), p. 176.

53 Report from the Landrat des Kreises Eiderstedt to the Regierungspräsident in Schleswig concerning "Notbewegung" Tönning, 6.3.1929, LAS 309/22921.

dent), Hamkens. The milder spring weather, it scoffed, had failed to thaw out many people's brains.[54]

The Landvolkbewegung's insurgent ambitions crystallised over the following days. First, *Das Landvolk*, on March 6, referred to the Crisis Assistance committees as "self-government".[55] Then, on March 11, *Das Landvolk* published a proclamation from the committee in Steinburg informing the county's "suffering people" that it had been established to represent their interests. It declared that if the "administrative apparatus" issued orders without the committee's approval it would be in breach of the first article of the Reich Constitution. Administrative bodies which presumed to disregard the "people's decision" would be "put in their place" and members of the tax commissions and local authorities who violated the decisions of March 4 were to be reported to the committee.[56]

The Steinburg, South Tondern and Eiderstedt committees subsequently notified Schleswig-Holstein's Oberpräsident, Regierungspräsident, county administrators, land registry offices and finance office that the "suffering people" had instated them as their sole representatives. Claiming the first article of the Reich constitution as their sanction, the committees required the administrative bodies to approach them if they wished to issue orders.[57]

These declarations proscribed the local and provincial authorities from carrying out their executive functions without the committees' permission. They did not go so far as to seek legislative power. The Crisis Assistance committees, then, were not actually an attempt to overthrow "the system" or authorities. Rather, they were to function alongside the existing bodies, as an "additional government" (*Nebenregierung*).[58]

---

54 *Schleswig-Holsteinische Volkszeitung* No. 55, 6.3.1929, p. 7.

55 *Das Landvolk* No. 13, 6.3.1929, p. 6.

56 *Das Landvolk* No. 17, 11.3.1929, p. 5. The committees in South Tondern and Eiderstedt echoed this declaration. See the issue of the verdict of the Schöffengericht in Itzehoe from 9.8.1930, pp. 18 and 23, and the *Westküste* 19.3.1929, both in LAS 301/4700.

57 Declarations from the "Nothilfe des Kreises Eiderstedt" to the "Oberpräsidium (Verwaltungsapparat zu Händen des Herrn Oberpräsidenten Kürbis) in Kiel" Garding, 13.3.1929, from "Die Nothilfe des Kreises Südtondern" Leck, 13.3.1929 and from "Die Nothilfe des Kreises Steinburg" Itzehoe, 14.3.1929, all in LAS 301/4695; declarations from the Nothilfe des Kreises Steinburg to the "Verwaltungsapparat des Landratsamtes Itzehoe" Itzehoe, 14.3.1929, from "Die Nothilfe des Kreises Steinburg" to the "Verwaltungsapparat Regierungspräsident Abegg, Schleswig" Itzehoe, 14.3.1929, from "Die Nothilfe des Kreises Südtondern" to the "Verwaltungsapparat der Provinz Herrn Regierungspräsidenten Abegg in Schleswig" Leck, 13.3.1929 and to the "Verwaltungsapparat des Kreises Herrn Landrat Skalweit in Niebüll" Leck, 13.3.1929 and, also see, the report from the Landrat des Kreises Eiderstedt to the Oberstaatsanwalt in Flensburg concerning "Nothilfe" Tönning, 25.3.1929, all in LAS 309/22921.

58 Report from the Regierungspräsident to the Preuß. Minister des Innern concerning "Politische Lage im Reg. Bez. Schleswig" Schleswig, 5.6.1929, LAS 309/22668. Also see *Das Landvolk* No. 1, 3.1.1929, p. 3.

After such bold beginnings, it appeared the Crisis Assistance committees would significantly facilitate the Landvolkbewegung's expansion. In respect of tax obstruction, it did help to consolidate the movement's position. In Eiderstedt, for example, it was associated with a dangerous increase in "tax sabotage".[59] As "additional governments", however, it was only the Steinburg and Eiderstedt committees which persisted, and then only for a short time, with frivolous attempts to assert authority, while other committees took no further such action. The Steinburg committee issued a notice reprimanding the deployment of 100 riot police in Altona and demanding that the Crisis Assistance committee was to be consulted before any such measures were taken.[60] From Eiderstedt, Hamkens wrote to the Chamber of Agriculture, forbidding further confiscations and the compulsory collection of taxes from "suffering farmers".[61] Soon, however, even the Steinburg and Eiderstedt committees faltered, and for reasons now to be discussed, the Crisis Assistance withdrew from public activity.

The Crisis Assistance's prospects as an "additional government" were poor from the beginning. Certainly they might have proven stronger had the committees in other counties joined them. Yet leaders in other counties, if entertaining political ambitions in the first place, may well have felt they could not count on their followers in such an undertaking: there were many among the Landvolk's own supporters who – though they liked the idea – doubted that the committees were workable.[62] Little consideration had been given to the committees' operations. Even some committee members were not entirely sure as to how they were to function; for instance, it was not even clear whether the collected tax notifications were simply to be returned en masse in a demonstration of the farmers' incapacity to pay or sent back with annotations, advising the finance office where reductions were necessary.[63] Haphazard, even in this respect, it is difficult to imagine the Crisis Assistance committees gaining and maintaining executive power.

A further weakness was that the committees were limited vocationally. Nominally, the Crisis Assistance was to represent "all in distress" (*alle Bedrängten*).[64] In

---

59 Preußisches Katasteramt Tönning to the Regierungspräsident in Schleswig concerning "Landvolkbewegung im Kreise Eiderstedt" Tönning, 16.3.1929, LAS 309/22921.

60 Issue of the verdict of the Schöffengericht in Itzehoe from 9.8.1930, p. 17, LAS 301/4700.

61 For Hamkens' letter to the Chamber of Agriculture, see the issue of the verdict of the Schöffengericht in Itzehoe from 9.8.1930, p. 19, LAS 301/4700.

62 For example, in Steinburg and Eckernförde. See the issue of the verdict of the Schöffengericht in Itzehoe from 9.8.1930, pp. 13 and 25, LAS 301/4700; *Eckernförder Zeitung* No. 54, 5.3.1929, p. 2.

63 Issue of the verdict of the Schöffengericht in Itzehoe from 9.8.1930, pp. 12-13, LAS 301/4700.

64 Report from the Regierungspräsident to the Oberpräsident in Kiel concerning "Landvolkversammlungen am 4.3.1929" Schleswig, 20.3.1929, LAS 301/4695; undated report titled "Zusammenfassung der Berichte der Landräte Niebüll, Husum, Tönning, Eckernförde,

reality, they largely represented farmers, and it was farmers who constituted the great majority of the committee representatives. Steinburg's county administrator was convinced that the inclusion of members from other occupations was merely intended to lend the movement the appearance of a universal crisis movement.[65] In Eiderstedt, by contrast, the appointment to the committee of the chairmen of the Business Association and of the Tradesman Association appeared more genuine, since these organisations had actually resolved to send them to the meetings.[66] Nevertheless, Eiderstedt's administrator also considered this a contrived measure on the part of the Landvolkbewegung to make it appear that the movement not only encompassed farmers but also those working in business and trade.[67]

The Landvolkbewegung's failure to mobilise a genuine cross-section of rural society points to the relatively greater susceptibility to radicalism among farmers than other occupations in the spring of 1929. Most shopkeepers and tradesmen rejected the Crisis Assistance committees, as they considered the Landvolkbewegung too radical.[68] Few even attended the Crisis Assistance meetings.[69] Quite simply, they had little cause to do so. Although shopkeepers and tradesmen had increasingly begun to suffer under deteriorating economic circumstances, their distress was more recent compared with those working in agriculture. In February they had held their own Reich-wide rallies to draw attention to their difficulties, but theirs were calls for assistance and reform much like the farmers' earlier rallies of 1928. They were not like the Landvolk's contemporaneous protests directed against "the system".[70]

Bolstering the farmers' readiness to explore alternative forms of association was the disarray of their organisations, which had persisted since January 1928. The townspeople had no such inducement, nor did the farmers supply others. The townspeople's sympathy for farmers had declined soon after the rallies of January 28, 1928 because agriculture had put its interests first, insisting on their primary importance for the economy.[71] At the Crisis Assistance meetings, the speeches and

---

Rendsburg und Bordesholm über die Landvolkkundgebungen am 4.3.1929", LAS 309/22921; *Rendsburger Tageblatt* No. 54, 5.3.1929, p. 8.

65 Landrat des Kreises Steinburg to the Regierungspräsident in Schleswig concerning "Aufruf der Interessenvertretung des notleidenden Volkes des Kreises Steinburg" Itzehoe, 16.3.1929, LAS 309/22921.

66 Issue of the verdict of the Schöffengericht in Itzehoe from 9.8.1930, p. 17, LAS 301/4700.

67 Report from the Landrat des Kreises Eiderstedt to the Regierungspräsident in Schleswig concerning "Notbewegung" Tönning, 6.3.1929, LAS 309/22921.

68 Wulf, *Die politische Haltung des schleswig-holsteinischen Handwerks*, pp. 72-75.

69 Issue of the verdict of the Schöffengericht in Itzehoe from 9.8.1930, p. 17, LAS 301/4700. Also see the *Lecker Anzeiger* No. 54, 5.3.1929, p. 3; *Itzehoer Nachrichten* No. 55, 6.3.1929, p. 4.

70 See the demands voiced at the rallies as reported in the *Heider Anzeiger* No. 40, 16.2.1929, p. 4.

71 See chapter 2, p. 62.

resolutions indicated that the townspeople's involvement would essentially be perfunctory since they did not articulate their specific concerns. Once again it appears the farmers considered the townspeople's interests secondary. Where workers, shopkeepers or tradesmen had been arranged to speak, it was generally only to declare that solidarity with the farmers was essential.[72]

Though limited regionally and vocationally, poorly organised and with no common resolve, it was not only internal weaknesses but also external factors which inhibited the Crisis Assistance's development. One incident, in particular, diverted the rural population's interest from the Landvolkbewegung to the NSDAP, a clash between the SA and communists in Wöhrden, which resulted in the deaths of two young SA men and a communist. As later became an established party tactic, the NSDAP skilfully exploited the funerals for the purpose of publicity.[73] Six thousand people, including uniformed SA units, a brass band and a drummer corps, heard Hitler and other senior party leaders speak at the burials. Hitler's appearance, in particular, made a strong impression on the rural people.[74]

The burials were conducted in an atmosphere of considerable agitation. Both National Socialist and Landvolk cries – "Heil!", "Germany awake!" and "Rather dead than a slave!" – were to be heard at the graves, while the appearance of police was met with angry shouts of "Prussian-Socialist Police" and demands for explanations for their absence in Wöhrden. Some local farmers were so enraged by their presence that storm troopers had to drag them into the nearest pub to avoid conflict. Afterwards speeches were made at gatherings in all the local pubs. There appears to have been a considerable entangling of the two movements. Some farmers called for the SA to help protect them from confiscations; some farmers called for the NSDAP to spread from village to village. Others, however, demanded civil war, arguing that waiting for the NSDAP to take control of the parliaments would take too long.[75]

During the following weeks the NSDAP accrued almost two thousand new members, nearly a quarter of its membership in the province at that time.[76] In Dithmarschen entire villages were reported to have joined or at least sympathised with the party, and its agitators were also better received in regions where they had not previously enjoyed support, such as in Husum County.[77] The actual extent to

72 See p. 164, fn. 35 above.
73 See reports in the *Schleswig-Holsteinische Tageszeitung* No. 61, 14.3.1929, p. 1 and No. 62, 15.3.1929, p. 1.
74 Report from the Landeskriminalpolizeiamt Berlin concerning "Ermittelungen und Beobachtungen in den Kreisen Norder- und Süderdithmarschen" Heide, 16.3.1929, LAS 309/22668.
75 Report from the Landeskriminalpolizeiamt Berlin concerning "Ermittelungen und Beobachtungen in den Kreisen Norder- und Süderdithmarschen" Heide, 16.3.1929, LAS 309/22668.
76 See the report from the Landeskriminalpolizeistelle Flensburg to the Regierungspräsident in Schleswig concerning "NSDAP" Flensburg, 25.4.1929, LAS 309/22998.
77 See the report from the Landeskriminalpolizeistelle Flensburg concerning "Die im Kreise Husum geführten Ermittelungen" Flensburg, 24.3.1929, the addendum to it titled "Zusatz-

which the Wöhrden incident and the NSDAP's resultant gains limited the potential development of the Crisis Assistance cannot be measured satisfactorily. Their impact, however, should not be underestimated, for among the party's new members were Landvolk followers, including many residual members of the movement's Defence Association. According to one police report, the forty-strong Büsum group joined the party as one.[78] Another police report noted that Defence Association members had gone over to the NSDAP after the NSDAP had continually criticised it.[79] The Defence Association's dissolution at the beginning of March, therefore, came at a propitious moment for the NSDAP, but in coinciding with the great sympathy the Wöhrden incident generated for the party, was unfavourable for the spread of the Landvolk's Crisis Assistance.

The authorities also played a significant role in containing the Crisis Assistance. Absurd as the Landvolk's notifications were, the authorities could not predict their implications and soon showed their resolve to suppress the committees.[80] The Wöhrden incident provided a welcome opportunity to prohibit all public processions as well as meetings of the National Socialists and Landvolkbewegung.[81] This was a much greater hindrance for the Landvolk than the NSDAP since the latter were able to use the funerals of the SA men to circumvent the ban. That the communists were initially spared from the ban strongly suggests that the provincial authorities were motivated more by political considerations – a wish to restrain the radical right – than concern for public safety. The omission incensed the greater part of the west coast population, not least because it was the communists who had attacked the National Socialists.[82]

In view of the lenient treatment of the communists, the ban on Landvolkbewegung meetings might be described as unfair. Yet it was not completely unjustified. The *Schleswig-Holsteinische Tageszeitung* emphasised that the NSDAP had nothing to do with the Landvolkbewegung, that the Oberpräsident had banned

---

Bericht" Flensburg, 27.3.1929 and the reports from the Landeskriminalpolizeiamt Berlin, Abteilung IA, Heide, 21.3.1929, Itzehoe, 26.3.1929 and Berlin, 2.4.1929, all in LAS 309/22668.

78 Report from the Landeskriminalpolizeiamt Berlin, Heide, 21.3.1929, LAS 309/22668.

79 Report from the Landeskriminalpolizeiamt Berlin, 2.4.1929, in both LAS 301/4695 and 309/22668.

80 Report from the Landrat des Kreises Eiderstedt to the Regierungspräsident in Schleswig concerning "'Nothilfe' im Kreise Eiderstedt" Tönning, 15.3.1929, LAS 309/22921.

81 Script of a telegram from the Regierungspräsident, Schleswig, 9.3.1929, appended to the report from the Regierungspräsident to the Oberpräsident in Kiel concerning "Versammlung und Umzugsverbot aus Anlaß der Vorgänge in Wöhrden" Schleswig, 9.3.1929, LAS 301/4695. For the ban on *Das Landvolk*, see chapter 4.

82 *Das Landvolk* No. 18, 12.3.1929, p. 2 and No. 19, 13.3.1929, p. 1; *Schleswig-Holsteinische Tageszeitung* No. 60, 13.3.1929, p. 1; *Heider Anzeiger* No. 11.3.1929, p. 1 and No. 62, 14.3.1929, p. 1. For further complaints in the nationalist press, see *Der Tag* No. 80, 3.4.1929 and the *Neue Preußische Kreuz Zeitung* No. 135, 4.4.1929, both in LAS 301/4695.

Landvolk meetings simply to shroud the entire affair.[83] This was not quite the truth. At a Landvolkbewegung meeting in Wöhrden on February 18, the NSDAP's local group from Heide had provided a guard (*Saalschutz*) of 20 SA men replete with party flags.[84] During the meeting, the communist leader, Christian Heuck, had challenged Paul Adam Ross to a fight. Ross drew a pistol before driving off in his car but the remaining Landvolk followers, Stahlhelm members and National Socialists banded together and gave Heuck and his friends a severe beating.[85] It was this occasion that had led Heuck and his friends to seek revenge and fall upon the unarmed SA procession in Wöhrden three weeks later.

The meeting ban, which coincided with the first ban on *Das Landvolk*, prevented the Crisis Assistance committees from assembling their followers and mustering public support. Consequently, it hindered the movement's expansion. Although lifted on March 18, the ban was in effect promptly restored by order of Prussian Interior Minister Grzesinski, who authorised the prohibition of all open-air processions and meetings which posed a threat to public safety, as well as all indoor meetings of an "unpeaceful" nature.[86] The fact that a Landvolk meeting in Heide in North Dithmarschen at the end of March was prevented on these grounds, while a nearby National Socialist meeting was allowed, indicates that at this time, the authorities gave greater priority to stemming the Landvolkbewegung than they did the NSDAP.[87]

Public warnings supplemented the meeting bans. Dismayed at the Landvolkbewegung's escalating incendiary character and hostility to the state, and fearing an extensive proliferation of tax obstruction, the president of the State Finance Office issued a press release warning against the Crisis Assistance and of the legal consequences – jail and fines – which awaited those who encouraged actions like those the committees espoused. He affirmed that the finance offices recognised the agricultural crisis and would therefore process all well-founded individual applications for postponement and remission benevolently, but cautioned that the return of taxation and confiscation notifications would only result in farmers having to pay arrears.[88]

---

83 *Schleswig-Holsteinische Tageszeitung* No. 58, 10.3.1929, p. 1 and No. 43, 21.2.1929, p. 5.

84 Report from the Landeskriminalpolizeiamt Berlin concerning "Süderdithmarschen" Itzehoe, 26.3.1929, LAS 309/22668.

85 Report from the Landeskriminalpolizeiamt Berlin concerning "Ermittelungen und Beobachtungen in den Kreisen Norder- und Süderdithmarschen" Heide, 16.3.1929, LAS 309/22668.

86 Regierungspräsident to the Polizeipräsident in Altona, Schleswig, 16.3.1929, LAS 309/22921; 'Runderlaß des preußischen Innenministers Grzesinski betreffend das Verbot von öffentlichen Versammlungen und Umzügen unter freiem Himmel vom 21.3.1929' in Huber, *Dokumente zur deutschen Verfassungsgeschichte,* pp. 442-443.

87 Copy of a report from the Landrat to the Regierungspräsident in Schleswig concerning "Landvolkversammlung in Heide" Heide, 30.3.1929, LAS 301/4695.

88 Präsident des Landesfinanzamtes Schleswig-Holstein to the Oberpräsident in Kiel concerning "Kundgebungen des 'Landvolks' " Kiel, 9.3.1929, in both LAS 301/4695 and

Next, Oberpräsident Kürbis declared that the committees were neither authorised nor qualified to represent the public and warned that anyone who approached them would face legal proceedings.[89] He admonished the committees a second time during an address to open the provincial parliament (*Landtag*) at the beginning of April. There he underlined the state's efforts to assist agriculture, asserted that the committees could only lead agriculture further into crisis because they were directed against the state and lamented, in reference to Wöhrden, that "political agitation" had led to bloodshed.[90] Two weeks later Reich Interior Minister Severing added a further warning during a visit to the province.[91]

These bans and warnings fuelled the Landvolk's resentment toward the authorities and led to an intensification of denunciations of them. In fact, it almost appeared as if the Landvolk was more engaged in this quarrel than actually pursuing the development of its committees. Oberpräsident Kürbis was the chief target of censure in March and April. From a publicised order issued by the Crisis Assistance committees to lift the meeting ban, to a letter of complaint to Reich President Hindenburg published in *Das Landvolk*, Kürbis was maligned as a "non-Schleswig-Holsteiner", accused of serving "dark powers" and labelled "detrimental to the people" (*volksverderbend*), as well as apportioned responsibility for destroying profitability.[92] Calls for his removal from office and departure from the province actually pointed to the Crisis Assistance's powerlessness, given that their only means of enforcement was to threaten petitioning his superiors in Berlin. These did not, however, detract from the Landvolk's efforts to damage the Oberpräsident's standing in agricultural circles, nor lessen the polarisation of relations between the authorities and farmers.

309/22921; Präsident des Landesfinanzamts Schleswig-Holstein to the Oberpräsident in Kiel concerning "Kundgebung des 'Landvolks' " Kiel, 11.3.1929 and the appended press release titled "Das Landesfinanzamt Schleswig-Holstein zu den Kundgebungen des 'Landvolks' in der vorigen Woche", LAS 301/4695; *Eiderstedter Nachrichten* No. 8225, 14.3.1929, p. 2; *Itzehoer Nachrichten* No. 61, 13.3.1929, p. 4; *Kieler Neueste Nachrichten* No. 61, 13.3.1929, p. 3; *Nordischer Kurier* No. 61, 13.3.1929, p. 4.

89 Regierungspräsident to the Oberpräsident in Kiel, Schleswig, 22.3.1929, LAS 301/4695; *Kieler Zeitung* No. 84, 25.3.1929, p. 3; *Nordischer Kurier* No. 72, 26.3.1929, p. 1.

90 Press release of the Amtlicher Preußischer Pressedienst titled "Eröffnung des Schleswig-Holsteinischen Provinziallandtags. Ernste Mahnworte des Oberpräsidenten Kürbis" 2.4.1929, LAS 309/22921. Also see, for example, the *Kieler Neueste Nachrichten* No. 78, 4.4.1929, p. 9.

91 *Nordischer Kurier* No. 63, 16.4.1929, p. 1.

92 Letter addressed to the "Verwaltungsapparat für Schleswig-Holstein z. Hd. des Herrn Oberpräsidenten Kürbis, Kiel" Schleswig-Holstein, 11.3.1929, LAS 301/4695, also published, for example, in *Das Landvolk* No. 19, 13.3.1929, p. 3 and *Rendsburger Tageblatt* No. 62, 14.3.1929, p. 2; copy of a letter addressed to the "Verwaltungsapparat von Schleswig-Holstein z. Hd. des Herrn Oberpräsidenten Kürbis" Tetenbüll, 7.4.1929, LAS 301/4695; letter to Reichspräsident Hindenburg in *Das Landvolk* No. 35, 17.4.1929, p. 3.

This dispute also signified a new level of hostility towards the authorities because it contained a personal element. Malice reached a peak in this episode with an edited version of Kürbis's Landtag speech in *Das Landvolk*. One might concede that criticism of Kürbis's claims that the state had done much to aid agriculture was not entirely unfounded. Yet this was entirely secondary to the deliberate misspellings, punctuation errors and sarcastic comments, as well as a charge of arrogance, that supplemented a further demand for his departure.[93]

The authorities would like to have banned the Crisis Assistance or indeed the Landvolkbewegung in its entirety. The impulse to do so came from the Prussian Interior Ministry, which while speculating whether the Crisis Assistance amounted to "high treason", recommended the Landvolkbewegung be dissolved if it could be defined as an "organisation" or "association" on the grounds that it encouraged disobedience against the laws by calling for the concealment of the dairy earnings.[94] The provincial authorities' endeavours were to prove fruitless on this account, but their investigations and examinations of the legal grounds for a ban cast light on the structure of the movement.

A ban on the basis of the Reich's Association Laws from April 19, 1908 proved impracticable because by definition an "association" (*Verein*) was "a lasting, at least for a prolonged period, calculated, voluntary coalition of several people for the purpose of a definite and common aim under common leadership and organisation."[95] The county administrators, however, concurred that the Landvolkbewegung was a "temporary junction of various personalities, a loose structure without tight connections".[96] Consequently, the Oberpräsident conceded that the requirements of the definition were not fulfilled, while the Interior Ministry confirmed that police measures could not be employed unless the Landvolkbewegung or Crisis Assistance exhibited these requisite features.[97]

The provincial authorities subsequently attempted to pin down an organisational body in order to define the Landvolkbewegung as an "association". One step involved the state criminal police in Flensburg sending an agent to approach the

93 *Das Landvolk* No. 23, 3.4.1929, p. 1.

94 Regierungspräsident to the Oberstaatsanwalt in Kiel and to the Preuß. Minister des Innern in Berlin concerning "Nothilfebewegung innerhalb der schlesw. – holst. Bauernschaft" Schleswig, 16.3.1929, LAS 309/22921; Preuß. Minister des Innern to the Oberpräsident in Kiel, Berlin 16.3.1929, LAS 301/4695.

95 Oberpräsident to the Minister des Innern in Berlin concerning "Verbot der Landvolkbewegung" Kiel, 20.3.1929, LAS 301/4695.

96 Oberpräsident to the Minister des Innern in Berlin concerning "Verbot der Landvolkbewegung" Kiel, 20.3.1929, LAS 301/4695.

97 Reply to the Oberpräsident's report from 20.3.1929 from the Preuß. Minister des Innern to the Oberpräsident in Kiel, Berlin, 2.4.1929, LAS 301/4695.

editors of *Das Landvolk* to inquire how the movement was structured and, in particular, whether the Landvolkbewegung had statutes.[98] They replied to his queries,

> "The Landvolkbewegung.... is not an organisation and consequently has no statutes, membership-lists and similar such things. It also does not have a leader or chairman.... Wilhelm Hamkens, Tetenbüll, is merely the man who has brought this idea to life. ... All who follow him on this path are not in any way bound or obligated to him by any kind of organisational form. ... The Landvolkbewegung's intellectual link is our non-partisan, independent daily newspaper, *Das Landvolk* ..."[99]

Landvolk followers denied the existence of statutes and ruled out their formulation in future. Members were not accepted by written declaration; rather all fellow farmers could consider themselves part of the movement. They were, however, expected to subscribe to *Das Landvolk* and make donations by contributing to the so called "fighting funds".[100] Despite thorough investigations, the authorities failed to uncover any statutes or membership lists. The authorities, therefore, were compelled to admit that the Landvolkbewegung and its Crisis Assistance were not "organisations" or "associations" but a "movement".[101]

Though requiring a digression from the examination of the authorities' endeavours to restrain the Crisis Assistance, it would be appropriate here to discuss the reasons for the Landvolkbewegung's rejection of conventional organisation, structures and procedures. Rudolf Heberle observed that the Landvolk avoided rigid organisation partly in order to escape the control of the police and partly for fear of bureaucratic stagnation. In addition, he suggested that the Landvolk were also making a virtue of necessity, as the movement was unlikely ever to encompass broad masses of farmers.[102] First, though it by no means prevented entanglements with the law, the absence of organisational structure did help to avoid the prohibition of the movement, as the case of the Crisis Assistance showed. Second, it is true the

---

98 Report from the Regierungspräsident to the Oberpräsident in Kiel concerning "Nothilfe" Schleswig, 28.4.1928, LAS 301/4695.

99 "Die Landvolkbewegung .... ist keine Organisation und besitzt infolgedessen auch keinerlei Satzungen, Mitgliederlisten und ähnliche Sachen. Sie hat auch keinen Führer bzw. Vorsitzenden, ... Wilhelm Hamkens - Tetenbüll ist lediglich der Mann, der diese Idee zu wirklichem Leben gebracht hat .... Alle diejenigen, die ihm auf diesem Wege folgen, sind nicht etwa in irgendeiner organisatorischen Form an ihn gebunden bzw. verpflichtet .... Das geistige Bindeglied der Landvolkbewegung ist unsere in Itzehoe erscheinende überparteilich unabhängige Tageszeitung 'Das Landvolk' ..." Copy of a letter from *Das Landvolk's* editors to Herrn A. Nicolaisen (Flensburg), Itzehoe, 19.4.1929, LAS 301/4695.

100 Reports from the Landeskriminalpolizeistelle Flensburg to the Regierungspräsident in Schleswig concerning "Nothilfe" Flensburg, 22.4.1929, and from the Polizeipräsident Altona-Wandsbek to the Regierungspräsident in Schleswig concerning "Satzung der 'Nothilfe' " Altona, 20.4.1929, both in LAS 309/22921.

101 Reports from the Regierungspräsident to the Oberpräsident in Kiel concerning "Nothilfe" Schleswig, 28.4.1928, and from the Oberpräsident to the Minister des Innern in Berlin concerning "Verbot der Landvolkbewegung" Kiel, 4.5.1929, both in LAS 301/4695.

102 Heberle, *From Democracy to Nazism*, p. 69.

Landvolk feared that organisational structure spelt bureaucratic ossification and diminished stimulus for individuals to take initiative. That is, Landvolk leaders argued that avoiding organisational structures facilitated dynamism and versatility, demanded enduring vitality from the leaders, compelled their followers to assume responsibility and strengthened their will to fight.[103] There is no evidence, however, to verify Heberle's assumption that the movement's size might have given its leaders additional incentive to avoid organisational structure. Only in respect of membership lists could this possibly have applied. Arguably, if low membership figures had been publicised, they might have dissuaded some farmers from supporting the movement or been perceived as a measure of disapproval of it.

Distinctly ideological motivations, hitherto overlooked in the literature on the Landvolkbewegung,[104] are also extremely important when examining the movement's abhorrence of organisational structures. The Landvolkbewegung's rejection of organisation, which set it apart from the multitude of farmers' associations, political parties and paramilitary groups, was an integral element of its protest against "the system". The Landvolk made a point of proclaiming this abstinence, highlighting that it was a "movement" independent of party-politics, "... a closed front of German women and men, without organisational ties, sharing a common fate like 1914..."[105] Far from aspiring to parliamentary participation, the Landvolk argued that to fight "the system" required a refusal to adopt the methods and structures of the parties and other groups working within it. The failure of Johannsen's deputation following the demonstrations of January 1928 was seen by some as evidence that no help could be expected from Berlin.[106] But more importantly, the resolve to remain outside the system reflected the Landvolk's aversion to the factionalism the Weimar party system allowed.

In the words of Hamkens and other Landvolk leaders, cooperation with the system was pointless because the structure of parties and organisations sapped their

---

103 *Das Landvolk* No. 226, 20.12.1929, p. 1. Also the speech of Meinke (Sieverstedt) cited in the transcript of the Landesthing from the Polizeiverwaltung in Rendsburg titled "Landesthing des Schleswig-Holsteinischen Landvolkes, am 10. November 1929 in Rendsburg im 'Schützenhof' nachmittags 3 Uhr", in both LAS 301/4696 and 309/22590.

104 Heberle, *From Democracy to Nazism*, p. 69 and for Le Bars' and Stoltenberg's cursory descriptions of the movement's organisation, Le Bars, *Le Mouvement Paysan*, pp. 72-74 and Stoltenberg, *Politische Strömungen im schleswig-holsteinischen Landvolk*, p. 123.

105 Cited from a postcard issued at the Landesthing appended to the transcript from the Polizeiverwaltung in Rendsburg titled "Landesthing des Schleswig-Holsteinischen Landvolkes, am 10. November 1929 in Rendsburg im 'Schützenhof' nachmittags 3 Uhr" Rendsburg, 12.11.1929, in both LAS 301/4696 and 309/22590. For similar assertions, see the letter published in the *Bramstedter Nachrichten* No. 280, 3.8.1929, LAS 301/4696 and Hamkens' speech at a meeting in Burg in Dithmarschen cited in *Das Landvolk* No. 161, 4.10.1929, p. 5.

106 See the letter in *Das Landvolk* No. 154, 26.9.1929, p. 9 and also Wulf's speech at a meeting in Soltau reported in the *Soltauer Nachrichten* No. 279, 27.11.1929, LAS 301/4696.

strength, turned them against one another and allowed them to become the instruments of other forces. Political parties and even other right-wing organisations were said to be incapable of effecting betterment because they had succumbed to a "struggle among brothers" (*Bruderkampf*) and were played off against each other by "international powers". The leaders of organisations, parties and associations, they claimed, were mere puppets, "dancing to the whistle of Jewish high-finance".[107]

Returning to the examination of the Crisis Assistance, it should be noted that it is difficult to weigh up the comparative influences of the factors which inhibited the committees' development. It is, however, difficult to envisage their success as "auxiliary governments" because of their own weaknesses even before they drew the authorities' attention. The authorities may not have been able to ban the committees due to the absence of organisational structure, but their warnings and certainly the resultant rumours of charges of high treason did succeed in intimidating the Landvolk leaders on this occasion. Hamkens' bitter objections to these particular charges at a meeting in Heide in May confirmed this.[108] Yet as early as the first week of April the Oberpräsident was already satisfied that other measures were not imperative since the committees had withdrawn from public activity.[109]

The Crisis Assistance committees surrendered as "additional governments" but they did not disband or dissolve. They continued to expand but were restricted to pursuing established Landvolk tactics. This retreat was confirmed at a meeting held to organise a further committee in the County of Plön at the end of May. There was no mention of the new committee supplementing or replacing the authorities and the resolutions at the meeting were modified accordingly.[110] Nonetheless, some 400 rural people attended the meeting, which suggests enduring support for the Crisis Assistance idea.[111]

---

107 For examples in Hamkens' speeches, see: *Kaltenkirchener Zeitung* No. 127, 3.6.1929, p. 2; *Marner Zeitung* No. 73, 27.3.1929, LAS 301/4695; *Segeberger Kreis- und Tageblatt*, No. 199, 26.8.1929, LAS 301/4696; report from the Kriminal- und Grenzkommissariat, Niebüll, 6.1.1930, LAS 309/22590. Also see Kühl, cited in the *Schleswiger Nachrichten* No. 154, 4.7.1929, p. 8; Kenstler's article in *Das Landvolk* No. 153, 25.9.1929, pp. 1-2; Hönck's article in *Das Landvolk* No. 226, 20.12.1929, p. 1.

108 *Heider Anzeiger* No. 116, 21.5.1929, p. 2.

109 Report from the Oberpräsident to the Regierungspräsident in Schleswig, Kiel, 6.4.1929, LAS 309/22921.

110 Undated resolution titled "Entschließung" from the Landvolknotausschuss in County Plön, LAS 301/4695.

111 Report from the Landrat to the Oberpräsident in Kiel concerning "Landvolkkundgebung in Preetz am 25.5.1929" Plön, 29.5.1929, LAS 301/4695; Preuß. Minister des Innern to the Oberpräsident in Kiel "Auf die Berichte vom 10./11. Juni 1929" Berlin, 22.6.1929 and Regierungspräsident to the Oberpräsident in Kiel concerning "Landvolkbewegung" Schleswig, 5.8.1929, both in LAS 301/4700.

The Landvolkbewegung now took care to conceal the Crisis Assistance committees' activities. Offices, first set up in Steinburg, South Tondern and Eiderstedt, were never the brazen displays of autonomy which one might have expected after the initial blunt assertions of "self-government". Rather, they were makeshift quarters for the collection of mail and complaints. The offices in Itzehoe and in Preetz (Plön) shared premises with the Rural League, which gave them a legitimate cover and made it especially difficult for the authorities to monitor their activities in these counties.[112]

For example, in Plön County, even though the authorities knew from the Crisis Assistance meetings' resolutions of the Landvolk's intention to conceal dairy earnings, they were still unable to confiscate the farmers' earnings because the method of evasion escaped their detection. Peter Petersen's post-war testimony revealed that farmers in Plön County who were unable to gain postponement of payment of interest or taxes transferred their dairy earnings to Petersen's bank account so that the bailiffs could not locate the farmers' liquid assets. Petersen then gave the farmers cheques so that later they could withdraw their money. Of course, as Petersen admitted, this tactic sometimes ran the risk of payment by compulsory auctions.[113] Nevertheless, it was said to have successfully helped farmers to conceal their earnings like this on hundreds of occasions.[114]

Although preliminary inquiries began as early as March 1929, proceedings against the Landvolk on account of the Crisis Assistance did not take place until the summer of 1930.[115] Numerous committee members were then tried, principally on charges of unauthorised assumption of office, at what became known as the "Great Landvolk Trial". It also incorporated the prosecution of other Landvolk leaders. These included *Das Landvolk's* printer and publisher, Pramor, and others for defamation of the Republic in *Das Landvolk* and for breaches of the newspaper bans in

---

112 Oberpräsident to the Preuß. Minister des Innern in Berlin, Kiel, 24.4.1929, LAS 301/4695; report from the Landeskriminalpolizeistelle Flensburg to the Regierungspräsident in Schleswig, Flensburg, 20.3.1929, LAS 309/22668; reports from the Landrat to the Regierungspräsident in Schleswig concerning "Landvolknotbewegung" Husum, 15.4.1929, LAS 309/22921; reports from the Landrat to the Regierungspräsident in Schleswig concerning "Büro der Nothilfe" Niebüll, 15.4.1929, from the Landrat des Kreises Eiderstedt to the Regierungspräsident in Schleswig concerning "Die Nothilfe" Tönning, 18.4.1929 and from the Landrat to the Regierungspräsident in Schleswig concerning "Landvolknotbewegung" Rendsburg, 22.4.1929, all in LAS 309/22921.

113 Interview with Peter Petersen, Zeitzeichen Archive (1990).

114 Beyer, 'Die Landvolkbewegung Schleswig-Holsteins und Niedersachsens', p. 186.

115 Report from the Oberstaatsanwalt to the Regierungspräsident in Schleswig concerning "Ermittlungssache gegen den Johannes Gravert in Landweg und Genossen wegen Vergehens gegen § 110 StGB" Altona, 20.3.1929, LAS 309/22921; report from the Oberpräsident to the Minister des Innern in Berlin concerning "Verbot der Landvolkbewegung" Kiel, 20.3.1929, LAS 301/4695; indictments issued by the Staatsanwaltschaft Altona, 29.10.1929 and 31.12.1929, LAS 301/4700; indictment issued by the Oberstaatsanwaltschaft Altona, 27.6.1930, GStA PK I. Rep. 84a No. 55410.

March and July 1929. The prosecutions also included Hamkens, Kühl and Weschke for some 17 instances of incitement to tax-strikes, defamation of the Republic and slander of the police at Landvolk meetings.[116] Hamkens was also tried for slander, having claimed that Grzesinski and other ministers had deliberately caused inflation in 1923 in order to put Germany at the mercy of international capital.

Ultimately, 14 committee members were found guilty of continued unauthorised assumption of office, as were two others for aiding and abetting on this charge. All 14 received fines, barring Hamkens. Contrasting testimonies and the difficulty of interpreting insidious language and ascertaining the context – in which, for example, Hamkens' calls to the farmers to continue paying taxes until they "bled white" had been made – cleared him on all but two counts of public incitement to tax-strikes and three of slander. In total, he was sentenced to two months prison and a RM 250 fine.[117] Weschke received a six-week prison term and an RM 80 fine, Kühl two months prison on bail for RM 250 and a RM 30 fine and Pramor four months prison and a fine of RM 50, subsequently converted to bail on parole.[118]

Held shortly before the "Great Bombing trial" (see Chapter Seven), the "Great Landvolk Trial" aroused scant public interest.[119] The press coverage differed little to that of other Landvolk trials; *Das Landvolk*, for example, once again contended that it was the system and not the accused that was really on trial, lauded the virtuous motives of their actions and complained of judicial severity at the bidding of the system.[120] The trial was interesting, however, for what it revealed about the Crisis Assistance: how it had lawfully helped farmers on numerous occasions. In fact, the court acknowledged that the Crisis Assistance had often proven itself a "blessing".[121]

In Steinburg and Eiderstedt the Crisis Assistance committees had approached the authorities on behalf of their colleagues and had sometimes been able to prevent confiscations or compulsory auctions by negotiation. On one occasion, Hamkens

116 Issue of the verdict of the Schöffengericht in Itzehoe from 9.8.1930, pp. 27-63, LAS 301/4700.

117 Issue of the verdict of the Schöffengericht in Itzehoe from 9.8.1930, pp. 3 and 47-50, LAS 301/4700.

118 Issue of the verdict of the Schöffengericht in Itzehoe from 9.8.1930, pp. 3-5, LAS 301/4700. Due to an oversight, Pramor received a summons to prison in mid-1931, prompting further derision of the authorities in *Das Landvolk.* Oberstaatsanwalt to the Generalstaatsanwalt in Kiel concerning "Artikel: 'Jeder blamiert sich so gut er kann' in Nr. 111 d. Tageszeitung 'Das Landvolk' vom 1. Juni 1931" Altona, 8.6.1931, GStA PK I. HA Rep. 84a No. 55410.

119 Regierungspräsident to the Oberpräsident in Kiel concerning "Anreizung zum Steuerstreik" Schleswig, 11.10.1930, LAS 301/4700.

120 See, for example, *Das Landvolk* No. 176, 29.7.1930, p. 1, No. 182, 5.8.1929, pp. 1-2, No. 186, 9.8.1930, p. 1 and No. 187, 11.8.1930, pp. 1-2.

121 Issue of the verdict of the Schöffengericht in Itzehoe from 9.8.1930, pp. 17 and 21, LAS 301/4700.

had indeed gone too far, threatening the finance office in Husum that the farmers would resort to "self-help" if it did not remit one impoverished farmer's taxes.[122] Nevertheless, Hamkens continued throughout 1929 and 1930 to assist needy colleagues. So too did the Crisis Assistance in Rendsburg and South Tondern. There individual committee members provided information and advice and assisted their colleagues, for example, in applying for bridging finance or by accompanying them to discuss their taxation difficulties with a lawyer.[123]

With the object of obtaining more lenient sentences for the accused, the defence emphasised these virtuous undertakings.[124] Yet even if it overstated their importance, this constructive, philanthropic aspect of the Landvolkbewegung cannot be ignored. Underlining the spirit of unity among the Landvolk, genuine concern for their colleagues' welfare was clearly an important motivation for the movement's leaders notwithstanding all their rebellious aspirations.

The Crisis Assistance was one important step in the Landvolkbewegung's radicalisation. The movement's increasing militancy in 1929 was another. All the more remarkable was that this militancy often had a playful, almost carnival quality. As suggested briefly in Chapter Three, part of the Landvolkbewegung's attraction was the enjoyment farmers derived from their involvement. Sometimes it seemed that the farmers all but forgot the movement's serious cause and purpose, as they carried out what often can only be described as pranks. In contrast to the Crisis Assistance, which was at least in part an attempt to diminish the farmers' losses, these prank-like actions frequently had little to do with tax obstruction or the agricultural crisis. Continuing into 1930, they highlight the movement's ever-increasing absorption in simply defying the authorities. More and more, opposition itself became the movement's essential objective.

While some of these actions might be described as relatively harmless, others were considerably more spiteful. On one occasion farmers pretended to call a meeting to trick the authorities into despatching police.[125] On another occasion Hamkens led a march through Eiderstedt to such fanfare that the local fire brigade mistook it for an alarm.[126] Without making their motives clear, farmers sent Eiderstedt's administrator a package containing a rancid rabbit carcass wrapped up in copies of a communist newspaper.[127] More explicable was the prank played on the new director of the agricultural school in Garding, Hjuler. Enraged by the

122 Issue of the verdict of the Schöffengericht in Itzehoe from 9.8.1930, pp. 20-21, LAS 301/4700. Also see *Das Landvolk* No. 14, 17.1.1930, p. 5.

123 Issue of the verdict of the Schöffengericht in Itzehoe from 9.8.1930, pp. 17-26, LAS 301/4700. Also see the *Nordischer Kurier* No. 164, 15.7.1930, p. 4.

124 Issue of the verdict of the Schöffengericht in Itzehoe from 9.8.1930, p. 17, LAS 301/4700.

125 *Das Landvolk* No. 25, 30.1.1930, p. 5; *Rendsburger Tageblatt* No. 27, 1.2.1930, p. 3.

126 *Holsteinischer Courier* No. 3, 4.1.1930, p. 9.

127 Landrat des Kreises Eiderstedt to the Regierungspräsident in Schleswig concerning "Anonyme Übersendung eines Paketes" Tönning, 6.5.1929, LAS 309/22921.

Chamber of Agriculture's transfer of the former director, Schotten, who was an energetic Landvolk supporter, farmers tricked Hjuler into getting into a car under pretence of collecting soil samples for a fertiliser test. They drove him through the countryside and set him out across the Eiderstedt County border.[128]

Even the courts were not spared such mischief. Fined RM 70 for insulting police infiltrators at a meeting, Paul Adam Ross appealed and with the assistance of a professor from Kiel, who gave evidence on the historical linguistics of the lower German dialect, persuaded a judge that the expression he had used, "*Ick shiet op den ganzen Verwaltungsapparat und siene Abgesandten*" (I shit on the whole administrative apparatus and its emissaries) was not an insult but a general expression of rejection.[129] The laughter of Ross' friends and fellow farmers in the courtroom showed they considered the hearing a great joke. Their impudence and Ross' impunity meant that he actually succeeded in affronting the authorities, not once, but twice.[130]

The carnival aspect of the Landvolkbewegung held attraction and provided adventure for both farmers and those who joined the movement from outside agriculture. Bruno von Salomon compared the "fun" he was having as a "putschist" in Schleswig-Holstein to the society of the "Salon putschists" in the cities.[131] In *Der Fragebogen*, his brother Ernst not only recalled the "fun" young farmers had planting bombs, he later remembered his own experiences in the Landvolkbewegung as "coarse rural amusement" (*ein grober bäuerlicher Spaß*).[132]

Adventure became dangerous when mixed with violence. In part, the growing willingness to adopt more extreme methods can be attributed, as the Crisis Assistance meetings witnessed, to pressure from the movement's rank and file. In Husum, a member of the audience proposed that the local fire brigades chase the bailiffs and policemen out of the villages, and at the meeting in Rendsburg there were cries of "hanging" and "execution by firing squad" in response to Wulf's question as to what was to be done about individuals who had assisted the authori-

---

128 Report from the Landrat des Kreises Eiderstedt to the Regierungspräsident in Schleswig concerning "Landvolkbewegung" Tönning, 19.4.1930 and Hjuler's appended statement, Garding, 17.4.1930, in both LAS 301/4697 and 309/22590. For Schotten's role at *Das Landvolk*: report from the Landeskriminalpolizeistelle Altona, Abteilung IA, titled "Bericht über die Landvolkbewegung" Altona, 5.10.1929, in both LAS 301/4696 and 309/22784; Landwirtschaftskammer für die Provinz Schleswig-Holstein to the Oberpräsident in Schleswig-Holstein, Kiel, 25.10.1929, LAS 301/4697; report from the Landeskriminalpolizeistelle, Abteilung IA, concerning "Landvolkbewegung" Altona, 8.3.1930, LAS 301/4697.

129 *Nordischer Kurier* No. 285, 5.12.1929, p. 4.

130 *Das Landvolk* No. 44, 21.2.1930, p. 5.

131 See the lists of confiscated material, "42. Durchschlag eines Schreibens des Bruno v. Salomon an Hartmut Plaas in Berlin vom 9.3.1929" in "IV. In der Redaktion der Landvolk-Zeitung am 9.9.1929 beschlagnahmtes Material", LAS 309/22784.

132 v. Salomon, *Der Fragebogen*, p. 237 and *Deutschland deine Schleswig-Holsteiner*, p. 177.

ties in confiscations and compulsory auctions.[133] Spontaneous though such cries were, they could only have raised expectations of more militant action from the movement's leaders. Soon after the establishment of the Crisis Assistance, they found an opportunity to react and what followed was one of the most remarkable episodes in the Landvolkbewegung's history: the attempted liberation of a prisoner from the county courthouse prison in Hohenwestedt.

A painter, Georg Struve, was imprisoned in Schenefeld on the morning of March 5 after he failed to pay outstanding taxes and refused to give an oath of disclosure.[134] News of his imprisonment soon reached Landvolk leader and fellow NSDAP party member, Heinrich Hansen, who was asked if Steinburg's Crisis Assistance committee could help.[135] He and other local National Socialists visited the prison and requested Struve's release. Rumours quickly circulated around the village that Struve would be liberated by force. In the afternoon, however, he was transferred by train to the county court prison in Hohenwestedt.[136] At the station in Schenefeld a number of National Socialists then attempted to intervene but were blocked by police.[137]

In the meantime Hansen had an appointment with Hamkens and Weschke.[138] Relating Struve's predicament, Hansen stressed that at Landvolk meetings it had often been said that the time had come to take action and he impressed upon them that this was such an opportunity and also a chance for the Landvolk to do something for tradespeople. On their way to further engagements, Hamkens and Weschke passed through Hohenwestedt, met with the village's "trusted man", Rathjen, and reconnoitred. They determined a rendezvous point for 2 a.m. at a cutting some 200 metres from the courthouse, before driving on to Nortorf and Eisendorf near Rendsburg and meeting with the Landvolk leaders there, Wulf and Thun. On learning of the plan, Wulf and Thun also resolved to take part and they found

133 Police report titled "Bericht über die öffentliche Versammlung der Notbewegung am 4. März 1929" Husum, 4.3.1929, LAS 309/22921; *Rendsburger Tageblatt* No. 55, 6.3.1929, p. 5; meeting transcript from the Polizeiverwaltung Rendsburg titled "Willenskundgebung zur Selbsthilfe im Schützenhof zu Rendsburg am 4.3.1929" Rendsburg, 4.3.1929, in both LAS 301/4695 and 309/22921.

134 Landrat to the Regierungspräsident in Schleswig concerning "Befreiungsversuch eines Inhaftierten durch Nationalsozialisten" Rendsburg, 13.3.1929, LAS 309/22668; *Das Landvolk* No. 13, 6.3.1929, p. 6.

135 Issue of the verdict of the Schöffengericht in Neumünster from 25.3.1929, p. 3, LAS 309/22668.

136 Issues of the verdicts of the Schöffengericht in Neumünster from 25.3.1929, pp. 4-5 and from 26.4.1929, pp. 3-4, both in LAS 309/22668.

137 Landrat to the Regierungspräsident in Schleswig concerning "Befreiungsversuch eines Inhaftierten durch Nationalsozialisten" Rendsburg, 13.3.1929, LAS 309/22668.

138 The account here follows the issues of the verdicts of the Schöffengericht in Neumünster from 25.3.1929, pp. 4-9 and 26.4.1929, pp. 4-8, both in LAS 309/22668.

further volunteers among their acquaintance, just as Hansen did among his colleagues.

Four cars carrying perhaps 20 farmers met at the rendezvous point and proceeded to the courthouse by foot.[139] Some had blackened their faces and others carried axes and picks. They gathered in the small courtyard in front of the building where they smashed open the outer door, ran up the stairs to the prison entrance and turned on the lights. The prison warden, who had been asleep in his adjacent apartment, woke at the noise and ran to his front door. The men outside called out to him that they were 100 in number and that they wouldn't harm him if he handed over the keys. With no telephone to call for help, the warden fired a shot through the door into the courtyard and further warning shots out of his bedroom window. The farmers, on determining that they were not equipped to break through the sturdy prison door, soon returned to their cars and departed. Struve yielded and gave an oath of disclosure the following day.[140]

Eight of the intruders as well as Hans Dammann, who had written threatening letters to the county court demanding Struve's release, were tried at a Schöffengericht in Neumünster on March 25. They were found guilty of "aggravated breaking and entering", attempted liberation of a prisoner, property damage and intimidation. Hansen, identified as the principal instigator and judged to have misused his influence as a leader among the rural people, was sentenced to six months' prison. Six others received prison terms ranging between one and three months. Hans Dammann was fined RM 100, as was his son, Heinrich, who was found guilty of aiding and abetting the raid.[141]

The trial exposed Hamkens and Weschke's role in preparing the raid and pointed strongly to their direct involvement.[142] They and eight others were tried at a second hearing in Neumünster on April 26. Hamkens received four months, Weschke three and most of the others, including Wulf, between one and two

139 Issue of the verdict of the Schöffengericht in Neumünster from 25.3.1929, p. 10, LAS 309/22668.

140 See the report from the Landrat to the Regierungspräsident in Schleswig concerning "Befreiungsversuch eines Inhaftierten durch Nationalsozialisten" Rendsburg, 13.3.1929, LAS 309/22668; *Kieler Neueste Nachrichten* No. 73, 27.3.1929, p. 14.

141 Issue of the verdict of the Schöffengericht in Neumünster from 25.3.1929, p. 2, LAS 309/22668; Polizeiverwaltung to the Regierungspräsident in Schleswig concerning "Verurteilung von Einwohnern aus Schenefeld und Umgebung wegen der Vorgänge am 6.3.1929 im Amtsgerichtsgefängnis in Hohenwestedt" Neumünster, 26.3.1929, in both LAS 301/4695 and 309/22668.

142 Polizeiverwaltung to the Regierungspräsident in Schleswig concerning "Verurteilung von Einwohnern aus Schenefeld und Umgebung wegen der Vorgänge am 6.3.1929 im Amtsgerichtsgefängnis in Hohenwestedt" Neumünster, 26.3.1929, in both LAS 301/4695 and 309/22668; Regierungspräsident to the Preuß. Minister des Innern in Berlin concerning "Befreiungsversuch eines Inhaftierten durch Nationalsozialisten" Schleswig, 4.4.1929, LAS 301/4695.

months' imprisonment.[143] Like Hansen, Hamkens was punished more severely because he was a respected Landvolk leader: he could have prevented the raid, but instead his participation had encouraged others to take part.[144] A number of the farmers appealed, but the verdicts were upheld with the exception of a revision of Hansen's sentence from six to four months.[145]

Initially, the NSDAP was believed to be behind the raid in Hohenwestedt because the first intruders identified were all members of the party.[146] The trials confirmed that most had indeed been party members but established that the party as such had not been involved. Rather it was judged that individual members had acted under the sway of the Landvolkbewegung.[147] The courts identified the origins of the incident in the disquiet the Landvolkbewegung had spread among the rural population and, in particular, concluded that the intruders had been motivated by the Crisis Assistance meetings, which several had attended, the day before they had formed their plan.[148]

Claus Heim remembered that the farmers were incensed at Struve's imprisonment because they perceived it was the authorities' answer to the formation of the Crisis Assistance committees.[149] It was, of course, Struve's financial difficulties that had led to his imprisonment and his refusal to pay outstanding taxes that linked him to the farmers' struggle. The crisis was the underlying reason for the raid. Heim's view, however, indicates that the overriding impulse was defiance. Ultimately, the raid was, both in intent and execution, yet a further manifestation of the Landvolk's increasing engrossment in rebelling against the authorities.

---

143 Polizeiverwaltung to the Regierungspräsident in Schleswig concerning "Verurteilung von Einwohnern aus der Schenefelder, Nortorfer und Eisendorfer Gegend wegen der bekannten Vorgänge am 6.3.1929 im Amtsgerichtsgefängnis in Hohenwestedt" Neumünster, 27.4.1929 and the issue of the verdict of the Schöffengericht in Neumünster from 26.4.1929, p. 2, both in LAS 309/22668; report from the Regierungspräsident to the Preuß. Minister des Innern concerning "Befreiungsversuch eines Inhaftierten durch Nationalsozialisten und Anhänger der Landvolkbewegung" Schleswig, 1.5.1929, LAS 301/4695.

144 Issue of the verdict of the Schöffengericht in Neumünster from 26.4.1929, pp. 10-11, LAS 309/22668.

145 Issue of the verdict of the Landgericht in Kiel from 27.6.1929, LAS 309/22668; report from the Regierungspräsident to the Preuß. Minister des Innern concerning "Befreiungsversuch eines Inhaftierten durch Nationalsozialisten und Anhänger der Landvolkbewegung" Schleswig, 4.7.1929, in both LAS 301/4696 and 309/22668.

146 Landrat to the Regierungspräsident in Schleswig concerning "Befreiungsversuch eines Inhaftierten durch Nationalsozialisten" Rendsburg, 13.3.1929, LAS 309/22668.

147 Issues of the verdicts of the Schöffengericht in Neumünster from 25.3.1929, p. 13 and of the Landgericht in Kiel from 27.6.1929, p. 5, both in LAS 309/22668. Also see the *Flensburger Nachrichten* No. 72, 26.3.1929, p. 5.

148 Issues of the verdicts of the Schöffengericht in Neumünster from 23.3.1929, pp. 3 and 13 and from 26.4.1929, p. 4, both in LAS 309/22668. Also see the *Kieler Neueste Nachrichten* No. 73, 27.3.1929, p. 14.

149 Heim, 'Claus Heim berichtet', LAS EI/4434, p. 22.

A network of local NSDAP and Landvolk connections as well as ties of friendship had constituted the raiding party. Hamkens' and Weschke's involvement was not premeditated, and most others also joined the scheme unexpectedly. One of the conspirators learned of the plan over a game of cards, and Wulf's cousin only joined when the Rendsburg party stopped at his house on their way to Hohenwestedt, waking him because they needed radiator water. The spontaneous willingness of all these men to take part in such a haphazard and desperate plot is startling, not least because their actual objectives remained unclear. The scheme was of dubious assistance to Struve and there was no evidence of a plan once they had freed him: it seems the matter was scarcely given a thought. In explaining the men's reckless willingness to take part, various factors must be taken into account including their distress as a consequence of the agricultural crisis, the impact of the Landvolkbewegung's agitation, Hansen and Hamkens' prestige and, presumably, an inner group dynamic which boosted the men's resolve. These factors combined with an enthusiasm for militant opposition to the authorities. Prudence and the purpose of the raid were forgotten in the enthusiasm for escapades that had become an accepted part of the movement's culture.

The raid in Hohenwestedt nevertheless proved too extreme for some Landvolk followers. At a meeting afterwards, Hamkens' leadership was questioned on account of the "stupidity" of the action.[150] Significantly, it was also far too wild for the liking of the NSDAP, which promptly expelled those members who had participated.[151] It did so for good reason. After the abortive putsch in 1923, Hitler increasingly steered his party onto a legal path to power. While various lesser displays of camaraderie between the two movements – principally in the form of common enmity toward the authorities and communists[152] – passed unhindered, the party leadership endeavoured to prevent open association with the Landvolkbewegung. Such unruly acts as the Hohenwestedt raid in particular could only

150 Landrat des Kreises Eiderstedt to the Oberstaatsanwalt in Kiel concerning "Angriff auf das Hohenwestedter Gefängnis" Tönning, 23.1.1930, LAS 309/22783.

151 Issue of the verdict of the Schöffengericht in Neumünster from 25.3.1929, p. 13 and of the verdict of the Landgericht in Kiel from 27.6.1929, p. 5, both in LAS 309/22668. Also see the *Flensburger Nachrichten* No. 72, 26.3.1929, p. 5.

152 For examples of *Das Landvolk* and *Schleswig-Holsteinische Tageszeitung* each printing notices for the other movement, see the *Schleswig-Holsteinische Tageszeitung* No. 50, 29.2.1929, p. 8 [sic], LAS 309/22921; *Das Landvolk* No. 17, 11.3.1929, p. 3. Also see Johannes Kühl's complaints of police protection for communists who attacked National Socialists and the *Schleswig-Holsteinische Tageszeitung*'s objections to the treatment of Hamkens in prison as a "common criminal" rather than a political prisoner: report from the Landeskriminalpolizeistelle Altona, Abteilung IA, titled "Bericht über die Landvolkversammlung am 2.10.1929 ... in Wiemersdorf bei Bad Bramstedt" Altona, 3.10.1929, in both LAS 301/4696 and 309/22784; *Schleswig-Holsteinische Tageszeitung* No. 106, 6.5.1930, p. 2.

damage the law-abiding image the party desired and give the authorities a pretext to ban it.[153]

Bodo Uhse's account of Hitler's visit to Schleswig-Holstein on the occasion of the Wöhrden funerals indicates that the Landvolkbewegung's radicalism put pressure on the province's National Socialists to take a more extreme course. According to Uhse, Hitler warned him to be more circumspect in his writings in the *Schleswig-Holsteinische Tageszeitung*. Uhse defended his work, arguing that the Landvolkbewegung was compelling him to use more militant language, but Hitler remained unmoved. He insisted on abiding by lawful methods, declaring that the Republic had to be defeated "with its own weapons".[154]

In order to counter the appeal of the Landvolk's activism, party leaders, such as the SA leader, Major Dincklage, argued that the Landvolk's methods were futile, that storming a finance office, for example, would change nothing, that the removal of "pests" was pointless when others would simply take their place.[155] Uhse, too, publicly attacked the Landvolk's activism as ineffective, describing it as "putschist" rather than genuinely revolutionary.[156] Secretly, however, he admired the farmers for it, judging his party as soft by comparison.[157] This was precisely Bruno von Salomon's opinion. He held the SA in high esteem but was "very disappointed" in Hitler and believed that the NSDAP had become "feeble".[158] In particular, he blamed the provincial leadership for limiting the party's freedom of action, calling it "narrow-minded bigwigedness" (*engstirniges Bonzentum*) and dubbing Gauleiter Lohse the "head-bigwig".[159]

The Landvolkbewegung always denied any links with the NSDAP, just as it professed its autonomy from all parties and organisations. To the frequently asked question from their audiences, Landvolk speakers such as Hamkens replied, "No, the Landvolkbewegung has nothing to do with the National Socialist German

153 Report from the Landeskriminalpolizeiamt, Abteilung IA, Berlin, 2.4.1929, LAS 309/22668.

154 Uhse, *Söldner und Soldat*, pp. 201-202.

155 Report from the Landeskriminalpolizeiamt Berlin concerning "Süderdithmarschen" Itzehoe, 26.3.1929, LAS 309/22668.

156 See the list of confiscated material titled "XIV. Bei Rittmeister a.D. Bohm aus Hamburg am 14. September 1929 beschlagnahmtes Material", LAS 309/22784.

157 Uhse, *Söldner und Soldat*, pp. 274-275.

158 See the letters from Bruno von Salomon to F. W. Heinz, 23.3.1929 and 17.5.1929, both cited in the list of confiscated material titled "V. Auslese wichtigen Materials aus den von der Redaktion der Landvolk – Zeitung beschlagnahmten Ordnermappen", LAS 309/22784.

159 Letter from Bruno von Salomon to F. W. Heinz, 17.5.1929 and the undated letter from Landtag Deputy Lohse to Bruno von Salomon, both cited in the list of confiscated material titled "V. Auslese wichtigen Materials aus den von der Redaktion der Landvolk – Zeitung beschlagnahmten Ordnermappen", LAS 309/22784.

Worker's Party."[160] Yet until at least the spring of 1929, there seems to have been some confusion between the two movements. Ignorance greatly contributed to this. Police and provincial government reports observed that "simple people" were ill-informed about the political aims of both the Landvolkbewegung and the NSDAP, that what attracted farmers to the NSDAP was simply its opposition to the Republic and the parties of the left.[161] One earlier report maintained, "It has not yet been possible ... to determine just how far the political parties have gained as a result of its [the Landvolkbewegung's] propaganda. ... The rural population, for the most part, is only very superficially informed about the dispositions of the individual parties and describes all possible directions as National Socialist."[162] Among the urban working population the persistence of this mistake would not have been especially surprising, given that the left-wing press made little effort to distinguish between the Landvolkbewegung and the NSDAP.[163] Even in the countryside the regional press, to some degree, also contributed to this erroneous association when reporting on the NSDAP's and Landvolkbewegung's activities and developments as if they were a single item.[164] But if Schleswig-Holstein's rural population sometimes confused the Landvolkbewegung and NSDAP, they were not alone, for as late as June 1929, a police infiltrator observing a Landvolk meeting in Kaltenkirchen wrote in his report that he had attended a meeting of the NSDAP.[165]

The main reason for the confusion between the two movements, as these reports indicated, was the similarities between their völkisch and anti-Semitic ideologies. Moreover, these reports observed that the NSDAP was cleverly using the dissatisfaction in the countryside to gain new members and that many Landvolk followers had already joined the NSDAP as well. The reports also indicated that at the

160 Hamkens cited at the meeting in Plön in May 1929, *Kieler Neueste Nachrichten* No. 121, 28.5.29, p. 10. For a further example, see Ernst von Salomon's denials that the National Socialists had influence on the Landvolkbewegung in *Das Landvolk* No. 82, 14.6.1929, p. 1.

161 Reports from the Landeskriminalpolizeiamt Berlin concerning "Ermittelungen und Beobachtungen in den Kreisen Norder- und Süderdithmarschen" Heide, 16.3.1929 and Heide, 21.3.1929, both in LAS 309/22668; report from the Landeskriminalpolizeiamt, Berlin, 2.4.1929, in both LAS 301/4695 and 309/22668; report from the Regierungspräsident to the Oberpräsident in Kiel concerning "Landvolkbewegung im Kreise Südtondern" Schleswig, 29.4.1929, in both LAS 301/4695 and 309/22783.

162 Report commissioned by the Außendienst II. G. St. des Berliner Polizeipräsidiums, 27.12.1928, LAS 301/4695.

163 For examples of the Landvolkbewegung and its leaders identified as National Socialist, see *Ein Blatt am Montag* No. 53, 31.12.1928, LAS 301/4695; *Vossische Zeitung* No. 75, 28.3.1929, 309/22668; *Schleswig-Holsteinische Volkszeitung* No. 151, 2.7.1929, p. 7 and No. 152, 3.7.1929, p. 6, and for a description simply as "fascist", see the communist *Norddeutsche Zeitung* No. 149, 2.7.1929, LAS 309/23017.

164 See, for example, *Der Schleswiger* No. 225, 24.9.1929, LAS 301/4696.

165 Report from the Landjägereisamt Nahe, Nahe, 3.6.1929, sent to the Landrat in Bad Segeberg, in both LAS 301/4695 and 309/22668.

very least, the Landvolkbewegung had prepared the way for a radicalisation of rural voters at future elections.[166]

Though they did not confuse the two movements, the provincial authorities speculated whether there were close ties between the Landvolkbewegung and the NSDAP, and even conjectured that an internal working-committee might coordinate each groups' activities.[167] An investigation into the links between the Landvolkbewegung, NSDAP and DNVP involved discussions with the Jungdeutscher Orden leader, Hogrefe, who believed that the Landvolkbewegung was a product of a DNVP and NSDAP alliance. He presumed that Bruno von Salomon was in fact the National Socialist, Franz Pfeffer von Salomon.[168] The authorities' search for evidence continued until August but yielded no positive results.[169]

The NSDAP and Landvolkbewegung continued to compete for the farmers' favour in the spring of 1929, but unlike the Landvolkbewegung, the NSDAP was determined to undermine its rival. It claimed the Landvolkbewegung's successes for its own, emphasised its perceived weaknesses and, whenever possible, drew attention to moments of inner Landvolk discord. Revealing no small measure of envy, the NSDAP responded to Stahlhelm leader Eduard Stadtler's praise of the strength of the Landvolkbewegung, insisting that this was solely the result of the long years of its own dogged and bitter work.[170] Conversely, it downplayed the Landvolkbewegung's successful meetings as merely the result of the crisis and claimed to have made the same speeches in the province itself five years beforehand.[171] Fur-

---

166 Report from the Landeskriminalpolizeiamt, Berlin, 2.4.1929, in both LAS 301/4695 and 309/22668; report from the Regierungspräsident to the Oberpräsident in Kiel concerning "Landvolkbewegung im Kreise Südtondern" Schleswig, 29.4.1929, in both LAS 301/4695 and 309/22783.

167 Report from the Regierungspräsident to the Preuß. Minister des Innern in Berlin concerning "Landvolkbewegung und die politischen Sprengstoffanschläge" Schleswig, 30.9.1929 in LAS 301/4696 and 28.9.1929 in LAS 309/22784.

168 Report from the Landeskriminalpolizeistelle Flensburg to the Regierungspräsident in Schleswig, Flensburg, 20.3.1929, LAS 309/22668; report from the Regierungspräsident to the Oberpräsident in Kiel concerning "Landvolkbewegung, DNVP, Stahlhelm und NSDAP" Schleswig, 15.4.1929, LAS 301/4695; report from the Landeskriminalpolizeistelle Flensburg to the Regierungspräsident in Schleswig concerning "Landvolkbewegung und Deutschnat. Volkspartei" Flensburg, 16.4.1929, LAS 309/22668.

169 Report from the Regierungspräsident to the Oberpräsident in Kiel concerning "Das Verhältnis Hugenburg - Hitler zur Landvolkbewegung" Schleswig, 3.5.1929, in both LAS 301/4695 and 309/22668; report from the Landeskriminalpolizeistelle Flensburg to the Regierungspräsident in Schleswig concerning "Hugenberg-Hitler-Landvolk" Flensburg, 13.8.1929, LAS 309/22668.

170 *Schleswig-Holsteinische Tageszeitung* No. 54, 6.3.1929, p. 4. For further evidence, see the letter from Bruno von Salomon to Ernst von Salomon, 22.2.1929, cited in the list of confiscated material titled "IV. In der Redaktion der Landvolk-Zeitung am 9.9.1929 beschlagnahmtes Material", LAS 309/22784.

171 *Schleswig-Holsteinische Tageszeitung* No. 58, 10.3.1929, p. 4.

ther, it condemned the Landvolkbewegung for its want of organisation, programme, statutes and leader. The *Schleswig-Holsteinische Tageszeitung* argued that only the NSDAP with its clear programme, organisational structure, ideological aims and leader could deliver Germany from all ills.[172]

The NSDAP also noted the Landvolkbewegung's failure to find a following among other occupational groups and encouraged the belief that it alienated potential supporters. For example, the *Schleswig-Holsteinische Tageszeitung* published a letter pointing out the frequent harsh and unjustified criticisms of public servants at Landvolkbewegung meetings where they were often depicted as "the beneficiaries of the current state". The letter maintained that by attacking officials like this, the Landvolkbewegung was purportedly destroying the sympathy individual officials had for the movement. Predictably, the *Schleswig-Holsteinische Tageszeitung* claimed that it was the only newspaper which had defended public servants.[173]

Significantly, the NSDAP also sought to drive a wedge between the Landvolkbewegung and the Farmers' League. Leaders of the Landvolk and Farmers' League did criticise each other, but far stronger than this tension was an overriding spirit of cooperation.[174] Discussing the relationship at a Farmers' League meeting in Husum in March, the chairman, Köhler, had attempted to clarify their respective domains, stressing that the Landvolkbewegung should leave economic-political affairs to the Farmers' League but commending the Landvolk's fight against "international powers" and "Jewry" as well as its endeavours to change the system for the benefit of agriculture. His call for the Landvolkbewegung and Farmers' League to "march together" exacted emphatic avowals of cooperation from Landvolk leaders.[175] At the Crisis Assistance meeting in Husum, Hönck made it clear that the Landvolkbewegung and the Farmers' League were working in unison, that the Landvolkbewegung merely consisted of the more "active forces" – read militant forces – among the farmers.[176]

However, the NSDAP mentioned nothing of the cooperation between the Farmers' League and Landvolkbewegung and where other newspapers passed over matters of conflict, the *Schleswig-Holsteinische Tageszeitung* dwelt upon them.[177]

---

172 *Schleswig-Holsteinische Tageszeitung* No. 54, 6.3.1929, p. 4 and No. 58, 10.3.1929, p. 4.

173 *Schleswig-Holsteinische Tageszeitung* No. 56, 8.3.1929, p. 7.

174 *Heider Anzeiger* No. 55, 6.3.1929, p. 5; *Schleswig-Holsteinische Tageszeitung* No. 55, 7.3.1929, p. 5.

175 *Husumer Nachrichten* No. 53, 4.3.1929, pp. 5-6; *Kieler Neueste Nachrichten* No. 54, 5.3.1929, p. 6; *Lecker Anzeiger* No. 54, 5.3.1929, p. 5. Also see the meeting transcript from the Polizeiverwaltung Rendsburg titled "Willenskundgebung zur Selbsthilfe im Schützenhof zu Rendsburg am 4.3.1929" Rendsburg, 4.3.1929, in both LAS 301/4695 and 309/22921.

176 Report from the Landeskriminalpolizeistelle Flensburg to the Regierungspräsident in Schleswig concerning "Landvolkbewegung" Flensburg, 6.3.1929, LAS 309/22921; *Rendsburger Tageblatt* No. 55, 6.3.1929, p. 5.

177 *Schleswig-Holsteinische Tageszeitung* No. 48, 27.2.1929, p. 4 and No. 53, 5.3.1929, p. 4.

Pointing to the party's future strategy to secure rural support – infiltrating rather than competing with existing farming organisations – the NSDAP courted Köhler. It demanded that he take a clear position on the Landvolkbewegung and promised to support him if he recognised the NSDAP's position on the crisis. If he did not, the NSDAP vowed to create its own agricultural organisation and smash the unity organisation in the course of one year.[178]

When considering the factors driving the Landvolkbewegung's growth and growing extremism from the spring of 1929, it is important to remember that the initial causes of unrest in the countryside remained. There were neither signs that the crisis would abate, nor any indication of progress in the agricultural associations' efforts to unify. Moreover, the movement's expansion and increased extremism discussed in this chapter built on the groundwork of earlier developments in the Landvolkbewegung's history that were of continuing significance in this period. To recall the most important: the Landvolk's obstruction at confiscations and compulsory auctions was continuing to spread; *Das Landvolk's* propaganda was reaching a growing readership during this period; and legal action against Landvolk leaders and followers reached a climax with the Beidenfleth trial in April, intensifying the differences between the farmers and authorities and also producing additional publicity.

Assisting the Landvolkbewegung's expansion and giving it even greater momentum from March 1929 were the more concentrated promotional efforts at meetings and, in particular, the endeavours to establish the Crisis Assistance committees. The Crisis Assistance meetings alone attracted the interest of thousands of farmers and were a key factor in the spread of the movement from the west coast into the province's central and eastern counties. Furthermore, additional meetings held throughout the province during the year consistently drew substantial audiences.

The movement's rejection of organisation, of membership and statutes, was a strength in allowing the participation and support of all fellow farmers. Since farmers did not have to abandon other parties and organisations, this also facilitated cooperation with agricultural organisations such as the Farmers' League. Moreover, the absence of membership underlined the considerable and continuing appeal of the Landvolkbewegung's meetings since farmers were under considerably less duress to attend them than if they had been obliged as members to do so. While curiosity, the meetings' social allure and perceived expectations or pressure from colleagues may also have played a role, clearly what continued to draw crowds was the great attraction of the movement's speakers and their speeches.

This points to the enduring appeal of the Landvolk's ideology, a body of nationalistic, anti-democratic, anti-Semitic, anti-capitalistic and völkisch ideas, which as the Crisis Assistance meetings showed, were shared by leaders across the prov-

178 *Schleswig-Holsteinische Tageszeitung* No. 58, 10.3.1929, p. 4.

ince. From its inception, the movement's rhetoric had quickly moved from discussion of the crisis toward vehement opposition, to the Weimar system and the authorities – or as the Landvolk put it – to "the system", its "administrative apparatus" and "lackeys". Gradually, the movement's speakers, foremost Hamkens, developed a routine in which they both promoted resistance to the payment of taxes, and indoctrinated audiences with an enticingly simple explanation for the complex circumstances that had led to their misfortune, as well as all other real and supposed ills afflicting Germany. This was an updated version of the older anti-Semitic conceptions of the radical right, tailored to the farmers' plight. After the spring of 1929, there was little significant variation in the content of Landvolk ideology beyond an increase in völkisch elements from the autumn of that year. For the most part these came from a small number of extremists who used the movement as a forum to promote their fanatical racial views.

The Crisis Assistance committees threatened to increase considerably the actions which had given the Landvolkbewegung momentum in the winter: resistance to confiscations and the prevention of compulsory auctions. However, in driving the committees underground, the authorities hindered the institutionalisation of obstructive actions, leaving Landvolk leaders to content themselves with a less ambitious scope of operations. Nevertheless, throughout 1929 and 1930, they continued to help farmers conceal their dairy earnings and supported them in their regular pecuniary dealings with the authorities and financial institutions.

On the other hand, the Landvolkbewegung's plan of setting up Crisis Assistance committees as "additional governments" was a complete failure. The authorities' continued strategy against the movement in the form of bans, warnings and threats of legal action put an end to this attempt to gain greater control over local and provincial affairs. Even beforehand, however, the Crisis Assistance had little chance of success without a decided and decisive strategy and sufficient support both from other occupations and the farmers themselves. Here, the weakness of the movement's rejection of organisation became apparent. As much as it allowed all farmers to take part in its actions, the freedom of action that the absence of organisation permitted meant they were under less obligation to go along with their leaders' plans.

While the Crisis Assistance may have had little prospect of institutionalising tax obstruction or gaining executive power, it was an alarming expression of desperation and of a lack of confidence in the authorities and Weimar system. Indeed it was all the more striking precisely because the Landvolk's leaders had even contemplated such a hopeless and haphazard venture. Even in failure the Crisis Assistance furthered the radicalisation of the countryside. That is to say, when the Crisis Assistance stumbled, the Landvolk channelled its frustration against the authorities' bans and warnings. This intensified antagonisms culminating in the malicious denunciations of Oberpräsident Kürbis and provincially-patriotic demands for his departure.

Pranks and militant actions were an additional manifestation of the Landvolkbewegung's radicalisation. One measure of this was their unbridled and aggressive nature, to be seen most prominently in the raid on the courthouse prison in Hohenwestedt. Another measure was their purpose. The movement was acquiring an oppositional character that soon went far beyond impertinent refusals to follow police instructions.[179] Pranks, such as the abduction of the new director of the agricultural school in Garding and Ross's appeal against his conviction for insulting police officers, were clearly a source of spiteful amusement for the farmers, but thwarting the authorities was their essential objective. Viewed alongside the Crisis Assistance committees' attempts to gain greater control over local and provincial affairs, it seems that these actions, though driven by despair as a result of the crisis, also reflected the farmers' deeply-rooted unwillingness to accept the rule of outsiders, a consequence of their former independence and self-administration as discussed in Chapter One. Yet these actions, unlike the obstructive actions at confiscations and auctions, and the Crisis Assistance too, were only indirectly, if at all, connected with the agricultural crisis and the tax issue. Opposition to the authorities and the system had become central.

Despite enjoying a significant surge of growth after the Wöhrden incident, the NSDAP was envious of the Landvolkbewegung's success and continued to view it as a rival. The Landvolk's unruly and unpredictable actions gave the NSDAP good reason to be wary of being associated with it, especially because similarities, particularly in their demagogy, made it difficult for many in the wider community to differentiate between them. Abiding by the Reich leadership's instructions that the party was on a legal path to power, the NSDAP resisted pressure to become more extreme and took every opportunity to undermine the farmers' movement, stressing that its tactics were ineffective and that it failed to attract the support of other occupational groups. In attempting to win over the farmers, the NSDAP was dogged and repetitive, claiming that the Landvolkbewegung's ideas merely mirrored its own and stressing that only it had a programme, organisational structure and leader who could save Germany. Nonetheless, the Landvolk's popularity appeared to continue unabated, and its activism had not yet reached its peak.

---

179 Examples of this, as discussed in the previous chapter, were to be seen at the rally outside the Husum courthouse on January 3 and at the gathering at Husum's finance office on February 7. See chapter 4, pp. 120 and 128.

Chapter Six

The Landvolk's popularity peaks

Disturbances at two Landvolk rallies, in Husum and Neumünster, in the summer of 1929 marked the height of open conflict between Schleswig-Holstein's farmers and the authorities. Coinciding with the movement's bombing campaign, which will be considered in the following chapter, they made for the most turbulent period in the Landvolk's history. The rallies were held in honour of the commencement and conclusion of Hamkens' first prison term and both culminated in violent clashes with the police. The dramatic episode in Neumünster provided the inspiration for Hans Fallada's novel, *A Small Circus* (*Bauern, Bonzen und Bomben*). Police reports and abundant witness testimony enhance court findings, allowing a reconstruction of the events that reveals how the authorities lost control of the situation. This is fortunate since the details of the story – seemingly trivial decisions, moments and coincidences – were to have significant consequences. Yet just as important for the unrelenting radicalisation of the rural population was the rallies' aftermath, which led to the boycott of Neumünster. Relations between the authorities and the Landvolk were already bad, but the course of the rallies and the subsequent boycott exacerbated their differences and drew wider rural circles into their quarrel. While these developments had a radicalising effect on the province's agricultural associations, they aggravated tensions between the Landvolk and the NSDAP. Additional points of agreement and variance between the two movements came to light in this period as the NSDAP continued its efforts to profit from the Landvolkbewegung.

The Landvolk rally in Husum on July 1 marked the beginning of Hamkens' first prison term for inciting tax-strikes and took the county administrator and local police by surprise. Firstly, and much to their annoyance, the penal administration had failed to notify them that it had ordered Hamkens to Husum to begin the four-week sentence to which he had been sentenced in March.[1] Secondly, they had received no warning signs from the countryside because the Landvolk had kept preparations for the meeting and subsequent march to the county court prison underground.[2] The county administrator and local police, therefore, were quite unprepared to deal with the 300 to 400 farmers who gathered in the town in order to bid Hamkens farewell.

In the early afternoon the farmers met at a tavern where Paul Adam Ross led proceedings. Hamkens made a characteristic speech, which was described by the

1 See chapter 4, p. 120.

2 Landrat to the Regierungspräsident in Schleswig concerning "Landvolkdemonstration in Husum" Husum, 2.7.1929 and report from the Regierungspräsident to the Preuß. Minister des Innern concerning "Landvolkbewegung" Schleswig, 11.7.1929, both in LAS 301/4696 as well as 309/23017.

senior constable who watched the meeting as his "cursing at the Jewish-parliamentary system".[3] Ross, Johannes Kühl and Weschke followed, adding further bitter criticisms of the system and paying tribute to Hamkens. They praised him for his honesty, ideals and leadership, threatened vengeance and declared his prison term as establishing an Order Pour le Mérite for the movement.[4] Roused by these speeches and then singing the national anthem accompanied by Kiel's Stahlhelm band, the farmers were agitated even before they began their march.

Just ten police officers could be mustered from the town and neighbouring areas, too few to disperse the crowd and enforce a ban on a gathering of such size, too few, even, to cordon off the streets to the prison.[5] Their sole success was to prevent the band from playing outside. Otherwise, the farmers simply ignored or ridiculed their orders in typically defiant Landvolk fashion. When informed that the mayor had banned the march, Ross belittled the officers, "Oh, don't talk rubbish. Today, no one is going to tell us what to do!" and he added, menacingly, "Now stop kicking up a stink, otherwise something just might happen here today."[6] Then, underscoring the farmers' militancy, he turned to the crowd and told them the march had been banned, only to order them, "Into formation, march!" As they strode off in song, some of the farmers, evidently inspired by Ross's example, warned the police to refrain from any "imprudent acts".[7]

While the farmers passed through the town, cheering Hamkens and singing patriotic songs, the mayor directed the police to fall back in order to protect the courthouse at least, for he feared that the farmers might attempt to enter the building.[8] The ten officers hurried to Theodor Storm Street where they formed two columns blocking the farmers' path. On the farmers' approach, the sergeant in charge raised his arm for the crowd to halt. Initially, the first rows slowed a little, but they were pushed forward by those shoving from behind. With ever more force, the farmers pressed the police back, causing some of the officers to fall to the ground.[9] Driven some 20 meters, the officers from the country constabulary (*Landjäger*) drew their

3 Untitled police report signed "Polizeihauptwachtmeister Indorf" Husum, 1.7.1929, in both LAS 301/4696 and 309/23017.

4 Untitled police report signed "Polizeihauptwachtmeister Indorf" Husum, 1.7.1929, in both LAS 301/4696 and 309/23017; *Husumer Nachrichten* No. 132, 2.7.1929, p. 4; *Rendsburger Tageblatt* No. 152, 2.7.1929, p. 10.

5 Oberstaatsanwalt to the Preuß. Justizminister in Berlin concerning "Ausschreitungen von Landleuten in Schleswig-Holstein" Flensburg, 4.7.1929, LAS 309/22783.

6 Report titled "Bericht über die Vorgänge in der Theodor-Stormstraße am 1. Juli 1929 nachm. 15/40 Uhr" Husum, 1.7.1929, in both LAS 301/4696 and 309/23017.

7 *Vorwärts* No. 304, 2.7.1929 and *Norddeutsche Zeitung* No. 149, 2.7.1929, both in LAS 309/23017.

8 Report titled "Bericht über die Vorgänge in der Theodor-Stormstraße am 1. Juli 1929 nachm. 15/40 Uhr" Husum, 1.7.1929, in both LAS 301/4696 and 309/23017.

9 *Schleswig-Holsteinische Volkszeitung* No. 152, 3.7.1929, p. 4.

sabres, and with the communal police using their truncheons, finally brought the crowd to a stop directly in front of the courthouse entrance.[10]

Before commencing his first prison term, Hamkens bids the crowd farewell, Husum, July 1, 1929[11]

The farmers revelled in the uproar and their insubordination. Again they struck up the national anthem and at the end of the second verse Hamkens bade them farewell, declaring, "I'm now going to prison for four weeks and hope that in that time you will not rest in the fight against the Jewish leech system!"[12] As he passed through the gate, the farmers cheered and proceeded with the last verse. While they sang, a young member of the crowd approached the police officers shouting, "In the service regulations it is stated that civil servants are to remove their head wear when the national anthem is sung. I see that not one of you has done so, boo!" And with that the farmers booed the policemen. When they had quietened, the sergeant

10 Report from the Regierungspräsident to the Preuß. Minister des Innern concerning "Landvolkbewegung" Schleswig, 11.7.1929, in both LAS 301/4696 and 309/23017; *Rendsburger Tageblatt* No. 152, 2.7.1929, p. 10.

11 Photo: Stadtarchiv Neumünster.

12 Hamkens as quoted in the *Husumer Nachrichten* No.152, 2.7.1929, p. 4 and *Das Landvolk*, No. 92, 16.7.1929, p. 5. Also see the report titled "Bericht über die Vorgänge in der Theodor-Stormstraße am 1. Juli 1929 nachm. 15/40 Uhr" Husum, 1.7.1929, in both LAS 301/4696 and 309/23017.

instructed Ross to call on the crowd to disperse but Ross simply swore at him.[13] Nevertheless, after shouting and singing a few moments longer, the crowd began to disband of its own accord and thereby probably avoided further confrontation as some ten to fifteen minutes later a riot police squad, requested by the county administrator, arrived from Kiel.[14]

The events in Husum received considerable coverage in the provincial press. While most refrained from editorial comment, some of the middle-class and left-wing newspapers disapproved of the farmers' impertinence and provocative behaviour and argued that only the prudence of the police had prevented a more serious conflict.[15] Predictably, *Das Landvolk* countered, claiming that Hamkens' imprisonment during the hay harvest was chicanery. It blamed the "administrative apparatus" for the clash and referred to the police as the "soldateska".[16] Fearing that those who had taken part would face legal action, it blustered that if just one "of perhaps 1,000 participants" were put on trial, then "... we'll all testify that we marched despite the police's ban and then the administrative apparatus will have to take all 1,000 of us to court!"[17] They did not. The leaders, Ross, Weschke and Kühl, were later found guilty of resisting authority and sentenced to six months suspended prison terms and fines of RM 200 and RM 50, while Ross was also fined RM 30 for insulting the police officers.[18]

Under pressure from Husum's county administrator as well as the left-wing press, the provincial authorities sought to have Hamkens transferred from Husum to a location far from the grazing centres and livestock markets where there was less danger of further "sympathy rallies" or an attempt to break into the prison.[19]

---

13 Ross is reported to have said "Ich scheiß ihnen 'was!" (a fairly unusual but very offensive expression): report titled "Bericht über die Vorgänge in der Theodor-Stormstraße am 1. Juli 1929 nachm. 15/40 Uhr" Husum, 1.7.1929, in both LAS 301/4696 and 309/23017.

14 Landrat to the Regierungspräsident in Schleswig concerning "Landvolkdemonstration in Husum" Husum, 2.7.1929, in both LAS 301/4696 and 309/23017.

15 See, in particular, the *Itzehoer Nachrichten* No. 153, 3.7.1929 and *Norddeutsche Zeitung* No. 149, 2.7.1929, both in LAS 309/23017; *Nordischer Kurier* No. 153, 3.7.1929, p. 3. For further reports, see *Der Schleswiger* No. 154, 3.7.1929, LAS 309/23017; *Flensburger Nachrichten* No. 152, 2.7.1929, p. 4; *Husumer Nachrichten* No. 132, 2.7.1929, p. 4; *Rendsburger Tageblatt* No. 152, 2.7.1929, p. 10; *Schleswiger Nachrichten* No. 152, 2.7.1929, p. 6; *Schleswig-Holsteinische Tageszeitung* No. 154, 3.7.1929, LAS 309/23017.

16 *Das Landvolk* No. 92, 16.7.1929, p. 5 and No. 93, 17.7.1929, p. 5.

17 *Das Landvolk* No. 93, 17.7.1929, p. 5.

18 Oberstaatsanwalt to the Preuß. Justizminister in Berlin concerning "Ausschreitungen von Landleuten in Schleswig-Holstein" Flensburg, 4.7.1929 in LAS 309/22783; issue of the verdict of the Schöffengericht in Husum, 13.2.1931 and Oberstaatsanwalt to the Preuß. Justizminister in Berlin, Flensburg, 14.2.1931, GStA PK I. HA Rep. 84a No. 55410.

19 Regierungspräsident to the Preuß. Minister des Innern concerning "Landvolkbewegung" Schleswig, 11.7.1929, in both LAS 301/4696 and 309/23017; Landrat to the Regierungspräsident in Schleswig concerning "Landvolkdemonstration in Husum" Husum, 2.7.1929, in both LAS 301/4696 and 309/23017; *Volkszeitung* No. 152, 3.7.1929, p. 4.

The penal administration consequently transferred Hamkens to Neumünster and arranged for his release at a time of day not conducive to large rallies.[20] The transfer took place smoothly and several weeks passed without incident.[21] Concerns heightened, however, after detectives on July 27 intercepted a confidential notice sent out from *Das Landvolk's* business administrator, John Johnsen, to the movement's "trusted men", calling on the rural people to gather in Neumünster on August 1 to greet Hamkens on his release.[22]

The plans for a rally in Neumünster became public after the *Schleswig-Holsteinische Volkszeitung* printed a letter written by Hamkens from prison to Johannes Kühl, requesting that a crowd meet him on August 1. It is not clear how the *Schleswig-Holsteinische Volkszeitung* came by the letter which it claimed "favourable winds" had blown on to its editorial desk, but it delighted in its find, mocking Hamkens' instructions as those of a "film director" and scoffing at the farmers' inability to pay taxes when entrance fees at meetings were being used to finance "Landvolk spectacles".[23] The *Schleswig-Holsteinische Volkszeitung's* assessment of the crisis was unsympathetic, but it certainly got to the heart of the matter by highlighting the Landvolk's great capacity for creating a commotion.

Alarmed at this news, the provincial authorities took steps to avoid another disturbance. First, they arranged for Hamkens to be secretly moved to Flensburg as a precautionary measure.[24] Then, on the day before his release, representatives of the Regierungspräsident travelled to Neumünster to meet with the town's mayor, Lindemann, and police commander, Chief Inspector Bracker, seeking to prevent the anticipated demonstration.[25] As the town's police administrator, it was Mayor Lindemann who had the power to enforce the prohibition of open rallies and even of indoor meetings if they were deemed to pose a danger to the public peace, safety and order. In answer to the Regierungspräsident's representatives' pleas, however,

20 Regierungspräsident to the Präsident des Strafvollzugsamtes in Kiel concerning "Strafvollstreckung gegen den Hofbesitzer Hamkens in Tetenbüll" Schleswig, 2.7.1929, LAS 309/23017.

21 Präsident des Strafvollzugsamtes to the Regierungspräsident in Schleswig concerning the "Strafvollzug gegen den Hofbesitzer Hamkens in Tetenbüll" Kiel, 4.7.1929, in both LAS 301/4696 and 309/23017. Also see *Der Schleswiger* No. 156, 5.7.1929, LAS 309/23017.

22 Copy of a notice from John Johnsen, Itzehoe, sent from the Abteilung IA to the Regierungspräsident in Schleswig, Kiel, 27.7.1929, LAS 309/23017.

23 *Schleswig-Holsteinische Volkszeitung* No. 173, 27.7.1929, p. 2; Regierungspräsident to the Preuß. Minister des Innern in Berlin concerning "Vorfälle in Neumünster am 1.8.1929 aus Anlaß der Entlassung des Landvolkführers Hofbesitzers Wilhelm Hamkens in Tetenbüttel [sic] aus dem Gefängnis" Schleswig, 5.8.1929, LAS 309/23017.

24 Undated report titled "Niederschrift über die polizeilichen Maßnahmen anläßlich der Entlassung des Landvolk-Führers, Landwirt Hamkens aus Tetenbüll", LAS 309/23017.

25 Report from the Bürgermeister to the Regierungspräsident in Schleswig concerning "Landvolkbewegung in Neumünster" Neumünster, 12.8.1929 and the undated report titled "Niederschrift über die polizeilichen Maßnahmen anläßlich der Entlassung des Landvolk-Führers, Landwirt Hamkens aus Tetenbüll", both in LAS 309/23017.

Lindemann declared that he viewed a Landvolk rally as harmless and explained that rallies of the communists and the Republican-friendly paramilitary corps, the *Reichsbanner*, were always peaceful. Despite warnings that the Landvolkbewegung was more dangerous than the communists, especially because it was a movement without organisation, definite membership or leaders, Lindemann was unmoved.[26] He could see no reason why the rally should be banned and was adamant that such events should be left to run their course.[27]

Doubting that Neumünster's police force, a chief inspector and 27 officers, were sufficient to maintain order, the provincial government representatives pressed Bracker and Lindemann to accept the assistance of a riot police contingent.[28] In fact, so great was their concern, they even offered to put a further unit on call in Kiel.[29] Bracker, however, was of the opinion that a riot police contingent presence would simply be a provocation.[30]

It was at this point in the discussion that Lindemann was called away to meet with Franz Schwarzloh (Tungendorf), a local Landvolk leader who had come to the town hall on Lindemann's invitation. Seeking Lindemann's consent, Schwarzloh confirmed that no demonstration would take place in front of the prison and explained that the farmers intended to wait for Hamkens' arrival in pubs on the town square before marching to the town's livestock auction hall where they would hold a meeting.[31] Reassured, Lindemann gave Schwarzloh his permission for these plans and returned to the provincial government representatives, who in the meantime had continued fruitlessly to warn Bracker of the dangers the Landvolkbewegung posed.[32] Lindemann reaffirmed his decision and he refused, despite repeated urgings, to allow the deployment of riot police in Neumünster. Only after much discussion did he agree to have the contingent stationed on call in the village of Einfeld, on Neumünster's outskirts.[33]

---

26 Issue of the verdict of the Schöffengericht in Neumünster, 12.11.1929, in both LAS 301/4701 and 309/23017.

27 Undated statement by Bürgermeister Lindemann, Neumünster, LAS 309/23017.

28 Statement by Polizei-Major Bocatius, Schleswig, 8.8.1929, LAS 309/23017.

29 Regierungspräsident to the Preuß. Minister des Innern in Berlin concerning "Vorfälle in Neumünster am 1.8.1929 aus Anlaß der Entlassung des Landvolkführers Hofbesitzers Wilhelm Hamkens in Tetenbüttel [sic] aus dem Gefängnis" Schleswig, 5.8.1929, LAS 309/23017.

30 Copy of a report titled "Zusammenfassende Darstellung der Neumünsteraner Vorfälle" Schleswig, 10.8.1929, in both LAS 301/4701 and 309/23017.

31 See the report titled "Abschrift des in Neumünster erhaltenen Lageberichtes" Neumünster, 31.7.1929, appended to the report from the "3. Polizei-Bereitschaft" concerning "Einsatz am 1. u. 2. August 1929 in Neumünster" Kiel, 2.8.1929, LAS 309/23017.

32 Undated statement by Bürgermeister Lindemann and statement by Polizeioberinspektor Bracker, Neumünster, 6.8.1929, both in LAS 309/23017.

33 Statement by Polizei-Major Bocatius, Schleswig, 8.8.1929 and undated statement by Bürgermeister Lindemann, both in LAS 309/23017; issue of the verdict of the Schöffengericht in Neumünster, 12.11.1929, in both LAS 301/4701 and 309/23017.

On the evening of July 31, Hamkens was transferred to Flensburg without incident and without detection.[34] That the provincial government went to great lengths to keep this a secret is puzzling.[35] Doing so might have served to prevent the farmers from proceeding to Flensburg to welcome Hamkens there, but it did not deter them from demonstrating in Neumünster. At best, the belated discovery of the transfer could merely cause the farmers inconvenience and, at worst, lead to unpredictable changes of plans with unknown consequences for the authorities.

On August 1, farmers began to gather in the pubs around Neumünster's main square throughout the course of the morning, but so few had arrived by midday that Mayor Lindemann and Chief Inspector Bracker must have felt vindicated in their belief that the provincial government's fears were inflated.[36] Yet even after news arrived in the early afternoon that trucks packed with farmers were on their way from the west coast, Lindemann saw no cause to call on the assistance of the riot police contingent.[37] By three o'clock, the number of farmers in the pubs had swelled to some 2,000 and curious citizens of Neumünster began to line the streets.[38] There was still no sign of trouble. However, after learning of Hamkens' transfer to Flensburg and that he would not arrive for some time, the farmers soon resolved to march to the auction hall immediately and wait for him there. In response, Chief Inspector Bracker hastily made arrangements for his officers to take up positions along the march route while he made his way to the head of the procession.[39]

As Bracker hurried across the town square, Walther Muthmann, the agronomist and member of the Wehrwolf who assisted Landvolk leader Peter Petersen in Plön County, emerged from a pub carrying a flag. Reminiscent of the standard of the German Peasant Wars, it was black, but completing the old national colours, in its centre were symbols of the farmers' toil and struggle: a white plough and a red sword. Although it had been on display in the pub throughout the afternoon, the

---

34 Undated report titled "Niederschrift über die polizeilichen Maßnahmen anläßlich der Entlassung des Landvolk-Führers, Landwirt Hamkens aus Tetenbüll", LAS 309/23017.

35 Report from the Abteilung IA 2 titled "Bericht" Kiel, 2.8.1929, in both LAS 301/4701 and 309/23017.

36 Issue of the verdict of the Schöffengericht in Neumünster, 12.11.1929, in both LAS 301/4701 and 309/23017.

37 Undated statement by Bürgermeister Lindemann and the report from the "3. Polizei-Bereitschaft" concerning "Einsatz am 1. u. 2. August 1929 in Neumünster" Kiel, 2.8.1929, both in LAS 309/23017.

38 Undated statement by Polizeioberinspektor Bracker, LAS 309/23017.

39 Statement by Polizeioberinspektor Bracker, Neumünster, 6.8.1929 and the undated report titled "Niederschrift über die polizeilichen Maßnahmen anläßlich der Entlassung des Landvolk-Führers, Landwirt Hamkens aus Tetenbüll", both in LAS 309/23017; issue of the verdict of the Schöffengericht in Neumünster, 12.11.1929, in both LAS 301/4701 and 309/23017.

flag caused a stir among the bystanders, particularly on account of a sickle Muthmann had attached to the top of the pole.[40]

The Landvolk flag: black with a white plough and red sword[41]

Bracker, who did not realise the sickle was not nearly as dangerous as it appeared, as its blade had been dulled, proceeded directly toward Muthmann.[42]

40 Issue of the verdict of the Schöffengericht in Neumünster, 12.11.1929, in both LAS 301/4701 and 309/23017. For the bystanders' reactions, see the statements of Rudolf Jürgensen, Neumünster, 19.8.1929, Ludwig Stender, Neumünster, 5.8.1929 and Hans Tietgen, Neumünster, 11.8.1929, all in LAS 309/23017.

41 Photo of the Landvolk flag: Stadtarchiv Neumünster.

Bracker reached Muthmann just as he joined the farmers standing at the front of the procession and asked him politely to return the flag to the pub. When Muthmann did not respond and then subsequently refused an order, the chief inspector declared the flag confiscated. But as Bracker stretched out to seize the flag, the procession suddenly set in motion, bumping him aside.

Accompanied by two of his officers, Bracker quickly caught up with the front of the procession, where he ordered once more that the flag be handed over. As Muthmann again refused, the officers moved in, attempting to grab hold of the flagpole. Muthmann held tight. Led by Paul Adam Ross, the farmers in the first rows lent him assistance, surging forward, pushing and shouting at the officers, before beginning to strike them with their fists and sticks.[43] In the ensuing melee, Bracker and one of the officers were almost impaled on the sickle as the flagpole was thrust back and forth. Meanwhile, another officer who had rushed to his colleagues' assistance tripped and fell to the ground in the turmoil and was momentarily knocked unconscious by the farmers' unrelenting assault.[44]

Overwhelmed, Bracker ordered sabres to be drawn. Regrettably, the one remaining officer able to help him used his against Johannes Thiess, a Landvolk follower who had not been involved in the fray. Struck on the shoulder, Thiess caught hold of the blade, pushed the officer against a lamppost and held him in a deadlock.[45] Bracker was left surrounded and the farmers succeeded in grabbing his raised sabre. Struck from all sides, Bracker drew his unloaded pistol and ordered the other officers to do the same. "Make way!" he shouted. Finally, the farmers fell back.[46]

---

42 Indictment issued by the Oberstaatsanwalt, Kiel, 28.9.1929 and the report from the Magistrat der Stadt Neumünster to the Regierungspräsident in Schleswig concerning "Die Bauerndemonstration am 1. August, Neumünster" Neumünster, 22.8.1929, both in LAS 309/23017.

43 See the statements by Polizeihauptwachtmeister Nickel, Neumünster, 1.8.1929, Friedrich Strohbach, Neumünster, 7.8.1929, Hans Tietgen, Neumünster, 11.8.1929 and Polizeihauptwachtmeister Gussmann, Neumünster, 19.10.1929 and the indictment issued by the Oberstaatsanwalt, Kiel, 28.9.1929, all in LAS 309/23017.

44 Statements by Polizeihauptwachtmeister Wilhelm Peters, Neumünster, 2.8.1929 and 13.8.1929, by Wilhelm Heller, Neumünster, 1.9.1929, Hermann Drößler, Neumünster, 4.8.1929, Agnes Körner, Neumünster, 4.8.1929 and Polizeihauptwachtmeister Gußmann, Neumünster, 19.10.1929, all in LAS 309/23017; issue of the verdict of the Schöffengericht in Neumünster, 12.11.1929, in both LAS 301/4701 and 309/23017.

45 Statements by Johannes Markus Thiess, Neumünster, 1.8.1929, Polizeihauptwachtmeister Nickel, Neumünster, 1.8.1929 and Kriminalsekretär Zimmny, Neumünster, 21.10.1929, all in LAS 309/23017. Also see the issue of the verdict of the Schöffengericht in Neumünster, 12.11.1929, in both LAS 301/4701 and 309/23017.

46 Issue of the verdict of the Schöffengericht in Neumünster, 12.11.1929, in both LAS 301/4701 and 309/23017; statements by Polizeioberinspektor Bracker, Neumünster, 6.8.1929, both in LAS 309/23017.

Threatening the farmers with his pistol, Bracker created enough space for himself to hurry in search for more officers. He did not have to go far, as some of those who had been posted at intervals along the march route had been alerted by a passing cyclist and consequently came running toward him.[47] Without delay, Bracker turned back with his reinforcements to face the oncoming procession.[48]

Bracker called on the crowd to stop and again ordered Muthmann, still at the procession's head, to hand over the flag. Twice he repeated this order, but the farmers continued forward.[49] There was a cry of "That is our symbol. We will not surrender it!" as Muthmann dropped back into the crowd and Ross urged the marchers on.[50]

When the officers moved to seize the flag, the farmers once more beat them with their sticks.[51] The ferocious but disciplined manner in which they attacked astonished the chief inspector who later described the front rows as an "assault squad" (*Stoßtrupp*).[52] The police truncheons were poor defence against the farmers' longer, heavier makeshift weapons and when one of his men collapsed with blood running out from under his helmet, Bracker gave the command, "Sabres out, strike!"[53]

Falling to the ground under the officers' attack, Muthmann tried desperately to retain the flag, pushing and kicking wildly as he lay supine. He yelled at the officers, "The flag is our symbol and we will defend it with our blood and our lives! I will die for it. I will not give it up!"[54] The officers continued to strike him, slashing at his hands, but only after cutting off one of his fingers and severing the muscles

47 Issue of the verdict of the Schöffengericht in Neumünster, 12.11.1929, in both LAS 301/4701 and 309/23017 and the statements by Polizeihauptwachtmeister Peters, Neumünster, 2.8.1929 and Rudolf Jürgensen, Neumünster, 19.8.1929, all in LAS 309/23017.

48 Report from the Abteilung IA 2 titled "Bericht" Kiel, 2.8.1929, in both LAS 301/4701 and 309/23017; statements by Kriminalsekretär Zimmny, Neumünster, 21.10.1929 and Polizeihauptwachtmeister Petersen, Neumünster, 19.10.1929, both in LAS 309/23017.

49 Statements by Polizeihauptwachtmeister Nissen, Neumünster, 10.10.1929, Polizeihauptwachtmeister Münnich, Neumünster, 17.10.1929 and Polizeihauptwachtmeister Petersen, Neumünster, 19.10.1929, all in LAS 309/23017.

50 Statements by Polizeihauptwachtmeister Afflerbach, Neumünster, 16.10.1929 and Hans Tietgen, Neumünster, 11.8.1929, both in LAS 309/23017.

51 Statements by Polizeihauptwachtmeister Afflerbach, Neumünster, 2.8.1929 and Polizeihauptwachtmeister Münnich, Neumünster, 17.10.1929, both in LAS 309/23017.

52 Statement by Polizeioberinspektor Bracker, Neumünster, 6.8.1929, LAS 309/23017.

53 Statements by Polizeihauptwachtmeister Fritsche, Neumünster, 16.10.1929 and Polizeihauptwachtmeister Roos, Neumünster, 19.10.1929, both in LAS 309/23017.

54 Issue of the verdict of the Schöffengericht in Neumünster, 12.11.1929, in both LAS 301/4701 and 309/23017; statements by Polizeihauptwachtmeister Afflerbach, Neumünster, 2.8.1929, Polizeihauptwachtmeister Fritsche, Neumünster, 16.10.1929 and Polizeihauptwachtmeister Peters, Neumünster, 2.8.1929, all in LAS 309/23017.

of his right forearm, slicing them to the bone, did Muthmann finally let go of the flag.[55]

Walther Muthmann is helped from the ground, Neumünster, August 1, 1929[56]

The police accounts of this incident are extensive and detailed, but conspicuously deficient in explaining the use of their sabres.[57] Perhaps the officers feared

---

55 Indictment issued by the Oberstaatsanwalt, Kiel, 28.9.1929, LAS 309/23017.

56 Photo of Walther Muthmann: Stadtarchiv Neumünster.

57 See, for example, the undated statement by Polizeioberinspektor Bracker and the statements by Polizeihauptwachtmeister Afflerbach, Neumünster, 2.8.1929, Polizeihauptwachtmeister Peters, Neumünster, 2.8.1929, Polizeihauptwachtmeister Nissen, Neumünster, 10.10.1929, Polizeihauptwachtmeister Münnich, Neumünster, 17.10.1929, Polizeihauptwachtmeister

they had overreacted and did not wish to implicate one another for heavy-handedness. Significantly, three farmers were seriously injured who had not been involved in the melee, but who were, in fact, quite some distance from it. One received a cut to the shoulder and another was struck on the head from behind as he climbed the stairs to enter a pub.[58] The third, Ernst Behr (Mettenhof), was an elderly farmer, who on crossing the street, did not hear or perhaps ignored an officer's order to stop. He received a sabre swipe across the face that left him lying in a pool of blood with his nose severed from his face.[59]

Kiel's Stahlhelm band leads the procession as the march continues. Behind, on the picture's right, march Claus Heim and Paul Adam Ross, Neumünster, August 1, 1929[60]

Cries for revenge could be heard as the march continued to the auction hall, now with Kiel's Stahlhelm band at its head.[61] There, some 2,000 Landvolk follow-

Fritsche, Neumünster, 16.10.1929 and Polizeihauptwachtmeister Diercks, Neumünster, 21.10.1929, all in LAS 309/23017.

58 Issue of the verdict of the Schöffengericht in Neumünster, 12.11.1929, in both LAS 301/4701 and 309/23017.

59 Report from the Abteilung IA 2 titled "Bericht" Kiel, 2.8.1929, in both LAS 301/4701 and 309/23017; statements by Friedrich Strohbach, Neumünster, 7.8.1929, Polizeimeister Lütjens, Neumünster, 18.10.1929 and Polizeihauptwachtmeister Storkebaum, Neumünster, 18.10.1929, all in LAS 309/23017.

60 Photo of the march in Neumünster: Stadtarchiv Neumünster.

ers gathered, intending to make speeches and sing patriotic songs until Hamkens arrived.[62] Schwarzloh, and in particular, Kühl and then Max Bestmann (Hohn) inflamed the audience, telling of the flag's bloody confiscation and calling for retribution. Kühl, for example, told the farmers it was their duty to rebel and Bestmann called on them to retrieve the flag and demanded that Neumünster's police chief be forced to apologise.[63]

However, before any action could be undertaken, the Landvolk discovered that the police, together with the riot police detachment from Kiel, were assembling outside the building. On learning that bystanders had seen a number of farmers loading revolvers during the procession and that they were presently listening to inflammatory speeches, Mayor Lindemann had ordered that the farmers be disarmed and the meeting dissolved if they failed to cooperate.[64]

Appointed to head a Landvolk delegation to request the return of the flag and the release of those arrested, Bestmann left the auction hall, only to be sent back in by Chief Inspector Bracker under instructions to inform the farmers to surrender their sticks and other weapons. When Bestmann returned soon afterwards to tell the chief inspector that the farmers had demanded the police come into the hall themselves if they wanted anything, Bracker dispatched a sergeant to declare the meeting closed.[65] His entrance was greeted with thunderous screams, shouts and stamping, while the Stahlhelm band struck up the national anthem. Waiting until it was finished, the sergeant hoped to make himself understood, but the crowd immediately began to sing the Schleswig-Holstein anthem, leaving him little choice but to withdraw.[66]

The police now proceeded to clear the hall by force, breaking in through the main entrance and removing the band, which had begun to play the national anthem

---

61 Indictment issued by the Oberstaatsanwalt, Kiel, 28.9.1929 and the undated statement by Polizeioberinspektor Bracker, both in LAS 309/23017.

62 Issue of the verdict of the Schöffengericht in Neumünster, 12.11.1929, in both LAS 301/4701 and 309/23017.

63 Report from the Abteilung IA 2 titled "Bericht" Kiel, 2.8.1929, in both LAS 301/4701 and 309/23017.

64 Undated statements by Polizeioberinspektor Bracker and Bürgermeister Lindemann and the report from the "3. Polizei-Bereitschaft" concerning "Einsatz am 1. u. 2. August 1929 in Neumünster" Kiel, 2.8.1929 in LAS 309/23017. Bystanders' estimates of the number of farmers with firearms ranged between four and eight. See statements by Hugo Schlemmer, Neumünster, 3.8.1929, Ludwig Stender, Neumünster, 5.8.1929 and Friedrich Strohbach, Neumünster, 1.8.1929 and 7.8.1929, all in LAS 309/23017.

65 Issue of the verdict of the Schöffengericht in Neumünster, 12.11.1929, in both LAS 301/4701 and 309/23017.

66 Report from the Abteilung IA 2 titled "Bericht" Kiel, 2.8.1929 and the issue of the verdict of the Schöffengericht in Neumünster, 12.11.1929, both in LAS 301/4701 as well as 309/23017.

again.[67] Bestmann and Rudolf Jens, who had been making a speech as the officers entered the hall, allegedly encouraged the farmers to resist the police.[68] Those who did were driven out under the blows of the officers' truncheons, while others climbed through the windows. Outside, the police confiscated some 100 oak sticks and conducted body searches. Though they found just one revolver and a steel rod, it was assumed that the other farmers with guns were among the many who had managed to escape.[69]

Farmers leaving the auction hall in Neumünster, August 1, 1929[70]

Most of the farmers made their way to the train station, where they regrouped at the square outside and began to cheer for Hamkens and to resume their singing

67 Report from the Abteilung IA 2 titled "Bericht" Kiel, 2.8.1929, in both LAS 301/4701 and 309/23017; report from the "3. Polizei-Bereitschaft" concerning "Einsatz am 1. u. 2. August 1929 in Neumünster" Kiel, 2.8.1929, LAS 309/23017.

68 Indictment issued by the Oberstaatsanwalt, Kiel, 28.9.1929, LAS 309/23017.

69 Report from the "3. Polizei-Bereitschaft" concerning "Einsatz am 1. u. 2. August 1929 in Neumünster" Kiel, 2.8.1929 and report from the Regierungspräsident to the Preuß. Minister des Innern in Berlin concerning "Vorfälle in Neumünster am 1.8.1929 aus Anlaß der Entlassung des Landvolkführers Hofbesitzers Wilhelm Hamkens in Tetenbüttel [sic] aus dem Gefängnis" Schleswig, 5.8.1929, both in LAS 309/23017. For bearing the weapons, Max Thun and Johann Nagel, were convicted in October in separate trials and fined RM 100 and RM 50 respectively. See the reports from the Generalstaatsanwalt in Kiel to the Preuß. Justizminister in Berlin, Kiel, 29.10.1929 and Kiel, 6.11.1929, both in GStA PK I. HA Rep. 84a No. 55389.

70 Photo: Stadtarchiv Neumünster.

of the national anthem.[71] They defied police orders to disperse but when they attempted to form ranks they were driven into the nearby pubs.[72] Despite rumours of an attempt to recover the flag, most of the farmers departed towards evening, as there was still no sign of Hamkens' arrival. Several hundred farmers proceeded to Rendsburg, hoping to intercept him there, but this meant that when he finally did arrive in Neumünster, his appearance was an anti-climax, for just 60 farmers had remained to welcome him.[73]

One might assume that the main questions asked in the aftermath of the events in Neumünster were whether the conflict could have been avoided had Mayor Lindemann taken greater heed of the Regierungspräsident's warnings and ensured more police were in place, or if the march would have passed uneventfully had Chief Inspector Bracker not chosen to remove the flag, or at least done so either with greater composure or greater force from the beginning. These were not, however, the questions that dominated public discourse afterwards, particularly in the countryside.

The Landvolk, of course, firmly believed that the police intervention was completely unjustified. Yet there is some evidence that a significant section of the rural public in Schleswig-Holstein also shared this view. According to Thygge Thyssen, the contemporary director of the Chamber of Agriculture, the entire countryside stood behind the Landvolk.[74] Outraged that the farmers should not have been allowed to carry a flag when they had clearly been permitted to march, the rural population overlooked the fact that the Landvolk had failed to comply with police instructions. The rural population discounted the Landvolkbewegung's share of responsibility for the bloodshed in attacking the police, but was convinced the police's conduct was heavy-handed and directed against the farmers as an occupation.

Large attendances at Landvolkbewegung meetings in August denoted a surge in the movement's popularity. After his recovery, these meetings often featured

71 Issue of the verdict of the Schöffengericht in Neumünster, 12.11.1929, in both LAS 301/4701 and 309/23017.

72 Report from the "3. Polizei-Bereitschaft" concerning "Einsatz am 1. u. 2. August 1929 in Neumünster" Kiel, 2.8.1929 in LAS 309/23017; issue of the verdict of the Schöffengericht in Neumünster, 12.11.1929, in both LAS 301/4701 and 309/23017.

73 Report from the Abteilung IA 2 titled "Bericht" Kiel, 2.8.1929, in both LAS 301/4701 and 309/23017; undated report titled "Niederschrift über die polizeilichen Maßnahmen anläßlich der Entlassung des Landvolk-Führers, Landwirt Hamkens aus Tetenbüll", LAS 309/23017; Polizeiverwaltung to the Regierungspräsident in Schleswig and the Landrat in Rendsburg concerning "Landvolkbewegung" Rendsburg, 2.8.1929, LAS 309/23017; report from the Landrat des Kreises Eiderstedt to the Regierungspräsident in Schleswig concerning "Entlassung des Landmannes Wilhelm Hamkens aus dem Gefängnis" Tönning, 2.8.1929, in both LAS 301/4696 and 309/23017.

74 Thyssen, *Bauer und Standesvertretung*, p. 408.

Muthmann, the "flag-bearer", as a secondary speaker.[75] Johannes Thiess, detained in Neumünster after wrestling with an officer, received a hero's welcome on his return home to Wiemersdorf. A reception committee, the fire brigade's brass band and flags in the imperial colours awaited him at the station and almost the entire village attended a subsequent meeting where the "reckless" conduct of the police was vehemently condemned.[76] Symptoms of unabating anger in the following weeks included fervent poems printed in *Das Landvolk* as well as the expedition of three drunken farmers, two of them community chairmen, who drove to the Neumünster town hall and demanded that the truncheons with which they had been struck be handed over.[77]

Although none could match its rage, other newspapers joined *Das Landvolk* in criticising the police conduct and sympathising with the farmers.[78] Agricultural organisations and political parties, including the Farmers' League and the province's DNVP, demanded a thorough investigation into the police's conduct.[79] Public opinion was so hostile that the Public Servants Association deemed it necessary to call on Regierungspräsident Abegg for his support against the unceasing public agitation towards the police.

From the beginning, the public was bound to side with the farmers' representation of the events and question the police's conduct because initial reports were brief and did not elucidate the police's motives for striking the farmers.[80] More importantly, the authorities' official report which followed patently overstated the threat the farmers had posed: exaggerating the danger of the farmers' walking sticks, falsely claiming there had been altercations between the townspeople and the farmers and insisting the flag with its sickle was a "dangerous instrument".[81]

75 Report from the Regierungspräsident to the Oberpräsident concerning "Landvolkversammlung in Angeln" Schleswig, 5.9.1929, LAS 301/4696; *Das Landvolk* No. 120, 17.8.1929, p. 5, No. 121, 18.9.1929, p. 5, No. 180, 26.10.1929, p. 9 and No. 211, 3.12.1929, p. 4.

76 Report from the Landrat to the Regierungspräsident in Schleswig concerning "Landvolk-Kundgebung in Wiemersdorf, Kreis Segeberg" Bad Segeberg, 6.8.1929, in both LAS 301/4696 and 309/23017; *Bramstedter Nachrichten* No. 280, 3.8.1929, LAS 301/4696.

77 *Nordischer Kurier* No. 198, 24.8.1929, p. 7 and for examples of poems in *Das Landvolk*, see No. 108, 3.8.1929, p. 1, No. 112, 8.8.1929, p. 5, No. 115, 11.8.1929, p. 6 and No. 127, 25.8.1929, p. 5.

78 *Das Landvolk* No. 108, 3.8.1929, p. 1 and No. 109, 4.8.1929, p. 1, No. 110, 6.8.1929, p. 6 and No. 112, 8.8.1929, p. 5. Also see the *Schleswig-Holsteinische Tageszeitung* No. 182, 4.8.1929, p. 4 and *Flensburger Nachrichten* No. 357, 3.8.1929, p. 2.

79 Letter from the Schleswig-Holsteinischer Bauernbund to the Regierungspräsident in Schleswig, 2.8.1929 and letter from the DNVP Landesverband Schleswig-Holstein to the Regierungspräsident in Schleswig, Kiel, 5.8.1929, both in LAS 309/23017; *Flensburger Nachrichten* No. 357, 3.8.1929, p. 2; *Kieler Neueste Nachrichten* No. 182, 7.8.1929, p. 1.

80 See, for example, the report in the *Kieler Neueste Nachrichten* No. 179, 3.8.1929, p. 3.

81 Report from the Regierungspräsident to the Preuß. Minister des Innern in Berlin concerning "Vorfälle in Neumünster am 1.8.1929 aus Anlaß der Entlassung des Landvolkführers Hofbesitzers Wilhelm Hamkens in Tetenbüttel [sic] aus dem Gefängnis" Schleswig, 5.8.1929,

Far from regaining public support, these blatantly calculated embellishments undermined the authorities' credibility. In particular, the discovery that the sickle was blunt soon became public.[82]

According to the Rural League, the authorities' report was simply designed to justify the police's conduct. It declared its resolve not to participate in any meetings, shows or other functions in Neumünster until amends were made.[83] At the same time, a farmer, Claus Harms, placed a notice in Neumünster's daily newspaper calling for the immediate dismissal of Bracker and the officers who had attacked the farmers and driven them from their own auction hall. If Neumünster's authorities did not comply with this demand he threatened that farmers would resort to "self-help" and move their annual gymkhana, auctions and agricultural association meetings to other locations.[84]

Das Landvolk eagerly embraced the idea, which it soon expanded to a call for a general boycott of the town. Though conscious that it could harm many citizens undeservedly, *Das Landvolk* nevertheless considered a boycott the most effective means of putting pressure on the town's authorities.[85] One after another, agricultural organisations moved their business from the town. The provincial gymkhana was transferred to Bad Segeberg, where it has remained to this day. The Schleswig-Holstein Horse Breeders' Association moved its crossbreed foal show and stallion ratings elsewhere. The cattle auctions, the pig market and all other agricultural association meetings followed. Then the farmers began to forego shopping in the town.[86] It was a boycott that was to last a full year.

Both the Rural League and the Farmers' League readily endorsed the boycott.[87] The Farmers' League, moreover, now took the opportunity to support the Landvolkbewegung openly and welcome cooperation. It conceded that many of its members were in the movement and it forgave the occasional critical comments made about it by individual Landvolk followers. Apparently these gave no cause for concern for they were considered a consequence of the movement's open struc-

---

LAS 309/23017; *Vossische Zeitung* No. 184, 3.8.1929, p. 4; *Kieler Neueste Nachrichten* No. 180, 4.8.1929, p. 10.

82 Magistrat der Stadt Neumünster to the Regierungspräsident in Schleswig concerning "Die Bauerndemonstration am 1. August, Neumünster" Neumünster, 22.8.1929, LAS 309/23017.

83 *Landbote für Schleswig-Holstein* No. 62, 8.8.1929, p. 1; *Kieler Neueste Nachrichten* No. 182, 7.8.1929, p. 1.

84 *Holsteinischer Courier* No. 183, 7.8.1929, p. 4. For the subsequent legal proceedings against Harms, see the indictment issued by the Schöffengericht Kiel, undated, GStA PK I. HA Rep. 84a No. 55389; *Holsteinischer Courier* No. 26, 31.1.1930, p. 9.

85 *Das Landvolk* No. 113, 9.8.1929, p. 5.

86 *Flensburger Nachrichten* No. 368, 9.8.1929, p. 2; *Kieler Zeitung* No. 221, 11.8.1929, p. 14; *Das Landvolk* No. 126, 24.8.1929, p. 5; report from the Landrat to the Regierungspräsident in Schleswig concerning "Boykott Neumünster" Bad Segeberg, 8.9.1929, LAS 309/23017. Also see Thyssen, *Bauer und Standesvertretung*, p. 408.

87 For a further call from the Rural League, see *Das Landvolk* No. 116, 13.8.1929, p. 5.

ture which allowed all to participate.[88] Hailing what it saw as an end to the Farmers' League's reserve, *Das Landvolk* pointed to the complementary nature of their relationship. It judged the Landvolkbewegung's "fundamental struggle against the Jewish-parliamentary system" as the "prerequisite for [the] successful economic political work" of the Farmers' League.[89]

Even the Chamber of Agriculture seemed to side with the Landvolkbewegung. Incensing Neumünster's town council and the provincial authorities, the Chamber endorsed a press notice which affirmed the boycott and claimed that the Neumünster incident had united agriculture.[90] Oberpräsident Kürbis' enquiries revealed that a member of the Chamber's staff had released the notice without the board's consent and on Kürbis' insistence the Chamber agreed to publish a declaration disclaiming its responsibility for the notice. Much to the authorities' chagrin, however, the Chamber published the repudiation not in the province's daily papers but only in its inconsequential agricultural weekly. Additionally, the Chamber refused to take a position on the incident and the boycott until there was an official "clarification".[91] The Chamber thus found itself in an awkward position, displeasing the provincial authorities, yet failing to join its members in condemning the police and supporting the boycott.

As the extent of the boycott increased, so too did the terms demanded for its end. At a meeting in Meldorf in South Dithmarschen, the Landvolk called for the return of the flag in an official, honourable and public manner; Mayor Lindemann's dismissal; a payment of RM 10,000 for damages to the rural people, as well as a life-long pension for Muthmann.[92] The DNVP indirectly supported these demands in the Prussian Landtag, inquiring if the injured farmers had received compensation and whether disciplinary measures had been taken against Neumünster's police administration.[93]

---

88 See the letter from South Dithmarschen and Neumünster's Rural League and the extract from the Bauernbund in *Das Landvolk* No. 116, 13.8.1929, p. 5; *Schleswiger Nachrichten* No. 187, 12.8.1929, p. 7.

89 *Das Landvolk* No. 123, 21.8.1929, p. 5.

90 *Holsteinischer Courier* No. 193, 19.8.1929, p. 5. Letter from the Magistrat der Stadt Neumünster to the Oberpräsident in Kiel, Neumünster, 20.8.1929 and letter from the Oberpräsident to the Landwirtschaftskammer z. Hd. des Herrn Direktors Dr. Thyssen in Kiel, Kiel, 20.8.1929, both in LAS 301/4701.

91 See the letters from the Landwirtschaftskammer für die Provinz Schleswig-Holstein to the Oberpräsident in Schleswig, Kiel, 21.8.1929 and 9.9.1929, both in LAS 301/4701; report from the Regierungspräsident to the Oberpräsident in Kiel concerning "Veröffentlichung im Amtsblatt der Landwirtschaftskammer über die Vorfälle in Neumünster" Schleswig, 30.9.1929, LAS 301/4701.

92 *Nordischer Kurier* No. 193, 18.8.1929, p. 3.

93 *Itzehoer Nachrichten* No. 193, 18.8.1929, p. 7; *Rendsburger Tageblatt* No. 195, 21.8.1929, p. 4; *Flensburger Nachrichten* No. 195, 21.8.1929, p. 2.

Fearful of serious repercussions for the town even before the boycott began, Neumünster's citizens had hoped to restore amicable relations with the Landvolk by calling for an investigation to establish whether the farmers had acted illegally or the police had been at fault.[94] Once the boycott began, these calls amplified. Espousing the prevailing rural view that it was the authorities who were to blame, the town's trade and commercial associations appealed to Regierungspräsident Abegg for disciplinary measures against those responsible, echoing the farmers in their criticisms of the police's "incomprehensible" and "reckless" conduct.[95] Then, at a meeting at the end of August, citizens established a committee to initiate negotiations with the rural people, repeated demands for an investigation and took up the Landvolk's call for compensation.[96]

Although investigations were underway, the longer the public had to wait for their findings, the more it appeared as if the authorities wished to conceal the truth. As Hans Fallada suggested, the authorities' indecisiveness aggravated the situation.[97] Even the *Segeberger Kreis- und Tageblatt*, a newspaper the authorities had considered calm and responsible, questioned the delay. It dismissed as a "fairytale" claims that the flag was dangerous and demanded that if the investigations had shown that the police's intervention was "... overhasty and completely unjustified, which can not be doubted from the state of affairs, then one should at least finally find the courage to tell the truth."[98]

Conceal the truth is precisely what Regierungspräsident Abegg chose to do. While he informed the Prussian Interior Ministry that the sickle was blunt and not dangerous, he lied to the Farmers' League. Rather than back down and simply defend the police's actions by contending they could not have known at the time that the sickle posed little risk, Abegg insisted that it was "regarded as a weapon".[99] Furthermore, he exaggerated the threat the farmers posed in order to justify the dissolution of the meeting held after the march, and he falsely gave the impression that

---

94 *Holsteinischer Courier* No. 183, 7.8.1929, p. 5.

95 Letter from the Bürgerverein e.V. Neumünster, Industrieverband, Kreishandwerksverband, Großhandelsverband and Wirteinnung to the Regierungspräsident in Schleswig, Neumünster, 13.8.1929 and telegram from the Handelsbund Neumünster to the Regierungspräsident, Neumünster, 14.8.1929, both in LAS 309/23017.

96 Resolution from a protest meeting of the "Bürgerschaft Neumünsters" in Neumünster on 26.8.1929 to the Regierungspräsident in Schleswig, 27.8.1929, LAS 309/23017. Also see *Das Landvolk* No. 130, 29.8.1929, p. 8.

97 H. Fallada, 'Bauern-Krieg wider Neumünster' *Das Tagebuch*, 10, No. 37, 1929, cited in Bredohl, 'Some thoughts on the Political Opinions of Hans Fallada', p. 528.

98 *Segeberger Kreis- und Tageblatt* No. 191, 16.8.1929 and the report from the Landrat to the Regierungspräsident in Schleswig, Bad Segeberg, 16.8.1929, both in LAS 309/23017.

99 Regierungspräsident to the Preuß. Minister des Innern in Berlin concerning "Zusammenstöße zwischen Landvolk und Polizei in Neumünster am 1.8.1929" Schleswig, 24.8.1929 and Regierungspräsident to the Schleswig-Holsteinischer Bauernbund in Kiel, Schleswig, 24.8.1929, both in LAS 301/4701.

the investigations had been even-handed, claiming untruthfully that many farmers had been heard. It should be noted, however, that he briefly mentioned misgivings about the manner of the police's actions too, and announced that Chief Inspector Bracker was to be suspended from service.[100]

Abegg's duplicitous manoeuvring was also evident in his report to the Interior Ministry. On the one hand, he criticised Bracker's conduct, for his initially insufficient and then excessive use of force.[101] On the other hand, he advised the Ministry of the Interior to respond to questions in parliament by denying that there was any evidence to indicate the authorities' conduct against the Landvolk demonstrators was "uncommonly severe".[102]

Under pressure from the agricultural associations, the press, Neumünster's trade, business and citizen organisations, but in particular from the Interior Ministry – which spotted deficiencies and discrepancies in the Regierungspräsident's reports[103] – Abegg, Lindemann and Bracker were compelled to clarify their roles. None, of course, wished to be held accountable; prestige and careers were at stake. So instead of providing a unified, consistent and mutually supportive response, they blamed one another. In this way, the authorities ensured that the farmers' boycott was more detrimental to them than if they had simply yielded to the farmers' demands. For Hans Fallada, their bickering furnished the essence of his novel, *A Small Circus*: the conflict of all against all in a Northern German town as a microcosm of the social and political divisions of the Weimar Republic.

Mayor Lindemann maintained that he had been directed only to prevent a rally in front of Neumünster's prison.[104] Abegg, however, insisted that he had given clear instructions to prevent a procession under all circumstances. He thrust all responsibility on Bracker and Lindemann. He reproached them for assuming they

---

100 Regierungspräsident to the Schleswig-Holsteinischer Bauernbund in Kiel, Schleswig, 24.8.1929, LAS 301/4701. There is no evidence in the files that any of the rallies' participants, apart from those who were arrested, were questioned other than one mathematics student. See the statement by Friedrich Möller, Neumünster, 21.9.1929, LAS 309/23017. The indictment listed 44 witnesses, almost all of whom were policemen or citizens of Neumünster: indictment issued by the Oberstaatsanwalt, Kiel, 28.9.1929, LAS 309/23017.

101 Regierungspräsident to the Preuß. Minister des Innern in Berlin concerning "Zusammenstöße zwischen Landvolk und Polizei in Neumünster am 1.8.1929" Schleswig, 24.8.1929, LAS 301/4701.

102 Report from the Regierungspräsident to the Preuß. Minister des Innern in Berlin concerning "Zusammenstöße zwischen Landvolk und Polizei in Neumünster am 1.8.1929" Schleswig, 24.8.1929, LAS 301/4701.

103 Replies to the Regierungspräsident in Schleswig from the Preuß. Minister des Innern "zum Bericht vom 5.8.1929" Berlin, 17.8.1929 and "zum Bericht vom 24.8.1929" Berlin, 30.9.1929, both in LAS 309/23017.

104 The Bürgermeister to the Regierungspräsident in Schleswig concerning "Landvolkbewegung in Neumünster" Neumünster, 12.8.1929, LAS 309/23017; copy of a report titled "Zusammenfassende Darstellung der Neumünsteraner Vorfälle" Schleswig, 10.8.1929, both in LAS 301/4701 as well as 309/23017.

could supervise the rally with just 30 officers and blamed the police's tactically inappropriate conduct for the bloodshed.[105] Abegg argued that Lindemann should have deployed all forces from the outset to prevent conflict.[106] Bracker, on the other hand, vindicated his actions, arguing that the Regierungspräsident 's representatives had led him to believe he should proceed with severity against the Landvolk, and that Abegg considered the Landvolkbewegung more dangerous than the communists. A further complicating element was that communists had held a rally with flags in Neumünster on the evening of August 1. That no attempt had been made to prevent it was perceived as a further injustice in rural circles.[107] Abegg denied all knowledge of the communist rally and that consequently he had not afforded them special treatment.[108]

The Landvolk's anger toward the authorities showed no sign of abating. When Abegg asserted that the authorities had been "generous" in tolerating the rally, he was attacked as a non-native Schleswig-Holsteiner, and as an "outstanding agent" of the "Jewish leech system" who had come to office because he was Jewish.[109] Lindemann and Oberpräsident Kürbis were also renounced as foreigners who should leave the province and be replaced by native Schleswig-Holsteiners, while Bracker's return to service gave rise to rumours in the countryside that he might be attacked, perhaps even murdered.[110]

At the end of August, Abegg called a meeting in Schleswig to discuss the restoration of business relations between Neumünster and the rural people.[111] To this end, he invited the province's Chambers of Commerce and Trade, representatives from Neumünster's town council, its citizens', business and industry associations,

---

105 Reports from the Regierungspräsident to the Preuß. Minister des Innern in Berlin concerning "Vorfälle in Neumünster am 1.8.1929 aus Anlaß der Entlassung des Landvolkführers Hofbesitzers Wilhelm Hamkens in Tetenbüttel [sic] aus dem Gefängnis" Schleswig, 5.8.1929 and from the Regierungspräsident to the Preuß. Minister des Innern in Berlin concerning "Vorfälle in Neumünster am 1. August 1929" Schleswig, 31.10.1929, both in LAS 309/23017.

106 Regierungspräsident to the Polizeiverwalter in Neumünster concerning "Zusammenstöße zwischen Landvolk und Polizei am 1.8.1929" Schleswig, 24.8.1929, LAS 301/4701.

107 Letter from the DNVP Landesverband Schleswig-Holstein to the Regierungspräsident in Schleswig, Kiel, 5.8.1929, LAS 309/23017.

108 Regierungspräsident to the Oberstaatsanwalt in Kiel, Schleswig, 31.10.1929, LAS 309/23017.

109 *Das Landvolk* No. 129, 28.8.1929, p. 1 and No. 131, 30.8.1929, p. 5. A further example was Hamkens' speech at a meeting in Gelting in August. See the untitled report, Flensburg, 24.8.1929, in both LAS 301/4696 and 309/22668.

110 *Das Landvolk* No. 113, 9.8.1929, p. 5; report from the Landrat des Kreises Steinburg to the Regierungspräsident in Schleswig concerning "Wiedereinstellung des Polizeioberinspektors Bracker in Neumünster" Itzehoe, 10.10.1929, LAS 309/23017.

111 Letter from the Regierungspräsident to Landtagsabgeordneter Jürgensen, Schleswig, 30.8.1929 and the undated document titled "Zur Wiederherstellung des Wirtschaftsfriedens um Neumünster", both in LAS 309/23017.

as well as the county administrators neighbouring Neumünster. He also invited the three agricultural associations, the Chamber of Agriculture and representatives from individual local organisations.[112] He was not prepared, however, to invite Landvolk leaders, reasoning that the movement was incapable of negotiation because it was not an organisation but a politically revolutionary current, which rejected the state and opposed the authorities.[113]

Even if they had been invited, the Landvolkbewegung's leaders would not have attended, as a refusal in *Das Landvolk* clearly showed.[114] The Rural League and Farmers' League excused themselves from the meeting, first under the pretext that the investigations were not complete and then, after Abegg postponed the meeting, revealing their true motives.[115] They believed the meeting would offer no opportunity for negotiation and that the authorities would continue to portray the farmers in a negative light. Moreover, they were not prepared to endorse the conduct of the police.[116]

The meeting promised little prospect of ending the boycott without the participation of the Rural League and Farmers' League. Yet remarkably, Abegg placed no hope in the meeting even before they refused his invitation. Reporting to the Interior Ministry, he described the meeting merely as an opportunity to clarify the situation and to help ease the existing tension. In fact, it appears his proposed meeting was a sham, for he actually believed the boycott would end of its own accord.[117] It seems, therefore, that Abegg's purpose was merely to create the appearance the state was taking appropriate action.

At the meeting, Abegg attempted to isolate the Landvolk by differentiating between the movement and other farmers, insisting the police had acted not against

---

112 Undated document titled "Zur Wiederherstellung des Wirtschaftsfriedens um Neumünster", LAS 309/23017.

113 Report from the Regierungspräsident to the Preuß. Minister des Innern in Berlin concerning "Zusammenstöße zwischen Landvolk und Polizei in Neumünster am 1.8.1929" Schleswig, 28.8.1929, in both LAS 301/4701 and 309/23017; letter from the Regierungspräsident to Landtagsabgeordneter Jürgensen, Schleswig, 30.8.1929, press release from the Korrespondenzbüro "Nordschleswig", Flensburg, 2.9.1929, Bl. 3 and the undated document titled "Zur Wiederherstellung des Wirtschaftsfriedens um Neumünster", all in LAS 309/23017.

114 *Das Landvolk* No. 138, 7.9.1929, p. 5.

115 Letters to the Regierungspräsident in Schleswig from the Landbund Schleswig-Holstein, 27.8.1929, and from the Schleswig-Holsteinischer Bauernbund, 25.8.1929, both in LAS 309/23017.

116 Letters to the Regierungspräsident in Schleswig from the Schleswig-Holsteinischer Bauernbund, Kiel, 2.9.1929, and from the Landbund Schleswig-Holstein, Kiel, 2.9.1929, both in LAS 309/23017.

117 Report from the Regierungspräsident to the Preuß. Minister des Innern in Berlin concerning "Zusammenstöße zwischen Landvolk und Polizei in Neumünster am 1.8.1929" Schleswig, 28.8.1929, in both LAS 301/4701 and 309/23017.

the farmers but only against the Landvolk.[118] Determined to deny the movement any influence, there was no discussion of its demands.[119] Town and agricultural representatives present stated their desire for the return to amicable relations, but rather than drive a wedge between the Landvolk and the province's "prudent" farmers, as Abegg viewed the Farmers' Association[120], the meeting in Schleswig only underlined the farmers' unity in supporting the boycott. In spite of derisive claims from the extreme right over its unrelenting cooperation with the authorities, shared animosity concerning the events in Neumünster had given the Farmers' Association renewed impetus to consider unification with the Farmers' League. The Farmers' Association now declared that it would continue to boycott the town until the situation was clarified and voiced its displeasure with what it too perceived to be the partial treatment of the communists.[121]

At best, the meeting in Schleswig was futile, serving only to produce a common declaration containing little more than a call for the trial of the rally leaders to be held as soon as possible, which – it was argued – would provide clarification and ease tension between Neumünster and the countryside. At worst, the meeting drew attention to the powerlessness of the provincial authorities to resolve the conflict and highlighted their inability even to gain the support of the moderate Farmers' Association.

Abegg realised he was losing control of the situation. Reporting on the meeting to the Prussian Interior Ministry, he reflected, for instance, on the radicalisation of the Rural League. He believed its leaders had not attended the meeting because they had succumbed to a "psychosis" as a result of what had happened in Neumünster. Schleswig-Holstein's farmers, he warned, were gradually losing their respect for the state, as it failed to prevent the bombings (see the following chapter) and failed to hinder the increasing activities of the "agitators" who were "at least morally responsible" for the attacks. As a consequence, Abegg recommended that

118 Undated document titled "Zur Wiederherstellung des Wirtschaftsfriedens um Neumünster", LAS 309/23017; memorandum "über die vom Herrn Reg. Präs. Schleswig am 3.ds. Mts. anberaumte Besprechung über die Neumünsteraner Boykottbewegung ..." Kiel, 5.9.1929, LAS 301/4701.

119 Report from the Regierungspräsident to the Preuß. Minister des Innern in Berlin concerning "Boykott gegen die Stadt Neumünster" Schleswig, 5.9.1929, in both LAS 301/4701 and 309/23017.

120 Report from the Regierungspräsident to the Preuß. Minister des Innern in Berlin concerning "Boykott gegen die Stadt Neumünster" Schleswig, 5.9.1929, in both LAS 301/4701 and 309/23017.

121 Memorandum "über die vom Herrn Reg. Präs. Schleswig am 3.ds. Mts. anberaumte Besprechung über die Neumünsteraner Boykottbewegung ..." Kiel, 5.9.1929, LAS 301/4701. According to Otto Johannsen, leader of the rallies of January 28, 1928, moves by the Farmers' League and Farmers' Association to fuse were the "fruit of Neumünster": *Nordischer Kurier* No. 193, 18.8.1929, p. 3.

extraordinary measures be considered, including greater control over the press, in order to prevent the radicalisation of the entire agricultural community.[122]

Once again suppression appeared to be Abegg's favoured strategy. He continued to closely monitor *Das Landvolk*, hoping to instigate further bans and confiscations.[123] He banned a further Landvolk rally with musical accompaniment and flag in Rendsburg at the end of September and he instructed the county administrators to ban all further rallies on the grounds they posed a threat to public safety. He also demanded that indoor meetings be controlled rigorously and closed the moment they became disorderly.[124]

Of further significance in Abegg's report on the meeting in Schleswig was a belated acknowledgment that constructive undertakings were required to counterbalance the repressive measures aimed at containing the radicalisation of the rural population. This realisation was expressed in his hope of winning the farmers back by calmly educating them.[125] Contemplating a press campaign, Abegg arranged for the Reich Office for Regional Services to provide clichés and copies sufficient for numerous articles on the agricultural crisis and the measures taken to combat it, and on the Haag conference, the Dawes and Young Plans and the liberation of the Rhineland.[126] Yet, as various county administrators were quick to indicate, the material

122 Report from the Regierungspräsident to the Preuß. Minister des Innern in Berlin concerning "Boykott gegen die Stadt Neumünster" Schleswig, 5.9.1929, in both LAS 301/4701 and 309/23017.

123 Report from the Regierungspräsident to the Preuß. Minister des Innern in Berlin concerning "Landvolkbewegung und die politischen Sprengstoffanschläge" Schleswig, 30.9.1929, in LAS 301/4696 and 28.9.1929 in 309/22784.

124 Note in the files titled "2) An Bestmann. Antrag vom 25.9.1929 auf Genehmigung öffentlicher Kundgebungen, die am 30.9.29 in Rendsburg stattfinden sollen" and the copy of a letter from Max Bestmann to the Polizeiverwaltung der Stadt Rendsburg, Rendsburg, 25.9.1929, both in LAS 309/22784; report from the Regierungspräsident to the Preuß. Minister des Innern in Berlin concerning "Landvolkbewegung und die politischen Sprengstoffanschläge" Schleswig, 30.9.1929, in LAS 301/4696 and 28.9.1929 in 309/22784; Regierungspräsident to the Landräte, the Polizeipräsidenten in Altona and Kiel, the Polizeiverwaltungen in Flensburg, Neumünster and Rendsburg, Schleswig, 28.9.1929, and Regierungspräsident to the Landräte, Polizeipräsidenten and the Polizeiverwaltungen in Neumünster and Rendsburg concerning "Auflösung von Versammlungen" Schleswig, 9.10.1929, both in LAS 309/22784.

125 Report from the Regierungspräsident to the Preuß. Minister des Innern in Berlin concerning "Boykott gegen die Stadt Neumünster" Schleswig, 5.9.1929, in both LAS 301/4701 and 309/23017.

126 Report from the Regierungspräsident to the Preuß. Minister des Innern in Berlin concerning "Rechtsradikale Bewegung in der Provinz Schleswig-Holstein" Schleswig, 27.9.1929, and Regierungspräsident to the Landräte, Schleswig, 27.9.1929, both in LAS 309/22783; Reichszentrale für Heimatdienst, Landesabteilung Schleswig-Holstein, to the Regierungspräsident in Schleswig, Kiel, 4.10.1929 and the copy of the report titled "Bericht über die politische Lage in Schleswig-Holstein und Vorschläge der Landesabteilung betreffend notwendiger Maßnahmen" sent from the Reichszentrale für Heimatdienst, Landesabteilung

had to be concise if it was to suit the regional press format. Obtaining the bulk of their news ready for printing from large press agencies, newspapers such as the *Heider Anzeiger* and *Husumer Nachrichten* merely took care of the local reports themselves and were loathe to undertake the extensive editing that the material required.[127] Consequently, it seems that this well-intentioned endeavour was abandoned.[128]

The trial of six of the demonstrators began at the end of October with hopes set that it would clarify the Neumünster incident. The principal charges brought against Muthmann, Ross, Thiess, Bestmann, Hell and Jens concerned resistance to public authority (*Widerstand gegen die Staatsgewalt*) and encouraging resistance to public authority.[129] Muthmann, Ross and Thiess received suspended sentences on probation of one month, three weeks and two weeks respectively while Jens was fined RM 50 for insulting a police officer.[130] Bestmann and Hell were acquitted because of insufficient evidence and questionable testimony, as was Ross on a further charge of intimidating a public servant (*Beamtennötigung*). He had allegedly called for the forceful removal of a detective from the auction hall.[131]

Lenient sentences were given on the grounds of various mitigating circumstances. Despite the fact that it was unable to obtain a "completely clear picture" of the events, the court determined that though Chief Inspector Bracker had made a legitimate mistake, the confiscation of the flag was not permissible in law because the sickle neither constituted a weapon nor was it proven that it had been used as

---

Schleswig-Holstein, to the Regierungspräsident in Schleswig, Kiel, 4.10.1929, both in LAS 309/22783. The Reich Office for Regional Services (*Reichszentrale für Heimatdienst*), founded in 1918 to foster calm on the home front, was used during the Weimar Republic to propagate republican-democratic ideas and support the state's institutions, which in practice often comprised the promotion of acceptance of the governments' course on the various difficulties which beset the nation. See K. W. Wippermann, *Politische Propaganda und staatsbürgerliche Bildung. Die Reichszentrale für Heimatdienst in der Weimarer Republik* (Cologne, 1976).

127 Landrat to the Regierungspräsident in Schleswig concerning "Pressenachrichten über Probleme der Gegenwart" Meldorf, 30.9.1929, Landrat des Kreises Stormarn to the Regierungspräsident in Schleswig concerning "Beeinflussung der Presse" Wandsbek, 4.10.1929 and Landrat to the Regierungspräsident in Schleswig concerning "Einwirkung auf die Presse" Husum, 21.10.1929, all in LAS 309/22783.

128 No further evidence of this undertaking is to be found in the Regierungspräsident's files in Schleswig.

129 Indictment issued by the Oberstaatsanwalt, Kiel, 28.9.1929, LAS 309/23017.

130 Issue of the verdict of the Schöffengericht in Neumünster, 12.11.1929, pp. 2 and 39, in both LAS 301/4701 and 309/23017; report from the Regierungspräsident to the Preuß. Minister des Innern in Berlin concerning "Zusammenstöße von Landvolkanhängern mit Polizei in Neumünster am 1.8.d.Js." Schleswig, 21.11.1929, in both LAS 301/4701 and 309/23017.

131 Issue of the verdict of the Schöffengericht in Neumünster, 12.11.1929, pp. 38-40, in both LAS 301/4701 and 309/23017.

such.[132] The court criticised the police conduct, in particular, for continuing to strike Muthmann with their sabres while he was lying on the ground.[133] No evidence was found that an "assault squad" had led the march, or that farmers had loaded revolvers, or that the participants had deliberately armed themselves with sticks before the rally.[134] The grounds on which the police had subsequently dissolved the farmers' meeting were, therefore, dubious. The court also took Muthmann's injuries, the farmers' agitated state and their belief that the police had acted unjustly into account. The court's clemency toward Ross was less comprehensible. Despite evidence to the contrary, the court judged that he had had a calming effect on the procession, and that his presence had ensured that the conflict with the police had not been worse.[135]

The "Neumünster trial" in October 1929. From the left: Max Bestmann, Johannes Thiess, Paul Adam Ross (standing), Walther Muthmann, and behind him, Walter Luetgebrune[136]

The soft sentences incensed the left-wing press, which complained of partial treatment for the anti-Republican farmers. In its wish for the suppression of the farmers' movement as a "rebellion against the state", it arguably neglected the legal

132 Issue of the verdict of the Schöffengericht in Neumünster, 12.11.1929, pp. 3, 18-22 and 24-25, in both LAS 301/4701 and 309/23017.

133 Issue of the verdict of the Schöffengericht in Neumünster, 12.11.1929, pp. 32-37, in both LAS 301/4701 and 309/23017. Also see the *Kieler Zeitung* No. 311, 9.11.1929, p. 15.

134 Regierungspräsident to the Preuß. Minister des Innern in Berlin concerning "Zusammenstöße von Landvolkanhängern mit Polizei in Neumünster am 1.8.d.Js." Schleswig, 21.11.1929, in both LAS 301/4701 and 309/23017.

135 Issue of the verdict of the Schöffengericht in Neumünster, 12.11.1929, pp. 42-43, in both LAS 301/4701 and 309/23017.

136 Photo: Stadtarchiv Neumünster.

substance of the case.[137] In this respect, the left and right-wing press were alike, though the former elevated the defence of the Republic and the latter the "people's sense of justice".[138] *Das Landvolk* exulted in the verdict, which, it claimed, had proven the "police disgrace".[139] Though less impassioned, other newspapers also observed that the trial had proven the public's misgivings about the legitimacy of the police conduct.[140]

The court's findings were also agreeable to Regierungspräsident Abegg. Freed of blame himself, he was able to report to the Prussian Interior Ministry that the trial confirmed Lindemann's responsibility and that Bracker had acted "completely incorrectly, rashly and nervously".[141] Abegg believed the police had lost prestige as a result of their poor conduct, something that was further accentuated, when Behr, who had had his nose cut from his face, later successfully sued the town of Neumünster for the compensation of his medical costs and exemplary damages.[142] Nevertheless, Abegg's awkward intervention during the trial did his office no service either. Disapproving of his attempt to influence the choice of an expert witness, the *Rendsburger Tageblatt* observed that the trial had highlighted the gulf between the authorities and the farmers.[143] More critical still, Hans Fallada, who in a report on the trial for the *Weltbühne*, wrote of the "absurd governance" (*unsinnige Regiererei*) that was simply creating new martyrs and further resentment.[144]

The trial brought no respite for Neumünster as the farmers continued to boycott the town. Persisting anger over the incident and perhaps pressure from colleagues help to explain the farmers' fortitude. Admittedly, there is no evidence to confirm that farmers did coerce their neighbours to abide by the boycott as Fallada's novel suggests.[145] Yet the need for such pressure should be questioned because the boycott demanded little sacrifice on the farmers' part. Unlike the planned purchase strike of the Rural Leagues in a number of Germany's central and eastern provinces in May 1929, which failed because their members were not prepared to forfeit their own

137 See the excerpts from reports in the *Schleswig-Holsteinische Volkszeitung* and *Berliner Tageblatt* cited in *Das Landvolk* No. 197, 15.11.1929, p. 5; Werner, *Die Prozesse gegen die Landvolkbewegung*, pp. 131-134.

138 Werner, *Die Prozesse gegen die Landvolkbewegung*, p. 134.

139 *Das Landvolk* No. 196, 14.11.1929, p. 1.

140 *Rendsburger Tageblatt* No. 267, 13.11.1929, p. 1.

141 Regierungspräsident to the Preuß. Minister des Innern in Berlin concerning "Zusammenstöße von Landvolkanhängern mit Polizei in Neumünster am 1.8.d.Js." Schleswig, 21.11.1929, in both LAS 301/4701 and 309/23017.

142 Issue of the verdict of the Landgericht in Kiel from 4.7.1930, LAS 309/23017. NB the court did not determine the final sum to which Behr was entitled. Also see the *Schleswig-Holsteinische Volkszeitung* No. 215, 16.9.1930, p. 11.

143 *Rendsburger Tageblatt* No. 267, 13.11.1929, p. 1.

144 H. Fallada, 'Landvolkprozeß', *Die Weltbühne* No. 49, 3.12.1929, p. 835, GStA PK I. HA Rep. 84a No. 55389.

145 Fallada, *Bauern, Bonzen und Bomben*, pp. 297-301.

gain, the Neumünster boycott was bound to be more effective because it was directed at a single town.[146] The rural people merely had to take their custom elsewhere at little or no cost to themselves.

The cost to Neumünster, or at least the perceived cost, was substantial. As the Regierungspräsident's office noted, it was difficult to tell whether the alleged falls in turnover were entirely a consequence of the boycott or the result of the general unfavourable economic conditions.[147] Aware that it was considered accountable, Neumünster's police administration expressed doubts about the boycott's severity.[148] Neumünster's businesses, however, whose calls for assistance could only benefit from a gloomy portrayal of the situation, claimed the boycott was growing worse.[149] They complained that trade relations with the surrounding villages and countryside had been severed. They believed that not only were rural people avoiding the town, but that townspeople were also being shunned when they travelled the countryside.[150] Some businesses tried to circumvent the problem by using private connections or paying agents in the villages a provision to take their goods, but such measures did not offset the shortfalls. Businesses in all domains from manufacturers to retailers reported significant falls in turnover, some of 40 percent and more over the previous year, and businesses specialising in agricultural wares fared even worse. The cancelled agricultural events also occasioned losses, from the rent on stalls and the printing of catalogues to the patronage of cigar shops, restaurants, inns and hotels.

Das Landvolk was only too happy to publicise the town's misfortune as well as repeat the requirements for lifting the boycott.[151] The rural people were determined to humble the town and were not perturbed, for example, by moves from the SPD in the Prussian Landtag to terminate public funding of a provincial cattle-breeding association because it supported the boycott.[152] If anything, such discussion was

146 In order to exact greater political attention to agriculture's demands, the Rural Leagues in East Prussia, Pomerania, Mecklenburg-Schwerin, Brandenburg, Hanover, Thüringia and Hessen directed their members to refrain from buying anything they could do without, including fertilisers and new machinery. Merkenich, *Grüne Front gegen Weimar*, pp. 257-260; Pyta, *Dorfgemeinschaft und Parteipolitik*, pp. 206-208.

147 Report from the Regierungspräsident to the Preuß. Minister des Innern in Berlin concerning "Vertretung der Polizeiverwaltung Neumünster im Strafprozeß" Schleswig, 5.9.1929, LAS 301/4701.

148 Reports from the Polizeiverwaltung to the Regierungspräsident in Schleswig concerning "Boykottbewegung gegen Neumünster" Neumünster, 15.10.1929 and to the Regierungspräsident in Schleswig, Neumünster, 10.1.1930, both in LAS 309/23017.

149 Report from the Handelsbund Neumünster e.V. titled "Bauernboykott und Wirtschaftsschaden von Neumünster" Neumünster, 5.12.1929, LAS 309/23017.

150 Report from the Handelsbund Neumünster e.V. concerning "Bauernboykott und Wirtschaftsschaden von Neumünster" Neumünster, 5.12.1929, LAS 309/23017.

151 *Das Landvolk* No. 47, 25.2.1930, p. 5 and No. 221, 14.12.1929, p. 5.

152 *Das Landvolk* No. 46, 23.2.1930, p. 2 and No. 78, 2.4.1930, p. 1.

counterproductive, for it ensured that the Landvolk knew their endeavours were being taken seriously.

Negotiations to end the boycott commenced in mid-January when Hamkens, Behr and other Landvolk leaders met with the heads of the business organisations and representatives from the town, seeking the return of the flag on January 28, 1930 to mark the second anniversary of the great rallies. Newspapers reported that the farmers and townspeople appeared to be approaching an understanding, but Mayor Lindemann would not agree to the Landvolk's demands. Certainly in answer to Abegg's enquiries, he denied having considered allowing the flag to be carried through the town. But in any case the flag's return was impossible as long as it remained in the possession of the public prosecutor.[153]

Slowly hurdles were removed. Neumünster's police administration and then the public prosecutor withdrew appeals against the verdict for Muthmann, Ross and their fellow protestors.[154] Lindemann became mayor of a small town in Brandenburg, a promotion *Das Landvolk* happily interpreted as a "disgraced departure" induced by Neumünster's citizens.[155] At the end of June, once the public prosecutor had released the flag to the Landvolk's lawyer, Luetgebrune, and Neumünster agreed to a celebration, a declaration signed by Landvolk leaders and agricultural associations announced the rural people's peace with the town. The farmers were to march through the town to music and a town representative was to present the flag.[156]

Just days beforehand, however, Abegg derailed the plan, banning an outdoor procession and gathering on the grounds that it constituted a danger to the public peace and order.[157] He declared that he welcomed all measures to restore peaceful relations between town and country but rejected an outdoor procession because "people like Claus Heim and Wilhelm Hamkens" had heralded it and because it was to take place under Landvolk leadership, "... a political direction which re-

---

153 Magistrat der Stadt Neumünster to the Regierungspräsident in Schleswig concerning "Boykottbewegung gegen Neumünster" Neumünster, 17.12.1929, Regierungspräsident to the Polizeiverwaltung zu Neumünster, Schleswig, 17.1.1930, reports from the Polizeiverwaltung to the Regierungspräsident in Schleswig, Neumünster, 18.1.1930, from the Polizeiverwaltung to the Regierungspräsident in Schleswig concerning "Boykottbewegung" Neumünster, 18.2.1930 and the report titled "Bericht über Verhandlung mit führenden Landleuten am Montag den 13. Januar 1930" Neumünster, 18.1.1930, all in LAS 309/23017. Also see the *Holsteinischer Courier* No. 12, 15.1.1930, p. 7 and *Das Landvolk* No. 14, 17.1.1930, p. 5.

154 Magistrat der Stadt Neumünster to the Regierungspräsident in Schleswig concerning "Bauernboykott in Neumünster" Neumünster, 27.2.1930, LAS 309/23017; Luetgebrune to the Preuß. Justizminister in Berlin, Sankt Peter, 18.5.1930 and Vermerk IV f311 B/30, both in GStA PK I. HA Rep. 84a No. 55389.

155 *Das Landvolk* No. 90, 16.4.1930, p. 5.

156 *Holsteinischer Courier* No. 148, 27.6.1930, p. 8; *Das Landvolk* No. 149, 28.6.1930, p. 1; *Kieler Neueste Nachrichten* No. 148, 28.6.1930, p. 7.

157 Regierungspräsident to the Preuß. Minister des Innern in Berlin concerning "Landvolkbewegung" Schleswig, 7.7.1930, LAS 301/4701.

jected the state, its constitution, laws and institutions, which tolerated and heroised bombings". As a further pretext, Abegg pointed to a clash between communists and National Socialists in the town, claiming that it was often impossible to differentiate between the Landvolkbewegung and NSDAP.[158]

The farmers felt compelled to cancel the celebration.[159] *Das Landvolk* vented its anger by once more assailing Abegg as Jewish and foreign to the province.[160] In a perhaps unsurpassed rage, it answered Abegg's declaration,

> "Arselic (*Schleimsch*[sic]) .... to the front! Who wants to belong to those who are better as [sic] us! Who wants to belong to those whom the Regierungspräsident would gladly tolerate in Neumünster's streets? He speaks of responsible organisations which he favourably differentiates from the Landvolk. Which of these organisations wants Abegg's recognition? We hope – none! .... The entire manner of expression of this knight in inverted commas, Abegg, is spattering hatred, almost criminal hatred against a valuable portion of the population ... The hatred is Jewish, really typically Jewish ..."[161]

Hatred can only describe *Das Landvolk's* response, not Abegg's pettiness in banning the procession. However, the farmers were not alone in their rage. Neumünster's citizens – the social democrats excepted – were also furious.[162] The town's businesses had collected RM 4,000, which had already been paid out to the Landvolk as compensation, and it was Abegg, they believed, who had placed the peace in jeopardy.[163] A delegation from the town pleaded with him to lift the ban but Abegg would not be moved, arguing that the state's interests were more important than the risk to the town's economy.[164] This was to inspire the animosity of the town's businesses and it was rumoured that the citizens even threatened to sue him should the boycott continue.[165]

Although the farmers made no effort to continue the boycott, it was not until fresh plans were made for a celebration at Neumünster's cattle yards that *Das Landvolk* reiterated its call for farmers to return to the town.[166] Abegg dispatched a detachment of riot police as a precautionary measure, but this time stopped short of

---

158 *Holsteinischer Courier* No. 156, 7.7.1930, p. 5; *Das Landvolk* No. 157, 8.7.1930, p. 1.

159 *Das Landvolk* No. 152, 2.7.1930, p. 2 and No. 153, 3.7.1929, p. 1.

160 See *Das Landvolk* No. 154, 4.7.1930, p. 1 and No. 157, 8.7.1930, p. 3.

161 *Das Landvolk* No. 157, 8.7.1930, p. 1.

162 *Holsteinischer Courier* No. 151, 1.7.1930, p. 5.

163 Regierungspräsident to the Preuß. Minister des Innern in Berlin concerning "Landvolkbewegung" Schleswig, 7.7.1930, LAS 301/4701.

164 *Kieler Neueste Nachrichten* No. 153, 4.7.1930, p. 3.

165 *Schleswig-Holsteinische Tageszeitung* No. 154, 3.7.1930, pp. 4 and 7.

166 Report from the Landeskriminalpolizeistelle, Abteilung IA, titled "Bericht" Kiel, 25.10.1930, LAS 301/4698; *Das Landvolk* No. 233, 3.10.1930, p. 1, No. 240, 11.10.1930, p. 1 and No. 260, 4.11.1930, pp. 5 and 8.

banning the celebration because the farmers refrained from planning a procession.[167]

As many as 4,000 people attended the celebration on November 7, 1930. A Stahlhelm band from Rendsburg provided musical interludes and to great jubilation, a representative of Neumünster declared the town's solidarity with the farmers and presented the flag, which was given to Johannes Thiess for safekeeping. Schwarzloh, Hamkens and others made speeches, in which they declared the rural people's peace with the citizens of Neumünster but not with "the system." It was subject to the customary opprobrium, as was Regierungspräsident Abegg. Closing the celebration with characteristic Landvolk symbolism, the members of the crowd sang the national anthem and during the last verse raised their fists in solidarity.[168]

The Stahlhelm band looks on, while Hamkens (on the right) delivers a speech and the Landvolk flag is returned, Neumünster, November 7, 1930[169]

The Neumünster affair was not quite over as it generated one last dispute. An unhindered KPD march through the town that evening was an echo of August 1, 1929, and gave further occasion to assail Abegg. Though the authorities had been

167 Report from the Regierungspräsident concerning "Strafantrag gegen die Tageszeitung 'Das Landvolk' " Schleswig, 13.11.1930, in both LAS 301/4698 and 309/22761.

168 *Holsteinischer Courier* No. 263, 8.11.1930, p. 4; *Das Landvolk* No. 264, 8.11.1930, p. 1.

169 Photo: Stadtarchiv Neumünster.

unaware of the communists' intentions, *Das Landvolk* was appalled at what it felt was preferential treatment.[170] Consequently, Johannes Kühl accused Abegg of preventing the Landvolk from marching unlawfully but allowing the communists to march, and maligned him once again, this time as "just a clerk of the party coalition which currently governed Prussia", an "enemy of the nation," of Jewish descent and therefore "incapable of understanding us".[171]

The Neumünster incident and its aftermath did not serve to bring the Landvolkbewegung and NSDAP any closer together. Anticipating the spurt of popularity the farmers' movement would gain on Hamkens' release from prison, Gauleiter Lohse responded pre-emptively on August 1, 1929 with a damaging full page leading article in the *Schleswig-Holsteinische Tageszeitung*. He claimed that Landvolk speakers and their newspaper borrowed their practicable ideas from National Socialist thinking but failed to help agriculture effectively for lack of a prominent leader with a clear purpose who could deliver decisive agricultural policy. Lohse maintained that Landvolk aspirations to a "farmers' revolution" – a term he maliciously ascribed to the Landvolk – could only fail because of the resistance of other occupations. These, he explained, were of commensurate importance and it was only united under Adolf Hitler's leadership that they could effect the German revolution.[172] Lohse's intentions were transparent: he sought to take credit for the Landvolk's spreading of radical right ideas, highlight the Landvolk's failure to influence agricultural policy and exploit the Landvolk's inability to appeal to other occupational groups.

Landvolk and NSDAP leaders, their followers and the authorities were well aware that little separated them in their ideological outlooks. It was no secret that the National Socialists hoped to profit from the Landvolk and that each was helping to pave the way for the other, since there was little difference in, what one provincial government report termed, their "phraseology". The NSDAP had keenly observed the Landvolkbewegung's activities in the hope that it would later be able to

---

170 Report from the Regierungspräsident concerning "Strafantrag gegen die Tageszeitung 'Das Landvolk' " Schleswig, 13.11.1930, in both LAS 301/4698 and 309/22761.

171 This resulted in a series of corrections and further accusations in *Das Landvolk*, (see chapter 4, p. 137, fn. 105), and to Abegg suing Kühl for defamation, for which Kühl ultimately received a fine of RM 100. See, in particular, *Das Landvolk* No. 264, 8.11.1930, p. 1 and also No. 269, 14.11.1930, p. 1; Regierungspräsident to the Oberstaatsanwalt in Altona, Schleswig, 10.11.1930, Regierungspräsident to the Oberstaatsanwalt in Altona concerning "Strafantrag gegen den verantwortlichen Schriftleiter der Tageszeitung 'Das Landvolk' " Schleswig, 14.11.1930 and Regierungspräsident to the Oberpräsident in Kiel concerning "Strafantrag gegen die Tageszeitung 'Das Landvolk' " Schleswig, 9.12.1930, all in LAS 301/4698 as well as 309/22761; issue of the verdict of the Schöffengericht in Itzehoe from 15.5.1931, LAS 309/22761.

172 *Schleswig-Holsteinische Tageszeitung* No. 179, 1.8.1929, p. 1.

take advantage of its "groundwork".[173] Specifically, as the *Vossische Zeitung* wrote, the Landvolk was an "involuntary path-maker" for the NSDAP because it had encouraged hostility to the state.[174]

The NSDAP recognised the potential of the Neumünster incident for unabashed self-promotion. Among its most blatant undertakings, it adopted the very same tactic it had used after Beidenfleth, luring an audience to a gathering advertised as a meeting on the affair. But the meeting, held in Neumünster in late August, was one the socialist *Schleswig-Holsteinische Volkszeitung* accurately described as "brazen" for little was said about the Neumünster affair other than that it was the product of the system. Otherwise, it was no different from any other NSDAP promotional meeting.[175]

The Landvolk would have welcomed greater cooperation with the NSDAP. Its avoidance of membership and bureaucratic procedure allowed members of other groups to participate in its activities without compelling them to abandon their other party or organisational commitments. Indeed it encouraged this explicitly.[176] However, the National Socialist leadership's instructions – that its members not take part in organisations which adopted illegal methods – meant this openness was not reciprocated. The party feared this might give the authorities a pretext to impose bans. It also tacitly feared that such association would frighten away potential supporters. Shortly after the Neumünster incident, Hitler repeated these instructions, this time with special emphasis that the ban applied to the Landvolkbewegung.[177] Accordingly, both Friedrich Wilhelm Heinz, the editor of the party's *Niedersächsischer Beobachter* in Hanover, and Otfried Rademacher were expelled from the party. They had collaborated with Bruno von Salomon, both supplying him with material for *Das Landvolk* and printing and disseminating Landvolk literature.[178]

---

173 Report from the Regierungspräsident to the Preuß. Minister des Innern in Berlin, Schleswig, 30.9.1929 in LAS 301/4696, and 28.9.1929 in 309/22784. This was also Bruno von Salomon's assessment. See the letter from Bruno von Salomon to F. W. Heinz, 17.5.1929, cited in the list of confiscated material titled "V. Auslese wichtigen Materials aus den von der Redaktion der Landvolk – Zeitung beschlagnahmten Ordnermappen", LAS 309/22784.

174 *Vossische Zeitung* No. 189, 9.8.1929, LAS 301/4696.

175 *Schleswig-Holsteinische Volkszeitung* No. 203, 31.8.1929, LAS 309/23017.

176 See Hamkens speech at a meeting in Uelzen in October 1929. Report from the Landeskriminalpolizeiamt, Abteilung IA, titled "Bericht über die öffentliche Versammlung der Landvolkbewegung am 6.10.1929 in Uelzen auf dem Schützenplatz" Altona, 7.10.1929, LAS 309/22784.

177 *Schleswig-Holsteinische Tageszeitung* No. 181, 3.8.1929, p. 12.

178 *Schleswig-Holsteinische Tageszeitung* No. 183, 6.8.1929, p. 7; *Rendsburger Tageblatt* No. 183, 7.8.1929, p. 2. Friedrich Wilhelm Heinz (1899-1968) joined the Freikorps after the First World War and was active in various organisations of the extreme right, working in cooperation with Hermann Ehrhardt. Working as a publicist and author, Heinz found work as editor of the NSDAP's newspaper in South-Hanover from September 1928 by way of connections with the Strasser brothers until his expulsion from the party. After the seizure

The Landvolk responded to the NSDAP's hostility with a warning to the Reich leadership in Munich, written anonymously by Rademacher and printed in *Das Landvolk.* Dismayed at the party's rejection of the farmers' activism, which had been repeated at the NSDAP's party day in Nuremberg, Rademacher accused Lohse and other provincial leaders of stabbing the Landvolk in the back and of treason against the Landvolk and even against National Socialism itself. Mistakenly blaming them for the party's enmity, and for its deficient activism and "unwillingness really to oppose the administrative apparatus", Rademacher assumed that Hitler was ignorant of their treachery. Though desiring the party's open support, he saw room for coexistence with the Landvolk at the very least. "Even if the existence of an activist movement in Holstein impeded the efficacy of a party, which has similar ideas but can not itself pursue an active struggle without risking its dissolution, the party must allow room for action, if it is to take the ideas of resistance seriously."[179]

Though the Reich leadership's instructions were clear, there were extensive violations of them at the local level. As Paul Kurt Rogers wrote to Hartmut Plaas in August 1929, it seems that Hitler's orders were simply ignored.[180] Perhaps some National Socialist farmers felt it permissible to participate in the Landvolk's obstructive actions at compulsory enforcement proceedings, on the basis of Werner Willikens' essay in the 1929 *National Socialist Yearbook* in which he had written that farmers had to decide themselves whether they wished to use strikes, including tax-strikes, to defend their interests.[181] NSDAP members, like Peter Ludwig Petersen in Husum County, Erich Paprotny who worked for *Das Landvolk* in the Flensburg area and Lendt (Alt-Lipke), one of the movement's "trusted-men" in Brandenburg, were not disciplined by the party, ostensibly because their involvement

---

of power, he joined the army and worked in counter-intelligence where he was involved in the conspiratorial nationalist resistance against Hitler. Most notably, Heinz led the detachment that had intended to arrest and eliminate Hitler in the aborted Oster Plot of September 1938. This was discovered after the plot of July 20, 1944 and Heinz subsequently went into hiding. Continuing to work in espionage after the war, Heinz led his own intelligence office, the "FWH-Dienst", from 1950 to 1953, which was affiliated to the Federal Chancellor's Office. See S. Meinl, *Nationalsozialisten gegen Hitler: Die nationalsozialrevolutionäre Opposition um Friedrich Wilhelm Heinz* (Berlin, 2000), esp. pp. 137-165, and S. Meinl and D. Krüger, 'Der politische Weg von Friedrich Wilhelm Heinz: Vom Freikorpskämpfer zum Leiter des Nachrichtendienstes im Bundeskanzleramt', *Vierteljahrshefte für Zeitgeschichte*, 42 (1994), pp. 36-69.

179 *Das Landvolk* No. 113, 9.8.1929, p. 1. According to Susanne Meinl, the article was written by Otfried Rademacher: Meinl, *Nationalsozialisten gegen Hitler*, pp. 154-155.

180 Report titled "Vorläufiges Ergebnis der Sichtung des bei Plaas im Büro Ehrhardt, Berlin W.35, Steglitzerstr. 28, beschlagnahmten Materials", LAS 309/22784.

181 W. Willikens, 'Nationalsozialismus und Landvolk' in the *Nationalsozialistisches Jahrbuch* (Munich, 1929), p. 197.

was not flaunted and did not harm the party's image.[182] Nor did National Socialists feel inhibited from donating to the Landvolk. The leader of the local in Niebüll contributed no less than RM 200 for Luetgebrune's legal fees.[183]

One of the most blatant instances of fraternisation after Hitler repeated his prohibition was a joint meeting in Oelixdorf in October 1929. On discovering they had called a meeting at the same time and in the same pub, leaders of both groups agreed to hold their meetings together, acknowledging they essentially pursued the same aims. Little separated the speakers ideologically. Where they differed was in Kühl's rejection of organisation and the SA leader Schoene's (Lockstedter Lager) insistence on the necessity of tight organisation for the maintenance of discipline and the preservation of unified purpose.[184]

Though it always welcomed collaboration, the Landvolkbewegung rejected suggestions made at local meetings that it amalgamate with the NSDAP.[185] Landvolk speakers reproached the NSDAP for its aloofness, but the chief reason they gave concerned their own aversion to organisational structure and party politics. They declined joining with the NSDAP, asserting their wish to remain a "movement", a claim which, when used in contradistinction to the term "party", upset the National Socialists, for they too saw themselves as a movement.[186] Landvolk speakers argued that the formation of a party or an organisation would halt the movement's drive, and that involvement in Weimar party politics was inescapably corrupting, for its factionalism engendered "brotherly struggle" and the neglect of the people's interests in the pursuit of remaining in office.[187] Nevertheless, they usually softened these rejections with conciliatory remarks, for example, wishing the NSDAP success in its struggle to deliver its promises.[188]

---

182 Report from the Landeskriminalpolizeistelle Altona-Wandsbek, Abteilung IA, concerning "Sprengstoffanschläge" Itzehoe, 20.3.1930, in both LAS 301/4697 and 309/22590; Heinacher, *Der Aufstieg der NSDAP im Stadt- und Landkreis Flensburg*, p. 292.

183 Report from the Polizeipräsident, Landeskriminalpolizeistelle, to the Regierungspräsident in Schleswig, Flensburg, 3.3.1930, LAS 301/4697.

184 Report from the Landeskriminalpolizeistelle Altona, Abteilung IA, titled "Bericht über eine Landvolkversammlung in Oelixdorf bei Itzehoe" Itzehoe, 16.10.1929, in both LAS 301/4696 and 309/22784.

185 Two such instances were to be found in reports on Landvolk meetings, in Schafflund in October and in Glückstadt in December 1929. Report from the Landjägerei Kreis Steinburg titled "Bericht über die Landvolkversammlung in Glückstadt am 13.12.1929 im Lokal 'Englischer Garten' Itzehoe, 15.12.1929, in both LAS 301/4697 and 309/22783; report from the Polizeipräsident to the Regierungspräsident concerning "Landvolkbewegung" Flensburg, 20.10.1929, LAS 309/22784.

186 For an example of the National Socialist objections to the Landvolk labelling it a "party", see the *Schleswig-Holsteinische Tageszeitung* No. 129, 4.6.1929, p. 6.

187 See, for example, Johannes Kühl's speech at a meeting in Husby in July 1929 as reported in *Der Schleswiger* No. 156, 5.7.1929, LAS 301/4695.

188 See, for example, Hamkens' response to a young National Socialist at a meeting in Wrist in October 1929 and Wulf's response to National Socialist criticism at a Landvolk meeting in

This was characteristic of the Landvolkbewegung's overriding eagerness to promote amity in the nationalist camp. It was said that Claus Heim was especially critical of the NSDAP. However, for the sake of harmonious relations, he did not express his opinions openly.[189] Blinded perhaps by fraternisation at the local level and by the desire for unity overall, the Landvolk failed to comprehend fully that the NSDAP's provincial and Reich leaderships would never tolerate a rival, but simply wished wherever possible to capitalise and undermine the farmers' movement for their own benefit.

The rallies in Husum and Neumünster were among the most significant manifestations of radicalisation in rural Schleswig-Holstein, both as impressive demonstrations of the farmers' support for Hamkens and the Landvolkbewegung, as well as displays of their enthusiasm for defiance and for some, violence. What saw these clashes have an ever more radicalising effect was farmers and their organisations siding with the Landvolk in sympathy and anger, and subsequently joining the movement in its boycott of Neumünster. This, and not simply the fact that the demands for compensation were largely met, was what made the boycott such a success for the Landvolk. It drew the province's squabbling agricultural associations together in a rare display of unity. Alarmingly, it was farmers' defiance of the authorities that achieved this as can clearly be seen by the refusals of the Farmers' League and Rural League to attend the Regierungspräsident's meeting in Schleswig. Although the Farmers' Association accepted Abegg's invitation, it too took sides with the Landvolk, by openly supporting the boycott at the meeting.

There can be no doubt the Landvolk's aggression was the main ingredient in the clashes at both rallies. Yet neither may have transpired but for avoidable mistakes on the part of the authorities. The most significant of these were the want of warning which left the county administrator and police unprepared in Husum; and the conflicting views on the strategy required to deal with the movement in Neumünster, not to mention the police chief's hasty decision-making and improvised tactics in confiscating the flag.

While these were unfortunate mistakes, Abegg's conduct during the boycott was completely unsuccessful. Though it should be acknowledged that he was compelled to take a more forceful approach in dealing with the Landvolk due to its involvement in the bombings, his strategy did nothing to halt the course of radicalisation in the province. Perhaps he believed he was doing the wrong thing for the right

Burg in Dithmarschen in October 1929. Report from the Landeskriminalpolizeistelle Altona, Abteilung IA, titled "Bericht über eine Landvolkversammlung in Wrist" Itzehoe, 18.10.1929, in both LAS 301/4696 and 309/22784; *Das Landvolk* No. 161, 4.10.1929, p. 5.

189 Undated extract from the examination transcript of Claus Heim titled "Klaus Heim, 24.3.1884 in Oesterfeld", LAS 309/22784; examination transcript of Johannes Kühl, Altona, 16.9.1929, LAS 309/22701; Helene Hönck's manuscript sent from the Staatspolizeistelle in Altona to the Geheimes Staatspolizeiamt in Berlin, 14.12.1933, LAS 301/4699.

reasons by overstating the danger the farmers had posed in Neumünster, supposing it necessary in order to convince the public that it really was the farmers who had been to blame. Delays in the investigations and then the exposure of various details meant this approach backfired. It diminished the prospects of prevailing over the Landvolk's popular view that the authorities' treatment had been unfair and heavy-handed, and ultimately damaged the state's authority because Abegg's version was so patently misleading. Soon afterwards he put integrity to the test again, attempting to influence the choice of an expert witness during the trial of the rally leaders. Moreover, his reluctance to be held accountable, though quite understandable in itself, was divisive when it entailed resentful reproaches of Mayor Lindemann and Chief Inspector Bracker. Finally, Abegg's attempts to prevent the Landvolk from returning to march in Neumünster adhered to principle instead of pragmatism, and this could only appear officious. His intervention simply impeded the resumption of peaceful relations between the farmers and the townspeople. It earned him the ill will of the town's tradespeople and businesses on top of the Landvolk's existing enmity toward him, which through the affair had become even shriller and eventually degenerated into vicious personal and anti-Semitic attacks.

The NSDAP was neither involved in the rallies in Husum and Neumünster, nor did it have any influence on the Landvolkbewegung's management of the boycott. However, it took every opportunity to join in the censure of the authorities, while continuing in its efforts to capitalise on the unrest the Landvolk had generated. Unlike the Farmers' League, which came out in full support of the Landvolkbewegung after the Neumünster incident, the NSDAP persisted in its criticisms of its perceived rival, accusing it of appropriating its ideas, of lacking a genuine leader and of the one-sided pursuit of farmers' interests. Ever hoping for cooperation, the Landvolk took umbrage that a fellow right-wing organisation would seek to do it harm, although its own abhorrence of organisational structure and party politics also caused tension. Though fraternisation at the local level continued, nothing impaired relations more than the NSDAP's fear of being associated with the Landvolk's illegal tactics. The Landvolkbewegung was most angered at what it understood as a rejection of its activism and showed no inclination of changing its course. Concerned for its image, the NSDAP did not refrain from expelling members who disobeyed the Reich leadership's orders and joined the farmers' movement. To understand fully the party's fears, however, requires an examination of the bombings, which will be the focus of the next chapter.

The Neumünster rally and its aftermath marked the peak of the Landvolkbewegung's popularity. This was revealed by the turn out at the rally, public approval for those who were arrested, large attendances at meetings, widespread sympathy in the provincial press and the united and outspoken support of the boycott by the agricultural organisations. At this stage, neither Abegg's warnings nor the NSDAP's criticisms appeared to have any detrimental impact on support for the movement. Beyond the boycott, however, the movement continued as before. It did not seek to harness the support it had gained in any new way. The success did not induce the movement to consider any practical undertaking that might have improved the farmers' circumstances or advanced its subversive ends. Remarkably, the confrontations with the police did not even lead to any further instances of mass violence. Yet by this time the Landvolk bombing campaign was well under way.

Chapter Seven

Bombs

The Landvolkbewegung is perhaps best remembered for its bombing campaign. It was the most extreme manifestation of radicalisation in the countryside during the Weimar Republic. The bombings commenced in November 1928 with a number of small attacks on Schleswig-Holstein's west coast and in the province of Hanover, and in the summer of 1929 were followed by a series of stronger and more spectacular explosions in Schleswig-Holstein, Hanover, Oldenburg and Berlin. The bombings raise a number of significant questions in respect of their planning, preparation and execution as well as their legal consequences and impact on public opinion. Possibly even more interesting is why a group within the movement came to embrace terrorist methods at the risk of bringing down the full force of the authorities and of losing many supporters.

The Landvolkbewegung's adoption of bombings hinged on the remarkable personality of Claus Heim, the architect of the attacks. Later, and therefore apparently romantic, depictions of Heim as the embodiment of his ancestors who had fiercely struggled for Dithmarschen's independence in the Middle Ages were actually not much embellished upon perceptions of him at the time.[1] He was known as an unusually assertive and obstinate person, as imposing in character as he was in stature.[2] A police profile described him as a wealthy farmer who had suffered financial losses, as strictly conservative and strongly nationalistic, something Heim himself attributed to having spent many years overseas. He was reportedly viewed as noble (*vornehm*) but industrious and his opinions were authoritative in the farming community.[3]

Heim was among those farmers, who in the weeks after the rallies of January 28, 1928 and the subsequent deputation in Berlin, began to doubt that these had made sufficient impression on the Reich and Prussian governments. Together with Hamkens he soon began to voice the need for more extreme methods. These views were not shared by all the fellow organisers of the rallies, including Otto Johannsen, and the leader of the Unity Organisation, Otto Köhler. Heim and Hamkens collaborated, promoting their plans for stronger protests and tax-strikes.[4] Speakers at Landvolk meetings, including Hamkens, received rousing audience responses when

---

1 Volck, *Rebellen um Ehre*, p. 346; v. Salomon, *Deutschland deine Schleswig-Holstiener*, p. 176.

2 See p. 270 below.

3 See the police profile on Heim from the Abteilung IA, Außendienst VII, Geschäftsstelle, Berlin, 5.6.1930, LAS 309/22701.

4 See chapter 3, pp. 84 ff; Heim, 'Claus Heim berichtet', LAS EI/4434, p. 2.

they talked of resorting to violence to defend their farms and secure profitability.[5] From the beginning of the Landvolkbewegung, however, it was Heim in particular who wished to put the rhetoric into practice.[6] The earliest manifestation of Heim's more extreme inclinations was the Defence Association. Though Herbert Volck, Hans Nickels and, to a lesser extent, Hamkens assisted him, it should be remembered that it was Heim who was the driving force behind this Landvolk vigilante network.[7]

Publicly, Defence Association members were to arm themselves with truncheons. Secretly, however, they were to defend their farms with firearms. Heim and Volck arranged for the delivery of weapons for the Defence Association with the assistance of Gustav Kurtze, a member of Organisation Consul. The remodelled military rifles which Kurtze found for them, and which were delivered to farmers in Schleswig-Holstein, Hanover and East Prussia from April to June 1929, numbered in the hundreds.[8] The rifles were never used, and it was not until much later that anything but rumours of their existence reached the authorities.[9] One can only speculate on the purpose Heim and his followers had in mind for the weapons, and for which eventualities they may have contemplated using them. Nonetheless, the

---

5 See, for example, Paul Adam Ross's speech in Husum on October 17, 1928; Frauen and Kelting's speeches in Itzehoe on November 26 and Hamkens' speech in Garding on December 18. See chapter 3, pp. 83, 98 and 106.

6 Hamkens' examination transcript, Altona, 15.9.1929, LAS 309/22701; Hamkens' examination transcript, Berlin, 20.9.1929, GStA PK I. HA Rep. 84a No. 55381.

7 For the development of the Defence Association, see chapter 3.

8 Undated extracts from the examination transcript of Gustav Kurtze titled "Kurzer Auszug aus der Vernehmungsniederschrift des Beschuldigten Gustav Willi Adolf Kurtze..." of Amandus Vick Jr. titled "Amandus Vick jun, 24.7.1897 in Rönne, Hofbesitzer in Winsen" and of Walther Muthmann titled "Geschäftsführer des Landbundes Walther Paul Muthmann, 22.4.1902 ...", all in LAS 309/22784.

9 For the authorities' responses to rumours of dealings in arms, control of weapon licences and searches of farms: Regierungspräsident to the Landräte des Bezirks, Schleswig, 7.12.1928, LAS 309/22569; Landrat to the Regierungspräsident in Schleswig, Heide, 25.3.1929, LAS 309/22921; reports from the Landrat to the Oberpräsident in Kiel concerning "Landvolkbewegung" Niebüll, 8.4.1929 and from the Regierungspräsident to the Oberpräsident in Kiel concerning "Landvolkbewegung im Kreise Südtondern" Schleswig, 29.4.1929, both in LAS 301/4695 and 309/22783; report from the Landeskriminalpolizeistelle Flensburg to the Regierungspräsident in Schleswig concerning "Landvolkbewegung und Deutschnat. Volkspartei" Flensburg, 16.4.1929, LAS 309/22668; report from the Landeskriminalpolizeistelle Flensburg to the Regierungspräsident in Schleswig concerning "Landvolkbewegung" Flensburg, 22.4.1929, LAS 309/22783; report from the Regierungspräsident to the Oberpräsident in Kiel concerning "Landvolkbewegung im Kreise Südtondern, Schleswig, 14.5.1929, in both LAS 301/4695 and 309/22783. For the later discoveries of two arsenals, the first in Winsen an der Luhe in Hanover, the second in Schlotfeld in County Steinburg in Schleswig-Holstein, see p. 238 below and the reports from the Landeskriminalpolizeistelle Altona, Abteilung IA, concerning "Angebliches Waffenlager" Itzehoe, 16.10.1929 and Itzehoe, 23.10.1929, both in LAS 309/22784.

dealings in arms clearly highlight the willingness to employ violence within Defence Association circles and it was from these circles that Heim was to recruit some of his principal fellow bombing conspirators.

It is not known whether Heim was inspired by any particular precedent in his choice to adopt the method of bombings. If so, the most evident possibility was the resistance against the French occupation of the Ruhr where activists such as Schlageter and Kölpin had used explosives in sabotaging the railways.[10] The fact that Heim and his accomplices later sought explosives in the Ruhr for their campaign would appear to bolster this supposition.

Heim's motivations, on the other hand, were quite clear. As well as impressing more emphatically upon the authorities the measure of the farmers' distress, he intended to intimidate those who supported "the system", neglected the farmers' concerns and opposed the Landvolkbewegung. In fact, this was his objective even before he took up the method of bombing. In November 1928 Heim and two fellow farmers smashed the windows of the offices of the *Heider Anzeiger* and notified the paper anonymously that it had failed to stand up for the rural people's interests.[11] In embracing bombings, Heim may have been encouraged by perceptions that the movement's followers expected more radical action. According to Volck, Heim declared that the farmers always wanted to see new "actions".[12] Further incentives for bombing was his belief that the rural population wanted invigorating and the Landvolkbewegung required greater publicity. He thought a series of explosions would provide the "loudest advertisement possible"[13] to rouse the rural population from its apathy, maintain the movement's momentum and encourage donations.

Heim also seemed to believe he could use bombs to overthrow the Weimar system, although it is unclear how he thought this possible. In this respect, Bodo Uhse later reflected that Heim himself was not sure about the reasons that drove him.[14] The clearest statement he ever made on the matter was recorded later, while he was in prison. On a visit by the SA leader of East Hanover, Hasse, in late 1931, Heim said that it was impossible to destroy the system by legal means and "pretty speeches" and that he had chosen illegal methods to "wear the system down". Heim claimed he was convinced that no supporter of the system would have dared leave his doorstep had he been able to continue his actions for a few months longer.[15] On the basis of this evidence, it appears Heim hoped he might be able to generate such

---

10 For an account of the active resistance during the French occupation of the Ruhr, see C. Fischer, *The Ruhr Crisis, 1923-1924* (Oxford, 2003), pp. 43 and 165-181.

11 Heim, 'Claus Heim berichtet', LAS EI/4434, p. 2.

12 Issue of the verdict of the Schwurgericht Altona, 31.10.1930, p. 20, in both LAS 301/4702 and 309/22764; Volck, *Rebellen um Ehre,* p. 343; Heim, 'Claus Heim berichtet', LAS EI/4434, pp. 3-5; Mrs Hamkens' interview, Zeitzeichen Archive (1990).

13 Issue of the verdict of the Schwurgericht Altona, 31.10.1930, p. 20, in both LAS 301/4702 and 309/22764.

14 Uhse, *Söldner und Soldat*, p. 276.

15 Report signed "Krim. Ass. Peters", Hanover, 14.11.1931, LAS 309/22701.

fear so as to paralyse political life in the Weimar Republic. Of course, whether this was a realistic expectation is very much debatable and it would be a further step still to suppose that it could by itself lead to the Republic's downfall.

In October and November 1928 Heim discussed his plans with Volck who willingly agreed to assist him.[16] Hamkens, on the other hand, was utterly opposed to the idea and voiced his objections from the outset. In what ultimately spelled a parting of ways that would later have significant consequences for the movement's development, Hamkens rejected their plans, making clear that he meant to continue employing rhetoric to defeat "the system", to convince the people to abandon the political parties and organisations and refuse all positive cooperation with them.[17] In his view, bombs would do the Landvolk much more harm than good. He feared that they would only lead to the arrest of the movement's leaders without helping to remove the system.[18] In fact, he felt so strongly on the matter that he interrupted a publicity trip in Oldenburg in a fruitless effort to dissuade Heim.[19]

Heim always remained the central figure behind the bombings although Volck was instrumental in the initial stages, offering Heim advice on choosing targets and helping to seek suitable fellow conspirators. But Heim later began to grow wary of Volck's talkativeness, and fearing he might expose their plans, encouraged him to leave Schleswig-Holstein.[20] It was Heim who came up with the idea of a bombing campaign and commissioned the construction of the bombs and provided the necessary funds. It was Heim who was ultimately responsible for arranging the delivery of materials and for the bombs' collection and storage.[21] Further, Heim supervised the implementation of the bombings, he determined the locations where the devices were to be planted and selected the personnel involved.

The first bombs were more or less improvised devices. A farmer, Johannes Rathjen (Bargfeld), prepared packages from the remainder of explosives he had been authorised to use to remove rocks and stumps from his fields. Volck helped to transport the bombs and Heim chose the targets, the homes of community chairmen: Mahlstedt in Beidenfleth, Thomsen in Hollingstedt and Max Ross in Fleder-

---

16 Issue of the verdict of the Schwurgericht Altona, 31.10.1930, p. 20, in both LAS 301/4702 and 309/22764.

17 Hamkens' examination transcript, Berlin, 20.9.1929, GStA PK I. HA Rep. 84a No. 55381.

18 Volck, *Rebellen um Ehre*, pp. 344-345.

19 Hamkens' examination transcript, Altona, 15.9.1929, LAS 309/22701; undated extracts from Hamkens' examination transcript titled "Kurzer Auszug aus der Vernehmungsniederschrift des Hofbesitzers Wilhelm Cornelius Hamkens, ...", LAS 309/22784; issue of the verdict of the Schwurgericht Altona, 31.10.1930, p. 21, in both LAS 301/4702 and 309/22764.

20 Issue of the verdict of the Schwurgericht Altona, 31.10.1930, pp. 20, 35 and 58-60, in both LAS 301/4702 and 309/22764. Also see the *Rendsburger Tageblatt* No. 265, 9.11.1929, p. 2; *Vossische Zeitung* No. 271, 13.11.1929, p. 3; *Kieler Zeitung* No. 314, 12.11.1929, p. 3.

21 Issue of the verdict of the Schwurgericht Altona, 31.10.1930, p. 51, in both LAS 301/4702 and 309/22764.

wurth.[22] Reprisal provided the immediate motive, for they were among those who at the Itzehoe meeting just the night before had been described as "dishonest, cowardly and listless colleagues" who had endorsed compulsory enforcement and supported "the system".[23] The bomb planted in Beidenfleth on the night of November 27, 1928 by Landvolk leader Weschke and Matthes, a farmer from Bücken near Itzehoe, made a loud noise but caused little material damage as did the bomb planted in Hollingstedt. That placed in Max Ross's garage in Flederwurth failed to explode, as did another planted by John Johnsen and farmers Amandus Vick and Franz Luhmann on Heim's instructions as a decoy at the finance office in Winsen an der Luhe in the province of Hanover.[24]

Heim was disappointed with the attacks which he believed had made no impression on the authorities. Reportedly, the rural population perceived them as a prank rather than an act of intimidation.[25] Heim complained to Volck that the farmers were simply amused, and that it was time to employ "stronger methods".[26] By stronger methods, Heim meant a series of attacks using much more potent bombs. His aims remained essentially the same. Bigger bombs were to provide publicity for the Landvolkbewegung, animate the rural population, lend the farmers courage and frighten the authorities and the Landvolk's opponents.[27]

To these ends, Heim charged Hans Nickels, who he had initially recruited to organise the Defence Association, with the task of obtaining explosives. Nickels had been involved in the Ruhr resistance and knew of reserves of explosives in a quarry on the outskirts of Mülheim an der Ruhr, where he had lived during the French occupation.[28] At the end of January he and Heinrich Wiborg, who ran a transport company in Lunden, drove a truck to Mülheim, and with other unknown

---

22 Issue of the verdict of the Schwurgericht Altona, 31.10.1930, pp. 23-25, in both LAS 301/4702 and 309/22764.

23 See the resolution from the Itzehoe meeting of November 26, 1928 as cited in chapter 3, p. 102.

24 Issue of the verdict of the Schwurgericht Altona, 31.10.1930, pp. 25-33, in both LAS 301/4702 and 309/22764; Weschke's examination transcript, Altona, 12.9.1929, LAS 309/22701; undated extracts from Weschke's examination transcript titled "Syndikus Guido Hermann Hans Weschke, 6.9.1893 in Pichelsdorf ...", LAS 309/22784; *Wilstersche Zeitung* No. 280, 28.11.1928, p. 2; *Rendsburger Tageblatt* No. 290, 28.11.28, p. 2; *Kieler Neueste Nachrichten* No. 275, 24.11.1929, p. 10; *Nordischer Kurier* No. 275, 23.11.1929, p. 1.

25 Report from the Landeskriminalpolizeistelle Flensburg concerning "Bauernbewegung in der Provinz Schleswig-Holstein" Flensburg, 24.1.1929, in both LAS 301/4589 and 309/22696; *Flensburger Nachrichten* No. 281, 29.11.1928, p. 4; *Rendsburger Tageblatt* No. 290, 28.11.1928, p. 2.

26 Issue of the verdict of the Schwurgericht Altona, 31.10.1930, p. 33, in both LAS 301/4702 and 309/22764. Also see Heim, 'Claus Heim berichtet', LAS EI/4434, p. 3.

27 Issue of the verdict of the Schwurgericht Altona, 31.10.1930, p. 34, in both LAS 301/4702 and 309/22764.

28 Issue of the verdict of the Schwurgericht Altona, 31.10.1930, pp. 35-36, in both LAS 301/4702 and 309/22764.

accomplices broke into the quarry store. Their exciting expedition, which also involved lodgings under false names, a rendezvous by a cemetery and a long and frosty, dangerous return drive, ended with the transfer of the explosives to Volck's car under the guise of a breakdown. Volck and Johnsen took the 52.5 kilograms of ammonite and 700 detonators to Peter Holländer's remote farm in South Tondern where they were stored until required.[29]

Volck urged Heim to seek technically qualified assistance for the construction of the new bombs.[30] Whether the two men recommended by Ehrhardt's office were qualified by previous experience or simply by their occupations is unknown. Yet Alfred Kaphengst, an electrician, and Herbert Schmidt, an artist, were both politically dependable, having belonged to various radical organisations including the Wehrwolf, Bund Wiking and the Nationalist Club in Hamburg. There they had met Bruno von Salomon and Walther Bohm and they occasionally visited Landvolk and Defence Association meetings.[31] In early spring they spent several weeks experimenting with explosives until they produced a reliable model, a device with an alarm clock that would give those who planted them a chance to flee and avoid detection.[32]

Heim did not immediately put this model to use. First, he resolved to frighten two opponents of the Landvolkbewegung, the farmers Loy and Huesmann in Wesselburen near Heide. To this purpose he was given hand-grenades by Amandus Vick, taken from a substantial arsenal Vick had hidden on his farm in Rönne in Hanover Province since the dissolution of the Baltic troops in 1920.[33] In the early hours of April 6, Heim, Kaphengst, Schmidt and another unknown accomplice drove by Huesmann and Loy's residences and threw a grenade at each.[34] The gre-

---

29 Issue of the verdict of the Schwurgericht Altona, 31.10.1930, pp. 35-45, in both LAS 301/4702 and 309/22764; Holländer's confession as reported in the *Kieler Neueste Nachrichten* No. 276, 26.11.1929, p. 2.

30 Issue of the verdict of the Schwurgericht Altona, 31.10.1930, p. 33, in both LAS 301/4702 and 309/22764.

31 Indictment against Alfred Kaphengst issued by the Oberstaatsanwalt Altona/Elbe, pp. 8-11, dated 1.11.1931 in LAS 309/22701 and 3.11.1930 in LAS 301/4703; issue of the verdict of the Schwurgericht Altona from 23.1.1932, pp. 4-7, in both LAS 301/4703 and 309/22701; issue of the verdict of the Schwurgericht Altona from 31.10.1930, pp. 46-47, in both LAS 301/4702 and 309/22764. Also see the report titled "Bericht" from the Landeskriminalpolizeiamt, Abteilung IA, Altona, 31.10.1929, LAS 309/22784; the examination transcripts of Herbert Schmidt, Altona, 30.10.1929 and 6.11.1929, both in LAS 309/22701; Stoltenberg, *Politische Strömungen im schleswig-holsteinischen Landvolk*, p. 137.

32 For descriptions of the bombs' construction, see the issue of the verdict of the Schwurgericht Altona, 31.10.1930, pp. 48-51, in both LAS 301/4702 and 309/22764; *Kieler Zeitung* No. 253, 12.9.1929, p. 1.

33 Issue of the verdict of the Schwurgericht Altona from 31.10.1930, pp. 53-54, in both LAS 301/4702 and 309/22764.

34 Issue of the verdict of the Schwurgericht Altona, 31.10.1930, pp. 55-57, in both LAS 301/4702 and 309/22764.

nades smashed windows but failed to detonate and simply rebounded into the gardens where police found them in the morning.[35] Though but a prelude to the campaign proper, these attacks showed that Heim was willing to use explosives and to risk injuring people in the pursuit of his objectives.

A bomb, constructed by Kaphengst and Schmidt, The inscription reads "K & S ALTONA. The dependable company for terror wares. Beware of imitations of K & S products[36]

The bombings using the powerful devices built by Kaphengst and Schmidt began in May 1929 and continued until early September. Heim's accomplices planted eight of these bombs in total: at the county administrator's office in Itzehoe on the night of May 22; Oldenburg's finance office on June 2; the residence of the county administrator, Skalweit, in Niebüll (Südtondern) on July 9; the residence of a lawyer, Emil Strauß, and the offices of the state health insurance department (*Landkrankenkasse*) in Lüneburg on July 31; the residence of the province's *Vize-Regierungspräsident* (vice-government president), Grimpe, in Schleswig on August

35 Landjägerabteilung Heide to the Landrat in Heide, 6.4.1929, LAS 301/4695; issue of the verdict of the Schwurgericht Altona, 31.10.1930, pp. 57-58, in both LAS 301/4702 and 309/22764; *Rendsburger Tageblatt* No. 81, 8.4.1929, p. 4. and No. 82, 9.4.1929, p. 4.

36 Photo: Stadtarchiv Neumünster.

29; the Reichstag in Berlin on the morning of September 1; and at the government offices in Lüneburg on September 5.

County administrator's office in Itzehoe after the bombing, May 1929[37]

All of the attacks were similar in their execution. The bombs were planted after nightfall and set to detonate in the early hours of the morning. The bombers took great care to avoid detection and in most cases they worked in groups of two or three and drove by car to their targets. For want of ballast, the attack on the Reichstag produced only a spectacular spurt of flame and a loud bang, but the other devices that exploded all caused considerable material damage.[38] The bomb in Itze-

37 Photo: Dohnke collection.

38 For details of the bombing in Itzehoe on May 23, see the issue of the verdict of the Schwurgericht Altona, 31.10.1930, pp. 61-62 in both LAS 301/4702 and 309/22764; *Itzehoer Nachrichten* No. 119, 24.5.1929, p. 1; *Kieler Neueste Nachrichten* No. 118, 24.5.1929, p. 1. For the bomb planted by the finance office in Oldenburg (Oldenburg) on June 3, see the issue of the verdict of the Schwurgericht Altona, 31.10.1930, p. 70, in both LAS 301/4702 and 309/22764. For details of the bombing in Niebüll on July 10, see the issue of the verdict of the Schwurgericht Altona, 31.10.1930, pp. 71-74, in both LAS 301/4702 and 309/22764; report from the Regierungspräsident to the Minister des Innern in Berlin concerning "Sprengstoffanschlag auf die Dienstwohnung des Landrats in Niebüll" Schleswig, 10.7.1929, LAS 301/4704; *Flensburger Nachrichten* No. 159, 10.7.1929, p. 1 and No. 160, 11.7.1929, p. 1; *Kieler Neueste Nachrichten* No. 159, 11.7.1929, p. 1. For the bombing of Dr Strauß's villa in Lüneburg on August 1, see the issue of the verdict of the Schwurgericht Altona, 31.10.1930, pp. 79-81, in both LAS 301/4702 and 309/22764; *Kieler Neueste*

hoe, for instance, detonated with a thundering boom that woke the town and did great harm to the county administrator's office and neighbouring buildings, while the resultant air-pressure shattered every window all the way to the train station 100 meters distant. In Itzehoe, as on several other occasions, the buildings were occupied at the time of the explosions but there were neither fatalities nor even serious injuries. In this respect, the daughter of the county administrator in Niebüll was very lucky to escape. Sleeping on the far side of the house, she awoke covered in glass shards but sustained only minor cuts.[39]

More fortunate still was Vize-Regierungspräsident Grimpe. The bomb, packaged in a box, and planted in a corner of the loggia of his villa would have been extremely dangerous, as it had been placed inside the building.[40] However, coming home late at night, his servant, Hunwardsen, saw the box as she passed, and since its appearance struck her as peculiar, she opened the lid and heard the ticking of the timing device. Realising she had found a bomb, she carried the box outside, down to the garden gate. Woken by the noise downstairs, Grimpe called the police. Timorously, he ordered Hunwardsen to carry the bomb further away, out over the street, where it was subsequently defused by the police.[41] Those later discovered to have participated in the attacks – driving, reconnoitring, keeping watch and planting the bombs – were Johnsen, Kaphengst, Schmidt and Heim's neighbour, Hennings, and, in Lüneburg, farmers Vick, Becker and Luhmann. In two cases, further accomplices were engaged without Heim's initiative. Fearful of being caught themselves, Kaphengst and Schmidt persuaded a friend and political colleague, Rieper, to help them plant a bomb in Oldenburg, just as Vick, Becker and Luhmann

---

Nachrichten No. 179, 3.8.1929, p. 2; *Volkszeitung* No. 178, 2.8.1929, p. 2; *Das Landvolk* No. 107, 2.8.1929, p. 1. For the Reichstag bombing on September 1, see the *Kieler Neueste Nachrichten* No. 205, 3.9.1929, p. 1 and *Das Landvolk* No. 134, 3.9.1929, p. 2. For the bombing of the government building in Lüneburg on September 6, see the issue of the verdict of the Schwurgericht Altona, 31.10.1930, pp. 90-91, in both LAS 301/4702 and 309/22764; *Kieler Neueste Nachrichten* No. 209, 7.9.1929, p. 1; *Das Landvolk* No. 138, 7.9.1929, p. 1.

39 *Flensburger Nachrichten* No. 160, 11.7.1929, p. 1; *Kieler Neueste Nachrichten* No. 160, 12.7.1929, p. 3. The villa of Dr. Strauß, for example, was occupied as were the government offices in Lüneburg on September 6. See the issue of the verdict of the Schwurgericht Altona, 31.10.1930, p. 81, in both LAS 301/4702 and 309/22764 and *Kieler Neueste Nachrichten* No. 209, 7.9.1929, p. 1.

40 *Kieler Neueste Nachrichten* No. 203, 1.9.1929, p. 1.

41 Issue of the verdict of the Schwurgericht Altona from 31.10.1930, pp. 85-86, in both LAS 301/4702 and 309/22764; report titled "Bericht" from the Polizeihauptwache, Schleswig, 30.8.1929, LAS 301/4704. Also see reports in the *Schleswiger Nachrichten* No. 204, 31.8.1929, p. 5; *Das Landvolk* No. 133, 1.9.1929, p. 2; *Kieler Neueste Nachrichten* No. 204, 1.9.1929, p. 14; *Rendsburger Tageblatt* No. 204, 21.8.1929, p. 4; *Volkszeitung* No. 202, 30.8.1929, p. 4.

engaged a young local Stahlhelm leader, Manecke, to help them carry out an attack in Lüneburg.[42]

With the exception of Christine Holländer's, who found out about the bombings after her husband agreed to store the explosives in their house, all who joined Heim in his conspiracy appear to have done so willingly.[43] This holds true whether they were fellow farmers or members of nationalist organisations of different occupations. Heim may, of course, have brought his imposing strength of character to bear in recruiting his accomplices. In the case of Volck, Nickels and Johnsen, moreover, their employment in the Defence Association and later at *Das Landvolk* may also have created a sense of obligation to their patron. For them, however, the opportunity to combat "the system", rather than drawing attention to the agricultural crisis, provided their principal motivation. Quite frank about this, Volck later wrote that he was not interested in the Landvolkbewegung as a struggle for better pork and butter prices, but only as a struggle against the system and for "blood and soil".[44] Yet it was the struggle against the authorities and system that provided the most important motivation for others too. Both Wiborg and Holländer agreed to lend assistance in the conviction that they were serving the national interest, while others, such as Weschke, liked the idea of intimidating those they considered adversaries of the rural people.[45]

Accounts in some histories of the Weimar Republic suggest that the targets of the bombings were finance offices, creating the impression that the Landvolkbewegung was principally a debtors' movement.[46] This is mistaken. As has been argued throughout this discourse, the Landvolkbewegung channelled the rural people's distress over the agricultural crisis against the Weimar system. Heim's specific choice of targets, too, reflected not only the farmers' financial distress but also their broader anger at the authorities said to serve the administrative apparatus. The chosen targets were easily recognisable symbols of "the system", and only in two instances, in Winsen and in Oldenburg, were the targets actually finance offices.[47] As noted above, in the first attacks, the community chairmen selected by Heim

42 Issue of the verdict of the Schwurgericht Altona, 31.10.1930, pp. 65 and 86-90, in both LAS 301/4702 and 309/22764.

43 Issue of the verdict of the Schwurgericht Altona from 31.10.1930, p. 45, in both LAS 301/4702 and 309/22764.

44 Volck, *Rebellen um Ehre*, pp. 301 and 315.

45 Issue of the verdict of the Schwurgericht Altona from 31.10.1930, pp. 37 and 42, in both LAS 301/4702 and 309/22764; the examination transcripts of Peter Holländer, Altona, 22.11.1929 and Guido Weschke, Altona, 12.9.1929, both in LAS 309/22701.

46 The mistake probably originated in Heberle's summary of the movement. See Heberle, *From Democracy to Nazism*, p. 67 but the mistake has been repeated elsewhere. See, for example, T. Childers, *The Nazi Voter: The Social Foundations of Fascism in Germany, 1919-1933* (Chapel Hill and London, 1983), p. 148.

47 Issue of the verdict of the Schwurgericht Altona, 31.10.1930, pp. 68-69, in both LAS 301/4702 and 309/22764.

were perceived to be particularly evil representatives of "the system" because they had endorsed compulsory enforcement proceedings.[48] In Niebüll the county administrator, Skalweit, was chosen evidently because he was considered to show no understanding for agriculture and had spoken out against the Landvolk's radical disposition.[49] The attack on the villa of the lawyer, Strauß, in Lüneburg was an exception in that personal motives and perhaps anti-Semitism also played a role. If so, it was also an exception in respect of the latter, because for all the movement's anti-Semitic propaganda, it did not otherwise target local Jewish people. Heim gave Vick the bomb but left Vick and his friends to plan and carry out the attack. Vick, Luhmann and Becker considered Strauß a government crony who had carried out a number of confiscations on behalf of the state health-insurance office. Yet Becker, who referred to him as "Strauß the Jew", also had a particular dislike, as he held him responsible for a Reichsbanner demonstration in 1924 in which he had been threatened and affronted on his farm.[50]

Following the first bombings in November 1928, the press and the authorities rightly suspected the Landvolkbewegung, or at least farmers who sympathised with the movement as the perpetrators. At the Itzehoe meeting, the farmers had declared that they were prepared to resort to "illegal methods" and the Regierungspräsident's office, police in Lunden and the *Flensburger Nachrichten* pointed to this apparent connection and also suggested they were a sequel to the obstructive actions at Lunden and Beidenfleth.[51] Since the National Socialists had expressed their disapproval of illegal methods and condemned the bombings, the Regierungspräsident doubted the party's involvement, though in a later report he wrongly claimed that the Landvolkbewegung was under the party's influence.[52] This report, however, recognised Heim's leading role in the Landvolkbewegung, and Flensburg detectives actually contended that Weschke was among those behind the attacks.[53]

48 See p. 237 above.

49 For criticisms of Skalweit, see chapter 3, p. 107 and the report from the Landeskriminalpolizeistelle Flensburg to the Regierungspräsident in Schleswig, Flensburg, 27.2.1929, LAS 309/22921. Also see W. Koops, 'Die Bombe von Niebüll: Sprengstoffanschlag als Bauern-Protest', *Nordfriesland*, 111 (1995), p. 20.

50 Issue of the verdict of the Schwurgericht Altona, 31.10.1930, pp. 75-79 and 87, in both LAS 301/4702 and 309/22764.

51 Report from the Regierungspräsident to the Preuß. Minister des Innern in Berlin concerning "Sprengstoffanschläge in Schleswig-Holstein" Schleswig, 30.11.1928, LAS 301/4695; *Flensburger Nachrichten* No. 281, 29.11.29, p. 4; report from the Landjägereiposten Lunden II, Kreis Norder Dithmarschen to the Oberstaatsanwalt in Flensburg, Lunden, 28.11.1928, LAS 301/4704. For further such rumours in the press, see the *Itzehoer Nachrichten* No. 120, 25.5.1929, p. 7 and the *Vossische Zeitung* No. 124, 25.5.1929, p. 2.

52 Report from the Regierungspräsident to the Preuß. Minister des Innern in Berlin concerning "Sprengmittelanschläge in Schleswig-Holstein" Schleswig, 15.12.1928, LAS 301/4704.

53 Report from the Landeskriminalpolizeistelle concerning "Bauernbewegung in der Provinz Schleswig-Holstein" Flensburg, 24.1.1929, LAS 309/22696.

The accuracy of these deductions was all the more remarkable given that it was to take so long before the bombers were finally apprehended.

The rural population also widely accepted that the Landvolkbewegung instigated the bombings, even if it was not known which individuals were responsible.[54] Only in the case of the hand-grenade attacks on the houses of Loy and Huesmann in Wesselburen was public opinion divided. The authorities were inclined to agree with those who believed the perpetrators were communists. Loy and Huesmann were wealthy farmers, communists had frequently gathered outside Loy's house and sung the Internationale and, evidently by coincidence, other farmers had received threatening letters that day attacking them as "overfed capitalists" profiting at the cost of the "working proletariat".[55] The socialist *Hamburger Volkszeitung*, however, correctly surmised that "big fascist farmers" had instigated the attack to compel Loy and Huesmann to join the Landvolkbewegung.[56]

The NSDAP, nevertheless, feared that it would be associated with the bombings. Writing to the party leadership in Munich, Gauleiter Lohse warned, "As previously the Landvolk has caused us a great deal of trouble. The danger that the stupidity of these people could be dangerous [sic] to us is immense. The methods with which they proceed do not matter at all to these people."[57] Consequently, the party insisted publicly that it had nothing to do with the Landvolkbewegung, condemned the bombings vehemently and argued that only politics, not bombings, could improve the farmers' lot.[58]

Yet dispelling rumours that the NSDAP was involved proved surprisingly difficult. In the Prussian state parliament, Lohse objected to a reported observation

---

54 Report from the Landeskriminalpolizeistelle Flensburg titled "Bericht" Itzehoe, 12.6.1929 and *Norddeutsche Zeitung* No. 117, 25.5.1929, both in LAS 301/4704. Also see Uhse, *Söldner und Soldat*, pp. 203-204; Volck, *Rebellen um Ehre*, p. 359; Schimmelreiter, *Unter der schwarzen Bauernfahne*, pp. 37-40; P. Petersen interview, Zeitzeichen Archive (1990).

55 Report from the Landrat to the Oberpräsident in Kiel, Heide, 13.4.1929, LAS 301/4695; report from the Regierungspräsident to the Oberpräsident in Kiel concerning "Handgranatenanschlag in Wesselburen" Schleswig, 10.4.1929, LAS 301/4704; copy of a report from the Landeskriminalpolizeistelle Flensburg titled "Bericht über die letzten Nachforschungen und Ermittelungen über den Handgranatenanschlag zum Nachteil Huesmann und Loy in Wesselburen" Flensburg, 18.4.1929, LAS 301/4704; copy of a letter to Herrn Otto Kröger (Süderdeich), 6.4.1929, appended to the report from the Landjägerabteilung Heide to the Landrat in Heide, Heide, 6.4.1929, LAS 301/4695.

56 *Hamburger Volkszeitung* No. 84, 12.4.1929, LAS 301/4695.

57 See the lists of material confiscated on the arrests of those suspected in the bombings for the extract from Gauleiter Lohse's correspondence in article 37 "Durchschlag eines Schreibens des Gauleiters Lohse an den Untersuchungs- und Schlichtungsausschuss der NSDAP München vom 27.5.1929" in section 4, "In der Redaktion der Landvolk-Zeitung am 9.9.1929 beschlagnahmtes Material", in both LAS 309/22784 and 301/4704.

58 See, for example, *Schleswig-Holsteinische Tageszeitung* No. 120, 24.5.1929, p. 1, and for the criticisms of Schoene, the SA leader, the report from the Landeskriminalpolizeistelle Flensburg to the Regierungspräsident in Schleswig, Flensburg, 2.6.1929, LAS 309/22569.

from Regierungspräsident Abegg that the culprits were undoubtedly a small group belonging to the "National Socialist Landvolk". Abegg subsequently insisted the *Vossische Zeitung* had misquoted him.[59] The confusion, however, continued. A short time later Lohse was compelled to resist Severing's references to National Socialists "creeping secretly through the streets at night with a few pounds of dynamite".[60] Lohse's indignation was understandable, of course, for the bombers were not National Socialists.

After the Reichstag bombing, the National Socialists found it even more difficult to refute involvement because a poster with a swastika proclaiming "Greater Germany awake!" was found close to the site of the attack.[61] Though the police made it clear that they could not confirm whether the poster was at all related to the incident, this time Hitler himself took action, denouncing the bombers as "enemies of our movement". In addition, he announced a reward of RM 2,000 for any party member who could expose the culprits and as much as RM 10,000 if they could provide evidence that they were in fact the "new-Prussian authorities" (*neupreußische Behörden*) themselves.[62]

The Landvolkbewegung, however, also vigorously resisted claims that it was responsible and *Das Landvolk* called on its readers to combat such rumours as strongly as possible.[63] Bruno von Salomon and his fellow editors knew very well that the bombers belonged to the movement. Nevertheless, it should not be assumed that they knew exactly who the bombers were.[64] If they did, they always guarded their knowledge carefully, even within their own circles, and in public went so far as to suggest that the bombings could simply be an invention of the "administrative apparatus" to compromise its opponents.[65]

Although it denied the movement's involvement, *Das Landvolk* reported the bombings in great detail from the beginning and grew increasingly enthusiastic about them as the campaign continued.[66] When the police released pictures and a description of the bomb found in Lüneburg, it took the liberty of printing them

---

59 *Flensburger Nachrichten* No. 160, 11.7.1929, p. 1; *Kieler Neueste Nachrichten* No. 160, 12.7.1929, p. 3; report from the Regierungspräsident to the Minister des Innern in Berlin concerning "Sprengstoffanschlag auf die landrätliche Dienstwohnung in Niebüll" Schleswig, 11.7.1929 and the copy of the "Staatstelegramm" from the Regierungspräsident to the *Vossische Zeitung* in Berlin, Schleswig, 11.7.1929, both in LAS 301/4704.

60 *Schleswiger Nachrichten* No. 167, 19.7.1929, p. 2.

61 *Das Landvolk* No. 134, 3.9.1929, p. 2.

62 See Hitler's article titled 'Irrsinnige Subjekte oder beamtete Provokateure', *Völkischer Beobachter*, 5.9.1929 in K. A. Lankheit (ed.), *Hitler: Reden Schriften Anordnungen: Februar 1925 bis Januar 1933,* vol. 3 (2) (Munich, 1994), pp. 371-373.

63 *Das Landvolk* No. 64, 24.5.1929, p. 1, No. 65. 25.5.1929, p. 2, No. 81, 13.6.1929, p. 5.

64 Heim, 'Claus Heim berichtet', LAS EI/4434, p. 4; issue of the verdict of the Schwurgericht Altona, 31.10.1930, p. 65, in both LAS 301/4702 and 309/22764.

65 *Das Landvolk* No. 133, 1.9.1929, p. 2.

66 See, for example, *Das Landvolk* No. 64, 24.5.1929, p. 1.

under the headline, "How does one construct a bomb? A police tutorial!" It explained its intentions were to assist the police in making this "well-liked" but "somewhat complicated game" accessible to the broader public, stating that thanks to the "administrative apparatus", planting bombs could soon become the "national sport".[67] The paper ventured even further after the final attack in Lüneburg, elaborating on this joke by contrasting the popularity of the bombings, as the term "Volkssport" implied, with the system, which it judged the most hated in Germany's history. Not only did it portray the bombings as innocuous by claiming that the bombers had planted their bombs with such precision that they could never have intended to risk lives, but it declared the system so contemptible, that there was "actually" no one who considered the bombings "particularly abhorrent or damnable".[68]

Support for the bombings was not as extensive as these reports suggested and certainly not as strong as Ernst von Salomon later conveyed in his writings.[69] Heim believed his fellow farmers were pleased, and to be sure, there were many people who applauded the attacks in the province's country pubs.[70] Moreover, after the bombing in Itzehoe, the county administrator in Steinburg, Göppert, distinguished between the declining number of calm and prudent people who were indignant about the attack and the many who had expressed amusement or pleasure, which was a reflection, he felt, of the effect of the Landvolkbewegung's agitation.[71] In other words, the public's response to the bombings was a sign of the radicalisation in the countryside. The victims, however, were not without sympathisers. Though particularly disliked in Landvolk circles, County Administrator Skalweit in South Tondern received avowals of support after the attack on his residence.[72] Furthermore, it should not be forgotten that there were strong reservations about the bombings from within the movement's own ranks, although these were due to views that they constituted an inappropriate strategy rather than empathy for the victims.[73] Above all, Hamkens, though he refrained from open censure, continued to object to Heim's plans. As late as June 30, when Heim and Volck visited him on his farm in Eiderstedt to take their leave before he went to prison, Hamkens quarrelled with

67 *Das Landvolk* No. 116, 13.8.1929, p. 2.

68 *Das Landvolk* No. 138, 7.9.1929, p. 1.

69 v. Salomon, *Der Fragebogen*, pp. 234-240. Volck also claimed that Landvolk meetings were full of farmers who approved of the attacks. Volck, *Rebellen um Ehre*, p. 323.

70 Heim, 'Claus Heim berichtet', LAS EI/4434, p. 4; *Schleswig-Holsteinische Volkszeitung* No. 229, 2.10.1930, p. 13; *Volkszeitung* No. 230, 2.10.1930, p. 5.

71 Report from the Regierungspräsident to the Oberpräsident in Kiel concerning "Bombenanschlag auf das Landratsamt in Itzehoe" Schleswig, 11.6.1929, LAS 301/4704.

72 According to Koops, *Südtondern in der Zeit der Weimarer Republik*, p. 253; Koops, 'Die Bombe von Niebüll', p. 20.

73 Report titled "Pressebericht. Landvolkbewegung des Kreises Rendsburg" Rendsburg, 22.12.1928, in both LAS 301/4695 and 309/22569.

them as he tried in vain to persuade them that bombings were counterproductive to achieving the movement's aims.[74]

Tremendous public excitement, numerous false confessions, false alarms, threats and mock attacks in Schleswig-Holstein and beyond did not help the police in their efforts to catch the bombers.[75] A further bomb that exploded in the garage of a Freemason in Hohenwestedt on May 29, cast suspicion on the type-setter at *Das Landvolk*, Gustav Dammann, but it did not lead them any closer to Heim and his accomplices since the attacks were unrelated.[76] Nevertheless, the police soon named Weschke, Muthmann, Hamkens, Volck, Hansen and Claussen (Hohenwestedt) as suspects and even presumed that Heim was complicit on the basis of his close contact with Muthmann and Volck.[77] Yet they were unable to find tangible

74 Interview with Mrs Hamkens, Zeitzeichen Archive (1990).

75 For false confessions, see Generalsstaatsanwalt to the Justizminister in Berlin concerning "Sprengstoffattentat auf das Reichstagsgebäude" Berlin, 10.9.1929, GStA PK I. HA Rep. 84a No. 55379; *Kieler Neueste Nachrichten* No. 209, 7.9.1929, p. 3 and No. 211, 10.9.1929, p. 2. Targets of attacks with firecrackers and mock bombs included a clinic in Hamburg, a post office by the Alexanderplatz in Berlin, the house of a police sergeant in Nahe in County Segeberg and the residence of a doctor in Rendsburg believed to be a Freemason. For the incident in Hamburg, see the *Kieler Neueste Nachrichten* No. 193, 20.8.1929, p. 2; for the post office in Berlin: *Kieler Neueste Nachrichten* No. 209, 7.9.1929, pp. 1 and 3 and No. 210, 8.9.1929, p. 1; for the incident in Nahe: reports from the Regierungspräsident to the Oberpräsident in Kiel concerning "Fingiertes Attentat in Nahe" Schleswig, 2.10.1929 and Schleswig, 17.1.1930 and the report from the Landjägersamt in Nahe, Nahe, 15.9.1929, all in LAS 301/4704; for the incident in Rendsburg: Polizeiverwaltung to the Regierungspräsident in Schleswig, Rendsburg, 4.10.1929, LAS 309/22784. For threats directed, for example, at the finance office in Elmshorn and Oberpräsident Kürbis, see the report from the Landrat to the Oberpräsident in Kiel concerning "Schutz der öffentlichen Gebäude vor Attentaten" Pinneberg, 21.6.1929, the anonymous and undated but anatomically unmistakable postcard marked "Manure hotbed – pumpkin inferior quality" (*Mistbeet – Kürbis II Sorte*), the report from the Polizeiverwaltung II to the Oberpräsident in Kiel concerning "Anonyme Drohbriefe" Flensburg, 18.7.1929, copy of a report from the Polizeiverwaltung II, Flensburg, 3.8.1929, report from the Regierungspräsident to the Preuß. Minister des Innern in Berlin, Schleswig, 7.8.1929, the examination transcript of Brottinger by the Polizeiverwaltung, Abteilung II, Flensburg, 4.8.1929 and the Polizeipräsident in Flensburg to the Oberpräsident in Kiel concerning "Strafverfahren gegen Brottinger (Schlageterbriefe)" Flensburg, 10.4.1930, all in LAS 301/4704.

76 Copies of reports from the Landjägerabteilung Hohenwestedt, Hohenwestedt, 30.5.1929, Polizeipräsident to the Regierungspräsident concerning "Sprengstoffattentat in der Nacht zum 30.5.29 in Hohenwestedt" Kiel, 19.7.1929 and the Landrat to the Regierungspräsident in Schleswig, Rendsburg, 30.5.1929, all in LAS 301/4704; reports from the Oberstaatsanwalt in Kiel to the Preuß. Justizminister, Kiel, 14.6.1929, 17.6.1929, 29.6.1929 and 25.4.1931, all in GStA PK I. HA Rep. 84a No. 55375; *Rendsburger Tageblatt* No. 124, 30.5.1929, p. 4; *Kieler Zeitung* No. 149, 31.5.1929, p. 7; *Das Landvolk* No. 70, 31.5.1929, p. 5, No. 72, 2.6.1929, p. 2 and No. 73, 4.6.1929, p. 1.

77 For the police's suspicions and what was apparently a false confession from Heinrich Wiborg intended to mislead the investigations, see the copy of the report titled "Bericht zum

evidence. Even by late July the provincial government had nothing new to report to the Prussian Interior Ministry other than speculation over possible connections to Organisation Consul in Berlin.[78]

Not once did the police find a trace of the bombers at the scene of the attacks, not even in Lüneburg where the government buildings were well illuminated at night and the police-station was situated directly opposite.[79] Analysis of the remnants of the bombs and of the devices that failed to explode confirmed that the attacks in Lüneburg and Berlin were related to those in Schleswig-Holstein.[80] However, the police found no informants in Heim and Hamkens' home villages of St. Annen and Tetenbüll, even though the rewards set for evidence leading to the culprits' arrests, which began at RM 5,000, rose to RM 25,000 in total after the Reichstag bombing. In late August, one informant came forward, claiming that Bruno von Salomon had intimated his brother, Ernst, and Guido Weschke were involved.[81] Otherwise the authorities were simply inundated with well-meant but ultimately worthless information.[82]

Not until they began a covert operation to monitor the movement of traffic, compiling lists of registration numbers as cars passed by and then comparing their lists when a detonation occurred, did the police begin to make progress. The process was elaborate, with roadworks simulated in numerous locations in order to ensure that cars slowed sufficiently for officers to read their licence plates, but gradually a large field of suspects was narrowed down until suspicion finally fell on

---

Verdacht gegen Hermann Claussen in Hohenwestedt" Itzehoe, 29.5.1929, and reports from the Landeskriminalpolizeistelle Flensburg titled "Bericht" Itzehoe, 12.6.1929, from the Kriminalpolizeistelle Flensburg to the Regierungspräsident in Schleswig, Itzehoe 5.6.1929 and from the Landeskriminalpolizeistelle Flensburg titled "Bericht" Itzehoe, 12.6.1929, all in LAS 301/4704.

78 Report from the Regierungspräsident to the Preuß. Minister des Innern in Berlin concerning "Sprengstoffanschläge" Schleswig, 22.7.1929, LAS 301/4704.

79 See, for example, reports in the *Kieler Neueste Nachrichten* No. 210, 8.9.1929, p. 1, No. 127, 4.6.1929, p. 5, No. 209, 7.9.1929, p. 1, No. 211, 10.9.1929, p. 2 and No. 210, 8.9.1929, p. 1.

80 Polizeipräsident in Berlin to the Generalstaatsanwalt, Landgericht I Berlin, Berlin, 4.9.1929, GStA PK I. HA Rep. 84a No. 55379; issue of the verdict of the Schwurgericht Altona, 31.10.1930, p. 82 in both LAS 301/4702 and 309/22764; *Kieler Neueste Nachrichten* No. 180, 4.8.1929, p. 2; *Volkszeitung* No. 205, 3.9.1929, p. 1.

81 Oberstaatsanwalt to the Preuß. Justizminister concerning "Sprengstoffattentate zu Itzehoe vom 23. Mai 1929" Altona, 20.8.1929 and the examination transcript of Otto Glamann, Altona, 20.8.1929, both in GStA PK I. HA Rep. 84a No. 55374.

82 Report from the Landeskriminalpolizeistelle Flensburg titled "Bericht" Itzehoe, 12.6.1929, LAS 301/4704; Reichsbahndirektion Altona to the Regierungspräsident in Schleswig concerning "Plakataushang" Altona, 9.8.1929, LAS 309/22668; *Schleswig-Holsteinische Tageszeitung* No. 123, 28.5.1929, p. 1; *Husumer Nachrichten* No. 122, 28.5.1929, p. 8; *Kieler Neueste Nachrichten* No. 205, 3.9.1929, p. 2; *Vossische Zeitung* No. 213, 6.9.1929, p. 6.

a Ford driven by Nickels.[83] A warrant was issued for his arrest after an officer spotted him transporting a "suspicious" package on September 6.[84] The officer had given chase but had had to abandon his pursuit after his motorbike broke down.

Three days later police saw Nickels in Hamburg, carrying the suspicious package into a friend's flat.[85] On the request of Schleswig-Holstein's police, they raided the flat the next morning, September 10, and found the package, a bomb in a cigar box wrapped in paper, on a bookshelf.[86] Nickels had already departed by train for Krempe, where he had left his car, but when he arrived there he found no fewer than ten policemen and detectives waiting for him.[87]

Nickels was known to be a regular visitor at *Das Landvolk* and had been observed stopping off in Itzehoe that very morning. Police now descended upon the town, raiding the newspaper's offices and arresting almost the entire staff: Johnsen, Kühl, Weschke, Bruno von Salomon and the type-setter, Dammann, among others.[88] Hamkens and Muthmann, who were on a publicity tour in Silesia, were arrested near Breslau on September 11, while the Ford that had provided the key to Nickels' apprehension also implicated Heim and Volck because police had also observed them driving it.[89] Heim was arrested immediately. Volck was still abroad. The search for him led to the discovery that he had resided for a time at Vick's farm in Rönne. When the police raided Vick's farm, they discovered his arsenal.[90] He was detained for a week, his role in the Lüneburg bombings not coming to light until later.

Almost the entire Landvolk leadership was behind bars but there was still no evidence of their involvement in the attacks.[91] Ignoring legal procedure, possibly

---

83 See, for example, the records of radio messages to alert the traffic controls sent to the "Pol. Funkstelle Schleswig": "Funkspruch Nr. 403" from the Polizeidirektion Lüneburg and "Funkspruch 410" from the Regierungspräsident in Lüneburg, Schleswig, 6.9.1929, both in LAS 309/22668.

84 For details of the operation as given by Altona's police president, Eggerstedt, see the *Kieler Neueste Nachrichten* No. 214, 13.9.1929, p. 1.

85 *Volkszeitung* No. 213, 12.9.1929, p. 2. For details of Nickels' visit to Hamburg, see the issue of the verdict of the Schwurgericht Altona, 31.10.1930, pp. 92-95, in both LAS 301/4702 and 309/22764.

86 Issue of the verdict of the Schwurgericht Altona, 31.10.1930, pp. 92-94, in both LAS 301/4702 and 309/22764.

87 *Schleswig-Holsteinische Tageszeitung* No. 215, 12.9.1929, p. 1; *Volkszeitung* No. 213, 12.9.1929, pp.1-2.

88 *Rendsburger Tageblatt* No. 213, 11.9.1929, p. 1; *Kieler Neueste Nachrichten* No. 213, 12.9.1929, p. 1 and No. 214, 13.9.1929, p. 1.

89 *Volkszeitung* No. 213, 12.9.1929, p. 2.

90 *Rendsburger Tageblatt* No. 215, 13.9.1929, p. 1; *Kieler Neueste Nachrichten* No. 214, 13.9.1929, p. 1 and No. 215, 14.9.1929, p. 2.

91 For the police's reservations as late as the beginning of September to arrest the suspects owing to a lack of evidence, see the Vermerk IV f444/29, 5.9.1929, GStA PK I. HA Rep. 84a No. 55374.

on the urging of the Prussian Interior and Justice Ministries, the police did not bring the arrested before a magistrate "without delay" as required, but rather waited for up to nine days while they collected evidence to justify the arrests.[92] However, this task became considerably easier when, on September 12, Weschke confessed his participation in the attack on the house of community chairman Mahlstedt in Beidenfleth.[93] Three days later Johnsen confessed to the bombing of the county administrator's office in Itzehoe and soon afterwards to his participation in the bombing in Niebüll and the attempted bombing of Grimpe's residence in Schleswig.[94] These confessions, of course, implicated other conspirators and led to further arrests.[95]

In the meantime Hamkens conceded that he had heard of Volck and Weschke's involvement, an admission of little consequence as they had already been incriminated by Weschke's own confession. He otherwise professed his innocence and ignorance. He emphasised that he did not condone the attacks and had avoided any discussion about them whenever he had heard them mentioned.[96] However, when he was examined again on September 15, he exposed Heim's central responsibility, speaking at length about their differences over preferred methods.[97] For himself, Hamkens admitted only to complicity. Reportedly, he told police that every farmer in Dithmarschen had known who was behind the bombings.[98]

Within a week of Nickels' arrest, the core of the conspiracy was unmasked and Heim, though he denied all responsibility, was recognised as the group's leader.[99]

---

92 Luetgebrune, *Neu-Preußens Bauernkrieg*, pp. 162-163; Werner, *Die Prozesse gegen die Landvolkbewegung*, pp. 234-237.

93 See Weschke's examination transcripts, Altona, 11.9.1929, Altona, 12.9.1929 and Altona, 13.9.1929, all in LAS 309/22701; *Rendsburger Tageblatt* No. 215, 13.9.1929, p. 1. NB there is no evidence to confirm later claims that the police tricked or intimidated Weschke, Johnsen, Volck or any of their accomplices into making confessions. For such claims, see Heim, 'Claus Heim berichtet', LAS EI/4434, pp. 6-7; Luetgebrune, *Neu-Preußens Bauernkrieg*, p. 170.

94 *Nordischer Kurier* No. 217, 15.9.1929, p. 1 and No. 218, 17.9.1929, p. 1; *Schleswiger Nachrichten* No. 221, 20.9.1929, p. 1; *Rendsburger Tageblatt* No. 217, 16.9.1929, p. 1; *Volkszeitung* No. 217, 17.9.1929, p. 2 and No. 219, 19.9.1929, p. 2. Also see the issue of the verdict of the Schwurgericht Altona, 31.10.1930, pp. 61-62, in both LAS 301/4702 and 309/22764.

95 *Rendsburger Tageblatt* No. 217, 16.9.1929, p. 1; *Nordischer Kurier* No. 218, 17.9.1929, p. 1.

96 Hamkens' examination transcript, Altona, 13.9.1929, LAS 309/22701.

97 Hamkens' examination transcript, Altona, 15.9.1929, LAS 309/22701.

98 *Kieler Zeitung* No. 262, 21.9.1929, p. 2. For the extensive speculation about Hamkens' statements, see *Vorwärts* No. 436, 17.9.1929, GStA PK I. HA Rep. 84a No. 55381; *Kieler Zeitung* No. 261, 20.9.1929, p. 2; *Rendsburger Tageblatt* No. 221, 20.9.1929, p. 1; *Das Landvolk* No. 192, 9.11.1929, p. 1; *Schleswiger Nachrichten* No. 263, 8.11.1929, p. 1.

99 *Nordischer Kurier* No. 218, 17.9.1929, p. 1; *Rendsburger Tageblatt* No. 217, 16.9.1929, p. 1. Also see the transcripts of examinations of Heim, Altona, 11.9.1929 and 13.9.1929, both in LAS 309/22701 and the transcripts of the interrogation of Heim, Berlin, 22.9.1929 and of Hamkens and Heim, Berlin, 25.9.1929, both in GStA PK I. HA Rep. 84a No. 55381.

Over the following months, further arrests and confessions followed so that by December, though Kaphengst eluded arrest until August 1930, all but the accomplices in Wiborg and Nickels' theft in Mülheim and the bombing of the Reichstag had been clarified.[100]

Correspondence confiscated during searches at *Das Landvolk* and its staff's residences, though ultimately containing nothing incriminating, was used as a basis for further interrogations and arrests in Berlin. Those questioned included high-profile proponents of national conservatism such as the writer, Ernst Jünger,[101] while most of those arrested were members of Organisation Consul, who police had suspected and monitored from the first attacks.[102] Organisation Consul leader Ehrhardt was not among them but he was questioned and his house was searched. He professed his friendship to the Landvolkbewegung but denied having anything to do with the affair.[103] Police did find a detonator similar to those from the bombings as well as 174 rounds of ammunition at the home of Plaas, his secretary.[104]

---

100 For Schmidt's arrest and confessions, see the examination transcripts of Schmidt, Altona, 30.10.1929 and Altona 31.10.1929, GStA PK I. HA Rep. 84a No. 55381; *Das Landvolk* No. 183, 30.10.1929, p. 2 and No. 184, 31.10.1929, p. 2; *Nordischer Kurier* No. 255, 30.10.1929, p. 1, No. 256, 31.10.1929, p. 1, No. 258, 2.11.1929, p. 1 and No. 260, 5.11.1929, p. 1. For Volck's confessions, see the copies of the examination transcripts of Volck, Stresa (Lago Maggiore) 24.10.1929 and 28.10.1929, GStA PK I. HA Rep. 84a No. 55381. For Volck's arrest in Lugano in November and the rumours surrounding his capture, see the *Vossische Zeitung* No. 269, 10.11.1929, p. 16; *Rendsburger Tageblatt* No. 265, 9.11.1929, p. 2; *Kieler Zeitung* No. 314, 12.11.1929, p. 3; *Nordischer Kurier* No. 180, 2.8.1930, p. 1; F. W. v. Oertzen, *Im Namen der Geschichte! Politische Prozesse der Nachkriegszeit* (Hamburg, 1934), pp. 79-101. For the arrests of Vick, Luhmann, Becker and Manecke and their confessions to the bombings in Lüneburg, see the *Nordischer Kurier* No. 276, 24.11.1929, p. 1 and No. 277, 25.11. 1929, p. 2; *Kieler Neueste Nachrichten* No. 276, 26.11.1929, p. 1. For Holländer's confession, see the *Kieler Neueste Nachrichten* No. 276, 26.11.1929, p. 1. For the arrests of Ross, Wiborg and Hönck, see the *Nordischer Kurier* No. 275, 23.11.1929, p. 1. For the robbery in Mülheim; Generalstaatsanwalt to the Preuß. Justizminister in Berlin concerning "Strafsache gegen Nickels und Genossen" Berlin, 18.12.1929, GStA PK I. HA Rep. 84a No. 55367.

101 Jünger was questioned owing to his acquaintance with Ernst and Bruno von Salomon. See the examination transcripts of Ernst Jünger, Berlin, 26.9.1929 and Berlin, 3.10.1929, GStA PK I. HA Rep. 84a No. 55381.

102 Report from the Regierungspräsident to the Preuß. Minister des Innern in Berlin concerning "Landvolkbewegung und die politischen Sprengstoffanschläge" Schleswig, Schleswig, 30.9.1929, in LAS 301/4696, and 28.9.1929 in 309/22784; the extensive lists of confiscated items, in both LAS 301/4704 and 309/22784; for the reports of the arrests in Berlin: *Rendsburger Tageblatt* No. 213, 11.9.1929, p. 1; *Kieler Neueste Nachrichten* No. 213, 12.9.1929, p. 3 and No. 214, 13.9.1929, pp. 1-2; *Kieler Zeitung* No. 253, 12.9.1929, pp. 1-2 and No. 254, 13.9.1929, p. 1; *Neue Preußische Kreuz Zeitung* No. 296, 12.9.1929, p. 1; *Volkszeitung* No. 213, 12.9.1929, p. 2 and No. 215, 14.9.1929, p. 6.

103 *Neue Preußische Kreuz Zeitung* No. 301, 17.9.1929, p. 7.

104 *Kieler Neueste Nachrichten* No. 214, 13.9.1929, p. 2 and No. 215, 14.9.1929, p. 1.

Further material confiscated there, however, confirmed only that Plaas had provided articles for *Das Landvolk*. There was no actual evidence of his or Ehrhardt's involvement in the attacks.[105]

There was considerably stronger circumstantial evidence, on the other hand, against the main suspects in the Reichstag bombing, Ernst von Salomon and Muthmann. A witness, who had seen two men, one with a large package under his arm, pass by the Brandenburg Gate heading toward the Reichstag on the night of the attack, later recognised von Salomon in a line up.[106] His subsequent claim, however, that the second man was Volck, who was known to be abroad, proved him unreliable. Another witness claimed to have seen Muthmann inspecting the Reichstag. Muthmann disappeared in November and rumours abounded as to his whereabouts. Later, a further witness testified that he had met Muthmann while travelling in Argentina and that he had confessed to the attack, but the public prosecutor failed to pursue this apparently dubious tale.[107] Though Ernst von Salomon's biographer, Markus Joseph Klein, asserted that von Salomon and Muthmann were responsible, the case remained unsolved, and even in his memoirs, Heim was equivocal, writing only that Muthmann probably planted the bomb.[108]

Fuelling claims in the nationalist press that the authorities were attempting to use the bombings to cast aspersions on their political opponents on the right, a

105 Plaas' examination transcript, Berlin, 21.9.1929, GStA PK I. HA Rep. 84a No. 55381; undated document titled "Ehrhardt Kreis" appended to a report from the Regierungspräsident to the Preuß. Minister des Innern concerning "Die Landvolkbewegung und die Sprengstoffanschläge" Schleswig, 17.9.1929 and the undated preliminary results of the investigation of the material found at Ehrhardt's office titled "Vorläufiges Ergebnis der Sichtung des bei Plaas im Büro Ehrhardt, Berlin W.35, Steglitzerstr. 28, beschlagnahmten Materials", both in LAS 309/22784.

106 Luck's examination transcripts, Berlin, 27.9.1929, GStA PK I. HA Rep. 84a No. 55381; *Vorwärts* No. 455, 28.9.1929 and *Berliner Volkszeitung* No. 458, 28.9.1929, both in GStA PK I. HA Rep. 84a No. 55382; *Schleswiger Nachrichten* No. 228, 28.9.1929, p. 1; *Nordischer Kurier* No. 229, 29.9.1929, p. 1.

107 Werner, *Die Prozesse gegen die Landvolkbewegung*, pp. 198-199 and 241-242; report titled "Bericht" concerning "Sprengstoffanschlag in Neumünster" Kiel, 21.3.1930, LAS 301/4702; report from the Landrat Kreis Steinburg to the Regierungspräsident in Schleswig concerning "Landvolkbewegung" Itzehoe, 22.3.1930, in both LAS 301/4697 and 309/22590; report from the Landeskriminalpolizeistelle Altona, Abteilung IA, concerning "Landvolkbewegung" Altona, 8.1.1930; report from the Polizeipräsident, Landeskriminalpolizeistelle, to the Regierungspräsident in Schleswig concerning "Landvolkbewegung" Flensburg, 12.7.1930, in both LAS 301/4697 and 309/22590; reports titled "Bericht" Flensburg, 6.2.1930 and Flensburg, 15.2.1930, sent from the Polizeipräsident, Landeskriminalpolizeistelle Flensburg, in both LAS 301/4697 and 309/22783; report from the Polizeipräsident, Landeskriminalpolizeistelle, to the Regierungspräsident in Schleswig, Flensburg, 18.3.1930, LAS 309/22590.

108 See Klein, *Ernst von Salomon*, p. 190 and Heim, 'Claus Heim berichtet', LAS EI/4434, p. 5.

number of National Socialists were also among the first to be arrested.[109] The link was tenuous, since all but Bodo Uhse were released the next day. Yet the party's fears for its image were warranted and there was even some truth in Lohse's claim that the authorities simply wished to draw the party into the affair.[110] The Regierungspräsident, at least, seemed to believe the Prussian Interior Ministry was this way inclined because he advised Berlin not to highlight evidence implicating individual National Socialists too strongly, fearing it might encourage the NSDAP and the Landvolk to draw closer together.[111] Although he had already identified the NSDAP as a greater longer-term danger than the Landvolkbewegung,[112] the unmasking of the bombing conspiracy soon after the Landvolk had enjoyed a burst of support following the Neumünster incident, led him to judge it "by far the most radical political movement" and requiring the greatest attention.[113] Isolating the Landvolkbewegung, he believed, would hasten its demise, for he anticipated that it would lose many followers as a consequence of its association with the bombings.

In any case, the material confiscated in connection with the arrests of the suspects did not support speculation over links between the Landvolkbewegung and the NSDAP leadership. It did show that some Landvolk personalities – including Walther Bohm who contributed to *Das Landvolk*, and Gustav Kurtze, who had briefly helped Heim with the Defence Association – belonged to the NSDAP. It also revealed the close relationship between editors Bruno von Salomon and Uhse, and indicated that an oppositional group within the NSDAP, containing Uhse and expelled members Heinz and Rademacher, sympathised with the Landvolk and admired its more extreme methods. Even so, there was no indication of any cooperation between Lohse and the Landvolk, nor that Hitler's warnings against the farmers' movement were anything but genuine.[114]

---

109 See, for example, *Neue Preußische Kreuz Zeitung* No. 302, 18.9.1929, p. 1. Those arrested included Uhse's fellow editor at the *Schleswig-Holsteinische Tageszeitung*, Ehlert, the Gau treasurer, Brix, the Gauleiter of Hamburg, Krebs, and Gregor Straßer. See the undated list titled "Festgenommene Personen anläßlich der Sprengstoffanschläge", LAS 309/22784; *Kieler Neueste Nachrichten* No. 215,14.9.1929, p. 2; *Schleswig-Holsteinische Tageszeitung* No. 216, 13.9.1929, p. 1.

110 *Schleswig-Holsteinische Tageszeitung* No. 217, 14.9.1929, p. 1.

111 Report from the Regierungspräsident to the Preuß. Minister des Innern concerning "Die Landvolkbewegung und die Sprengstoffanschläge" Schleswig, 17.9.1929, LAS 309/22784.

112 Regierungspräsident to the Preuß. Minister des Innern concerning "Politische Lage im Reg. Bez. Schleswig" Schleswig, 5.6.1929, LAS 309/22668.

113 Report from the Regierungspräsident to the Preuß. Minister des Innern in Berlin concerning "Landvolkbewegung und die politischen Sprengstoffanschläge" Schleswig, 30.9.1929, in LAS 301/4696 and 28.9.1929 in 309/22784.

114 Report from the Regierungspräsident to the Preuß. Minister des Innern in Berlin concerning "Landvolkbewegung und die politischen Sprengstoffanschläge" Schleswig, 30.9.1929, in LAS 301/4696 and 28.9.1929 in 309/22784. Also see the lists of confiscated material titled "XIII. Bei August Kurtze .... beschlagnahmtes Material, im Pol.-Präsidium eingeliefert am Abend des 14. September 1929" and "XIV. Bei Rittmeister a.D. Bohm aus Hamburg am 14

The arrests, however, had already caused sufficient confusion for the public to suppose the NSDAP was involved. As Goebbels noted in his diary, explanations in the press would not help much.[115] Indeed, reporting Johnsen's confession to the bombings in Itzehoe and Schleswig, the *Vossische Zeitung* erroneously identified him as a National Socialist.[116]

In the *Illustrierter Beobachter,* Hitler responded to the rumours by repeatedly condemning the bombings and denying his party's involvement. Though he criticised the "brutal and ruthless collection of taxes" and blamed the system for conditions in which "some obstinate, politically unenlightened fanatic" could end up in crime, he again insisted that bombs could not eliminate the system and he warned his followers that they would be playing into the hands of the Jews and their press should they participate in such activities.[117]

Uhse's release at the end of September should have put an end to speculation over the NSDAP's involvement, but the party's concerns about reports in some newspapers remained.[118] These were so strong that the Gau Hamburg pressed *Das Landvolk* to publish a declaration, affirming that these reports were false and that none of the "supposed" Landvolk bombers belonged to the NSDAP. Further, the declaration explained, "What is certain, is that these people belong to the Landvolkbewegung, to which the NSDAP stands in complete opposition".[119]

Das Landvolk agreed to print the declaration and though it disputed the NSDAP's use of the term "Landvolk-organisation", arguing that no such organisation existed, it did not impugn the statement's essential content. Nor did it appreciate the NSDAP's aloofness and what it understood as a further condemnation of Landvolk activism. Soon after *Das Landvolk* printed a report declaring conflict with the law acceptable and violence understandable when in the service of "glowing idealism". With what was unmistakably a slight at the National Socialists, it asked, "do these big talkers want to carry out their revolution with felt slippers and chocolate biscuits?"[120]

Ernst Jünger, who sympathised with the bombers, also publically rejected Hitler's and his party's condemnations.[121] Yet even proffering a qualified defence

---

September 1929 beschlagnahmtes Material", both in LAS 309/22784. For the expulsion of Heinz and Rademacher, see chapter 6 above.

115 J. Goebbels, *Tagebücher: 1924-1945, 1. Einführung, 1924-1929* (Munich, 1992), pp. 404-405.

116 *Vossische Zeitung* No. 221, 15.9.1929, p. 2.

117 A. Hitler, 'Politik der Woche', *Illustrierter Beobachter* 14.9.1929 and 21.9.1929 in Lankheit, *Hitler: Reden Schriften Anordnungen,* vol. 3 (2), pp. 378-382 and 382-385.

118 *Schleswig-Holsteinische Tageszeitung* No. 230, 29.9.1929, p. 1.

119 *Das Landvolk* No. 158, 1.10.1929, p. 6.

120 *Das Landvolk* printed the report which had appeared in the *Wehrwolf.* See *Das Landvolk* No. 162, 5.10.1929, p. 5.

121 H. Schwilk, *Ernst Jünger: Ein Jahrhundertleben. Die Biographie* (Munich and Zurich, 2007), pp. 341-342.

of the attacks was a surprisingly significant step for the Landvolkbewegung. For weeks after the arrests, *Das Landvolk* had stubbornly denied the movement's involvement. It feared a public backlash because opinion appeared to be widespread that the attacks were "a crime against the state and people".[122] Consequently, *Das Landvolk* claimed the arrests were a "bluff" by the authorities to attack the movement, vehemently and repeatedly dismissed reports to the contrary as lies and rumours spread by the "Jewish press" and employed anti-Semitic sentiment to encourage misgivings about the integrity of the investigations.[123]

As the bombers began to confess, *Das Landvolk* could only sustain this position by perpetuating misleading reports itself and by unashamedly rejecting the authorities' verification that the bombers belonged to the movement. Additionally, it touted the releases of individual Landvolk leaders, principally Hamkens and Kühl, as evidence of the movement's innocence.[124] In what amounted to a public expression of his long disagreement with Heim over their choice of methods, Hamkens directed *Das Landvolk* to dissociate itself from the bombers.[125] *Das Landvolk* claimed the movement had been falsely identified with a "small clique of adventurers and reckless fanatics" and had "never had anything to do with bombers and criminals".[126] Even after *Das Landvolk's* manager, Johnsen, confessed, the paper continued to assert that the Landvolkbewegung rejected terror, had always done so and had never had anything to do with violent actions.[127]

The Landvolkbewegung was compelled to relinquish this increasingly untenable position and slowly began to acknowledge that the bombers belonged to the movement after Luetgebrune spoke about the investigations at a meeting of 200 of

---

122 *Rendsburger Tageblatt* No. 220, 19.9.1929, p. 1 and further condemnations in No. 221, 20.9.1929, p. 1 and in a letter from a farmer from the Rendsburg area printed in No. 222, 21.9.1929, p. 1.

123 See, for example, *Das Landvolk* No. 142, 12.9.1929, pp. 1-3, No. 143, 13.9.1929, p. 1, No. 144, 14.9.1929, pp. 1-2, No. 145, 15.9.1929, pp. 1-2, No. 146, 17.9.1929, pp. 1-2 and No. 147, 18.9.1929, pp. 1-2.

124 See, for example, the deceptive headlines, "A gigantic police blunder" and "The complete innocence of most of the arrested already determined" in *Das Landvolk* No. 143, 13.9.1929, p. 1 and No. 144, 14.9.1929, p. 1. For *Das Landvolk's* attempts to distance itself from some of the suspects, including Nickels and Weschke, see *Das Landvolk* No. 142, 12.9.1929, p. 2, No. 145, 15.9.1929, p. 2, No. 146, 17.9.1929, p. 2, No. 147, 18.9.1929, p. 2 and No. 149, 20.9.1929, p. 1. For announcements of the releases of Dammann, Schade, Kühl and Hamkens, see *Das Landvolk* No. 145, 15.9.1929, p. 1, No. 149, 20.9.1929, p. 2, No. 150, 21.9.1929, p. 1 and No. 157, 29.9.1929, p. 1. Also see the *Vossische Zeitung's* objections to *Das Landvolk's* "unheard of assaults on the police." *Vossische Zeitung* No. 220, 14.9.1929, p. 2.

125 Letter from Hamkens, 16.9.1929, cited in the report from the Landeskriminalpolizeiamt, Abteilung IA, Außendienst, VII., Berlin, 5.6.1930, LAS 309/22701.

126 *Das Landvolk* No. 144, 14.9.1929, p. 1 and No. 145, 15.9.1929, p. 1, No. 146, 17.9.1929, p. 1.

127 *Das Landvolk* No. 147, 18.9.1929, p. 1. Also see *Das Landvolk* No. 149, 20.9.1929, p. 1.

the movement's "trusted people" in Rendsburg on September 30. For many, Luetgebrune's speech came as a shock because it was the first time someone who they considered reliable admitted that the Landvolkbewegung was implicated. Afraid that the details of the bombings, once public, might weaken support for the movement, Hamkens instructed those present to keep their knowledge to themselves, but signalling a change in tack, he and Kühl exhorted them not to leave the bombers in the lurch.[128]

Over the course of the next three months, the movement's leaders and newspaper gradually but dramatically adjusted their line of reasoning. Though continuing at first to claim that the movement had nothing more to do with the bombers, they added that they were not prepared to join in condemning them. Their disapproval of the bombers' mistaken choice of methods remained, but with ever more self-assurance they began to avow understanding for the bombers' actions, and emphasised that they had acted selflessly, according to their ideals and their love of the "fatherland".[129]

The transformation was not seamless and there were periods of silence, but bit by bit the Landvolk's leaders and newspaper went ever further.[130] By December, they excused the bombings as acts of desperation against the system. Indeed, they no longer just defended but effusively supported the bombers' motivations and hagiography replaced their previous disassociation. They described the bombers as "front soldiers" and "honourable men" and differentiated between criminal and il-

128 Report titled "Bericht" Heide, 1.10.1929 and the addendum to it dated Heide, 2.10.1929, LAS 301/4696; reports from the Abteilung IA titled "Bericht" concerning "Landvolkbewegung" Hohenwestedt, 2.10.1929 and titled "Bericht" Heide, 3.10.1929, both in LAS 309/22784.

129 See Kühl's speech at a meeting in Wiemersdorf and Wulf's speech at a meeting in Burg (Dithmarschen), both on October 2, 1929 and Kühl and Hamkens' speeches at a meeting in Wrist on October 17, 1929: report from the Landeskriminalpolizeistelle Altona, Abteilung IA, titled "Bericht über die Landvolkversammlung am 2.10.1929 ... in Wiemersdorf bei Bad Bramstedt" Altona, 3.10.1929, in both LAS 301/4696 and 309/22784; report from the Oberpräsident to the Preuß. Minister des Innern in Berlin concerning "Landvolkbewegung" Kiel, 9.10.1929, LAS 301/4696; report from the Polizei-Präsident Altona-Wandsbek, Abteilung IA, titled "Bericht über die am 2. Oktober 1929, ... in Burg i/Dith. stattgefundene Versammlung der Landvolkbewegung" Altona, 3.10.1929, in both LAS 301/4696 and 309/22784; report from the Landeskriminalpolizeistelle Altona, Abteilung IA, titled "Bericht über eine Landvolkversammlung in Wrist" Itzehoe, 18.10.1929, in both LAS 301/4696 and 309/22784. Also see the guidelines for leaders of smaller Landvolk meetings titled "Anweisungen für die Versammlungsleiter der Landvolkbewegung" sent by the Polizei-Präsident Altona-Wandsbek, Abteilung IA, to the Oberpräsident in Kiel, 7.10.1929, LAS 301/4696.

130 Report from the Abteilung IA titled "Bericht" concerning "Landesthing am 10. November 1929 in Rendsburg" Kiel, 11.11.1929, LAS 309/22783; report from the Landeskriminalpolizeistelle Altona, Abteilung IA, concerning the "Landvolkbewegung" Itzehoe, 6.11.1929, in both LAS 301/4697 and 309/22783.

legal methods. Although they acknowledged that the bombers had committed offences against the law, they deemed their actions as "pure and honourable".[131] By the beginning of the new year, Paul Adam Ross went as far as to claim that the bombers had acted out of "love", that they should not be viewed as criminals but as "the vanguard for the good cause".[132]

In all probability, the echo of the Landvolk leaders' views among their followers encouraged them to be bolder, while a further impetus seems to have come as a number of the bombers returned to the movement on their release from remand.[133] By March 1930 Steinburg's county administrator reported that in contrast to its initial stance, *Das Landvolk* had begun to give the bombers its full support and had completely reverted to its true character, that of "rabble-rousing" (*Hetze*).[134] By April *Das Landvolk* acknowledged the bombers as "our fellow fighters" and not only praised their motives but also began to argue that these justified their means.[135] Finally, in the months leading up to their trial, as the movement intensified its efforts to promote sympathy for the bombers, Landvolk leaders urged farmers to join the movement in displaying their support.[136] A flyer, of which over 40,000 copies were printed, called on farmers to show the "ruling powers" that they supported and acknowledged the accused as their "vanguard".[137]

Nothing, however, could surpass Bruno von Salomon's fanaticism. Despite the best efforts of the Prussian Interior Ministry and Regierungspräsident to prevent him from speaking, or at least have him punished by instigating legal action against him, von Salomon addressed a series of small meetings along the west coast after his release from remand. He voiced frank and unrestrained approval of the bombings, even responding to one incredulous police officer's enquiry that he did so

---

131 See the copy of the appeal to the Reich President, "Das Landvolk an Hindenburg!" in *Das Landvolk* No. 1, 1.1.1930, p. 1. Also see *Das Landvolk* No. 228, 22.12.1929, p. 2 and No. 230, 25.12.1929, p. 1. For an earlier, but unusual, example of the differentiation between illegal and criminal methods, see Otfried Rademacher's leading article in *Das Landvolk* No. 148, 19.9.1929, p. 1.

132 Report from the Landjägereikreisleiter, Kreis Steinburg, titled "Bericht über die Landvolkversammlung in Krempe am 12.d.Mts." Itzehoe, 14.12.1929, in both LAS 301/4697 and 309/22783.

133 For the rural population's response, see p. 262 below. For the releases of Vick, Becker and Wiborg, see *Das Landvolk* No. 36, 12.2.1930, p. 1; for Bruno von Salomon, No. 32, 7.2.1929, p. 1 and Hennings, No. 50, 28.2.1930, p. 1.

134 Report from the Landrat Kreis Steinburg to the Regierungspräsident in Schleswig concerning "Landvolkbewegung" Itzehoe, 22.3.1930, in both LAS 301/4697 and 309/22590.

135 *Das Landvolk* No. 77, 1.4.1930, p. 1.

136 Report from the Landeskriminalpolizeistelle Altona, Abteilung IA, titled "Bericht" Altona, 21.6.1930, in both LAS 301/4697 and 309/22590.

137 Report from the Landeskriminalpolizeistelle, Abteilung IA, titled "Bericht" Altona, 8.6.1930, in both LAS 301/4697 and 309/22590; report from the Landeskriminalpolizeistelle, Abteilung IA, Altona-Wandsbek, titled "Bericht über die Landvolkbewegung" Altona, 25.8.1930, LAS 309/22590.

"unconditionally and publicly". Von Salomon referred to Claus Heim as the "General", described him as a hero imprisoned for his love of the nation, explained the bombers' motives and made light of the danger the bombs had posed.[138]

It was testimony to the extensive trust in the Landvolk's propaganda that for weeks after the arrests many farmers had continued to believe that the suspected Landvolk bombers were innocent. Altona's police president and North Dithmarschen's county administrator appealed to the Regierungspräsident for a press release revealing the results of the investigations to counter the "clever propaganda" which misrepresented the facts.[139] Abegg, too, hoped that a press release would serve to weaken the movement but was disappointed in the outcome because it did not appear sufficiently "official".[140] Correspondence with his counterpart in Lüneburg indicated that he was reluctant to reveal details himself because it might give the Landvolkbewegung further opportunity to accuse him of attempting to influ-

138 For an overview of von Salomon's speeches in Lunden, Wesselburen and Hennstedt on June 12, 13 and 14, see the report from the Polizeipräsident, Landeskriminalpolizeistelle, to the Regierungspräsident in Schleswig concerning "Landvolkbewegung" Flensburg, 16.6.1930, LAS 301/4697. Also see the report from the Regierungspräsident to the Preuß. Minister des Innern in Berlin concerning "Landvolkbewegung" Schleswig, 27.6.1930, LAS 301/4698. For meetings in Itzehoe, Leck, Meldorf, Tellingstedt and Marne, see the copy of the report from the Polizeiverwaltung titled "Bericht über die am 21. Juli 1930 abends 8 Uhr stattgefundene und von dem früheren Schriftleiter Bruno von Salomon in Baumanns Gesellschaftshaus abgehaltene Versammlung" Itzehoe, 23.7.1930, in both LAS 301/4697 and 309/22590; reports from the Landrat to the Regierungspräsident in Schleswig concerning "Landvolkversammlung in Leck" Niebüll, 24.6.1930, from the Landrat to the Regierungspräsident in Schleswig concerning "Landvolk-Bewegung" Heide, 28.6.1930 and from the Landrat to the Regierungspräsident in Schleswig concerning "Landvolkversammlungen in Meldorf und Marne" Meldorf, 28.6.1930, all in LAS 309/22590. For the police's closure of a meeting in Hemme on June 23 and the provincial government's efforts to initiate legal action taken against Bruno von Salomon, see the reports from the Landrat to the Regierungspräsident in Schleswig concerning "Auflösung einer Versammlung" Heide, 24.6.1930 and from the Regierungspräsident to the Preuß. Minister des Innern in Berlin concerning "Landvolkbewegung" Schleswig, 27.6.1930; Regierungspräsident to the Oberstaatsanwalt in Flensburg, Schleswig, 18.6.1930 and report from the Regierungspräsident to the Preuß. Minister des Innern in Berlin concerning "Landvolkbewegung" Schleswig, 27.6.1930, all in LAS 301/4698.

139 Landrat to the Regierungspräsident in Schleswig concerning "Bombenattentate" Heide, 1.10.1929 and Polizeipräsident, Landeskriminalpolizeiamt, Abteilung IA, to the Regierungspräsident in Schleswig, Altona, 7.10.1929, both in LAS 309/22784.

140 Report from the Regierungspräsident to the Preuß. Minister des Innern in Berlin concerning "Landvolkbewegung und die politischen Sprengstoffanschläge" Schleswig, 30.9.1929, LAS 301/4696 and dated 28.9.1929, 309/22784; report from the Regierungspräsident to the Preuß. Minister des Innern in Berlin concerning "Landvolkbewegung" Schleswig, 15.10.1929, LAS 309/22784; the report on the investigations as published, for example, in the *Rendsburger Tageblatt* No. 236, 8.10.1929, p. 1.

ence the course of justice.[141] Yet he was eager for these to be disclosed, particularly in such a way that they damaged Hamkens in the eyes of his followers, and he harried the Ministry of the Interior to publish Hamkens' "confession".[142] Although Hamkens had not actually confessed at all, and his complicity was trivial in comparison to the active involvement of many others, Abegg's office singled him out simply because it viewed him as "one of the most dangerous leaders" of the movement.[143]

The publication of the transcripts of Hamkens' police examinations would not have discredited him among the movement's followers because they revealed his complicity in the bombings. They would have done so because they showed he had betrayed his companions.[144] Heim's lawyer, Hüttmann, learned of Hamkens' statement from Luetgebrune. Enraged, he informed other Landvolk leaders, denouncing Hamkens as an "egoist" and asserting that Hamkens' maxim "All for one and one for all" seemed nothing but an empty phrase. Hamkens appeared to regret his confession. He was reported to have suffered nervous problems and a present of RM 1,000 to Heim's wife was said to have been driven by his bad conscience.[145] This did not spare him his colleagues' censure. Responding to criticism at a meeting of "trusted men" in October, he played down the importance of his statements, pointing the finger at Weschke and Johnsen, who he described as "sissies" (*Waschlappen*), for their earlier confessions. However, two men in particular, noted as "noisy people" likely to generate opposition to Hamkens, were not placated. The first, Rudolf Peters (Dörpling), said that he had expected more from farmer and vanguard fighter Hamkens and he declared Heim a "true hero" who the authorities

---

141 Regierungspräsident to the Preuß. Minister des Innern concerning "Pressenotizen über die Landvolkbewegung" Lüneburg, 27.11.1929; Regierungspräsident in Schleswig to the Regierungspräsident in Lüneburg concerning "Pressenotizen über die Landvolkbewegung" Schleswig, 19.12.1929 and Regierungspräsident in Lüneburg to the Regierungspräsident in Schleswig, Lüneburg, 25.12.1929, all in LAS 309/22701.

142 Regierungspräsident to the Minister des Innern in Berlin, Schleswig, 18.10.1929 and Preuß. Minister des Innern to the Preuß. Justizminister, Berlin, 25.10.1929, both in GStA PK. I Rep. 84a No. 55367.

143 Regierungspräsident to the Minister des Innern in Berlin concerning "Presseveröffentlichungen über die Rolle, die der Hofbesitzer Wilhelm Hamkens aus Tetenbüll in der Bomben-Angelegenheit spielt" Schleswig, 12.11.1929, LAS 301/4704; Regierungspräsident to the Preuß. Minister des Innern in Berlin concerning "Sprengstoffanschläge" Schleswig, 28.1.1929, LAS 309/22783.

144 Hamkens' examination transcripts, Altona, 13.9.1929 and Altona, 15.9.1929, both in LAS 309/22701.

145 Report from the Landeskriminalpolizeistelle Altona, Abteilung IA, titled "Bericht über die Landvolkbewegung" Altona, 5.10.1929, in both LAS 301/4696 and 309/22784; report titled "Bericht" Heide, 10.10.1929, sent from the Polizeipräsident, Landeskriminalpolizeistelle, Abteilung IA, to the Regierungspräsident in Schleswig, Altona, 11.10.1929, LAS 309/22784.

had still not managed to wear down.[146] The second was Heinrich Frauen, who criticised Hamkens soon afterwards at a meeting of dairy members as a "big-mouthed, dumb and cowardly snooty-nosed fop" (*ein großschnauziger, dummer und feiger Schnösel*).[147] Even Hamkens' most loyal supporters, such as Kühl, were said to be angered by Hamkens' confession.[148]

Despite the leaders' fears that wider recognition of its involvement in the bombings would reduce support, until the year's end the movement continued to hold meetings on an almost daily basis in different locations throughout the province, generally attracting crowds of 300 to 500 people.[149] Publicity, however, should not be confused with good opinion. From the second half of October, there

146 Report titled "Bericht" Heide, 12.10.1929, LAS 309/22784.

147 Report dated Itzehoe, 15.10.1929, sent from the Polizeipräsident, Landeskriminalpolizeiamt, Abteilung IA, to the Regierungspräsident in Schleswig, Altona, 18.10.1929, in both LAS 301/4696 and 309/22784.

148 Report titled "Bericht" Altona, 24.10.1929, LAS 309/22784.

149 Reports from the Landeskriminalpolizeistelle Altona, Abteilung IA, titled "Bericht über die Landvolkbewegung" Altona, 5.10.1929 and concerning "Landvolkbewegung" Altona, 17.10.1929, both in LAS 301/4696 as well as 309/22784. For a representative sample of meeting attendances in this period, see the reports on meetings in Wiemersdorf and in Burg in Dithmarschen on October 2, 1929, Wrist on October 17, Leezen in County Segeberg on October 30, Wesselburen and Schönberg in early November, Oldenburg in Holstein on December 6, Glücksburg on December 9, Süderbrarup on December 10, Krempe on December 12, Glückstadt on December 13 and Leck on January 5, 1930: report from the Abteilung IA titled "Bericht über die Landvolkversammlung am 2.10.1929 ... in Wiemersdorf bei Bad Bramstedt" Altona, 3.10.1929, in both LAS 301/4696 and 309/22784; report titled "Bericht über die am 2. Okt. 1929, 11 Uhr vorm. in Burg i/Dith. stattgefundene Versammlung der Landvolkbewegung" Altona, 3.10.1929, in both LAS 301/4696 and 309/22784; report from the Landeskriminalpolizeistelle Altona, Abteilung IA, titled "Bericht über eine Landvolkversammlung in Wrist" Itzehoe, 18.10.1929, in both LAS 301/4696 and 309/22784; report from the Landeskriminalpolizeistelle, Abteilung IA, titled "Bericht" concerning "Landvolkkundgebung am 30.10.29 in Leezen" Kiel, 31.10.1929, LAS 309/22783; report from the Landrat to the Regierungspräsident in Schleswig concerning "Landvolkbewegung" Heide, 4.11.1929, LAS 309/22590; report from the Landrat to the Regierungspräsident in Schleswig concerning "Auftreten Hamkens – Tetenbüll" Plön, 12.11.1929, in both LAS 301/4696 and 309/22783; report from the Polizeiverwaltung to the Landrat concerning "Landvolkversammlung in Oldenburg am 5. ds. Mts." Oldenburg in Holstein, 6.12.1929, LAS 309/22783; report from the Polizeipräsident, Landeskriminalpolizeistelle, concerning "Landvolkversammlungen" Flensburg, 11.12.1929, in both LAS 301/4697 and 309/22783; report from the Landjägereikreisleiter, Kreis Steinburg, titled "Bericht über die Landvolkversammlung in Krempe am 12.d.Mts" Itzehoe, 14.12.1929, in both LAS 301/4697 and 309/22783; report titled "Bericht" concerning "Versammlung des Landvolkbundes im hiesigen Lokal 'Franscher Garten' " Glückstadt, 14.12.1929 and copy of report from the Landjägerei Kreis Steinburg titled "Bericht über die Landvolkversammlung in Glückstadt am 13.12.1929 im Lokal 'Englischer Garten' " Itzehoe, 15.12.1929, both in LAS 301/4697 as well as 309/22783; report from the Landjägereiposten Klixbüll to the Landrat in Niebüll, Klixbüll, 6.1.1930, in both LAS 301/4697 and 309/22783.

were a number of reports of subdued responses or weak applause from audiences, of opinion divided over Hamkens and of doubts expressed as to whether the Landvolkbewegung could improve circumstances.[150]

Slowly, more farmers accepted that the Landvolkbewegung was behind the attacks and this was reported as having a "sobering effect" on the rural population.[151] At first, many people in Dithmarschen were incredulous that Hamkens and Heim were incriminated so strongly, but others wished to disassociate themselves from the movement.[152] By mid-October support for the movement was in decline and a negative opinion of Hamkens was growing.[153]

While it is difficult to isolate disenchantment over the bombings from other factors that weakened support – and these will be examined in full in the following chapter – the movement reportedly lost many followers in South Tondern because they did not wish to be connected with the bombings. As a consequence, new Landvolk speakers in the county were received much more critically than previously.[154] As 1929 drew to an end, more and more farmers in the province distanced themselves from the movement, many leaving to join the NSDAP. The party had successfully avoided association with the attacks. The subsequent surge of support it received followed on the conviction that it was now the only party offering rescue and that, unlike the Landvolkbewegung, it fought with "realistic methods".[155]

150 See the report on meetings in Immenstedt in County Husum and Hohn in County Rendsburg, Groß-Harrie near Bordesholm and in Schafflund in the province's north: reports from the Landeskriminalpolizeistelle, Abteilung IA, titled "Bericht" Neumünster, 17.10.1929, from the Landjägerciamt Tungendorf titled "Bericht" Tungendorf, 17.10.1929, from the Landrat to the Regierungspräsident in Schleswig concerning "Öffentliche Versammlung der Landvolkbewegung in Gr. Harrie" Bordesholm, 18.10.1929, report titled "Bericht" concerning "Landvolkbewegung" Flensburg, 19.10.1929, sent by the Polizeipräsident, Landeskriminalpolizeiamt, Abteilung IA, to the Regierungspräsident in Schleswig, Altona, 21.10.1929 and the report from the Landeskriminalpolizeistelle, Abteilung IA, titled "Bericht" Rendsburg, 20.10.1929, all in LAS 309/22784.

151 Reports titled "Bericht" Heide, 3.10.1929 and from the Landeskriminalpolizeiamt titled "Bericht" Altona, 7.10.1929, both in LAS 309/22784.

152 Report titled "Bericht" Heide, 10.10.1929, sent from the Polizeipräsident, Landeskriminalpolizeistelle, Abteilung IA, to the Regierungspräsident in Schleswig, Altona, 11.10.1929, LAS 309/22784.

153 Report dated Itzehoe, 15.10.1929, sent from the Polizeipräsident, Landeskriminalpolizeiamt, Abteilung IA, to the Regierungspräsident in Schleswig, Altona, 18.10.1929, in both LAS 301/4696 and 309/22784; report titled "Bericht" concerning "Landvolkbewegung" Altona, 17.10.1929, LAS 309/22784.

154 Report from the Polizeipräsident, Landeskriminalpolizeistelle, to the Regierungspräsident in Schleswig concerning "Landvolkversammlung in Niebüll" Flensburg, 10.1.1930 and report from the Landrat to the Regierungspräsident in Schleswig, Niebüll, 24.1.1930, both in LAS 309/22783.

155 Report from the Landeskriminalpolizeistelle Flensburg titled "Bericht" and concerning "Landvolkbewegung" Husum, 8.11.1929, LAS 309/22783; report titled "Bericht" Heide, 12.10.1929, LAS 309/22784.

Despite declining support, the Landvolkbewegung's propaganda continued to influence rural opinion as its changing stance on the bombings found an echo in the countryside. Just as Landvolk leaders and *Das Landvolk* had at first began to avow understanding for the bombers while continuing to reject their methods, in November a police report noted that the rural population too disapproved of the attacks, but broadly sympathised with the bombers themselves because they believed they had acted according to their ideals.[156]

Though many farmers had flinched on first hearing that the Landvolkbewegung was responsible for the bombings, a number of displays of solidarity with the bombers in the new year also reflected the success of the movement's endeavours to spread understanding for them. First, the community representatives of Karlum in South Tondern proposed Holländer as their new community chairman.[157] Second, some 1,000 farmers reportedly joined a plot to accuse themselves of being accessories to the attacks.[158] Then, on his release on bail from remand, Wiborg was welcomed by the local voluntary fire-brigade and a Stahlhelm brass band as his train arrived in Lunden, causing the Regierungspräsident and the Prussian Interior Ministry to take steps to prevent the repetition of such receptions for fear they would have a detrimental effect on the state's authority.[159]

Guido Weschke's release in March was greeted with much jubilation. To the authorities' exasperation, his "near pathological self-confidence" and "boasting" about the bombings at meetings thereafter met with no dissension but rather "particularly strong applause".[160] Those responsible for the bombings in the province of Hanover enjoyed no such success, however, and the contrast to the response to Weschke's reappearance underlines the persisting and significantly more radical

---

156 Report from the Landeskriminalpolizeistelle Altona, Abteilung IA, concerning "Landvolkbewegung" Itzehoe, 14.11.1929, in both LAS 301/4697 and 309/22783.

157 Landrat to the Regierungspräsident in Schleswig concerning "Bestätigung von Gemeindevorstehern" Niebüll, 3.1.1930, LAS 309/22783.

158 Report from the Landeskriminalpolizeistelle Altona, Abteilung IA, Altona, 8.1.1930, LAS 301/4697.

159 Landjägerei Lunden III, Kreis Norderdithmarschen, to the Landrat in Heide, Lunden, 1.2.1930, LAS 309/22783; report from the Regierungspräsident to the Oberpräsident in Kiel, Schleswig, 24.2.1930, LAS 301/4702; *Das Landvolk* No. 40, 16.2.1920, p. 5. Also see the Regierungspräsident to the Landräte concerning "Freiwillige Feuerwehr Lunden" Schleswig, 7.2.1930, in both LAS 301/4702 and 309/22783; Werner, *Die Prozesse gegen die Landvolkbewegung*, pp. 248-251.

160 For Weschke's release and reappearance at Landvolk meetings, see the reports from the Landrat Kreis Steinburg to the Regierungspräsident in Schleswig concerning "Landvolkbewegung" Itzehoe, 22.3.1930 and concerning "Versammlung des Bauern- und Landvolkbundes im Hammonia-Hotel in Itzehoe. Auftreten Weschkes" Itzehoe, 7.4.1930, both in LAS 301/4697 and 309/22590; report from the Landeskriminalpolizeistelle Altona-Wandsbek, Abteilung IA, titled "Bericht" Oldesloe, 22.3.1930, LAS 309/22590. Also see the report in the *Schleswig-Holsteinische Tageszeitung* No. 83, 7.4.1930, p. 4.

mood in Schleswig-Holstein, even if support for the Landvolkbewegung on the whole was in decline.[161]

The most sinister endorsement of Heim's methods came in the form of an attempt to bomb the town hall in Bad Oldesloe and the finance office in Neumünster on March 15, 1930.[162] A circle of former Wehrwolf members from Hamburg who wished to create the impression Heim and his accomplices were blameless – Kaphengst and Schmidt were members of their circle – planted the devices, which failed to explode due to a fault in their construction.[163] The perpetrators were soon arrested, and when tried in December 1930, the two ring-leaders were each sentenced to five years' imprisonment.[164] Staff at *Das Landvolk* were said to have cheered on hearing of the attacks but, as befitted their failure, they were barely mentioned in the newspaper and the whole episode generated little public interest.[165]

The most notable reaction to them, in fact, came from the Oberpräsident and Regierungspräsident, who in an attempt to put a stop to the growing sympathy for Heim and his accomplices, published an appeal in the provincial press calling on the population to dismiss claims that such acts were signals of economic distress and driven by idealism.[166] Yet their branding of the perpetrators as "dishonourable" and their acts "cowardly" was unlikely to sway opinion when the Landvolkbewegung had long since instilled the idea among the rural population that the bombers and their motives were pure and honourable. Moreover, the appeal's emphasis on the grave danger the bombs posed – the death of innocent parties – was so blatantly overstated as to deprive it of credibility. There were no fatalities. In

161 For meetings held by Vick, Luhmann, Becker und Manecke, see the copy of a report from the Landjägerposten Bardowick, Kreis Lüneburg, titled "Bericht über die am 15. März ds. Js. abgehaltene öffentliche Versammlung der Landvolkbewegung in der Gastwirtschaft vom Gemeindevorsteher Rerup in Obermarschhacht" Bardowick, 16.3.1930 and the transcript of a Landvolk meeting in Wittorf, Kreis Winsen/Luhe, on 3.4.1930, sent by the Polizeipräsident Altona-Wandsbek to the Regierungspräsident in Schleswig, both in LAS 309/22590.

162 *Rendsburger Tageblatt* No. 63, 15.3.1930, p. 1, No. 64, 17.3.1930, pp. 1 and 4, No. 65, 18.3.1930, p. 6 and No. 68, 21.3.1930, p. 2.

163 Report titled "Bericht! Zusammenstellung der Verdachtsgründe, die eine Täterschaft des von Wilamowitz und Hellmann wahrscheinlich machen" Kiel, 24.3.1930, LAS 301/4702; issue of the verdict of the Schwurgericht Altona from 16.12.1930, pp. 4-7, LAS 301/4703.

164 For details of the trial, named the "Small bombers' trial" (*Kleiner Bombenlegerprozeß*) and the sentences, see the issue of the verdict of the Schwurgericht Altona from 16.12.1930, LAS 301/4703; report from the Oberstaatsanwalt in Altona to the Preuß. Justizminister, Altona, 18.2.1933, GStA PK I. HA Rep. 84a No. 55372.

165 Report from the Landeskriminalpolizeistelle Altona-Wandsbek, Abteilung IA, titled "Bericht" Oldesloe, 22.3.1930, in both LAS 301/4697 and 309/22590.

166 See the appeal "Aufruf von Oberpräsident und Regierungspräsident" in the *Rendsburger Tageblatt* No. 65, 18.3.1930, p. 1.

fact, the investigating judge had already made public that the bombers had not intended to endanger lives.[167]

The Landvolk's efforts to promote sympathy for the bombers continued to bear fruit. Though attendances were small at the series of meetings at which Bruno von Salomon spoke in June, the responses at them were entirely enthusiastic. The Landvolk's speakers were received everywhere with tremendous applause and in Lunden Paul Adam Ross also led the assembly with three cheers for Claus Heim.[168] Ross and von Salomon were among a number of Landvolk leaders who established a "Heim Group" aiming to obtain the bombers' release. Their collection of donations received a generous response from the rural population.[169]

Seeking to boost support further before the bombers' trial, Landvolk leaders endeavoured to gain the assistance of the urban middle-class press and, on Luetgebrune's initiative, discussions were also held with the leaders of the NSDAP, the Stahlhelm, Wehrwolf, the DNVP and farming organisations to obtain their cooperation.[170] Kühl met with Gauleiter Lohse in June and the *Schleswig-Holsteinische Tageszeitung* subsequently came out in defence of the bombers.[171] Both Lohse and Luetgebrune were unwilling, however, to endorse plans favoured by Bruno von Salomon to hold a series of combined meetings and rallies, together with the Stahl-

---

167 For the press conference given by the investigating judge, Masur, see the *Rendsburger Tageblatt* No. 259, 4.11.1929, p. 1. For the growing sympathy for the bombers among the agricultural associations, see the report from the Regierungspräsident to the Minister für Landwirtschaft, Domänen und Forsten in Berlin concerning "Schleswig-Holsteinischer Land- und Bauernbund" Schleswig, 15.8.1930, in both LAS 301/4697 and 309/22765.

168 Reports from the Polizeipräsident, Landeskriminalpolizeistelle, to the Regierungspräsident in Schleswig concerning "Landvolkbewegung" Flensburg, 16.6.1930, in both LAS 309/22590 and 301/4697; report from the Landrat to the Regierungspräsident in Schleswig concerning "Auflösung einer Versammlung" Heide, 24.6.1930, LAS 301/4698; report from the Regierungspräsident to the Preuß. Minister des Innern in Berlin concerning "Landvolkbewegung" Schleswig, 27.6.1930, LAS 301/4698; reports from the Landrat to the Regierungspräsident in Schleswig concerning "Landvolkversammlungen in Meldorf und Marne" Meldorf, 28.6.1930, from the Landrat to the Regierungspräsident in Schleswig concerning "Landvolkversammlung in Leck" Niebüll, 24.6.1930 and from the Landrat to the Regierungspräsident in Schleswig concerning "Landvolk-Bewegung" Heide, 28.6.1930, all in LAS 309/22590.

169 Reports from the Polizeipräsident, Landeskriminalpolizeistelle, to the Regierungspräsident in Schleswig concerning "Landvolkbewegung" Flensburg, 16.6.1930 and Flensburg, 12.7.1930, both in LAS 301/4697 as well as 309/22590.

170 Report from the Polizeipräsident, Landeskriminalpolizeistelle, to the Regierungspräsident in Schleswig concerning "Landvolkbewegung" Flensburg, 12.7.1930, in both LAS 301/4697 and 309/22590.

171 Reports from the Landeskriminalpolizeistelle Altona, Abteilung IA, titled "Bericht" Altona, 21.6.1930 and from the Polizeipräsident, Landeskriminalpolizeistelle, to the Regierungspräsident in Schleswig concerning "Landvolkbewegung" Flensburg, 12.7.1930, both in LAS 301/4697 as well as 309/22590.

helm, in the week before the trial.[172] While Luetgebrune was concerned that the rallies, like further bombings, would only make things more difficult for Heim and his accomplices, Lohse evidently still feared for his party's image and did not wish his members to become involved in any potentially reckless Landvolk actions.[173]

In other ways, though, the NSDAP showed a new willingness to support the bombers. The party's changed approach revealed a sharp sensibility to public opinion and its attentions to Heim in this period were striking. Heinrich Himmler was among his many visitors in remand and in July the party attempted to secure him as a candidate for the upcoming Reichstag election.[174] For Lohse, who initiated the proposition, this necessitated allowing political interest to overcome a mutual personal dislike.[175] For Heim, the candidacy would have guaranteed him parliamentary immunity. He refused. This caused a sensation in the province's press and beyond; all explained that he preferred jail to the Reichstag.[176] Whether his refusal is to be interpreted as an expression of opposition to the parliamentary system, an act of solidarity with his accomplices or a rejection of the National Socialists, its integrity was admirable. Heim, it seemed, was a principled terrorist.

The "Great Bombers' Trial" (*Großer Bombenlegerprozeß*) began on August 26 to immense public interest. Various complications had delayed the commencement of the trial, including the magnitude of the preliminary investigations, Luetgebrune's preparations for the defence, the concentration of judicial proceedings in Berlin and then their transfer to the *Landgericht* (Land court) in Altona.[177] The Prussian Interior Ministry had contrived the centralisation of proceedings in Berlin, hoping the trial would be held in a politically dependable court. Evidence against the Berlin suspects, however, had been found insubstantial and all the offences determined to be within the Altona court's jurisdiction.[178] The severity of the offences, breaches

---

172 Report from the Polizeipräsident, Landeskriminalpolizeistelle, to the Regierungspräsident in Schleswig concerning "Landvolkbewegung" Flensburg, 12.7.1930, in both LAS 301/4697 and 309/22590. Also see Uhse, *Söldner und Soldat*, p. 315.

173 Luetgebrune warned against new attacks when addressing the movement's trusted men in May. See the report from the Landeskriminalpolizeistelle Altona, Abteilung IA, titled "Bericht" Altona, 21.6.1930, in both LAS 301/4697 and 309/22590.

174 Heim, 'Claus Heim berichtet', LAS EI/4434, p. 9.

175 Heim, 'Claus Heim berichtet', LAS EI/4434, p. 13.

176 *Hamburger Nachrichten* No. 351, 31.7.1930 and *Kölnische Zeitung* No. 414, 31.7.1930, both in LAS 309/22590; *Das Landvolk* No. 178, 31.7.1930, p. 1; *Rendsburger Tageblatt* No. 177, 31.7.1930, p. 1; *Nordischer Kurier* No. 178, 31.7.1930, p. 1 and No. 179, 1.8.1930, p. 13; *Flensburger Nachrichten* No. 177, 31.7.1930, p. 2; *Schleswig-Holsteinische Tageszeitung* No. 179, 1.8.1930, p. 7; *Volkszeitung* No. 177, 1.8.1930, p. 4.

177 For details concerning the delays, especially those of legal process, see Werner, *Die Prozesse gegen die Landvolkbewegung*, pp. 201-206.

178 The Prussian Interior Ministry had deliberately left the Ministry of Justice and public prosecutor in the dark, so that the weaker evidence pointing to the Landvolk's involvement in the bombing of the Reichstag had prevailed against the considerably weightier reasons for

of the explosives' law, required a trial by *Schwurgericht*, a court composed of three professional and six lay judges.[179] Despite indications that Regierungspräsident Abegg may have contemplated interfering in the selection of the lay judges, this needs to be tempered with the recognition that rural interests were well represented with the inclusion of three farmers.[180]

Expected to take four weeks, the trial lasted just over two months.[181] Forty seats were reserved exclusively for the press, and the general public queued for two hours on the first day to vie for the few remaining places.[182] The excitement about the trial was heightened by an attempted bombing in the city of Hanover just days beforehand, as well as rumours of further possible attacks during the hearings and even that some "Republican heroes" would be killed if Heim and friends were punished severely.[183]

---

the investigation's concentration in Altona in Schleswig-Holstein: Generalstaatsanwalt to the Preuß. Justizminister in Berlin, Berlin, 9.12.1929, GStA PK. I. Rep. 84a No. 55367; Werner, *Die Prozesse gegen die Landvolkbewegung*, pp. 237-248. For accusations in the right-wing press that the concentration of proceedings in Berlin was politically motivated chicanery, see *Das Landvolk* No. 148, 19.9.1929, p. 1; *Hamburger Nachrichten* No. 229, 26.9.1929, LAS 301/4696; *Germania* No. 432, 17.9.1929, p. 1; for Ernst Jünger's response in the *Deutsche Zeitung*, see S. O. Berggöte (ed.), *Ernst Jünger: Politische Publizistik 1919 bis 1933* (Stuttgart, 2001), pp. 499-501. With the exception of Ernst von Salomon, most of the Berlin suspects had quickly been released: *Rendsburger Tageblatt* No. 217, 16.9.1929, p. 1 and No. 219, 18.9.1929, p. 1; *Volkszeitung* No. 216, 16.9.1929, p. 1; *Schleswiger Nachrichten* No. 237, 9.10.1929, p. 1; *Rendsburger Tageblatt* No. 247, 21.10.1929, p. 2; *Das Landvolk* No. 176, 22.10.1929, p. 1. For the transfer to Altona, see the *Rendsburger Tageblatt* No. 76, 31.3.1930, pp. 1 and 3; *Das Landvolk* No. 77, 1.4.1930, p. 1.

179 After 1924 a "Schwurgericht" was composed of three professional judges and six members of the "jury" (*Geschworenen*) who had the power of lay judges, deciding on both the question of guilt and the sentence. See W. Sellert and H. Rüping, *Studien- und Quellenbuch zur Geschichte der deutschen Strafrechtspflege,* vol. 2 (Aalen, 1994), p. 180 and H. Rüping and G. Jerouschek, *Grundriß der Strafrechtsgeschichte* (Munich, 2002), p. 105.

180 Polizeipräsident Altona-Wandsbek to the Regierungspräsident in Schleswig, Altona, 22.8.1930 and the report from the Polizeipräsident Altona-Wandsbek to the Regierungspräsident in Schleswig concerning "Politische Zugehörigkeit der Beschworenen im Bombenprozeß" Altona, 9.9.1930, both in LAS 309/22764. Also see Werner, *Die Prozesse gegen die Landvolkbewegung*, p. 204.

181 See the *Nordischer Kurier* No. 198, 23.8.1930, p. 3 and No. 200, 26.8.1930, p. 9.

182 *Kieler Zeitung* No. 237, 27.8.1929, p. 8.

183 For the attempted bombing at the union offices in Hanover, see the report from the Polizeipräsident Altona-Wandsbek to the Regierungspräsident in Schleswig concerning "Sprengstoffattentate in Hannover" Altona, 22.8.1930, LAS 309/22590. Also see the report from the Polizeipräsident in Kiel to the Polizeipräsident, Landeskriminalpolizeiamt Berlin, Abteilung IA, concerning "Sprengstoffanschläge" Kiel, 29.8.1930 and the appended report from the Abteilung IA titled "Bericht" Kiel, 25.9.1930, LAS 301/4702; reports from the Polizeipräsident, Landeskriminalpolizeistelle, to the Regierungspräsident in Schleswig concerning "Landvolkbewegung" Flensburg, 12.7.1930, in both LAS 301/4697 and 309/22590; copy of the report from the Regierungspräsident in Lüneburg to the Preuß. Minister des In-

The trial concerned not only the larger-scale bombings of 1929 but also the earlier, smaller attacks in Beidenfleth, Hollingstedt, Flederwurth; the attempted attack in Winsen an der Luhe on November 27, 1928; the hand-grenade attacks in Wesselburen on April 5, 1929; as well as the theft of explosives from Mülheim in January. Twenty-one people were tried in total. Hamkens and Bruno von Salomon were tried for failing to alert the authorities of their knowledge of the attacks, while the latter and Weschke also faced additional charges of perjury, having denied their knowledge when first questioned.[184] Luetgebrune defended most of the accused but other lawyers defended Heim, Nickels, Rehling, Schmidt and Rieper. Volck was defended by Rüdiger von der Goltz.[185]

The court worked through the cases essentially in chronological order, but skipped to and fro where connections between the accused and other attacks had been established. With the exception of Volck, who soon proceeded to speak at length about the Landvolkbewegung's history and the purpose of the bombings, Heim and his accomplices remained silent.[186] Publicly, their silence was a protest against "the system". Their further undisclosed purpose was to prevent the incrimination of others who had assisted them.[187] Apart from Heim and Rehling, however, all of the accused had responded during the police and preliminary investigations and so the transcripts of their examinations were read to the court instead.[188]

---

nern in Berlin concerning "Bombenattentate" Lüneburg, 9.8.1930, Regierungspräsident to the Polizeipräsidenten, Landräte and Polizeiverwaltung in Neumünster, Schleswig, 20.8.1930, LAS 309/22590; reports from the Abteilung IA titled "Bericht" Kiel, 6.8.1930 and from the Polizeipräsident in Kiel, Abteilung IA, to the Regierungspräsident in Schleswig concerning "Angeblich beabsichtigtes Bombenattentat in Burg auf Fehmarn" Kiel, 13.8.1930, the report from the Polizeipräsident in Kiel to the Polizeipräsident, Landeskriminalpolizeiamt Berlin, Abteilung IA, concerning "Sprengstoffanschläge" Kiel, 29.8.1930 and the appended report from the Abteilung IA, titled "Bericht" Kiel, 25.9.1930, all in LAS 301/4702.

184 Indictment issued by the Oberstaatsanwaltschaft Altona, 6.5.1930, pp. 11 and 13, LAS 301/4702; issue of the verdict of the Schwurgericht Altona, 31.10.1930, pp. 101-102, in both LAS 301/4702 and 309/22764.

185 Rüdiger von der Goltz (1865-1946) came to prominence as commander of the Freikorps troops fighting against the Bolsheviks in the Baltic in 1919. Afterwards he became one of the leaders of the *Vereinigte Vaterländische Verbände Deutschlands*, an umbrella organisation for right-wing extra-parliamentary organisations, uniting them against the pro-Republican and "Marxist" parties. He later joined the committee of Seldte, Duesterberg, Hugenberg, Hitler and Class against the Young Plan. See Diehl, *Paramilitary Politics in Weimar Germany*, pp. 45, 53, 201, 268.

186 See the *Kieler Zeitung* No. 238, 28.8.1930, p. 7; *Schleswig-Holsteinische Tageszeitung* No. 202, 28.8.1930, p. 1.

187 *Nordischer Kurier* No. 200, 26.8.1930, p. 9 and No. 203, 29.8.1930, p. 9; *Kieler Zeitung* No. 237, 27.8.1930, p. 8, No. 240, 30.8.1930, p. 9 and No. 243, 2.9.1930, p. 7; Heim 'Claus Heim berichtet', LAS EI/4434, p. 9.

188 *Kieler Zeitung* No. 244, 3.9.1930, p. 7, No. 245, 4.9.1930, p. 7 and No. 246, 5.9.1930, p. 7.

During the following weeks, the court was engaged in lengthy debate and questioning of witnesses. First the prosecution sought to establish and ultimately succeeded in showing that the bombs could have endangered lives and that the bombers were aware of the potential consequences of their actions.[189] Then, the defence attempted to persuade the court that despair had driven the accused to their deeds; that the particularly severe agricultural crisis and the governments' deficient relief efforts were to blame; that the crisis was a "crisis above the law" (*Übergesetzlicher Notstand*); and that the accused believed that bombs were their only recourse.[190] A further, peculiar highlight, at least in terms of the public interest it raised, was Volck's defence, in which von der Goltz contended that the authorities had tolerated and indirectly encouraged the bombings, the police even employing an *agent provocateur* to this purpose.[191]

The debate over the severity of the agricultural crisis ended in a comprehensive victory for the defence. It produced countless witnesses, farmers and agricultural experts, all ready to testify to the desperate situation of Schleswig-Holstein's agriculture: increasing costs, inadequate market prices for agricultural goods, excessive interest rates and taxes and mounting debts.[192] The prosecution attempted to argue that conditions in Schleswig-Holstein were improving and better than elsewhere but could not contest the defence's claims.[193] Its efforts failed completely when its principal witness actually ended up confirming the defence's grim portrayal of the crisis.[194]

Once again, Luetgebrune had assisted the Landvolk in converting the courtroom into an effective means of drawing attention to the farmers' grievances. The witnesses for the defence not only testified about the crisis itself but also com-

---

189 *Kieler Zeitung* No. 245, 5.9.1930, p. 7, No. 251, 10.9.1930, p. 7 and No. 261, 20.9.1930, p. 8; *Nordischer Kurier* No. 211, 8.9.1930, p. 9 and No. 226, 25.9.1930, p. 5; *Schleswig-Holsteinische Volkszeitung* No. 223, 25.9.1930, p. 15; Landrat Göppert's witness statement titled "Zeugenvernehmung Landrat Göppert" 8.10.1930, LAS 309/22764.

190 See, for instance, the *Nordischer Kurier* No. 229, 19.9.1930, p. 4, but, in particular, the final summaries of Luetgebrune, *Kieler Zeitung* No. 293, 22.10.1930, p. 7, von der Goltz and Pohlmann, No. 294, 23.10.1930 and Hüttmann, No. 296, 25.10.1930, p. 8.

191 *Kieler Zeitung* No. 244, 3.9.1930, p. 7, No. 261, 20.9.1930, p. 8 and No. 273, 2.10.1930, p. 8; *Schleswig-Holsteinische Volkszeitung* No. 222, 24.9.1930, p. 11. As rumoured in the press at the time of his arrest, Volck had been tricked into returning to Germany by the alleged "agent provocateur", Mrs von Oertzen. See the series of reports in *Tempo* No. 22, 28.1.1930 - No. 31, 7.2.1930, LAS 309/22590, apparently based on a manuscript, Volck had written on his experiences in the Landvolkbewegung. A copy of the manuscript titled "Landvolk und Bomben" is to be found in LAS 309/22783. See further F. W. v. Oertzen, *Im Namen der Geschichte!*, pp. 93-94.

192 *Kieler Zeitung* No. 268, 27.9.1930, p. 8, No. 271, 30.9.1930, p. 8, No. 272, 1.10.1930, p. 7 and No. 274, 3.10.1930, p. 11.

193 *Nordischer Kurier* No. 226, 25.9.1930, p. 5.

194 *Nordischer Kurier* No. 233, 3.10.1930, p. 9; *Schleswig-Holsteinische Tageszeitung* No. 234, 4.10.1930, p. 2; *Schleswiger Nachrichten* No. 232, 3.10.1930, p. 7.

plained that the state had failed them, that promises of relief had amounted to nothing and that despite assurances to the contrary the "ruthless" collection of taxes had continued.[195] Exemplifying the political justice which characterised so many trials of the right during the Weimar Republic, the judges indulged the defence, by, for example, allowing known Landvolk sympathisers, such as the DNVP Reichstag deputy Max Soth, to articulate their political convictions. In his testimony, Soth denounced "the system" and applauded Hamkens and Heim who, in his view, were fighting not just to retain their own farms but against "the destruction of everything".[196]

By publicising the farmers' grievances, bringing attention to the crisis and broadcasting criticism of the authorities' response to it, the trial ensured that Heim achieved some of his objectives in a way the bombs alone could not have done. Witness testimony made it entirely clear that the rural population had understood that the bombings were a response to the agricultural crisis and a protest against the continued collection of taxes.[197] Moreover, if the bombings had made the authorities aware of the extreme bitterness among the rural population, the trial also articulated the nature and extent of that bitterness far better than had the bombings themselves. It ensured that those who had been targeted were made aware of the specific grievances against them.[198] Among them was County Administrator Skalweit who denied that he had pursued the collection of taxes mercilessly, maintaining that the opposite was true.[199] The farmers who were called as witnesses, however, painted a different picture. They maintained that Skalweit was most unpopular for having failed to keep promises made to a deputation of farmers in January 1928 and, in particular, for having authorised a compulsory auction of a farm valued at RM 70,000 due to outstanding taxes of RM 187.[200] In this way, the trial expanded on the prior debate about the agricultural crisis by elucidating further the farmers' grievances against the authorities. Additionally, the testimony of the victims divulged the intimidating effects of the bombings and thereby served as a confirmation for Heim that they had been a success in this respect.[201]

---

195 *Kieler Zeitung* No. 268, 27.9.1930, p. 8, No. 272, 1.10.1929, p. 2, No. 273, 2.10.1930, p. 8 and No. 274, 3.10.1930, p. 11.

196 *Kieler Zeitung* No. 268, 27.9.1930, p. 8.

197 *Kieler Zeitung* No. 275, 4.10.1930, p. 11.

198 *Kieler Zeitung* No. 273, 2.10.1930, p. 7 and No. 275, 4.10.1930, p. 11; *Nordischer Kurier* No. 234, 4.10.1930, p. 11; *Schleswiger Nachrichten* No. 231, 2.10.1930, p. 7.

199 *Kieler Zeitung* No. 250, 9.9.1930, p. 7.

200 See the transcript of Hansen's examination, "Vernehmung Hofbesitzer und Tierarzt Hansen 55 Jahre alt" Altona, 6.10.1930 and of County Administrator Skalweit titled, "Vernehmung Landrat Skalweit", both in LAS 309/22764. Also see the *Kieler Zeitung* No. 275, 4.10.1930, p. 11.

201 *Kieler Zeitung* No. 246, 5.9.1930, p. 7 and No. 273, 2.10.1930, p. 8; *Schleswiger Nachrichten* No. 231, 2.10.1930, p. 7.

The trial also offered interesting insight into Heim's leadership role and personality as the court heard a number of witnesses give evidence in order to measure the influence he had over his accomplices. In effect, this explained Hamkens' failure to discourage Heim from his plans and, together with his persisting silence in the courtroom, no doubt contributed to the near-legendary reputation he was coming to attain.[202] The witnesses all depicted Heim as an unusually wilful and assertive character, who was even more obstinate, they said, than the kind of Dithmarscher farmer known from history.[203] Among these witnesses was District Chairman Ross, a victim of the attacks of November 27, 1928, who had grown up with Heim. He declared that had he been a Landvolk follower himself, he too would have obeyed him and planted a bomb. According to Ross, Heim was such a strong personality, one simply did whatever he demanded.[204]

Claus Heim (left) with fellow defendants during the "Great Bombers' Trial"[205]

The court did not accept the defence's assertions that the bombers' actions were justified because they were a last recourse to a "crisis above the law", and so

202 This is found most clearly in Ernst von Salomon's writings: v. Salomon, *Die Stadt*; v. Salomon, *Der Fragebogen*, pp. 126-145 and v. Salomon, *Deutschland deine Schleswig-Holsteiner,* p. 176.

203 *Kieler Zeitung* No. 273, 2.10.1930, p. 8; *Schleswig-Holsteinische Volkszeitung* No. 229, 2.10.1930, p. 13.

204 *Schleswiger Nachrichten* No. 231, 2.10.1930, p. 7.

205 Photo: Luetgebrune, *Neu-Preußens Bauernkrieg*, p. 161.

did not accept that the accused had not broken the law.[206] Though it is debatable whether the government had done much to alleviate the crisis, as the court claimed it had, it judged quite rationally in doubting the use of bombs as an appropriate means of assisting agriculture and it also decided that the crisis had not been so severe that no other options had been open to the farmers.[207]

Further, the court was not persuaded that an *agent provocateur* had encouraged Volck to undertake the attacks. The supposed agent admitted she had suspected Volck from as early as July, 1929 but as she had only gone to the police after the Reichstag bombing, that is shortly before the arrests, the court concluded that the police could not have knowingly allowed the bombings to continue.[208]

In respect of the bombings of 1929, the court found that everyone involved had known very well what damage could have ensued from their actions.[209] As no lives were lost, the bombers were not given maximum sentences. Heim and Volck were both sentenced to seven years penitentiary (*Zuchthaus*). Volck's sentence was as severe as Heim's not only because he was held responsible for helping devise the campaign, but also because he had persuaded Schmidt, still in his early twenties, and the Holländers, considered entirely harmless, to take part.[210] The other bombers also received substantial sentences: Schmidt, five and a half years, Johnsen five years and three months, Hennings five years and three months, Rathjen one year, Vick six years, Luhmann five years, Becker five years and three months, Manecke one year, and Rieper one year and three months penitentiary. Nickels received a two and a half year prison (*Gefängnis*) term, Wiborg one year and three months, Weschke six months and a fine of RM 50, Peter Holländer four months, Bossen four months and 14 days, and Hamkens, instead of two months imprisonment, a RM 500 fine.[211] Hamkens, in admitting that he had attempted to convince Heim not to proceed with the bombings, had proven himself guilty of the charges against

---

206 Issue of the verdict of the Schwurgericht Altona, 31.10.1930, in both LAS 301/4702 and 309/22764. Also see the defence's final summaries in the *Kieler Zeitung* No. 292, 21.10.1930, p. 7, No. 293, 22.10.1930, p. 7, No. 294, 23.10.1930, p. 8, No. 296, 25.10.1930, p. 8 and No. 300, 29.10.1930, p. 8.

207 See the reasons for the judgement read by the presiding judge as reported, for example, in the *Rendsburger Tageblatt* No. 257, 1.11.1930, pp. 1-2.

208 For the failure of Volck's defence, see the issue of the verdict of the Schwurgericht Altona, 31.10.1930, pp. 112-116, in both LAS 301/4702 and 309/22764. Also see von der Goltz's final summary in the *Kieler Zeitung* No. 294, 23.10.1930, p. 8.

209 Luetgebrune nevertheless continued to foster the idea that the farmers had taken great care to avoid human casualties. See Luetgebrune, *Neu-Preußens Bauernkrieg*, pp. 192-193.

210 See the *Kieler Zeitung* No. 303, 1.11.1930, p. 1; *Nordischer Kurier* No. 258, 1.11.1930, p. 4; *Schleswig-Holsteinische Volkszeitung* No. 255, 1.11.1930, p. 1; *Schleswig-Holsteinische Tageszeitung* No. 259, 2.11.1930, pp. 1-2.

211 Issue of the verdict of the Schwurgericht Altona, 31.10.1930, pp. 7-9, in both LAS 301/4702 and 309/22764.

him: he was required by law to report his knowledge to the authorities.[212] Rehling, accused of involvement in the Mülheim robbery, was acquitted as the defence produced sufficient witnesses to confirm his alibi.[213] Bruno von Salomon was also acquitted because there was insufficient evidence to prove that he was complicit.[214] Finally, Kaphengst, not arrested until August 1930 and tried separately from his accomplices, was sentenced in January 1932 to three years penitentiary.[215]

To some extent, the press responded to the verdict and the sentences as the political positions of the various newspapers dictated. Republican-friendly newspapers lauded the verdict, though some objected to the defence and depiction of the bombers as martyrs, while others regretted that those they believed had fomented the attacks – particularly Weschke and Bruno von Salomon – had escaped the punishments they deserved.[216] Nationalist newspapers, on the other hand, dismissed the verdict as "political" and though they rejected the bombers' methods, they defended them as victims of a system that had driven them to despair.[217] This was also the view of the National Socialist press, which throughout the trial had continued to support the bombers as it had done since the summer. The *Völkischer Beobachter* and *Schleswig-Holsteinische Tageszeitung* rejected Heim's methods but portrayed him as a hero, driven by his love of the nation and his struggle against "the system". They denounced the verdict as "monstrous" and promised that Heim's endeavours would not be in vain, declaring that the NSDAP would overthrow the system by other means and then Heim and his comrades would be freed.[218]

Das Landvolk, surprisingly, conceded that the court had weighed up the arguments in reaching its verdict. However, it claimed that the letter of the law had pre-

---

212 Issue of the verdict of the Schwurgericht Altona, 31.10.1930, p. 221, in both LAS 301/4702 and 309/22764.

213 Issue of the verdict of the Schwurgericht Altona, 31.10.1930, pp. 107 and 119-128, in both LAS 301/4702 and 309/22764. Also see the *Kieler Zeitung* No. 257, 16.9.1930, p. 11, No. 258, 17.9.1930, p. 7 and No. 259, 18.9.1930, p. 11.

214 Issue of the verdict of the Schwurgericht Altona, 31.10.1930, pp. 128-133, in both LAS 301/4702 and 309/22764. See also the *Kieler Zeitung* No. 264, 23.9.1930, p. 11.

215 Issue of the verdict of the Schwurgericht Altona from 23.1.1932, in both LAS 301/4703 and 309/22701. NB subsequent appeals were dismissed by the Reichsgericht. See the issue of the verdict of the Reichsgericht, Dritter Strafsenat, from 9.11.1931, LAS 301/4703; issue of the verdict of the Reichsgericht, Dritter Strafsenat, from 30.6.1932, LAS 301/4703. Also see Luetgebrune, *Neu-Preußens Bauernkrieg*, pp. 157-168.

216 *Volkszeitung* No. 256, 1.11.1930, p. 1; *Schleswig-Holsteinische Volkszeitung* No. 251, 28.10.1930, p. 13 and No. 256, 3.11.1930, p. 11; *Vossische Zeitung* No. 262, 2.11.1930, p. 15.

217 *Hamburger Nachrichten, Lokalanzeiger, Der Tag* and *Deutsche Zeitung* as cited in the *Rendsburger Tageblatt* No. 257, 1.11.1930, p. 2.

218 *Völkischer Beobachter* No. 262, 4.11.1930, p. 1; *Schleswig-Holsteinische Tageszeitung* No. 259, 2.11.1930, p. 1. Also see the *Schleswig-Holsteinische Tageszeitung* No. 199, 24.8.1930, p. 4 and No. 234, 4.10.1930, p. 2.

vailed over the people's hearts, that it was, therefore, a "false verdict", a "punch in the face" for the farmers[219] and a victory for the opposition in the "struggle between two worlds".[220] *Das Landvolk* had constructed this dichotomy in its report from the opening day of the trial, contending that two worlds, the German and the Jewish, opposed one another in the courtroom, the Jewish represented specifically by the state attorney, Dr. Brandt.[221] In its reaction to the trial, however, *Das Landvolk* offered little but hackneyed, grandiose portrayals of the bombers as heroes who had acted selflessly, for the nation and the freedom of the people.[222]

The response of the middle-class press provided a much better barometer of the shift in attitude toward the bombers. Rural circles had already begun to sympathise with them before the trial but now the broad mass of the middle-class press followed, both in the province and beyond. Taken as a whole, they pronounced the verdict correct in the sense that it accorded with the "explosives' law" but agreed that the penalties were severe. Some in fact judged them to be so severe, they predicted they would be incomprehensible to the public.[223] Though these newspapers continued to condemn the bombers' methods, they offered them sympathy, emphasising that they had acted out of despair. Moreover, all of them, even Mosse's liberal *Berliner Tageblatt*, demanded that the bombers be granted a reprieve (*Begnadigung*).[224] One of the few newspapers to offer a more considered view, the *Kieler Neueste Nachrichten* suggested that the crisis alone was insufficient explanation for the farmers undertaking such imprudent acts, and that a psychological cause had to be sought in the fanaticism resulting from the unrest among the rural population.[225] Much more numerous, however, were the provincial papers that now appeared to have set out on a more extreme path themselves. Though considerably less hostile than *Das Landvolk*, some began to echo the Landvolkbewegung's line of reasoning. The *Rendsburger Tageblatt*, a newspaper that had initially condemned the bombings as a crime against the state and people, deemed the verdict a victory for the system.[226] So too did the *Itzehoer Nachrichten*, which together with

---

219 *Das Landvolk* No. 259, 3.11.1930, p. 1.

220 *Das Landvolk* No. 258, 1.11.1930, p. 1.

221 *Das Landvolk* No. 201, 27.8.1930, p. 1.

222 *Das Landvolk* No. 258, 1.11.1930, p. 1 and No. 259, 3.11.1930, p. 1.

223 *Kieler Zeitung* No. 303, 1.11.1930, p. 1 and No. 304, 2.11.1930, p. 5; *Schleswiger Nachrichten* No. 257, 1.11.1930, pp. 1 and 6.

224 *Flensburger Nachrichten* No. 257, 1.11.1930, p. 1; *Kieler Neueste Nachrichten* No. 257, 2.11.1930, pp. 1-2; *Rendsburger Tageblatt* No. 257, 1.11.1930, p. 1 and No. 258, 3.11.1930, p. 1; *Kieler Zeitung* No. 304, 2.11.1930, p. 5; *Hamburger Anzeiger*, *Hamburger Fremdenblatt* and *Hamburger Korrespondent* as cited in the *Flensburger Nachrichten* No. 257, 1.11.1930, p. 1; *Husumer Nachrichten, Hamburger Anzeiger* and *Itzehoer Nachrichten* as cited in the *Flensburger Nachrichten* No. 258, 3.11.1930, p. 2; *Berliner Tageblatt* as cited in the *Rendsburger Tageblatt* No. 258, 3.11.1930, p. 2.

225 *Kieler Neueste Nachrichten* No. 257, 2.11.1930, pp. 1-2.

226 *Rendsburger Tageblatt* No. 257, 1.11.1930, p. 1.

the *Heider Anzeiger*, the *Kieler Zeitung* and the *Schleswiger Nachrichten* all pointed to the bombers' "honourable" motives, in particular, their "love of the fatherland".[227] This did not constitute approval of violent acts, but alongside acknowledgement of the agricultural crisis and the despondent mood in the countryside, it amounted to the qualified acceptance of them. Overall, the widespread sympathy for the bombers must be interpreted as a sign of increasing tolerance for radical action and of the growing shift toward the right both in Schleswig-Holstein and beyond its borders.

The bombing campaign was the most extreme expression of the Landvolkbewegung's radicalisation. It represented the ultimate step in a crescendo-like process that began with the rallies of January 28, 1928. The initial, smaller attacks were the accompaniment to the farmers' first direct actions at confiscations, while the later, more potent attacks were the deafening climax in the turbulent summer of 1929, over and above the conflicts with police at the rallies in Husum and Neumünster. The bombings also represented a further step for Claus Heim and his fellow conspirators on the more militant path taken by Heim when he formed the Defence Association. While the Defence Association was designed to defend the rural people's farms, the bombings were attacks against their supposed enemies. They therefore represented an escalation both in intention and method.

Inclined to adopt more extreme methods from the very beginning of the Landvolkbewegung, it was Claus Heim's decision to resort to bombings and his determination that drove their preparation and execution. He would not be dissuaded from his plans once he had resolved on bombings. This is not to say he was indifferent to outside influence since his decision was partly a response to pressure from some of the movement's followers for more radical action. In the Defence Association Heim moved in a milieu in which such expectations were concentrated. Additionally, Heim believed that bombs would help to retain the movement's momentum, that the movement required the "loudest advertisement possible".

The bombings were a product of growing desperation over the agricultural crisis. Heim and the farmers who assisted him resorted to bombs because they had lost hope in the governments' relief efforts and were convinced their earlier protests had not made a sufficient impact in Berlin. The bombings, therefore, were also a symptom of agriculture's political impotence, or at least of the farmers' feeling of helplessness. In this sense, Heim, physically imposing, influential and widely respected among his peers, presented an interesting paradox. His resort to stronger methods stood in inverse relationship to the farmers' sense of power and capacity to better their situation.

---

227 *Heider Anzeiger* as cited in the *Rendsburger Tageblatt* No. 258, 3.11.1930, p. 1; *Itzehoer Nachrichten* as cited in the *Flensburger Nachrichten* No. 258, 3.11.1930, p. 1; *Kieler Zeitung* No. 303, 1.11.1930, p. 1; *Schleswiger Nachrichten* No. 257, 1.11.1930, p. 1; *Rendsburger Tageblatt* No. 257, 1.11.1930, p. 1.

As the *Kieler Neueste Nachrichten* suggested, the crisis alone cannot explain the bombings. They were not simply intended as more forceful demonstrations of the farmers' distress. Rather, they were acts of intimidation directed at the authorities and "the system". Hostility towards these was the common factor motivating Heim and all of his accomplices, whether they were farmers or members of other occupations. This hostility, bred by the Landvolkbewegung's propaganda and the mounting conflicts with the authorities, must be identified as the single most important element in explaining the attacks.

If the bombings are to be considered a genuine attempt to overthrow the Republic – and Heim's discussion in prison with SA leader Hasse suggest he had thought it possible – then his choice of this particular strategy must be questioned. Though a matter for speculation, it would appear that Heim felt there was no alternative to violence. Heim, and the Landvolk leaders in general, strongly rejected all cooperation with "the system" and this ruled out effecting change by parliamentary participation. As Heim told Hasse, he did not believe the Republic could be destroyed by legal means. This left revolution or a putsch as further options. The Landvolk, however, was not a paramilitary organisation, nor did it have mass support, certainly not beyond Schleswig-Holstein's countryside. In fact, the Landvolk leadership never showed any ambition to seize power. Bombs, therefore, may have appeared to Heim as the only avenue open, that is, apart from conceding his objective impossible. Heim's discussion with Hasse indicates that he believed the continuation of his campaign would have caused great fear among supporters of the Republic and it can be deduced from this that he hoped the bombs would eventually paralyse political life. Yet it is difficult to picture how a regionally limited campaign could achieve this, and then, even if the attacks had become more extensive, how it could have led to anything more than temporary disorder without other forms of action and support. There is no evidence that Heim had thought beyond this, that he had really thought his plans through. If these speculations are correct, then ultimately the bombing campaign simply underlined the Landvolk leadership's lack of concept in its struggle against the Weimar Republic.

The provincial authorities were aware that the bombings were the product of a tense political situation and they recognised that approval of the attacks was a sign of the radicalisation of the rural population.[228] Nor was the seriousness of the situation overlooked in Berlin. Shortly after the bombers were arrested, Rudolf Hilferding wrote to Hermann Pünder, then the secretary of the Reich Chancellery, "The bombings and other recent incidents have shown that an inner tension has permeated the nation, like it has not done for many a year."[229]

---

228 Report from the Regierungspräsident to the Oberpräsident in Kiel concerning "Bombenanschlag auf das Landratsamt in Itzehoe" Schleswig, 11.6.1929, LAS 301/4704.

229 Reichsminister der Finanzen to Staatssekretär Pünder, 20.9.1929 in K. D. Erdmann (ed.), *Akten der Reichskanzlei Weimarer Republik: Das Kabinett Müller II,* vol. 2 (Boppard/Rhein, 1970), p. 959.

In some ways the bombings were a throwback to the militant activities of the extreme right in the early 1920s and it seems probable that it is there Heim and his accomplices found inspiration for their campaign. The Defence Association, from which circles many of the bombers came, was a return to post-war rural vigilantism, the secret weapons' caches were a remnant of the volunteer forces' stockpiles, the theft of explosives harked back to the Ruhr Crisis, while the ties to Organisation Consul connoted a revival of earlier terrorist methods. Nevertheless, the bombings might also be interpreted as a prelude to the escalation of political violence towards the close of the Weimar Republic. In this way the bombings served as a beacon for the rising unrest in Germany just as the growing sympathy for the bombers in the press, from their arrests until the trial, should be interpreted as forewarning of the increasing leaning toward the political right.

It is not true as historian Wilhelm Koops has claimed that the bombings had "completely the opposite effect" to Heim's intentions.[230] The bombings called national attention to Schleswig-Holstein and the agricultural crisis. In this respect, the bombers' trial was an unforeseen reward, articulating the farmers' grievances where the bombs themselves had merely underscored the intensity of their anger. The court acknowledged the severity of the crisis in the province and the trial publicised criticism of the authorities and relief efforts, doing so all the more successfully because it secured publicity and sympathy for the farmers in the press.

Furthermore, there can be no question the bombings succeeded in frightening the movement's opponents. Indeed the trial, by considering the testimony of the victims, ensured their intimidating effects were acknowledged. This did not mean, however, that the authorities were necessarily swayed to take action for the farmers' benefit. Schleswig-Holstein's Social Democratic Oberpräsident Kürbis, in fact, once pointed to the bombings and "disquiet in the countryside" to support a request for assistance for industry in the city of Kiel.[231]

Neither the bombers' trial and the considerable press coverage that accompanied it, nor the success of the movement's efforts to promote extensive sympathy for them should obscure the fact that the movement was in decline in 1930. The movement's decline is the subject of the following chapter, but the bombings in this respect were significant. Though they drew attention to the Landvolkbewegung, the advertisement they provided was contrary to that envisioned by Heim. After the bombers' arrests, many farmers distanced themselves from the movement, not wishing to be associated with the attacks or because they questioned the efficacy of such methods. Further, the bombings created irreconcilable differences between Hamkens and Heim, leading to divisions among the Landvolk's leaders which the movement's loose organisation was ill-suited to contain. Under

---

230 Koops, *Südtondern in der Zeit der Weimarer Republik*, p. 254.

231 Oberpräsident der Provinz Schleswig-Holstein to the Reichskanzler, Kiel, 27.9.1929 in Erdmann, *Akten der Reichskanzlei Weimarer Republik. Das Kabinett Müller II,* vol. 2, p. 965.

such circumstances competent leaders were required, but imprisonment deprived the movement of many of its best. Behind prison walls, Heim and his accomplices were poorly placed to take advantage of the renown their imprisonment had generated or to help resolve the internal strife that was to fracture the movement.

The greatest beneficiary was perhaps the NSDAP. While under great pressure to dissociate itself from the bombings before the arrests, the investigations freed the party from suspicion soon after. With its claims to legality no longer compromised, the NSDAP was well placed to gain the support of those who approved of the bombers' aims but objected to their methods. It was quick to respond to changing rural opinion by sympathising with the bombers from the summer of 1930. Perhaps the party's promise to overthrow the system was premature but in respect of their promise to free Heim and his comrades on coming to power, the National Socialists would prove better than their word.

Chapter Eight

The decline of the Landvolkbewegung

After growing so rapidly from the autumn of 1928 and having continued strongly throughout most of the following year, the Landvolkbewegung suddenly began to falter in late 1929. Various internal and then external pressures contributed to what proved a tortuous process of decline that included fractures as well as other groups taking over its activities from 1931. Though much diminished, the movement struggled on even until the National Socialist seizure of power.

For most of 1929 the Landvolk prospered, and a change in its fortunes seemed unlikely. Though the absence of a formal membership makes gauging the Landvolk's strength difficult, the circulation of its newspaper and the frequency of and attendance levels at its activities indicate that the movement reached its peak at the beginning of October. At that time *Das Landvolk* printed up to 12,000 copies daily and the movement was collecting substantial sums by way of donations and the sale of shares in the newspaper.[1] Meetings were held almost every day in different locations, generally attracting audiences of 300 to 500 people, and indeed, continued to draw crowds of this size until the beginning of the new year, with as many as 750 people attending a meeting in Niebüll in early January 1930.[2]

It was from October 1929, however, that the earliest signs of decline appeared. The trial of the Landvolk leaders and followers in connection with the Neumünster incident generated considerably less public interest than the authorities anticipated, while the Landesthing, the symbolic gathering of the tribes in November, also fell short of expectations with many farmers from the west coast not willing to travel to Rendsburg for the occasion.[3]

Attendances at meetings fell dramatically from the beginning of 1930. Where halls had been overcrowded the previous year, by spring often just 50 to 100 far-

1 Reports titled "Bericht" Heide, 3.10.1929 and from the Landeskriminalpolizeiamt titled "Bericht" Altona, 7.10.1929, both in LAS 309/22784; reports from the Landeskriminalpolizeistelle Altona, Abteilung IA, titled "Bericht über die Landvolkbewegung" Altona, 5.10.1929 and concerning "Landvolkbewegung" Altona, 17.10.1929, both in LAS 301/4696 as well as 309/22784.

2 See chapter 7, p. 260, fn 149 and the report from the Polizeipräsident, Landeskriminalpolizeistelle, to the Regierungspräsident in Schleswig concerning "Landvolkversammlung in Niebüll" Flensburg, 10.1.1930, LAS 309/22783.

3 Report from the Landeskriminalpolizeistelle Flensburg titled "Bericht" and concerning "Landvolkbewegung" Husum, 8.11.1929 and report titled "Bericht" and concerning "Landesthing in Rendsburg" Altona, 11.11.1929, sent from the Polizeipräsident, Landeskriminalpolizeiamt, Abteilung IA, to the Regierungspräsident in Schleswig, both in LAS 309/22783.

mers turned up.[4] At a meeting in Leck on May 5, Bossen complained that only 100 people were present although the meeting had been scheduled to follow the livestock market that day.[5] By mid-year a number of reports recorded even lower attendances.[6] Though there were exceptions, audiences smaller than 50 people were common thereafter, and where figures are available, they appear to have remained at these low levels into 1931.[7]

The case of Max Bestmann provides a further compelling illustration of the Landvolkbewegung's diminishing capacity to mobilise the rural population in 1930. Convicted for inciting tax-strikes, the Landvolk organised a meeting for the commencement of his prison term at the end of the year and the authorities feared a rally to accompany him to the prison might be planned.[8] However, only 11 Landvolk followers, joined by some members of the NSDAP and a few other curious people, fewer than 20 in all, turned out to hear Bestmann's short farewell speech.[9]

Yet it was not only attendance levels at Landvolk activities that fell from the beginning of 1930 but also the level of activity itself that decreased. There were few reports of actions at confiscations and compulsory auctions in 1930 and the number of meetings also slumped significantly in the first half of the year.[10] In

---

4 Polizeipräsident, Landeskriminalpolizeistelle, to the Regierungspräsident in Schleswig concerning "Landvolk" Flensburg, 8.3.1930, LAS 309/22783.

5 Report from the Polizeipräsident, Landeskriminalpolizeistelle, to the Regierungspräsident in Schleswig concerning "Landvolkbewegung" Flensburg, 9.5.1930, LAS 309/22590.

6 Report from the Polizeipräsident, Abteilung IA, concerning "Landvolkbewegung" Altona, 27.5.1930, in both LAS 301/4697 and 309/22590; report from the Landrat to the Regierungspräsident in Schleswig concerning "Auflösung einer Versammlung" Heide, 24.6.1930, LAS 301/4698; reports from the Landrat to the Regierungspräsident in Schleswig concerning "Landvolkversammlungen in Meldorf und Marne" Meldorf, 28.6.1930, from the Landrat to the Regierungspräsident in Schleswig concerning "Landvolkversammlung in Leck" Niebüll, 24.6.1930 and from the Landrat to the Regierungspräsident in Schleswig concerning "Landvolk-Bewegung" Heide, 28.6.1930 and the Landjägereiamt Bordesholm, Kreis Bordesholm, to the Landjägereiabteilung Bordesholm, Bordesholm, 27.7.1930, all in LAS 309/22590.

7 Landrat to the Oberpräsident in Kiel concerning "Rund-Erlaß des Herrn Ministers des Innern vom 13. März 1931" Itzehoe, 14.4.1931 and the reports from the Landrat to the Oberpräsident in Kiel, Pinneberg, 2.6.1931 and from the Landrat to the Oberpräsident in Kiel, Husum, 1.7.1931, all in LAS 301/4698.

8 Transcript of the proceedings against Bestmann at the Schöffengericht Kiel titled "Strafprozess Bestmann, Hohn am 30. September 1929", in both LAS 301/4696 and 309/22784; *Rendsburger Tageblatt* No. 23, 28.1.1930, p. 3; Polizeiverwaltung to Major a.D. Koopmann in Hohn, Rendsburg, 5.12.1930 and report from the Polizeiverwaltung to the Oberpräsident in Kiel, Rendsburg, 5.12.1930, both in LAS 301/4698.

9 Report from the Polizeiverwaltung to the Oberpräsident in Kiel, Rendsburg, 10.12.1930, LAS 301/4698.

10 For a rare report on obstructive action at a compulsory auction in 1930, see the report from the Kirchspielgemeindevorsteher in Lunden to the Landrat in Heide, Lunden, 6.5.1930, LAS 309/22590. Also see the report from the Regierungspräsident to the Oberpräsident in

North Dithmarschen, Eiderstedt and Husum this coincided with increasing activity and support for the NSDAP. In North Dithmarschen, including Heim's home village of St. Annen, police reported in mid-January 1930 that a Landvolk meeting had not been held for some months, and that the movement's followers had almost entirely gone over to the NSDAP. Similarly in rural Husum and the south-west of County Schleswig, entire community committees were now made up of National Socialists. In Eiderstedt the NSDAP also established locals, although there Hamkens and his colleagues' influence remained very strong.[11]

The falling number of subscribers to *Das Landvolk* was another indication of the movement's decline. The county administrators in North Dithmarschen, Eiderstedt and Husum reported this trend at the end of January, although it seems the paper still retained many readers who liked its sensationalist style and particularly the "Little War" (*Kleinkrieg*) section, which covered farmers' altercations with the authorities across Germany.[12] From 12,000 copies in October 1929, the circulation had fallen to 8,800 by mid-June 1930.[13] Greater interest in the weekly *Kämpfendes Landvolk* offset this loss somewhat, but by mid-September the circulation of *Das Landvolk* had fallen to as little as 5,000.[14]

Ever more, the Landvolkbewegung struggled to deal with the growing demands on its finances. Even when the movement was strong, *Das Landvolk* had run at a loss, recording a shortfall of RM 30,000 in the period March to September 1929.[15] Contributions from the Volksmacht in the last two months of 1929 kept the paper afloat for a time and even allowed the purchase of an additional setting machine.[16] Yet despite generous donations from its followers, the paper's financial problems

---

Kiel concerning "Landvolkbewegung" Schleswig, 4.2.1930, in both LAS 301/4697 and 309/22783; report from the Landeskriminalpolizeistelle Altona-Wandsbek, Abteilung IA, titled "Bericht" Altona, 10.5.1930, in both LAS 301/4697 and 309/22590.

11 Report from the Polizeipräsident, Landeskriminalpolizeistelle, to the Polizeipräsident, Landeskriminalpolizeiamt Berlin, concerning "Landvolkbewegung" Flensburg, 14.1.1930, in both LAS 301/4697 and 309/22783.

12 Reports to the Regierungspräsident in Schleswig from the Landrat concerning "Landvolkbewegung" Heide, 27.1.1930, from the Landrat concerning "Landvolkbewegung im Kreise Eiderstedt" Tönning, 28.1.1930 and from the Landrat concerning "Landvolkbewegung" Husum, 31.1.1930, all in LAS 309/22783.

13 Oberpräsident to the Minister des Innern in Berlin concerning "Landvolkbewegung" Kiel, 17.6.1930, LAS 301/4697.

14 Report from the Abteilung IA titled "Bericht über die Landvolkbewegung" Altona, 15.9.1930, LAS 301/4697.

15 Report from the Regierungspräsident to the Preuß. Minister des Innern in Berlin concerning "Landvolkbewegung und die politischen Sprengstoffanschläge" Schleswig, 30.9.1929 in LAS 301/4696 and 28.9.1929 in LAS 309/22784.

16 For the "Volksmacht" see chapter 4. Also see *Das Landvolk* No. 3, 4.1.1930, LAS 301/4697 and the lists of donations titled "Herrn Hamkens Konto (privat) Zeitung" and dated "bis zum 1. April 1930", in both LAS 301/4697 and 309/22590.

were never overcome for long.[17] From October 1929 the sale of shares began to decline and many newly acquired investors failed to pay up. Moreover, the authorities discovered that Heim and Nickels had resorted to selling some shares under value.[18] With subscriptions falling, by August 1930 *Das Landvolk* was again in desperate financial trouble.[19]

The financial burdens of maintaining the paper as well as the mounting costs of legal proceedings and fines necessitated the constant pursuit of income. Appeals for financial assistance to other right-wing groups proved fruitless, so the movement remained dependent upon the generosity of the local population.[20] Landvolk leaders worked hard to ensure their colleagues continued to support the movement with Hamkens, Kühl and von Lamezan exhorting their "trusted men" to persevere in collecting donations.[21] But willingness to make donations declined and Landvolk leaders were therefore compelled to find new means of securing fresh sources of funds. The sale of postcards with photos from the march in Neumünster proved lucrative for a time.[22] So too did collections in which the "trusted men" swayed farmers to make donations by flaunting lists upon which well-known colleagues had generously signed for sums of approximately RM 100. Further techniques included the distribution of remittance orders, obliging followers to transfer prescribed amounts on a monthly basis, and the creation of "Collection Commandos" with two "trusted men" in each county responsible for transferring funds to Luetgebrune.[23]

17 Report from the Landeskriminalpolizeistelle Altona, Abteilung IA, concerning "Landvolkbewegung" Altona, 17.10.1929, in both LAS 301/4696 and 309/22784.

18 Reports titled "Bericht" Heide, 12.10.1929 and titled "Bericht" concerning "Landvolkbewegung" Altona, 17.10.1929 and titled "Bericht" Altona, 24.10.1929, all in LAS 309/22784.

19 Regierungspräsident to the Oberpräsident in Kiel, Schleswig, 19.8.1930, LAS 301/4697.

20 For the DNVP's refusal, see the lists of confiscated material, "29. Schreiben der deutschnationalen Volkspartei Berlin N.W.7 (Nagel) an Hamkens vom 11.7.1929" in "II Bei Guido Weschke – Itzehoe am 11.9.1929 beschlagnahmtes Material", LAS 309/22784 and a copy of the letter from DNVP board member Nagel to Wilhelm Hamkens, Tetenbüll, Berlin, 11.7.1929, LAS 301/4704. For Volck's failed appeals to the Pan German League and Ehrhardt, see the *Kieler Zeitung* No. 245, 4.9.1930, p. 7 and the copy of Volck's manuscript, "Landvolk und Bomben", p. 70, sent from the Polizeipräsident, Abteilung IA, to the Regierungspräsident in Schleswig, Berlin, 16.1.1930, LAS 309/22783.

21 Report from the Polizeipräsident, Landeskriminalpolizeistelle, to the Regierungspräsident in Schleswig concerning "Landvolkbewegung" 3.3.1930, in both LAS 301/4697 and 309/22783.

22 Reports from the Landeskriminalpolizeistelle Flensburg titled "Bericht" and concerning "Landvolkbewegung" Husum, 8.11.1929 and from the Landrat to the Regierungspräsident in Schleswig, Niebüll, 24.1.1930, both in LAS 309/22783; report from the Landeskriminalpolizeistelle Altona, Abteilung IA, titled "Bericht über eine Landvolkversammlung in Wrist" z. Zt. Itzehoe, 18.10.1929, in both LAS 301/4696 and 309/22784.

23 Report from the Landeskriminalpolizeistelle Altona, Abteilung IA, concerning "Landvolkbewegung" Altona, 5.2.1930, in both LAS 301/4697 and 309/22783; report from the Polizeipräsident, Landeskriminalpolizeistelle, to the Regierungspräsident in Schleswig con-

All of these initiatives, however, brought no more than brief periods of respite. Only Claus Heim's popularity continued to attract more resilient financial backing, firstly in 1930 for the bombers' trial and afterwards for the support of their families during their imprisonment.[24]

The Landvolkbewegung's decline was characterised by fluctuations. The movement enjoyed a partial revival in County Steinburg, for instance, when Guido Weschke was released from remand in March 1930.[25] The bombers' trial and the return of the movement's flag in Neumünster in October and November 1930 discussed in the previous chapters also temporarily revived interest. Moreover, growth in some parts of the province also appeared to compensate for losses elsewhere, at least in the short-term.[26] In early 1930 there was a burst of activity in the province's north-east accompanied by increasing sales of *Das Landvolk*.[27] Hamkens addressed an audience of some 500 people in Flensburg on February 22, and he, Bestmann, von Lamezan and a local carpenter, Meinke (Sieverstedt), spoke at numerous meetings in County Flensburg and the Angeln region thereafter.[28]

---

cerning "Landvolkbewegung" Flensburg, 9.5.1930, LAS 309/22590; report from the Polizeipräsident, Abteilung IA, concerning "Landvolkbewegung" Altona, 27.5.1930, in both LAS 301/4697 and 309/22590. For the "Collection Commandos", see the report from the Landeskriminalpolizeistelle, Abteilung IA, titled "Bericht" Altona, 8.6.1930 and the instructions titled "Anweisung für die Kommission zur Aufbringung der Gelder für den Landvolkkampf", both in LAS 301/4697 as well as 309/22590. Also see the report from the Landeskriminalpolizeiamt, Abteilung IA, Außendienst, Berlin, 26.8.1930, LAS 309/22590; report from the Abteilung IA titled "Bericht über die Landvolkbewegung in Schleswig-Holstein" Altona, 28.1.1931, LAS 301/4698.

24 Report from the Landeskriminalpolizeiamt titled "Bericht" Altona, 7.10.1929, LAS 309/22784; report from the Polizeipräsident, Landeskriminalpolizeistelle, to the Regierungspräsident in Schleswig concerning "Landvolkbewegung" Flensburg, 3.3.1930, LAS 309/22783; report from the Polizeipräsident, Landeskriminalpolizeistelle, to the Oberstaatsanwalt in Altona, Flensburg, 15.8.1930, LAS 309/22764; report from the Polizeiverwalter to the Landrat, Heide, 5.2.1931, LAS 309/22701. For collections on Heim's behalf, also see: copy of an undated letter from Claus Heim to Emil Brix in Altona, LAS 309/22701; *Schleswig-Holsteinischer Bauernbund* No. 23, 7.6.1931, p. 4.

25 Report from the Landrat Kreis Steinburg to the Regierungspräsident in Schleswig concerning "Landvolkbewegung" Itzehoe, 22.3.1930, in both LAS 301/4697 and 309/22590.

26 *Der Schleswiger* No. 256, 30.10.1929, LAS 309/22784; report from the Landrat des Kreises Herzogtum Lauenburg to the Regierungspräsident in Schleswig, Ratzeburg, 29.10.1929, LAS 309/22783.

27 Report from the Polizeipräsident, Landeskriminalpolizeistelle, concerning "Landvolkversammlungen" Flensburg, 11.12.1929, in both LAS 301/4697 and 309/22783; report from the Polizeipräsident, Landeskriminalpolizeistelle, to the Polizeipräsident, Landeskriminalpolizeiamt Berlin, concerning "Landvolkbewegung" Flensburg, 14.1.1930, in both LAS 301/4697 and 309/22783.

28 Report from the Polizeipräsident, Landeskriminalpolizeistelle, to the Regierungspräsident in Schleswig concerning "Landvolkversammlung in Flensburg" Flensburg, 24.2.1930, LAS 309/22783; *Kieler Zeitung* No. 56, 25.2.1930, p. 12; *Der Schleswiger* No. 46, 24.2.1930 and

Similarly, there were signs that the Landvolkbewegung had expanded into other regions of Germany. From mid-1929 Landvolk leaders including Hamkens, Weschke, Muthmann, Ross, Knudsen, Wulf and Meinke had promoted the movement at meetings in Pomerania, Oldenburg, Hanover, Saxony, Silesia and East Prussia.[29] These endeavours proved particularly effective in the north of the state of Oldenburg and in the Lüneburg region in the province of Hanover, where agitation and the prevention of compulsory auctions continued into 1930.[30] From early that year, Hamkens also succeeded in spreading the movement to East Prussia, where he worked closely with his brother-in-law, Friedrich Doepner, a farmer in the province.[31] Hamkens stirred audiences with his criticisms of the authorities and crafty incitements to break the laws. The meetings attracted large audiences and, as in Schleswig-Holstein, the movement developed an activist profile.[32] In Pomerania, farmers also adopted Landvolk tactics at protests against compulsory auctions, al-

---

No. 50, 28.2.1930, LAS 301/4697; reports from the Polizeipräsident, Landeskriminalpolizeistelle, to the Regierungspräsident in Schleswig concerning "Landvolkversammlung" Flensburg, 3.3.1930 and concerning "Landvolk" Flensburg, 8.3.1930, both in LAS 309/22783 and concerning "Bauernversammlung in Satrup" Flensburg, 16.5.1930, LAS 309/22590. Also see the report from the Polizeipräsident, Landeskriminalpolizeistelle, to the Regierungspräsident in Schleswig concerning "Landvolkversammlung in Neukrug" Flensburg, 24.5.1930, LAS 309/22590. See further, Heinacher, *Der Aufstieg der NSDAP*, pp. 292-293.

29 *Das Landvolk* No. 85, 18.6.1929, p. 5 and No. 89, 22.6.29, p. 4; Ministerium des Innern in Oldenburg to the Regierungspräsident in Schleswig, Oldenburg, 25.7.1929, LAS 309/22668; report from the Landeskriminalpolizeistelle Altona, Abteilung IA, concerning "Landvolkbewegung" Altona, 9.10.1929, in both LAS 301/4696 and 309/22784; reports from the Landeskriminalpolizeistelle Altona, Abteilung IA, both concerning "Landvolkbewegung" Itzehoe, 6.11.1929 and Itzehoe, 14.11.1929, both in LAS 301/4697 as well as LAS 309/22783; *Das Landvolk* No. 205. 26.11.1929, p. 5 and No. 29, 4.2.1930, p. 7; report titled "Bericht" Flensburg, 6.2.1930, sent by the Polizeipräsident, Landeskriminalpolizeistelle, Flensburg, 17.2.1930, in both LAS 301/4697 and 309/22783. For evidence of contact in regions as far away as Bavaria and Austria, see the report from the Landeskriminalpolizeistelle Altona-Wandsbek, Abteilung IA, concerning "Landvolkbewegung" Itzehoe, 20.3.1930, LAS 309/22590; report from the Abteilung IA titled "Bericht über die Landvolkbewegung" Altona, 20.5.1931, LAS 309/22701.

30 Schaap, *Die Endphase der Weimarer Republik*, pp. 88-94; reports from the Regierungspräsident in Lüneburg and the Regierungspräsident in Stade to the Preuß. Minister des Innern as well as police reports to these Regierungspräsidenten, published by D. Stegmann, *Politische Radikalisierung in der Provinz*, pp. 151-158, 181-182, 189-190, 200-201, 234, 241-244.

31 See the excerpts from the "Mitteilungen des Landeskriminalpolizeiamts IA Berlin vom 15.11.30 Nr. 22", LAS 309/22576.

32 Copy of a report from the Regierungspräsident in Königsberg concerning "Bauernnotbewegung. Auftreten Hamkens – Tetenbüll" Königsberg, 14.2.1930 and Regierungspräsident in Schleswig to Regierungspräsident Dr. v. Bahrfeldt in Königsberg, Schleswig, 10.3.1930, both in LAS 309/22783.

though there the actions remained under the control of the Rural League.[33] In Lower Silesia the Landvolk cooperated with the *Bundschuh*, a nationalist farmers' movement formed in 1926, for which Hartmut Plaas edited *Die schwarze Fahne* (*The Black Flag*).[34]

Paradoxically, expansion into other regions of Germany was actually another indication of the movement's decline. Regierungspräsident Abegg judged that the movement had passed its peak in Schleswig-Holstein and suspected that the real reason for Hamkens' travels was that he was compelled to hunt further afield in order to finance *Das Landvolk*. Many years later, in an interview with Hans Beyer, Hamkens conceded that this was indeed the case.[35]

Any signs of the movement's recovery in Schleswig-Holstein in 1930 ended at the beginning of May when Hamkens and Weschke returned to prison.[36] In Angeln, the NSDAP capitalised on their absence, establishing local groups among erstwhile Landvolk supporters.[37] Likewise, in East Prussia the NSDAP managed to entice many Landvolk followers with its programme.[38] There the Landvolk's newspaper, also named *Die schwarze Fahne*, endured into 1931 but the movement's leaders played no role in disturbances at compulsory auctions after the autumn of 1930 and held no rallies or meetings after December.[39]

There was a brief revival in the spring of 1931 when the movement returned to the source of its earliest successes, its campaign against compulsory tax collection. It did so at the most propitious possible moment, just as the farmers' fortunes took another sharp turn for the worse. On top of ever-mounting debts,[40] livestock prices fell dramatically in 1930, eventually dropping even below their pre-war levels, as

33 Baranowski, *The Sanctity of Rural Life*, pp. 130 and 134; Pyta, *Dorfgemeinschaft und Parteipolitik*, p. 209. Also see *Vorwärts* No. 129, 18.3.1930, LAS 309/22590.

34 Report from the Landeskriminalpolizeistelle Altona, Abteilung IA, titled "Bericht über die Landvolkbewegung" Altona, 15.9.1930, LAS 301/4697; excerpt from the "Mitteilungen des Landeskriminalpolizeiamts (IA) Berlin vom 1. Mai 1931 Nr. 9", LAS 309/22576.

35 Regierungspräsident to the Regierungspräsident Dr. v. Bahrfeldt in Königsberg, Schleswig, 10.3.1930, LAS 309/22783; Beyer, 'Die Landvolkbewegung Schleswig-Holsteins und Niedersachsens 1928-1932', p. 196.

36 Oberstaatsanwalt to the Gemeindevorsteher in Hohenaspe, Tetenbüll, Itzehoe and Gnutz concerning "Strafsache gegen Hansen und Gen. wegen Beamtennötigung usw. (Angriff auf das Gerichtsgefängnis in Hohenwestedt.)" Kiel, 12.4.1930, LAS 309/22590; report from the Polizeipräsident, Abteilung IA, concerning "Landvolkbewegung" Altona, 27.5.1930, in both LAS 301/4697 and 309/22590.

37 Stoltenberg, *Politische Strömungen im schleswig-holsteinischen Landvolk*, p. 147.

38 Copy of a report from the Regierungspräsident concerning "Bauernnotbewegung. Auftreten Hamkens – Tetenbüll" Königsberg, 14.2.1930 and Regierungspräsident to the Regierungspräsident Dr. v. Bahrfeldt in Königsberg, Schleswig, 10.3.1930, both in LAS 309/22783.

39 Copy of a report from the Regierungspräsident to the Preuß. Minister des Innern in Berlin concerning "Landvolkbewegung" Königsberg, 21.9.1931, LAS 309/22701.

40 See the results of the Rural and Farmers' League's investigations in the *Schleswig-Holsteinischer Bauernbund* No. 40, 9.10.1932, pp. 1-3.

the agricultural crisis was compounded by the world economic crisis.[41] In Schleswig-Holstein this misfortune was exacerbated when a dry summer ended with floods, destroying 40 percent of the harvest.[42] Once again farmers found it increasingly difficult to meet their financial obligations and the number of compulsory enforcement proceedings began to rise.[43] Giving them added incentive to look once more to the Landvolkbewegung, the province's many livestock farmers believed the Reich government had again neglected to provide adequate tariff protection for capital intensive production, and their anger was fuelled when the province was omitted from the Eastern Aid programme (*Osthilfe*), which provided considerably easier access to protection from compulsory enforcement as well as reductions in interest and capital burdens.[44]

Landvolk leaders spoke at several meetings in May and June, promoting the sabotage of tax collection as well as campaigning for Claus Heim's release.[45] Among their most successful endeavours to generate publicity was a meeting in Holnis in the province's north where they presented a replica of the Landvolk flag to the local community chairman and community representatives who had resigned as one rather than collect taxes in their indebted community.[46]

---

41 *Schleswig-Holsteinischer Bauernbund* No. 1, 4.1.1931, p. 4, No. 10, 8.3.1931, p. 3, No. 14, 5.4.1931, p. 4 and No. 17, 26.4.1931, p. 1. Also see Gerschenkron, *Bread and Democracy*, pp. 133-145.

42 Undated document titled "Material für die Besprechung bei dem Herrn Preuss M.d.I. am 25. Nov. 1931", LAS 309/22924; Thyssen, *Bauer und Standesvertretung*, p. 396.

43 Oberlandesgerichtspräsident to the Landgerichtspräsidenten in Altona, Flensburg and Kiel concerning "Durchführung von Zwangsvollstreckungsmaßnahmen" Kiel, 13.11.1931, LAS 309/22924.

44 *Schleswig-Holsteinischer Bauernbund* No. 50, 13.12.1931, p. 3, No. 23, 12.6.1932, p. 1, No. 26, 3.7.1932, pp. 1-3, No. 37, 18.9.1932, pp. 1-2 and No. 49, 11.12.1932, p. 1.

45 Report from the Abteilung IA titled "Bericht über die Landvolkbewegung" Altona, 20.5.1931 and the flyer titled "Ist der Haushalt deiner Gemeinde in Ordnung?", both in LAS 309/22701; Regierungspräsident to the Oberpräsident in Kiel, Schleswig, 30.5.1931, in both LAS 301/4698 and 309/22701. Also see the reports to the Oberpräsident in Kiel from the Landrat in Pinneberg, 2.6.1931, Landrat in Flensburg, 8.6.1931, Landrat in Heide, 16.6.1931, Landrat in Bad Segeberg, 22.6.1931, Landrat in Marne, 24.6.1931 and Landrat in Husum, 1.7.1931, all in LAS 301/4698; report from the Abteilung IA titled "Bericht über die Landvolkbewegung" Altona, 15.7.1931, in both LAS 301/4698 and 309/22701.

46 Report from the Landrat to the Regierungspräsident in Schleswig concerning "Kundgebung des Landvolks in Holnis" Flensburg, 8.6.1931 and the report from the Landjägereiposten Wees, Flensburg, titled "Bericht über das Landvolk-Treffen in Holnis am Sonntag, den 7. Juni 1931" Wees, Flensburg, 8.6.1931, both in LAS 309/22701; report from the Landeskriminalpolizeistelle, Abteilung IA, titled "Bericht über die Landvolkbewegung" Altona, 15.7.1931, in both LAS 301/4698 and 309/22701. Also see the *Volkszeitung* No. 126, 3.6.1931 and the *Flensborg Avis* No. 128, 5.6.1931 and No. 131, 8.6.1931, all in LAS 309/22701.

The Landvolk's "tax struggle" resonated with farmers and it appeared that the movement was again attracting a greater following.[47] *Das Landvolk* also recovered to some extent, its circulation was back up to 7,000 in the spring of 1931. Its financial affairs, however, continued to deteriorate. Several staff had to be dismissed and a financial rescue plan introduced. The movement's leaders resolved that if it did not meet tight budgetary restraints and still depended on pecuniary aid after July 1, the business would have to be liquidated.[48]

The newspaper's survival was hampered when it was banned for two and a half weeks from April 10 for breaching the Reich President's decree of March 28, 1931 combating political acts of violence and for breaching the Prussian bylaw of 30.3.1931.[49] A further ban for breaches of these laws for a period of eight weeks at the end of June spelled its death. The *Vize-Oberpräsident* (deputy governor), Heinrich Thon, banned *Das Landvolk* on account of an article describing the signing of the Versailles treaty as a day of "shame and disgrace".[50] At the same time he banned the weekly *Kämpfendes Landvolk* for three months for a series of articles which accused the authorities of chicanery in the treatment of Claus Heim and his imprisoned colleagues. These articles claimed they should have been treated as political prisoners and thereby entitled to concessions such as not being compelled to perform demeaning menial tasks.[51]

---

47 Report from Abteilung IA titled "Bericht über die Landvolkbewegung" Altona, 20.5.1931, LAS 309/22701; report from the Regierungspräsident to the Preuß. Minister des Innern in Berlin concerning "Landvolkbewegung" Schleswig, 30.5.1931, in both LAS 301/4698 and 309/22701.

48 Copy of the financial rescue plan titled "Sanierungsplan für 'Das Landvolk', Tageszeitung GmbH" and report from Abteilung IA titled "Bericht über die Landvolkbewegung" Altona, 20.5.1931, both in LAS 309/22701; report titled "Bericht über die Landvolkbewegung" from the Abteilung IA, Altona, 15.7.1931, sent by the Polizeipräsident Altona-Wandsbek in Altona to the Regierungspräsident in Schleswig, 21.7.1931, in both LAS 301/4698 and 309/22701.

49 Oberpräsident's decrees, Kiel, 9.4.1931 and Kiel, 10.4.1931 (the second extended the ban across the entire Reich), LAS 301/4705. For the offending article, "Ein glänzender Erfolg des Verwaltungsapparates. Wie deutsche Substanz verschleudert wird", see *Das Landvolk* No. 81, 8.4.1931, p. 3 and for the announcement of the ban, *Das Landvolk* No. 83, 10.4.1931, p. 1. Also see the Reichsminister des Innern to the Oberpräsident in Kiel concerning "Verbot der in Itzehoe erschienenen periodischen Druckschrift 'Das Landvolk' " Berlin, 25.4.1931, Luetgebrune's appeal, Göttingen, 15.4.1931 and the decision of the Reichsgericht 4. Strafsenat from 30.4.1931, all in LAS 301/4705.

50 Oberpräsident's decree, Kiel, 29.6.1931, LAS 301/4705 and the article "Nur die Taten zählen im Kampf!" in *Das Landvolk* No. 134, 27.6.1931, p. 1.

51 Oberpräsident's decree, Kiel, 29.6.1931, LAS 301/4705. For the offending articles, see the *Kämpfendes Landvolk* No. 24, 2.6.1931, No. 25, 3.6.1931 and No. 26, 4.6.1931, all in LAS 301/4705. For the Reichsgericht dismissal of *Das Landvolk's* appeal, see the order of the Reichsgericht 4. Strafsenat from 17.7.1931 and the note in the files, Kiel, 23.7.1931, both in LAS 301/4705.

Das Landvolk's inevitable liquidation was announced shortly before the ban ended.[52] It had been dealt a further blow when Hamkens and Kühl returned to prison in July and August. Consequently, meetings and collections also came to a standstill. Pramor and various Volksmacht members strove to establish a successor in Itzehoe, a weekly called *Der Landvolkkampf*, and so too, after his release, did Hamkens with a new weekly *Landvolk* in Husum.[53] Dividing the old readership was counterproductive to the newspapers' viability, and legal action added even greater pressure, particularly on *Der Landvolkkampf*, which was banned for eight weeks from October and then again for three weeks in the new year.[54] Yet Hamkens' new *Landvolk* also failed to attract many readers and a nationwide bulletin,

52 Report from the Abteilung IA titled "Bericht über die Landvolkbewegung" Altona, 24.8.1931, LAS 309/22701; reports from the Regierungspräsident to the Minister des Innern in Berlin concerning "Landvolkzeitung" Schleswig, 2.9.1931 and from the Landrat des Kreises Steinburg to the Regierungspräsident in Schleswig concerning "Die Zeitung 'Das Landvolk' " Itzehoe, 27.8.1931, both in LAS 301/4698; report from the Abteilung IA titled "Bericht über die Landvolkbewegung" Altona, 15.7.1931, in both LAS 301/4698 and 309/22701. The *Vossische Zeitung* reported the liquidation of *Das Landvolk* as a sign that the movement was coming to its end, *Vossische Zeitung* No. 201, 23.8.1931, LAS 309/22701. For Hamkens' attempts to postpone imprisonment, his flight, capture in East Prussia and delivery to prison in Flensburg: Oberstaatsanwalt to Wilhelm Hamkens, Tetenbüll. Altona, 16.6.1931 and the reports from the Oberstaatsanwalt to the Preuß. Justizminister concerning "Strafvollstreckung gegen den Landwirt Wilhelm Hamkens in Tetenbüll" Altona, 16.6.1931 and concerning "Strafvollstreckungssache gegen Landwirt Hamkens in Tetenbüll" Altona, 3.7.1931, GStA PK I. HA Rep. 84a No. 55410; memoranda from the Landrat Eiderstedt to the Regierungspräsident in Schleswig concerning "Strafverbüßung Hamkens" Tönning, 4.7.1931 and from the Oberstaatsanwalt to the Generalstaatsanwalt concerning "Strafvollstreckungssache gegen den Landwirt Wilhelm Hamkens – Tetenbüll" sent from the Generalstaatsanwalt to the Regierungspräsident in Schleswig, Kiel, 8.7.1931, both in LAS 309/22701. For Kühl's imprisonment, Oberstaatsanwalt to the Regierungspräsident in Schleswig, 6.8.1931, LAS 309/22701.

53 Report from the Abteilung IA titled "Bericht über die Landvolkbewegung" Altona, 24.8.1931, LAS 309/22701; report from the Landrat des Kreises Steinburg to the Regierungspräsident in Schleswig concerning "Die Zeitung 'Das Landvolk' " Itzehoe, 27.8.1931, LAS 301/4698. Also see the last issue of *Das Landvolk* No. 135, 26.8.1931, p. 1 announcing the launch of the *Landvolkkampf*.

54 *Das Landvolk* received a warning in December 1931, Oberpräsident to the publishers of the weekly *Das Landvolk* in Husum, Kiel, 14.12.1931, LAS 301/4705. For the offending article, see "Heim und Hindenburg" in *Das Landvolk* No. 13, 15.12.1931, LAS 301/4705. For the bans of *Der Landvolkkampf* from the end of October 1931 and again in February 1932, see the Oberpräsident's decree, Kiel, 27.10.1931 and the *Landvolkkampf* No. 8 (1) and the Oberpräsident's decree, Kiel, 24.2.1932, LAS 301/4705. Appeals to overturn the bans were rejected. See the appeal from the publishers of *Der Landvolkkampf* in Itzehoe to the Reichsminister des Innern in Berlin, Itzehoe, 28.10.1931 and to the Oberpräsidium in Kiel, Itzehoe, 9.11.1931 and the decision of the Reichsgericht 4. Strafsenat from 7.11.1931, appeal from the *Landvolkkampf* to the Reichsminister des Innern in Berlin, Itzehoe, 25.2.1932 and the decision of the Reichsgericht 4. Strafsenat from 4.3.1932, all in LAS 301/4705.

the *Deutscher Landvolk-Kampfbund*, published by Ernst Niekisch in Berlin on Hamkens and Doepners' initiative, only had a minimal impact in the province.[55]

Still the movement persevered and Hamkens, in particular, tried desperately to continue the "tax struggle". In the autumn of 1931 a flyer warned that farms purchased at compulsory auctions would "go up in flames" and then, in an endeavour to resurrect the Crisis Assistance and keep what was left of the movement together, socalled "Village Communities" (*Dorfgemeinschaften*) were established.[56] Hamkens succeeded in founding some 60 Communities in the province's north: in Eiderstedt, South Tondern, Flensburg and Eckernförde counties. They proved considerably less boisterous than the Crisis Assistance of 1929, did not attempt to assume authority and concentrated purely on financial matters.[57] In fact, for most of the "Village Communities", there is no further record of their activities after their being established. In Eiderstedt and Ladelund in South Tondern, however, Landvolk leaders continued to hold meetings, prevent compulsory enforcement and assist needy colleagues.[58]

The Landvolk's campaign against compulsory tax collection might have effected a more lasting rejuvenation, had not the movement's activities been subsumed by other groups. The first to do so were the Rural League and the Farmers' League. In January 1930 they had finally fused to form the "Rural and Farmers' League" (*Land- und Bauernbund*). With the forces of these two agricultural asso-

55 Report from the Abteilung IA concerning "Landvolkbewegung" Altona, 23.2.1932 and the *Deutscher Landvolk-Kampfbund. Mitteilungsblatt* No. 1, 17.12.1931, both in LAS 309/22701. Niekisch hoped to integrate the Landvolkbewegung into his resistance movement and also took part in a meeting tour in East Prussia. The many inaccuracies in the short account of the Landvolkbewegung in his memoirs, however, do not point to an intimate connection with the movement, beyond the attempts to secure Heim's release in 1932. See Niekisch, *Erinnerungen eines deutschen Revolutionärs*, vol. 1, pp. 166-169.

56 Report from the Abteilung IA titled "Bericht über die Landvolkbewegung" Altona, 24.8.1931, LAS 309/22701.

57 Report from the Abteilung IA titled "Bericht über die Landvolkbewegung und den Land- und Bauernbund in Schleswig-Holstein" Altona, 22.12.1931, LAS 309/22701; report from the Landrat des Kreises Eiderstedt to the Regierungspräsident in Schleswig concerning "Landvolkbewegung" Tönning, 29.10.1931 and the letter from the Landvolkvereinigung des Kreises Eiderstedt to the Landrat des Kreises Eiderstedt in Tönning, Garding, 24.10.1931, both in LAS 309/22924. Also see *Der Schleswiger* No. 279, 29.11.1931, LAS 309/22701.

58 For the formation of the "Crisis Assistance Ladelund" in September 1931 and the renewed efforts of Bossen, Petersen (Hogelund) and Hamkens in the county, see *Der Schleswiger* No. 224, 25.9.1931, LAS 301/4698; reports from the Landjägereiposten Sande, Kreis Südtondern, to the Landrat in Niebüll, Sande b. Stedesand, 10.11.1931 and from the Landrat to the Regierungspräsident in Schleswig concerning "Nothilfe Ladelund und Landvolkbewegung" Niebüll, 11.11.1931, the letter from the Landesbank der Provinz Schleswig-Holstein to the Regierungspräsident in Schleswig, Kiel, 5.12.1931 and the copy of a letter from the Nothilfe Ladelund to Herr Söncke Hansen, Ladelundmühle, 29.11.1931, all in LAS 309/22924.

ciations combined, the Rural and Farmers' League represented the majority of the province's farmers.[59] From the autumn of 1931 it adopted the renegade tactics that had previously been the Landvolk's exclusive domain.[60] As *Der Schleswiger* put it, "Landvolk-policy" became "official".[61]

In September the Rural and Farmers' League demanded a moratorium on tax and interest payments and compulsory enforcement. It directed its members to ensure that no bids were made at compulsory auctions and to ostracise those who did socially and economically.[62] Next, it took up local calls for the formation of a new "self-help" organisation with the blessing of the NSDAP and Stahlhelm.[63] This led to the creation of the Rural and Farmers' League's own variant of the Landvolkbewegung's Crisis Assistance.

Despite Regierungspräsident Abegg's attempts to prevent it,[64] from October 10 the Rural and Farmers' League held meetings across the province at which farmers

---

59 For the fusion of the Rural League and the Farmers' League, see p. 314 below.

60 For a more detailed account of the Rural and Farmers' League's activism from the autumn of 1931, see A. Otto-Morris, ' "Only united can we escape certain ruin": rural protest at the close of the Weimar Republic', *Rural History*, 20, 2 (2009), pp. 187-208.

61 *Der Schleswiger* No. 219, 19.9.1931 and No. 271, 20.11.1931, both in LAS 309/22765. Also see the report from the Regierungspräsident to the Minister des Innern in Berlin concerning "Landvolkzeitung" Schleswig, 2.9.1931, LAS 301/4698.

62 Report from the Regierungspräsident to the Preuß. Minister des Innern in Berlin concerning "Stimmung in der ländlichen Bevölkerung" Schleswig, 18.9.1931, LAS 301/4698. Also see the Rural and Farmers' League's provincial *Schleswig-Holsteinischer Bauernbund* No. 38, 20.9.1931, p. 1, No. 45, 8.11.1931, p. 1 and No. 1, 10.1.1932, p. 1.

63 The calls for a new self-help organisation came, above all, from Henry Pötter (Kükels) at a meeting in Bad Segeberg at the beginning of September. Report from the Abteilung IA titled "Bericht" Kiel, 3.9.1931, LAS 301/4698. For Pötter's notice see the *Schleswig-Holsteinische Tageszeitung* No. 211, 3.9.1931, *Kaltenkirchener Zeitung* No. 207, 4.9.1931, the *Segeberger Kreis- und Tageblatt* No. 208 and the *Eckernförder Zeitung* from 8.9.1931, all in LAS 301/4698. For the preparations, see the letters to Major Scheel from Henry Pötter, Kükels, 25.9.1931, from Gauleiter Lohse to Major Scheel, Land- und Bauernbund in Kiel, Altona, 24.9.1931 and from Major Scheel to Otto Köhler, Bünsdorf, Bad Wildungen, 25.9.1931, all in LAS 301/4698; report from the Abteilung IA concerning "bevorstehende Aktionen des Land- und Bauernbundes in der Provinz Schleswig-Holstein" Altona, 10.10.1931, Regierungspräsident to the Preuß. Minister des Innern in Berlin concerning "Stimmung in der ländlichen Bevölkerung" Schleswig, 29.9.1931 and the circulars from Stahlhelm-Landesführer Tüllmann to the Gaue, Kiel, 6.10.1931 and from the Schleswig-Holsteinischer Land- und Bauernbund to its county organisations, No. 183, Kiel, 8.10.1931 and No. 184, 8.10.1931, all in LAS 301/4698.

64 Report from the Regierungspräsident to the Preuß. Minister des Innern in Berlin concerning "Selbsthilfeaktion des Land- und Bauernbundes" Schleswig, 23.10.1931, LAS 301/4698. Also see the directive to ban the meetings issued by the Regierungspräsident in Schleswig to the Landräte, Schleswig, 8.10.1931, LAS 309/22924 and the cover note to the confiscated material from the Polizeipräsident in Kiel to the Oberpräsident in Kiel concerning "Durchsuchung der Hauptgeschäftsstelle des Land- und Bauernbundes in Kiel" Kiel, 14.10.1931, the report titled "Durchsuchungsbericht" Kiel, 10.10.1931 and the appended

signed as members of the "Crisis and Shared Fate Community" (*Not- und Schicksalsgemeinschaft*) and pledged to undertake "self-help" in order to protect their capital.[65] These meetings and collection of signatures continued over the following weeks and, by the beginning of November, the Crisis and Shared Fate Community had 65,000 members and, by the end of the month, no less than 70,000. Including not only the majority of the province's farmers but also many rural labourers and tradesmen, this constituted an unprecedented basis for implementing and enforcing "self-help".[66]

The Crisis and Shared Fate Community's notion of "self-help" as well as much of its rhetoric were appropriated directly from the Landvolkbewegung. Its pledge, for example, concluded with the Three Musketeers' maxim "all for one and one for all", which Hamkens had made popular. Similarly, League Chairman Köhler echoed the Landvolk, warning community chairmen from lowering themselves to "lackeys of the system" by collecting taxes from the suffering farmers.[67] Calls for the farmers who could afford to pay taxes to continue to do so also mimicked Hamkens' style.[68] The Crisis and Shared Fate Community's pledge and signatures, however, made it stronger than the Landvolkbewegung, establishing a basic bond with its members, which the Landvolkbewegung had never demanded from its followers.

The Crisis and Shared Fate Community also went further than the Landvolkbewegung by setting up a "Property Protection Cooperative" (*Besitzschutzgenossenschaft*) in December.[69] Like the Landvolk's Crisis Assistance, the cooperative aimed to safeguard farms by negotiating with creditors and the authori-

---

list of confiscated items which includes the instructions to the county organisations, all in LAS 301/4698.

65 The farmers circumvented the Regierungspräsident's ban, holding the meetings in their own homes and barns and collecting signatures house by house. See the *Schleswig-Holsteinischer Bauernbund* No. 42, 18.10.1931, pp. 2-4. Also see the "Anlage zum Erlaß des Pr. Ministeriums des Innern vom 23. Februar 1932" sent from the Preuß. Minister des Innern to the Oberpräsident in Kiel, Berlin, 23.2.1932 and from the Oberpräsident in Kiel to the Regierungspräsident in Schleswig, Kiel, 7.3.1932, in both LAS 301/4699 and 309/22924.

66 *Schleswig-Holsteinischer Bauernbund* No. 44, 1.11.1931, p. 2, No. 45, 8.11.1931, p. 1 and No. 47, 22.11.1931, p. 4. According to Thyssen, there were 80,000 in November. See Thyssen, *Bauer und Standesvertretung*, p. 396.

67 Report from the Landrat to the Regierungspräsident in Schleswig concerning "Steuerverweigerungsbewegung" Bad Segeberg, 31.10.1931, in both LAS 301/4698 and 309/22924.

68 Report from the Landrat Plön to the Regierungspräsident in Schleswig concerning "Auftreten des nationalsozialistischen Hofbesitzers Pötter-Kükels" Plön, 19.11.1931, LAS 309/22924.

69 See the article titled "Schleswig-Holsteins Bekenntnis zur Landwirtschaftl. Besitzschutz-Genossenschaft" in the *Schleswig-Holsteinischer Bauernbund* No. 50, 13.12.1931, p. 3.

ties, and by providing farmers with financial guidance.[70] The Rural and Farmers' League's aspirations for the cooperative were entirely pragmatic. Rather than confuse financial and political objectives as the Crisis Assistance had done, it recognised the cooperative's limitations and made clear from the beginning that its intervention intended to defer the farmers' losses until there was fundamental political change from the right.[71] To improve on the Landvolk's Crisis Assistance and its Village Communities in financial matters also, the cooperative required that members buy shares creating funds to help needy colleagues and to enable the cooperative itself to bid at auctions if necessary.[72] In its first year, the cooperative gained postponements of payment in more than 500 cases and intervened in 42 compulsory auctions, including eight in which it purchased properties for members.[73]

Where the Rural and Farmers' League was entirely in keeping with the Landvolkbewegung was in the second wave of obstructionist actions at confiscations and compulsory auctions that began in the autumn and winter of 1931 and continued throughout 1932. Though rarely noted in the literature, these actions were quite as prolific as the Landvolkbewegung's activities in 1928 and 1929.[74] The Landvolkbewegung's earlier activities provided the model for obstructionist tactics and sometimes former Landvolk leaders were directly involved.[75] However, the

---

70 "Anlage zum Erlaß des Pr. Ministeriums des Innern vom 23. Februar 1932", in both LAS 301/4699 and 309/22924.

71 The newspaper *Schleswig-Holsteinischer Bauernbund* described the cooperative as a bridge to better times and as a blue marginal note against a cutting of the article in the files indicates, the Regierungspräsident believed that the farmers expected Hitler to deliver them. *Schleswig-Holsteinischer Bauernbund* No. 50, 13.12.1931, p. 3, LAS 309/22924. Also see the speech of the Rural and Farmers' League chairman, Köhler, at a compulsory auction of a farm in Lentföhrden in December 1931, report from the Regierungspräsident concerning "Aufhebung eines Zwangsversteigerungstermins durch die Landschaft" Schleswig, 12.12.1931, in both LAS 301/4698 and 309/22924.

72 Report from the Polizeipräsident Altona-Wandsbek to the Regierungspräsident in Schleswig concerning "Vertrauensmännerversammlung des Kreis-Bauernbundes Stormarn in Bargteheide" Altona, 3.1.1932, LAS 309/22924 and dated 31.12.1931, LAS 301/4698. Also see the *Schleswig-Holsteinischer Bauernbund* No. 3, 24.1.1932, p. 3 and No. 10, 13.3.1932, pp. 5-6.

73 See the reports on the cooperative's activities in the *Schleswig-Holsteinischer Bauernbund* No. 25, 26.6.1932, pp. 5-6, No. 44, 6.11.1932, p. 3 and No. 52, 31.12.1932, p. 6.

74 For more on these actions and the second wave of rural protest in Schleswig-Holstein and other regions of Germany from 1931, see Otto-Morris, ' "Only united can we escape certain ruin" ', pp. 187-208. While Nils Werner dealt with the trials arising from some of these actions, they have been paid little heed elsewhere. Werner, *Die Prozesse gegen die Landvolkbewegung*, pp. 92-99. A further exception is H. W. Gondesen, 'Ende der Weimarer Republik, Ende des Landvolks', *Jahrbuch für die Schleswigsche Geest*, 37 (1989), pp. 16-21. Stoltenberg only mentioned the actions in a footnote. See Stoltenberg, *Politische Strömungen im schleswig-holsteinischen Landvolk*, p. 176, fn. 4.

75 Two salient examples involving former Landvolk leaders were the rally to prevent the compulsory auction of Heinrich Hansen's farm in Pöschendorf on October 19 and the assault of

actions now took part under the leadership of the Crisis and Shared Fate Community. For this reason, it seems, the actions remained focussed solely on the preservation of the farmers' property. They did not give rise to nearly as much tangential mischief, nor were they accompanied by bombings. But when it came to confiscations and auctions, the actions were every bit as intense, and conducted in precisely the same style, as they had been in the Landvolkbewegung.

Farmers were willing to impede compulsory enforcement, whether it was a matter of outstanding taxes or private debt.[76] To prevent confiscations, they intimidated bailiffs, their assistants, compliant transport companies and the police. Sometimes terror succeeded threats and abuse.[77] At compulsory auctions, it was not only bidders who were subjected to threatening crowds, threats and violence.[78] Creditors

---

a bailiff who attempted to confiscate a pub's till in Hohn near Rendsburg in January 1932, led by the organiser of the Landesthing, Bestmann. Oberstaatsanwalt to the Regierungspräsident in Schleswig concerning "Vorgänge gelegentlich einer Zwangsversteigerung am 19.X.1931 in Pöschendorf (Hansen in Breitenfelde und Genossen) ..." 7.6.1932, LAS 309/22924; "Anlage zum Erlaß des Pr. Ministeriums des Innern vom 23. Februar 1932", in both LAS 301/4699 and 309/22924; Landjägereiabteilung Rendsburg to the Landrat in Rendsburg, Rendsburg, 29.1.1932, LAS 309/22924.

76 Reports from the Landrat to the Regierungspräsident in Schleswig concerning "Stimmungsbericht" Husum, 21.3.1932 and concerning "Zwangsversteigerung in Winnertfeld" Husum, 23.8.1932, both in LAS 309/22924.

77 See for example the confiscations at Diekshörn in South Dithmarschen and in Großensee in Stormarn. Reports from the Landrat concerning "Verhinderung der Durchführung einer Pfändung in der Mühlenstraße durch eine Anzahl Landleute" Meldorf, 25.10.1931, from the Landjägereiamt Brunsbüttelkoog to the Landjägereiabteilung zu Meldorf, Brunsbüttelkoog, 25.10.1931, both in LAS 309/22924; issues of the verdicts of the Schöffengericht Itzehoe, 22.1.1932 and 10.6.1932, GStA PK I. HA Rep. 84a No. 55434; reports concerning "Ermittelungsverfahren gegen Rosenau und Gen. wegen Aufruhrs (Verhinderung der Zwangsversteigerung einer Kuh am 30.10.31 in Grossensee b/Trittau)" from the Oberstaatsanwalt to the Preuß. Justizminister in Berlin, Altona, 18.11.1931, Altona, 24.11. 1931, Altona, 4.12.1931 and Altona, 22.2.1932, all in GStA PK I. HA Rep. 84a No. 55430; issue of the verdict of the Schöffengericht Altona from 16.12.1931, LAS 309/22924; report from the Landrat des Kreises Stormarn to the Regierungspräsident in Schleswig concerning "Verhinderte Zwangsversteigerung in Großensee" Wandsbek, 7.3.1932 and Oberstaatsanwalt to the Regierungspräsident in Schleswig, Altona, 19.7.1932, all in LAS 309/22924.

78 Landrat to the Regierungspräsident in Schleswig concerning "Versteigerungen" Meldorf, 10.11.1931, report from the Landrat to the Regierungspräsident in Schleswig concerning "Störungen von Zwangsversteigerungen" Bad Segeberg, 12.11.1931, *Kaltenkirchener Zeitung* No. 252, 17.10.1931, report from the Landrat to the Regierungspräsident in Schleswig concerning "Zwangsversteigerungen" Heide, 12.11.1931, all in LAS 301/4698; report from the Landrat des Kreises Steinburg to the Regierungspräsident in Schleswig, Itzehoe, 5.1.1932, report from the Obergerichtsvollzieher Haack to the Aufsichtsrichter des Amtsgericht Itzehoe, Itzehoe, 19.1.1932, report from the Landrat to the Regierungspräsident in Schleswig concerning "Zwangsversteigerung in Wesselburen" Heide, 3.2.1932, report from the Landjägereiabteilung Bramstedt to the Landrat in Bad Segeberg concerning "Bericht zu dem anl. Artikel der Volkszeitung v. 13.11.31 'Unerträglicher Terror der Nazibauern' "

who pursued their liabilities also risked their debtors' fury, and farmers invariably found opportunities to frighten their lawyers into agreeing to postponements even when police were able to prevent heckling crowds.[79] Moreover, trouble did not desist even in those rare cases when farming properties were auctioned successfully: on several occasions tenants refused to leave or forcefully returned to their farms while new owners risked boycott and ostracism.[80] One of the most memorable instances involved the black-marking of two community chairmen from Nordstrand, said to have profited at the expense of two needy fellow farmers by acquiring the land they had leased.[81] When one of the community chairmen subsequently took those responsible for defaming him to court, lay judges sympathised with the accused and acquitted them on finding their reproaches legitimate.[82]

Under the supervision of the Rural and Farmers' League, resistance to the payment of taxes reached new heights as the crisis worsened. Though there are no really satisfactory figures available to ascertain the extent of tax-strikes in Schleswig-Holstein, an undated and untitled document from the period recorded that the province's 13 county finance offices estimated a minimum of 15,000 cases requiring immediate enforced payment of "state taxes". The finance offices were unwilling to auction properties on account of small sums and only in conjunction

---

Bramstedt, 3.12.1931, report from the Landrat to the Regierungspräsident in Schleswig, Bad Segeberg, 24.5.1932, report from the Landrat to the Regierungspräsident in Schleswig concerning "Versteigerung in Lunden" Heide, 1.6.1932, report titled "Bericht über den Einsatz in Ahrensburg am 13.6.1932" (date missing due to document damage), all in LAS 309/22924.

79 See the letter from Heinrich Tödter to the Landesbank der Provinz Schleswig-Holstein concerning "Nichtbilligung des Verkaufs der im Zwangsverfahren befindlichen Gastwirtschaft von H. J. Schröder – Albersdorf" Oesterrade, 20.11.1931, LAS 301/4699; report from the Landrat to the Regierungspräsident in Schleswig concerning "Steuerstreik" Heide, 11.1.1932, report from the Landrat to the Regierungspräsident in Schleswig, Heide, 29.1.1929, report from the Ortspolizeibehörde to the Regierungspräsident in Schleswig, Neumünster, 18.4.1932 and report from the Landrat to the Oberstaatsanwalt in Altona, Meldorf, 22.6.1932, all in LAS 309/22924.

80 Report from the Landrat to the Regierungspräsident in Schleswig concerning "Störungsversuche bei Zwangsversteigerungen von landwirtschaftlichen Besitzungen" Pinneberg, 10.11.1931, LAS 301/4698; "Anlage zum Erlaß des Pr. Ministeriums des Innern vom 23. Februar 1932", in both LAS 301/4699 and 309/22924; report from the Regierungspräsident to the Oberpräsident in Kiel concerning "Zwangsräumungen Lutzhorn" Schleswig, 23.11.1931, LAS 301/4698.

81 Reports from the Landrat to the Regierungspräsident in Schleswig concerning "Stimmung der Bevölkerung" Husum, 11.3.1932 and concerning "Öffentliche Brandmarkung eines Landwirtes durch den Land- und Bauernbund" 25.4.1932, both in LAS 309/22924; *Husumer Nachrichten* No. 59, 10.3.1932, p. 8; *Rendsburger Tageblatt* No. 95, 23.4.1932, p. 16.

82 "Vermerk für Dez. 6" Regierungspräsident, Schleswig, 3.9.1932, LAS 309/22924. Also see the *Husumer Nachrichten* No. 36, 2.9.1932, p. 5; *Schleswig-Holsteinische Tageszeitung* No. 208, 4.9.1932, p. 3; *Schleswig-Holsteinischer Bauernbund* No. 36, 11.9.1932, pp. 3-4 and No. 44, 6.11.1932, pp. 1-2.

with outstanding community taxes did they actually proceed.[83] Yet compulsory enforcement, the only means of collecting the outstanding payments, had become impossible all over the province.[84] Furthermore, the movement also spread with renewed vigour over the border, as the Rural Leagues in Oldenburg and Hanover, Mecklenburg, Pomerania, Brandenburg and Silesia embraced the Crisis and Shared Fate Community idea or adopted similar measures.[85]

A surprising feature of the new wave of unrest in the countryside was the involvement of communists. Sometimes coming from the towns to take part, they were willing to support farmers' obstructionist activities in Schleswig-Holstein and in Hanover. They did so both independently and in open collaboration with other groups, including the Landvolk and NSDAP.[86] In a bid to make up for past neglect

---

83 Undated and untitled document, LAS 309/22924.

84 "Anlage zum Erlaß des Pr. Ministeriums des Innern vom 23. Februar 1932", in both LAS 301/4699 and 309/22924. See the copy of a letter to the Landrat in Ratzeburg, Juliusburg, 2.11.1931, the letter from the Lauenburgischer Land- u. Bauernbund to the Landrat in Ratzeburg, Mölln i. Lbg, 4.11.1931, the reports from the Landrat des Kreises Herzogtum Lauenburg to the Regierungspräsident in Schleswig, Ratzeburg, 21.12.1931 and the Regierungspräsident to the Preuß. Minister des Innern in Berlin concerning "Urantrag der Sozialdemokratischen Landtagsfraktion" Schleswig, 21.12.1931 and report from the Landrat to the Regierungspräsident in Schleswig, Plön, 30.7.1932, all in LAS 309/22924.

85 Schaap, *Oldenburgs Weg ins "Dritte Reich"*, p. 45; Pyta, *Dorfgemeinschaft und Parteipolitik*, pp. 217-222; report from the Regierungspräsident to the Preuß. Minister des Innern in Berlin concerning "Selbsthilfemaßnahmen der Landwirtschaft" Stade, 25.11.1931, LAS 309/22924. Also see the "Anlage zum Erlaß des Pr. Ministeriums des Innern vom 23. Februar 1932", in both LAS 301/4699 and 309/22924; *Schleswig-Holsteinischer Bauernbund* No. 47, 22.11.1931, p. 2 and No. 50, 13.12.1931, p. 4; Regierungspräsident in Stralsund to the Regierungspräsident in Schleswig, Stralsund, 13.11.1931, LAS 309/22701.

86 Communists were involved in the actions at Pöschendorf on October 19, 1931 and at Lutzhorn on October 26, at a compulsory auction in Gülzow in County Lauenburg in November 1931, at a compulsory auction in Quickborn in County Pinneberg in January 1932, at the evictions of farmers in Lütjenburg in June 1932 and Süderhastedt in South Dithmarschen in August 1932. See the "Anlage zum Erlaß des Pr. Ministeriums des Innern vom 23. Februar 1932", in both 301/4699 and 309/22924; report from the Landjägereiposten Gülzow to the Landrat in Ratzeburg concerning "Bericht eines Umzuges in Juliusburg am 24.11.1931 zwischen 13¼ und 14¼ Uhr" Gülzow, 24.11.1931, report from the Landrat des Kreises Herzogtum Lauenburg in Ratzeburg to the Regierungspräsident in Schleswig, 11.12.1931, report from the Kreisleiter der Landjägerei des Kreises Pinneberg to the Landrat in Pinneberg, Pinneberg, 27.2.1932, report from the Landrat to the Regierungspräsident in Schleswig, 18.3.1932, reports from the Amtsgericht to the Landgerichtspräsident in Kiel, Lütjenburg, 22.6.1932 and from the Landrat Plön to the Regierungspräsident in Schleswig, Plön, 22.7.1932 and report from the Regierungspräsident to the Preuß. Minister des Innern concerning "Urantrag Nr. 625, Zwangsräumung in Süderhastedt, Kreis Süderdithmarschen" Schleswig, 20.8.1932, all in LAS 309/22924. For the KPD's rural activities in Hanover, see the report from the Regierungspräsident to the Preuß. Minister des Innern in Berlin concerning "Störungen von Zwangsversteigerungen und kommunistische Agitation unter der Landbevölkerung" Lüneburg, 25.11.1931, LAS 309/22924.

and win greater support in the countryside, the KPD also directed more propaganda toward the farmers, highlighting their economic concerns and issues on which there was consensus with the right, while avoiding discussion of the party's more general aims.[87] The KPD's efforts were driven, in large part, by the former editor of *Das Landvolk*, Bruno von Salomon, and the former editor of the *Schleswig-Holsteinische Tageszeitung*, Bodo Uhse, who joined the party in late 1931.[88] In January 1932 they led the "Reich Farmers' Congress" of the KPD's Reich Farmers' League (*Reichsbauernbund*), at which it resolved to set up local rural committees. Reminiscent of the Landvolk's Crisis Assistance, Bruno von Salomon managed to establish perhaps as many as 27 of these "Farmers and Workers' Action Committees" in Schleswig-Holstein. Attracting the support of many small farmers, especially in the province's south-west, the KPD subsequently gained 33.8 percent of the vote in the Kremperheide area in the first Reichstag elections of 1932.[89]

Though communist agitation was reported to have gained the interest of many farmers at first, it was soon checked by the resistance of the right-wing parties and the Rural and Farmers' League. They denounced individuals who crossed over to the KPD and they branded communist promises as demagogic and deceitful.[90] They countered the KPD's calls for the release of Claus Heim and endeavours to allay farmers' fears of expropriation by denying that there was any connection between Heim and the party, and by pointing to Soviet Russia, warning that collective farming was what the farmers could expect from communism.[91] The KPD never enjoyed great success in rural Schleswig-Holstein, but its emergence nevertheless signified a further competitor for the Landvolkbewegung, hastening the movement's demise.

---

87 "Anlage zum Erlaß des Pr. Ministeriums des Innern vom 23. Februar 1932", in both LAS 301/4699 and 309/22924.

88 For von Salomon and Uhse's shift to the communists, see p. 309 below.

89 "Anlage zum Erlaß des Pr. Ministeriums des Innern vom 23. Februar 1932", in both LAS 301/4699 and 309/22924. Ehrenfried Friedrich and Otto-Ernst Schüddekopf both gave the figure of 27 committees in Schleswig-Holstein. There is no evidence in the files, however, that they had tens of thousands of followers, as stated in *Die Rote Fahne* cited in Friedrich's thesis. See Friedrich, *Ursachen, Verlauf und agrarhistorische Bedeutung der Landvolkbewegung*, p. 175 and Schüddekopf, *Linke Leute von rechts*, p. 310. Also see Stoltenberg, *Politische Strömungen im schleswig-holsteinischen Landvolk*, p. 175.

90 "Anlage zum Erlaß des Pr. Ministeriums des Innern vom 23. Februar 1932", in both LAS 301/4699 and 309/22924; report from the Regierungspräsident to the Preuß. Minister des Innern in Berlin concerning "Agitation unter der Landbevölkerung und Störungen von Zwangsversteigerungen" Lüneburg, 24.12.1931, both in LAS 309/22924. Also see the *Deutscher Landvolk-Kampfbund* 29.1.1932, LAS 309/22701; *Landbote für Schleswig-Holstein* No. 47, 22.11.1931, p. 3.

91 *Schleswig-Holsteinischer Bauernbund* No. 3, 24.1.1932, p. 2 and No. 5, 7.2.1932, p. 6. Also see the "Anlage zum Erlaß des Pr. Ministeriums des Innern vom 23. Februar 1932", in both LAS 301/4699 and 309/22924.

Considerably more prolific and sensational than the actions involving communists were those of the NSDAP, which worked in increasingly close cooperation with the Rural and Farmers' League. Though it had long spurned the Landvolkbewegung's tactics in its endeavours to impress upon the public that it was on a legal path to power, the party now eagerly and openly embraced all opportunities of involvement in obstructing compulsory enforcement. While the dramatic increases in support for the NSDAP will be discussed later in the chapter, it must be noted here that the party was no longer compelled to tread softly where its popularity had reached unprecedented heights. Consequently, National Socialists played an increasingly important role in the farmers' actions and readily availed themselves of violence and terror.[92]

In 1932 the party was pivotal in the planning and execution of the most spectacular actions in the province. Strangely these remarkable incidents have been largely forgotten, overshadowed perhaps by the proliferation of political violence between the left and right extremes in Schleswig-Holstein in the summer of 1932; most notably the "Bloody Sunday" riot in Altona in which 18 people died.[93] The actions included demonstrations against the eviction of a farmer in Groß-Bremsburg in April at which police repeatedly and forcefully repelled violent protesters, who continued to return to the property.[94] The farm's new administrator was boycotted and the sergeant of the local constabulary, who had assisted in blocking the protesters, received death threats and had rocks thrown through his

---

92 Report from the Landrat to the Regierungspräsident in Schleswig concerning "Zwangsversteigerung in Winnertfeld" Husum, 28.4.1932, reports from the Landjägereiamt Süderdeich to the Landrat in Heide concerning "Zwangsversteigerung Grundstück Herwig – Wesselburen" Süderdeich, 20.6.1932, from the Justizrat Theodor Voss to the Landesbank der Provinz Schleswig-Holstein, Wesselburen in Holstein, 1.7.1932 and from the Polizeibereitschaft Schleswig concerning "Streifenfahrt durch den Schutzbezirk" Schleswig, 2.7.1932, and the report from the Landjägereiamt Bordesholm to the Oberstaatsanwalt in Kiel, Bordesholm, 20.8.1932, all in LAS 309/22924.

93 For an account of the political murders and disturbances which peaked in July 1932, see W. Kopitzsch, 'Politische Gewalttaten in Schleswig-Holstein in der Endphase der Weimarer Republik' in E. Hoffmann and P. Wulf (eds), *"Wir bauen das Reich." Aufstieg und erste Herrschaftsjahre des Nationalsozialismus in Schleswig-Holstein* (Neumünster, 1983), pp. 19-39.

94 Reports concerning "Einsatz eines Schutzpolizeikommandos in Stärke von 1 Polizeioffizier und 20 Beamten zum Schutze der Zwangsräumung bei dem Landmann Max Levsen in Groß-Bremsburg" Flensburg, 8.4.1932 and Flensburg, 28.4.1932, reports from the Landrat to the Regierungspräsident in Schleswig concerning "Zwangsräumung bei dem Hofbesitzer Levsen in Groß-Bremsburg" Husum, 8.4.1932, concerning "Durchführung der Zwangsräumung in Groß-Bremsburg" Husum, 28.4.1932 and concerning "Zwangsräumung in Groß-Bremsburg" Husum, 2.5.1932, all in LAS 309/22924; report from the Regierungspräsident to the Preuß. Minister des Innern in Berlin concerning "Zwangsräumung der Generallandschaft" Schleswig, 14.4.1932, all in LAS 309/22924.

apartment windows.[95] A further notable incident was the auction of a farm in South Dithmarschen at the courthouse in Eddelak in September, where the NSDAP Reichstag Member Matthiessen (Auhof) frightened the creditors into withdrawing their application to auction the property by threatening that the crowd of 800 National Socialists gathered outside would raze the building to the ground should they not comply.[96] More dramatic still was the disturbance at a compulsory auction of a farm at Kellinghusen's county courthouse earlier that month. Rivalling the Landvolkbewegung's clash with police in Neumünster in August 1929 in both scale and intensity, the Crisis and Shared Fate Community, SA and SS mobilised a crowd 2,000 strong which came to blows with the hopelessly outnumbered local police force, before warning shots were fired causing the crowd to disperse.[97]

Not only did the agricultural organisations, NSDAP and KPD take over the Landvolkbewegung's struggle against compulsory enforcement, they also played an increasing role in the Landvolk's popular campaign on behalf of Claus Heim and his accomplices. Calls for clemency had begun during the bombers' trial and burgeoned once the convicted commenced their prison terms.[98] They appeared in the form of notices, declarations and petitions from local farmers, rural communi-

---

95 Report from the Landrat to the Regierungspräsident in Schleswig concerning "Unruhen wegen der Zwangsräumung in Groß-Bremsburg" Husum, 4.5.1932 and concerning "Vorgänge in Groß-Bremsburg" Husum, 1.6.1932 and copy of a letter to "Landjägermeister a.D. Halweg. Sherlock Holmes in Zivil, Ostenfeld" from "Die Korruptionsbekämpfer" sent from the Landrat to the Regierungspräsident in Schleswig, Husum, 6.6.1932, both in LAS 309/22924.

96 Reports dated Kiel, 22.9.1932, sent from the Schleswig-Holsteinische Generallandschaftsdirektion to the Oberpräsident in Kiel, 23.9.1932 and from the Regierungspräsident to the Preuß Minister des Innern in Berlin concerning "Zwangsversteigerung in Eddelak" Schleswig, 25.10.1932, both in LAS 301/4699.

97 Reports from the Kreisleiter der Landjägerei des Kreises Steinburg titled "Bericht über die Vorgänge in Kellinghusen am 12.9.1932" Kellinghusen, 12.9.1932, from the Landrat des Kreises Steinburg to the Regierungspräsident in Schleswig concerning "Versuche zur Verhinderung einer Zwangsversteigerung vor dem Amtsgericht in Kellinghusen" Itzehoe, 13.9.1932, from the Regierungspräsident to the Preuß. Minister des Innern in Berlin concerning "Ausschreitungen während einer Zwangsversteigerung in Kellinghusen, Kreis Steinburg" Schleswig, 19.9.1932, Regierungspräsident to the Oberlandesgerichtspräsident in Kiel concerning "Ausschreitungen während einer Zwangsversteigerung in Kellinghusen, Kreis Steinburg" Schleswig, 20.9.1932 and 14.11.1932, all in LAS 309/22924. For the impact of the incident and subsequent legal proceedings against some of the leading demonstrators, *Rendsburger Tageblatt* No. 250, 24.10.1932, p. 4 and No. 251, 25.10.1932, p. 4; reports from the Landjägerei-Inspektion III to the Regierungspräsident in Schleswig titled "Bericht über die Verhandlung vor dem Sondergericht Kiel gegen Otto Niedrof u. Genossen" Wandsbek, 28.10.1932 and from the Landrat des Kreises Steinburg concerning "Gerichtsverfahren anläßlich der Vorgänge bei der Zwangsversteigerung Uhl in Kellinghusen" Itzehoe, 29.10.1932, both in LAS 309/22924.

98 The bombers were imprisoned in Celle, Rendsburg, Flensburg, Münster and Werl in Westphalia. Heim, 'Claus Heim berichtet', LAS EI/4434, pp. 8-9.

ties and nationalist organisations, but, above all, from the Rural and Farmers' League.[99] Most prominent was the appeal of Count Kalckreuth, the president of its national organisation, the Reich Rural League, to Reich President Hindenburg.[100] The North Dithmarschen branch, however, demonstrated its solidarity with the bombers in word and deed, collecting money to provide for their families, organising assistance to bring in their harvests and even making Heim an honorary member of its board.[101]

Further calls for clemency and their release came from the political parties of the right, principally the NSDAP and DNVP, but also from the communists.[102] After Bodo Uhse and Bruno von Salomon joined the party, the KPD supported Heim's release as part of its campaign to win farmers to its cause and ultimately

99 For petitions for Heim's release from the communities of Kleve, Wesselburen, Wesseln, Hagen, Friedrichsgabekoog, Wellerhop, Weddingstedt, Ostrohe, Weddinghusen, Borgholz, Stelle-Wittenwurth and Rederstall, see LAS 309/22701; for the resolution of the community of Karolinenkoog, *Heider Anzeiger* No. 154, 4.7.1931, p. 3. For the Prussian Justice Ministry's delays in dealing with petitions for clemency from the Rural and Farmers' League, its South Dithmarschen organisation and the board of the United Nationalist Associations of Germany until after the verdict was made valid in March 1931, see Preußischer Justizminister to the Preuß. Minister des Innern concerning "Strafsache gegen Heim u. Gen. wegen Verbrechens gegen das Sprengstoffgesetz" Berlin, 9.2.1931, in both LAS 301/4703 and 309/22764. For a procession held in sympathy for Bossen and Holländer on the commencement of their sentences in December 1931 and the prohibition of a mass meeting in Heim's honour in Heide on the anniversary of the rallies of January 28, 1928, see the Oberstaatsanwalt to the Regierungspräsident in Schleswig, Altona, 28.11.1931 and 29.11.1931 and the report from the Landrat to the Regierungspräsident in Schleswig concerning "Auflösung eines Umzugs unter freiem Himmel" Niebüll, 9.12.1931, both in LAS 309/22701; report from the Landeskriminalpolizeistelle, Abteilung IA, titled "Bericht über die Landvolkbewegung" Altona, 15.7.1931, in both LAS 301/4698 and 309/22701; *Heider Anzeiger* No. 22, 27.1.1932, p. 4 and No. 23, 28.1.1932, p. 4.

100 For Kalckreuth's petition of Reich President Hindenburg in April 1931, see the *Kieler Zeitung* No. 114, 25.4.1931, p. 1; *Das Landvolk* No. 84, 28.4.1931, p. 1. For the rejection of Kalckreuth's appeal by Prussian Premier, Braun, in whose jurisdiction the question of clemency lay, see Werner, *Die Prozesse gegen die Landvolkbewegung*, pp. 219-220; *Rendsburger Tageblatt* No. 124, 30.5.1931, p. 1; *Kieler Zeitung* No. 148, 30.5.1931, p. 1. Prussian Interior Minister Severing's later claim to have supported a shortening of Heim's sentence is at odds with these reports, see C. Severing, *Mein Lebensweg*, vol. 2 (Cologne, 1950), p. 311.

101 Report from the Regierungspräsident to the Minister für Landwirtschaft, Domänen und Forsten in Berlin concerning "Land- und Bauernbund" Schleswig, 28.11.1930, LAS 301/4702; report from the Regierungspräsident to the Preuß. Minister des Innern in Berlin concerning "Stimmung in der ländlichen Bevölkerung" Schleswig, 29.9.1931, LAS 301/4698; *Schleswiger Nachrichten* No. 273, 21.11.1930, p. 3; *Heider Anzeiger* No. 141, 19.6.1931, p. 16; *Schleswig-Holsteinischer Bauernbund* No. 18, 3.5.1931, p. 4, No. 23, 7.6.1931, p. 4, No. 14, 10.4.1932, p. 2 and No. 17, 1.5.1932, p. 5, No. 47, 22.11.1931, p. 4.

102 *Schleswig-Holsteinische Tageszeitung* No. 250, 23.10.1930, p. 2; *Völkischer Beobachter* No. 163, 12.6.1931, LAS 309/22764; *Kieler Zeitung* No. 141, 22.5.1931, p. 2.

hoped to gain him as a candidate.[103] Though Heim maintained amicable relations with all correspondents and visitors in prison in Celle (Hanover), including representatives of the KPD and NSDAP, not for a moment did he consider joining them, nor would he be drawn into any party-political matters.[104] He was expressly open to cooperation with all groups, but insisted on the Landvolkbewegung's independence.[105]

For all the talk in right-wing circles of the bombers' ill-treatment in jail, Heim was a contented prisoner, more concerned for his older collaborators, Bossen and Holländer, and with ensuring that unemployed labourers found work on his farm in his absence, than he was with his own welfare.[106] He and most of his companions proudly refused all offers of assistance.[107] Apparently happy to bide his time, Heim planned to revive the Landvolkbewegung on his release. Although aware of the NSDAP's advance in the countryside, he was convinced the party would fail once

103 For the KPD's efforts, including a delegation to Justice Minister Schmidt in June 1931 and Reichstag Member Ernst Putz's visit of Heim in Celle, see the *Norddeutsche Zeitung* No. 84, 4.6.1931 and No. 87, 8.6.1931, LAS 309/22764; report from the Regierungspräsident to the Preuß. Minister des Innern in Berlin concerning "Störungen von Zwangsversteigerungen und kommunistische Agitation unter der Landbevölkerung" Lüneburg, 8.12.1931, LAS 309/22924; "Anlage zum Erlaß des Pr. Ministeriums des Innern vom 23. Februar 1932", in both LAS 301/4699 and 309/22924.

104 Report concerning "Überwachung der Landvolkbewegung" Hanover, 19.1.1932, LAS 309/22701; letter from Claus Heim to Johannes Rathjen, Celle, 2.2.1932, StaNms MA 5360. Heim's neutrality was echoed in Becker's correspondence. See the letters from Ernst Becker to the Stahlhelm "Gau Lüneburg" Lüneburg, Celle, 14.10.1931 and to Otfried Rademacher in Itzehoe, Celle, 29.10.1931, both in LAS 309/22701. Also see the report from the Polizeipräsident in Altona concerning "In Celle inhaftierte Landvolk-Aktivisten" 19.6.1931 in Stegmann, *Politische Radikalisierung in der Provinz*, pp. 244-245.

105 See the transcript of Heim's discussion with the SA leader of East Hanover, Hasse. Report signed "Krim. Ass. Peters" Hanover, 14.11.1931, LAS 309/22701.

106 Richard Schapke described the chicanery in detail. Schapke, *Aufstand der Bauern*, pp. 73-75. Heim later remembered his experiences in prison with good humour. See Heim 'Claus Heim berichtet', LAS EI/4434, pp. 9-13. For Heim's anger at the Prussian Interior Minister, Severing, for refusing Bossen and Holländer parole, see the letters from Claus Heim to Margarete Heim, Celle, 6.12.1931 and to Hermann Kröger in Neumünster, Celle, 6.12.1931, both in LAS 309/22701 and Severing, *Mein Lebensweg*, vol. 2, p. 310. For Heim's efforts to provide work for the unemployed, see the report signed "Krim. Ass. Peters", Hanover, 14.11.1931, LAS 309/22701 and Thyssen, *Bauer und Standesvertretung.*, p. 410.

107 Report from the Strafanstaltsdirektor der Strafanstalt Celle to the Landrat des Kreises Celle concerning "Besuch des Bombenattentäters Claus Heim" Celle, 23.9.1931, report signed "Krim. Ass. Peters", Hanover, 14.11.1931, letter from Claus Heim to Carla Allihn in Berlin, Celle, 3.12.1931 and the copy of an untitled report from the Strafanstaltsinspektor in Celle, Celle, 26.9.1931, all in LAS 309/22701.

it participated in government, and then, or so he believed, the Landvolk's time would follow.[108]

In any event, the bombers could afford to be patient when others had taken it upon themselves to gain their release. The KPD's endeavours culminated with a motion in the Prussian Landtag which failed to pass through the legal committee in January 1932.[109] Diverse groups and individuals then came together to form the "Claus Heim Komitee." Established at the beginning of 1932, the "Komitee" comprised Landvolk leaders from across the Reich; communists such as Uhse and Ernst Putz; the leader of the "Black Reichswehr", Buchrucker; the National Bolshevist, Niekisch; other publicists and authors, principally conservative revolutionaries, Ernst Jünger, Hans Franck, Hans Zehrer and Albert Erich Günther; former National Socialists, Otto Strasser, Stennes and Dinter; the leader of the Christian Social movement, Vitus Heller; and the president of the Prussian Chamber of Agriculture, Brandes. On Niekisch's initiative, the "Komitee" sought to gain support for Heim's candidacy at the Reich President elections, seeing in him a means of unifying various rebellious political interests. A list with 25,000 signatures was collected within eight days. Yet discouraged by Uhse, who wished to prevent competition with Ernst Thälmann, Heim declined, though he acknowledged the plan's object in mustering revolutionary forces from the right and left.[110]

The primary undertaking of the "Komitee" was to promote support for the bombers among members of the Prussian Landtag.[111] On June 13, 1932 the Landtag's legal committee accepted a draft of the NSDAP and DNVP supported by the KPD for the remission of penalty of the convicted.[112] Though each saluted Heim as a symbol of opposition in their own way, the NSDAP and KPD united on June 16 when the Landtag passed the motion.[113] The KPD refused to support Volck's release, possibly due to rumours he had participated in atrocities against communists

---

108 See the letter from Carl Baade to Claus Heim in Celle, Berlin, 6.1.1932 and the transcript of Heim's discussion with Uhse in the report concerning "Überwachung der Landvolkbewegung" Hanover, 19.1.1932, LAS 309/22701; letter from Claus Heim to Johannes Rathjen, Celle, 2.2.1932, StaNms MA 5360.

109 "Urantrag Nr. 7661" in 'Sammlung der Drucksachen des Preußischen Landtags, 3. Wahlperiode 1. Tagung begonnen am 8. Juni 1928', vol. 15 (Berlin, 1932), p. 7606; *Heider Anzeiger* No. 11, 14.1.1932, p. 9.

110 Letter from Heim cited in the *Heider Anzeiger* No. 72, 26.3.1932, p. 4; Schapke, *Aufstand der Bauern*, pp. 77-78; Niekisch, *Erinnerungen eines deutschen Revolutionärs*, vol. 1, pp. 168-169. Also see Schüddekopf, *Linke Leute von rechts*, pp. 310-311.

111 Heim, 'Claus Heim berichtet', LAS EI/4434, p. 13.

112 *Kieler Neueste Nachrichten* No. 137, 14.6.1932, p. 3.

113 'Preußischer Landtag 9. Sitzung am 16. Juni 1932' in the 'Sitzungsberichte des Preußischen Landtags, 4. Wahlperiode 1. Tagung begonnen am 24. Mai 1932' (Berlin, 1932), pp. 468-481; *Kieler Neueste Nachrichten* No. 141, 18.6.1932, pp. 1-2; *Schleswig-Holsteinische Volkszeitung* No. 140, 17.6.1932, p. 13; *Rendsburger Tageblatt* No. 140, 17.6.1932, p. 4; *Schleswig-Holsteinische Tageszeitung* No. 140, 17.6.1932, pp. 1-2 and No. 141, 18.6.1932, p. 3.

in the Russian Civil War, but all the other bombers yet to complete their sentences were freed the following day.[114] Finally, on March 21, 1933, the day of Hitler and Hindenburg's symbolic meeting in Potsdam, the bombers were granted a full amnesty by Reich President decree.[115]

A number of factors explain the Landvolkbewegung's long and fitful decline. First among these were the detrimental effects of the bombing campaign and its aftermath. Heim's dispute with Hamkens over the methods best suited to achieving their goals created a rift within the movement when the arrests exposed the movement's responsibility. Further, as discussed in the previous chapter, Hamkens lost many supporters who saw in his police statements a betrayal of his companions in complete contradiction of his own vociferous demands for solidarity. As a result, the bombings damaged the Landvolkbewegung even more than if the arrests had simply deprived it of many of its most active leaders.

The bombings also frightened away many supporters, despite having met with much approval among the rural population at the time, and despite extensive sympathy for the bombers after their arrests. This was hardly surprising when fearing a public backlash, Hamkens and other Landvolk leaders themselves condemned the attacks as acts of terror, and as reckless and criminal. Yet the bombings also led to a decline in support because they disposed the rural people to question the movement's direction, the efficacy of its methods and its capacity to improve their lot.[116] Some observers had already predicted that the movement would find it difficult to retain support if it did not procure advantages for the farmers.[117] Indeed, from as early as the spring of 1929 there had been signs of doubts emerging over its capacity to effect relief.[118]

Closely related to the above was the growing disapproval of the movement's lack of a programme. Such disapproval was discernible across the political spectrum, from the Social Democrats to the National Socialists.[119]Additionally, provin-

114 Oberstaatsanwalt to the Regierungspräsident in Schleswig, Altona, 20.6.1932, in both LAS 301/4703 and 309/22701; Oberstaatsanwalt to the Oberpräsident in Kiel, Altona, 22.6.1932, LAS 301/4703; *Rendsburger Tageblatt* No. 142, 20.6.1932, p. 2; *Schleswig-Holsteinische Tageszeitung* No. 148, 26.6.1932, p. 3; Volck, *Rebellen um Ehre*, pp. 459-464.

115 Oberstaatsanwalt to the Oberpräsident in Kiel, Altona, 10.4.1933, LAS 301/4703. Also see J. Christoph, *Die politischen Reichsamnestien 1918-1933* (Frankfurt/M, 1988), pp. 358-366.

116 See chapter 7.

117 See, for example, *Der Schleswiger* No. 183, 6.8.1929, LAS 301/4701.

118 Report from the Landeskriminalpolizeiamt, Abteilung IA, Berlin, 2.4.1929, LAS 309/22668. For a rare early report of a farmer abandoning the Landvolkbewegung due to its failure to deliver financial assistance, see the report titled "Bericht" Itzehoe, 5.6.1929, LAS 301/4704.

119 For derision over the Landvolk's lack of a programme, see the report on a Landvolk meeting in Groß-Harrie in the *Schleswig-Holsteinische Volkszeitung* No. 244, 18.10.1929, LAS 301/4696. For National Socialist criticism see, for example, the report of a meeting in Burg in Dithmarschen in *Das Landvolk* No. 161, 4.10.1929, p. 5.

cial newspapers, such as *Der Schleswiger*, and defiant opponents, such as the leader of the Farmers' Association, Willy Iversen, were prepared to criticise the Landvolk for failing to explain its "practical" and "positive" aims, and even accuse Hamkens of simply "carrying on".[120]

Recognising the strength of these objections, the Landvolkbewegung countered by emphasising the absence of a programme as desirable, that in its place it had an established goal: its struggle against international capital and the parliamentary system. Programmes, it was argued, caused disunity because political parties simply squandered their energy by stressing their differences to one another while the farmers' situation continued to deteriorate.[121]

Nevertheless, criticisms of its deficient political purpose and fear of losing support appear to have spurred the Landvolk to make greater efforts to highlight its aims toward the end of 1929. At the Landesthing, the gathering of the tribes in Rendsburg in November, Hamkens read out the movement's aims which were also disseminated in the form of a postcard. While the calls for the preservation of the farmers' properties were already well known, and its opposition to international capital and refusal to cooperate with "the system" as strident and forthright as ever, what distinguished this from earlier declarations was the relatively clearer formulation of what should replace the "Jewish parliamentary system".[122] At the Landesthing, the Landvolk called for a state based on a "national foundation" in which precedence was accorded to attributes such as "German ownership", "national character" and "German culture and honour", "blood and race" , the "German family" and a firm attachment to one's home.[123]

---

120 *Der Schleswiger* No. 156, 5.7.1929, LAS 301/4695. For Iversen's criticisms at meetings in Neukrug on June 7, 1929 and at Sterup and Sterupbek on August 8, 1929, *Flensburger Nachrichten* No. 132, 8.6.1929, p. 7 and the report titled "Versammlungsbericht von Sterup und Sterupbek" sent from the Landeskriminalpolizeistelle Flensburg to the Regierungspräsident in Schleswig, Flensburg, 24.8.1929, LAS 309/22668.

121 *Das Landvolk* No. 232, 29.12.1929, p. 1, No. 233, 31.12.1929, p. 1 and No. 40, 16.2.1930, p. 1. Also see von Lamezan's speech at a meeting in Glückstadt, report titled "Bericht" concerning "Versammlung des Landvolkbundes im hiesigen Lokal Franscher Garten" Glückstadt, 14.12.1929, in both LAS 301/4697 and 309/22783.

122 The preservation of the farmers' properties and opposition to international capital and "the system" were also encapsulated in a leaflet distributed in June 1929, which the Schöffengericht at the Great Landvolk Trial in the summer of 1930 took as representative of the Landvolk's aims. See the flyer titled "SOS Schleswig-Holsteins Kämpfer ringen um Heimat und Scholle", appended to the report from the Polizeipräsident to the Oberpräsident in Kiel concerning "Landvolkbewegung, Versammlung am 27. Juni d. Js. in der 'Deutschen Wacht' " Kiel, 1.7.1929, in both LAS 301/4695 and 309/22668; issue of the verdict of the Schöffengericht Itzehoe from 9.8.1930, pp. 8-10, LAS 301/4700.

123 Transcript of the Landesthing from the Polizeiverwaltung in Rendsburg titled "Landesthing des Schleswig-Holsteinischen Landvolkes, am 10. November 1929 in Rendsburg im 'Schützenhof' nachmittags 3 Uhr" Rendsburg, 12.11.1929 and accompanying postcard, in both LAS 301/4696 and 309/22590. Also see the *Nordischer Kurier* No. 266, 12.11.1929, p.

This concept of a Volksgemeinschaft, though the most specific the movement ever articulated, was still very amorphous. Of course it denoted the "German people" as an ethnic entity that defined the nation's borders, implied the revision of the Treaty of Versailles and elevated the "German people" above divisive economic and political interests. However, it did not even hint at what form a future German state should take. There can be no doubt that the Landvolk meant it to be strong and authoritarian. This was evident from remarks made during various other meetings and in *Das Landvolk.* It is also to be inferred from praise of fascism in Italy and calls for a leader like Mussolini as well as avowals of abhorrence of the "administrative apparatus" but love of the nation.[124] Furthermore, the Landvolk was critical of parliamentary democracy because of its failure to rise above party and interest-group politics. Rather than a liberal, pluralistic society, the Landvolkbewegung called for an authoritarian presence – of indeterminate form – to unify German aspirations, resolve conflicts and determine policy.[125] Yet the Landvolk could offer no clearer vision than this.

The absence of a coherent programme, or at least more specific intermediate objectives, led to increasing disillusionment among the rural population. Even at the Landesthing, where an express effort was made to clarify its aims, a number of young NSDAP and Stahlhelm members in uniform openly voiced criticism of the movement's lack of purpose.[126] This view proliferated in the countryside in the first months of 1930.[127] There were a number of reports that Landvolk audiences were

---

7. The postcards were distributed long afterwards. See the report from the Polizeipräsident Altona, Abteilung IA, concerning "Landvolkbewegung" Altona, 27.5.1930, in both LAS 301/4697 and 309/22590.

124 For this sentiment as expressed by Hamkens, Peter Petersen (Ahrenholz), Johannes Kühl and Major a. D. Koopmann, see *Der Schleswiger* No. 46, 24.2.1930, in both LAS 301/4697 and 309/22783; *Ostholsteinisches Tageblatt* No. 48, 26.2.1929, LAS 301/4695; *Das Landvolk* No. 138, 7.9.1929, p. 1; untitled meeting report, Schleswig, 30.8.1929, LAS 309/22668; report from the Landeskriminalpolizeistelle IA titled "Bericht" concerning "Landvolkkundgebung am 30.10.29 in Leezen" Kiel, 31.10.1929, LAS 309/22783; letter from Major a. D. Koopmann to the Oberpräsident in Kiel concerning "Aufhebung des Verbotes einer Kundgebung der Landvolkbewegung am 10.11.29 in Rendsburg" Hohn, 26.10.1929, LAS 301/4697. For further praise of Mussolini, see *Das Landvolk* No. 92, 18.4.1930, p. 9.

125 *Das Landvolk* No. 233, 31.12.1929, p. 1 and No. 51, 1.3.1930, p. 1.

126 Report titled "Bericht" and concerning "Landesthing in Rendsburg" Altona, 11.11.1929, sent from the Polizeipräsident, Landeskriminalpolizeiamt, Abteilung IA, to the Regierungspräsident in Schleswig, both in LAS 309/22783.

127 Report from the Polizeipräsident, Landeskriminalpolizeistelle, to the Polizeipräsident, Landeskriminalpolizeiamt Berlin, concerning "Landvolkbewegung" Flensburg, 14.1.1930, in both LAS 301/4697 and 309/22783. For the Rural and Farmers' League meeting in Plön County, see the report from the Landrat to the Regierungspräsident in Schleswig concerning "Bewegung in der Landwirtschaft" Plön, 13.3.1930 and the appended copy of the *Preetzer Zeitung* No. 53, 4.3.1930, LAS 309/22765.

tiring of repetitive speeches and stagnant, negative criticism, and that the continued avowals of opposition to the "administrative apparatus" were insufficient to retain support without a positive programme.[128]

The movement's decline was greatly accelerated by the imprisonment of so many of its leaders. Indeed, it would be difficult to exaggerate the significance of their absence. From the beginning, the Landvolkbewegung was dependent on a small number of leaders for its development and momentum,[129] and since it had no organisational structure, their role was even more important. Though the Landvolk had claimed its rejection of organisation was a strength, maintaining that it demanded greater initiative and vitality from all of its followers, it was to discover it had little to fall back on in the leaders' absence.

With most of the bombers already in remand and then in jail, Hamkens and Weschke's imprisonment in the summer of 1930 proved disastrous. Hamkens' appearances at meetings attracted particularly large crowds, and since entry fees typically raised RM 1 per head, his popularity as a speaker meant that he was, in effect, a good source of income for the movement.[130] The imprisonment of the leaders, therefore, not only deprived the movement of speakers and organisers, but was also detrimental to its finances.

Differences between various Landvolk leaders, quarrels over their methods and ideas, and the ensuing struggles for influence did further injury to the movement's fortunes, particularly from the second half of 1930. One of the most significant sources of friction was the newspaper, *Das Landvolk*. This was evident in the spring of 1930 when Hamkens and Heim's supporters vied for control. They quarrelled, among other things, over the reinstatement of Bruno von Salomon as editor-in-chief, as Heim's followers desired a more militant style of reporting.[131] At Luetgebrune's behest, however, they agreed to put aside their differences and cooperate until the close of the bombers' trial.[132]

---

128 Reports from the Polizeipräsident to the Polizeipräsident, Landeskriminalpolizeiamt Berlin, concerning "Landvolkbewegung" Kiel, 16.1.1930 and from the Landrat to the Regierungspräsident in Schleswig concerning "Landvolkbewegung" Heide, 27.1.1930, both in LAS 309/22783; report from the Regierungspräsident to the Oberpräsident in Kiel concerning "Landvolkbewegung" Schleswig, 4.2.1930, in both LAS 301/4697 and 309/22783.

129 See chapter 3, pp. 116-117.

130 Report from the Polizeipräsident, Landeskriminalpolizeistelle, to the Regierungspräsident in Schleswig concerning "Landvolkversammlung in Flensburg" Flensburg, 24.2.1930, LAS 309/22783; report from the Regierungspräsident in Schleswig to the Generalstaatsanwalt in Kiel, Schleswig, 27.2.1930, LAS 301/4697.

131 Report from the Regierungspräsident to the Preuß. Minister des Innern in Berlin concerning "Landvolkbewegung" Schleswig, 12.6.1930 and reports from the Landeskriminalpolizeistelle Altona, Abteilung IA, the first concerning "Landvolkbewegung" Altona, 27.5.1930, and the second titled "Bericht" Altona, 21.6.1930, all in both LAS 301/4697 and 309/22590.

132 Report from the Landeskriminalpolizeistelle Altona, Abteilung IA, titled "Bericht" Altona, 21.6.1930 and report from the Polizeipräsident, Landeskriminalpolizeistelle, to the Regier-

Tensions at *Das Landvolk* persisted, however, with the emergence of a contingent of Tannenbergbund members within the movement.[133] Well known for its leader, Ludendorff, and its fanatical propaganda against Jews, Freemasons and Jesuits, the Tannenbergbund had three members among the paper's 36 shareholders: Wulf, Fehrs (Drage) and Hamkens' brother-in-law, Hönck. After Hönck took over from Hamkens and Johnsen as the paper's business administrator in September 1929, his allegiance to the Tannenbergbund became clear. He began to press Kühl to print articles from Ludendorff's *Volkswarte* as leading articles in *Das Landvolk*. Kühl's objections led to serious disagreements between them.[134] Despite being initially unsuccessful, the three Tannenbergbund members were determined to propagate their movement's ideas and, at a general meeting of the newspaper's shareholders on June 4, 1930, advocated a member of their organisation's board for the position of editor-in-chief. Their suggestions were rejected, Hönck was called upon to resign as business administrator and two others were appointed to manage the paper's financial affairs in his place.[135]

Rather than accept the decision, Hönck, Fehrs and Wulf – the last able to exploit his position acting as chairman during Hamkens' imprisonment – staged what was effectively a coup, calling a further meeting in the Tannenbergbund stronghold of Nortorf. Stacking the meeting in their favour, even by resorting to blocking the entry of some shareholders, they passed their resolutions successfully. However, their subsequent attempts to dismiss Pramor and the editors, Winter and Kühl, who refused to comply with their orders, led to a series of squabbles and a brawl at the paper's offices in Itzehoe.[136]

---

ungspräsident in Schleswig concerning "Landvolkbewegung" Flensburg, 12.7.1930, in both LAS 301/4697 and 309/22590; report from the Regierungspräsident to the Minister des Innern in Berlin concerning "Landvolkbewegung" Schleswig, 26.7.1930 and report from the Polizeiverwaltung, Itzehoe, 18.8.1930, both in LAS 301/4697.

133 Report from the Landeskriminalpolizeistelle Altona, Abteilung IA, concerning the "Landvolkbewegung" Itzehoe, 6.11.1929, in both LAS 301/4697 and 309/22783; reports from the Polizeipräsident, Abteilung IA, concerning "Landvolkbewegung" Altona, 27.5.1930, from the Regierungspräsident to the Preuß. Minister des Innern in Berlin concerning "Landvolkbewegung" Schleswig, 12.6.1930 and from the Polizeipräsident, Landeskriminalpolizeistelle, to the Regierungspräsident in Schleswig concerning "Landvolkbewegung" Flensburg, 16.6.1930, all in both LAS 301/4697 and 309/22590.

134 Report from the Polizeiverwaltung, Itzehoe, 18.8.1930, LAS 301/4697.

135 Report from the Landeskriminalpolizeistelle Altona, Abteilung IA, titled "Bericht" Altona, 21.6.1930, in both LAS 301/4697 and 309/22590; reports from the Polizeiverwaltung, Itzehoe, 18.8.1930 and from the Regierungspräsident to the Oberpräsident in Kiel concerning "Landvolkbewegung" Schleswig, 1.9.1930, both in LAS 301/4697.

136 Reports from the Landeskriminalpolizeistelle Altona, Abteilung IA, titled "Bericht über die Landvolkbewegung" Altona, 21.7.1930, from the Regierungspräsident to the Minister des Innern in Berlin concerning "Landvolkbewegung" Schleswig, 26.7.1930, from the Polizeiverwaltung, Itzehoe, 18.8.1930 and from the Regierungspräsident to the Oberpräsident in Kiel, Schleswig, 19.8.1930, all in LAS 301/4697. Also see the reports in the *Nordischer*

Only after Hamkens' release, and an extra general meeting in Heide on September 7, was the quarrel settled. The Tannenbergbund followers were swiftly isolated and Hönck and Wulf were replaced as shareholders. Yet the meeting did not pass without fierce debate over Hönck and Wulf's actions and between Hönck and Hamkens over the former's suspected misappropriation of donations, underlining just how intense their personal, professional and political differences had become.[137]

Landvolk leaders endeavoured to contain discussion about the discord within the movement, but it had already become subject matter for the wider press and it had also spilled over into a further bitter dispute between Luetgebrune and Wulf and Hönck.[138] Preferring to stand by their actions at the Great Landvolk trial, they turned the movement's own language against its luminary lawyer, objecting to what they branded as Luetgebrune's wrangling over points of "Jewish-Roman law".[139] As a consequence of these quarrels, the Landvolkbewegung lost two stalwart leaders and many followers as well. By the end of August 1930, the disputes were reported to have cost *Das Landvolk* 700 subscribers.[140]

The Landvolkbewegung was not alone in experiencing internal strife in Schleswig-Holstein at this time. The NSDAP was in turmoil following the secession of two revolutionary factions from the party in May 1930: the "National Socialist Working Group Dithmarschen" led by a veterinarian, Emil Grantz (Albersdorf),[141] and Otto Strasser's "Action Group of Revolutionary National

Kurier No. 169, 21.7.1930, p. 9 and No. 171, 23.7.1930, p. 4; *Schleswig-Holsteinische Tageszeitung* No. 171, 23.7.1930, p. 6. For the advertisement for the meeting in Heide, see *Das Landvolk* No. 179, 1.8.1930, p. 2.

137 Report from the Landeskriminalpolizeistelle Altona, Abteilung IA, titled "Bericht über die Landvolkbewegung" Altona, 15.9.1930, LAS 301/4697.

138 For the attempts of Kühl and Schwarzloh at a meeting in Bordesholm to contain discussion about Wulf and *Das Landvolk*, see the report from the Landjägereiamt Bordesholm to the Landjägereiabteilung Bordesholm, Bordesholm, 27.7.1930, LAS 309/22590. For discussion in the press of the discord within the movement and of its decline, see *Der Schleswiger* No. 188, 14.8.1930 and the *Hamburger Fremdenblatt* No. 193, 14.7.1930, both in LAS 309/22590.

139 Report from the Regierungspräsident to the Minister des Innern in Berlin concerning "Landvolkbewegung" Schleswig, 26.7.1930, LAS 301/4697; *Nordischer Kurier* No. 170, 22.7.1930, p. 9 and No. 171, 23.7.1929, p. 4. Also see von Bodungen's criticisms as reported in the *Rendsburger Tageblatt* No. 174, 28.7.1930, p. 4.

140 Report from the Landeskriminalpolizeistelle Altona-Wandsbek, Abteilung IA , titled "Bericht über die Landvolkbewegung" Altona, 25.8.1930, LAS 309/22590.

141 For the struggle between Gauleiter Lohse and SA leader Dr. Grantz which centred on the party's parliamentary aspirations and led to the secession of some 800 members, see the "Manifest" Albersdorf, 28.5.1930 and the report from the Polizeipräsident in Kiel to the Polizeipräsident, Landeskriminalpolizeiamt in Berlin, Kiel, 11.7.1930, both in LAS 309/22576; Uhse *Söldner und Soldat*, pp. 263-264.

Socialists" or "Black Front ".[142] Both groups were opposed to the NSDAP's parliamentary course and, preferring greater activism, were quick to collaborate with the Landvolk.[143] Foremost among Strasser's faction, Bodo Uhse was expelled from the NSDAP for breaches of discipline.[144] Having already endeavoured to initiate closer cooperation between the Landvolk and the NSDAP a year earlier, once free of party restraints, he openly expressed his enthusiasm for the movement and advocated combating "the system" with all means possible.[145] Together with Bruno von Salomon, he travelled extensively throughout Schleswig-Holstein both promoting Strasser's Action Group and rallying support for Heim and the bombers.[146]

These developments initially promised to benefit the Landvolkbewegung. However, like Uhse, von Salomon was out of work in 1930. With his patron, Claus

142 For the formation at the end of May 1930 of the "Action Group of Revolutionary National Socialists" (*Kampfgemeinschaft revolutionärer Nationalsozialisten*) or "Black Front" (*Schwarze Front*), see the copy of a report titled "Kurzer Bericht über die zur Gründung der GSRN in der Zeit vom 28. bis 31.5.1930 abgehaltenen Tagung in Berlin" Kiel, 8.7.1930, copy of a dispatch from the "Gruppe sozialrevolutionärer Nationalisten" titled "4. Rundschreiben" and dated "Anfang Juni 1930" sent from the Polizeipräsident Altona-Wandsbek in Altona to the Regierungspräsident in Schleswig, 10.7.1930, excerpts from the "Mitteilungen des Landeskriminalpolizeiamts (IA) Berlin vom 15. Juli 1930 No. 14" titled "Rechtsbewegung" and Preuß. Minister des Innern to the Oberpräsident, Regierungspräsident and the Polizeipräsident in Berlin, Berlin, 29.7.1930, all in LAS 309/22576. Also see the extracts of correspondence, particularly between Bruno von Salomon and Friedrich Wilhelm Heinz in the lists of material confiscated from the offices of *Das Landvolk* and residences of its staff in both LAS 301/4704 and 309/22701.

143 Report from the Polizeipräsident Altona-Wandsbek in Altona to the Regierungspräsident in Schleswig concerning "Kampfgemeinschaft revolutionärer Nationalsozialisten" Altona, 25.10.1930 and the excerpts from the "Mitteilungen des Landeskriminalpolizeiamts (IA) Berlin vom 15. Juli 1930 No. 14", both in LAS 309/22576; report from the Landeskriminalpolizeistelle Altona, Abteilung IA, titled "Bericht über die Landvolkbewegung" Altona, 15.9.1930, LAS 301/4697. Also see Meinl, *Nationalsozialisten gegen Hitler*, pp. 148-149.

144 For Uhse's expulsion from the party, see the *Rendsburger Tageblatt* No. 165, 17.7.1930, p. 2. In *Söldner und Soldat*, Uhse's recollection of his expulsion from the party and subsequent unemployment in 1929 was too early. See Uhse, *Söldner und Soldat*, pp. 298, 308 and 313. Also see P. D. Stachura, *Gregor Strasser and the Rise of Nazism* (London, 1983), pp. 67-81.

145 *Nordischer Kurier* No. 186, 9.8.1930, p. 15.

146 Report from the Landeskriminalpolizeistelle, Abteilung IA, Altona-Wandsbek, titled "Bericht über die Landvolkbewegung" Altona, 25.8.1930, LAS 309/22590. For the Action Group's publication of articles by Rademacher and Plaas, glorifying the Landvolk's bombings, that led to a month's ban of its newspaper, see the report from the Polizeipräsident, Landeskriminalpolizeistelle, to the Regierungspräsident in Schleswig concerning "Kampftruppe der revolutionären Nationalsozialisten" Flensburg, 8.10.1930 and Polizeipräsident, Abteilung IA, to the "Verlag und den Herausgeber der Zeitung 'Der nationale Sozialist' Berlin" and to the "Redaktion der Zeitung 'Der Nationale Sozialist' Berlin", *Der Nationale Sozialist* Folge 200, 15.11.1930 and the issue of the verdict from the "4. Strafsenat des Reichgerichts vom 31.3.31", all in LAS 309/22576.

Heim, in custody, and dogged by claims that his wife had earned a share of the reward for providing information which led to the bombers' apprehension, he could not return to *Das Landvolk*.[147] Otto Strasser's Action Group also proved a disappointment. Farmers in Schleswig-Holstein showed just as little interest in it as they did in Grantz's Working Group.[148] In the face of failure, von Salomon and Uhse transferred their hopes to the KPD. Effectively, this spelled their departure from the Landvolkbewegung. In June 1931 von Salomon first praised the communists' agricultural programme and by October he and Uhse had begun to refer to themselves as "National Communists".[149]

Kenstler's endeavours to establish a new *Großdeutsche Bauern- und Landvolkbewegung* (Greater German Farmers' and Rural People's Movement) from Thuringia dealt the movement a further blow. In a manifesto drafted in early 1931, Kenstler appealed to rural Germany to join a revolution for "blood and soil" . Going against the Landvolk's resolve to avoid a programme, the manifesto comprised a series of demands, codifying and adding to many of the movement's ideas with a marked anti-urban tone.[150] In opposition to Hamkens, the Großdeutsche Bauern- und Landvolkbewegung claimed to support Claus Heim and even took his family's coat of arms as its symbol, though there is no evidence that Heim agreed to this. By and large, the new movement found few supporters either in Schleswig-Holstein or Thuringia. It did, however, draw away some dissatisfied leaders from the Land-

---

147 Report from the Landeskriminalpolizeistelle Altona-Wandsbek, Abteilung IA, concerning "Sprengstoffanschläge" Itzehoe, 20.3.1930, in both LAS 301/4697 and 309/22590.

148 For Hitler's rejection of Grantz's Working Group, see the *Schleswig-Holsteinische Tageszeitung* No. 145, 22.6.1930, p. 1 and for Grantz's expulsion from the party, the report from the Landrat to the Regierungspräsident in Schleswig concerning "Landvolkversammlungen in Meldorf und Marne" Meldorf, 28.6.1930, LAS 309/22590. Also see the report from the Landrat to the Regierungspräsident in Schleswig, Meldorf, 6.10.1920 and the excerpts from the "Mitteilungen des Landeskriminalpolizeiamts IA Berlin vom 15.11.30 Nr. 22", LAS 309/22576 and *Rendsburger Tageblatt* No. 163, 15.7.1930, pp. 1-2. For the failure of Strasser's Action Group in Schleswig-Holstein, see the report from the Polizeipräsident, Landeskriminalpolizeistelle, to the Regierungspräsident in Schleswig concerning "Kampftruppe der revolutionären Nationalsozialisten" Flensburg, 8.10.1930, reports from the Polizeipräsident to the Regierungspräsident in Schleswig concerning "Kampfgemeinschaft revolutionärer Nationalsozialisten" Kiel, 6.11.1930 and Kiel, 27.2.1931 and excerpts from the "Mitteilungen des Landeskriminalpolizeiamts (IA) Berlin vom 1.6.1931 Nr. 11" and from the "Mitteilungen des Landeskriminalpolizeiamts IA (Berlin) vom 1. September 1931 Nr. 17", all in LAS 309/22576.

149 *Flensborg Avis* No. 128, 5.6.1931, LAS 309/22701; report from the Polizeipräsident in Berlin to the Regierungspräsident in Schleswig, Berlin, 22.10.1931, LAS 309/22576.

150 "Das nationalrevolutionäre Manifest der deutschen Bauernschaft", in both LAS 301/4698 and 309/22701. For Doepner's rejection of the programme in a letter to Hamkens, see the report from the Landeskriminalpolizeistelle, Abteilung IA, titled "Bericht über die Landvolkbewegung" Altona, 20.5.1931, LAS 309/22701.

volkbewegung, namely, Johannes Tramsen (Groß-Quern), Lorenz Lorenzen (Kathrinenherd) and Heinrich Frauen (Hemme).[151]

The final notable fracture within the movement came as a result of the struggle between Hamkens and Pramor to establish a successor to *Das Landvolk* in the autumn of 1931. Even before its liquidation, Pramor had fought against attempts to remove him as publisher. Much embittered over the sacrifices he had made on behalf of the movement, he felt Hamkens had abandoned him, leaving only debts, by establishing his own successor newspaper in Husum.[152] Both claiming theirs as the true successor to *Das Landvolk*, Hamkens and Pramor squabbled over the old masthead and competed for a readership. The NSDAP took advantage of the conflict, blaming Hamkens for the "chaos" and boasting of its own unity and "crystal clear objectives" in order to poach what remained of the movement's following.[153]

A quite different cause of the Landvolkbewegung's decline was rumours of financial irregularities. Already in 1929 there were growing misgivings among the rural population that too many people in the movement were seeking to "live off the dumb farmers' money".[154] Volck and Nickels, in particular, were discovered to have drawn sizeable incomes from Landvolk funds.[155]

Rumours of misappropriation began to damage the movement in mid 1930. One farmer was said to have taken some RM 3,800 from Volksmacht collections for private use and Heim's supporters also believed that Hamkens himself had embezzled funds. However, they themselves did not appear above suspicion, for police reported that Hennings and Wiborg were so successful in their collections for the bombers' trial that they were able to afford excursions to the beach with their families.[156]

---

151 Reports from the Polizeipräsident, Landeskriminalpolizeistelle, to the Regierungspräsident in Schleswig concerning "Großdeutsche Bauernbewegung" Flensburg, 9.4.1931 and concerning "Großdeutsche Bauern- und Landvolkbewegung" Flensburg, 7.5.1931, both in LAS 309/22701; reports from the Regierungspräsident to the Preuß. Minister des Innern in Berlin concerning "Landvolkbewegung" Schleswig, 16.4.1931, Schleswig, 21.5.1931 and Schleswig, 30.5.1931, all in LAS 301/4698 as well as 309/22701; copy of a report from the Regierungspräsident to the Preuß. Minister des Innern in Berlin concerning "Landvolkbewegung" Erfurt, 30.5.1931, sent to the Regierungspräsident in Schleswig from the Preuß. Minister des Innern, LAS 309/22701.

152 Report from Abteilung IA titled "Bericht über die Landvolkbewegung" Altona, 20.5.1931 and the copy of the letter from Ferdinand Pramor to Claus Heim in Celle, Itzehoe, 6.3.1932, both in LAS 309/22701.

153 *Schleswig-Holsteinische Tageszeitung* No. 239, 11.10.1931, LAS 309/22701.

154 Report titled "Bericht" Itzehoe, 28.3.1929, LAS 309/22921.

155 Report from the Landeskriminalpolizeiamt Berlin concerning "Süderdithmarschen" Itzehoe, 26.3.1929, LAS 309/22668; report titled "Bericht", 15.10.1929, LAS 309/22784; *Kieler Neueste Nachrichten* No. 215, 14.9.1929, p. 1.

156 Reports from the Polizeipräsident, Landeskriminalpolizeistelle, to the Regierungspräsident in Schleswig concerning "Landvolkbewegung" Flensburg, 21.6.1930 and 12.7.1930, in both LAS 301/4697 and 309/22590.

In 1931 rumours that the donations had been pocketed for personal enrichment became more persistent and widespread.[157] The two most active remaining leaders, Hamkens and Weschke, were both accused of misusing donations collected for the trials, and this had a lasting effect on the public's willingness to make further contributions.[158] There is no real evidence to substantiate these claims but it appears that more and more of the movement's followers believed them. For example, farmers dismissed Hamkens brusquely while he was collecting on several occasions and, increasingly, they withdrew their stakes from *Das Landvolk*.[159]

Without question, the most harmful affair concerned the payment of Luetgebrune. Not only did his services drain the movement's finances, but the open criticisms of his expense and the resultant discord also weakened support. In 1930 these condemnations were led by the disenchanted Tannenbergbund members within the movement, in particular Wulf and Hönck, to whom Luetgebrune's response, that he had refused all payment other than "compensation" for his expenses and shortfalls, only served as confirmation of their accusations.[160]

Considerably more scathing was the open letter of one of the Landvolk's bombers, Johannes Rathjen, in August 1931 who lamented Luetgebrune's payment was an embarrassing scandal. Rathjen divulged that Luetgebrune had received between RM 85,000 and 100,000 for the "Great Landvolk trial" and no less than RM 250,000 for his defence of the bombers, whereas the other lawyers, including von der Goltz, had appeared pro-bono. Apparently some half a million marks had been collected up until the trial but there were no records for what had happened to the remainder. According to Rathjen, the Landvolk had failed because "crass materialism and injustice" had prevailed over idealism and dedication to the movement's cause.[161]

It is difficult to ascertain whether Luetgebrune did receive such sums. Various, considerably smaller, amounts are noted in police reports. Two reports from mid 1930 stated that Luetgebrune was to receive RM 250 daily; two others noted fees of

157 Report from the Regierungspräsident to the Preuß. Minister des Innern in Berlin concerning "Landvolkbewegung" Schleswig, 16.4.1931, in both LAS 301/4698 and 309/22701.

158 Reports from the Polizeipräsident to the Regierungspräsident in Schleswig concerning "Großdeutsche Bauernbewegung" Flensburg, 9.4.1931 and from the Landeskriminalpolizeistelle, Abteilung IA, titled "Bericht über die Landvolkbewegung" Altona, 24.8.1931, both in LAS 309/22701.

159 Polizeipräsident, Landeskriminalpolizeistelle, to the Regierungspräsident in Schleswig concerning "Großdeutsche Bauern- und Landvolkbewegung" Flensburg, 7.5.1931, LAS 309/22701; report from the Regierungspräsident to the Minister des Innern in Berlin concerning "Landvolkbewegung" Schleswig, 21.5.1931, in both LAS 301/4698 and 309/22701.

160 *Rendsburger Tageblatt* No. 175, 29.7.1930, p. 3 and No. 177, 31.7.1930, p. 3.

161 *Rendsburger Tageblatt* No. 200, 27.8.1931, p. 6; *Der Schleswiger* No. 203, 1.9.1931, LAS 309/22701. Also see the letter from Claus Heim to Johannes Rathjen, Celle, 2.2.1932, StaNms MA 5360.

around RM 40,000 and another RM 65,000.[162] However, the particular expenses these payments were to cover were not specified, they all predated the bombers' trial and proceedings then took much longer than anticipated. Whatever the total amount, it was sizable and therefore there is almost certainly some truth in the depiction of Luetgebrune as the "real grave-digger" of the movement.[163]

The authorities, and Regierungspräsident Abegg in particular, tried to hasten the Landvolkbewegung's decline wherever possible. This included encouraging charges to be laid against the Landvolk leaders and ensuring that Hamkens was sent to prison at a time that would prevent his appearances at meetings.[164] Abegg combed the press laws and constantly sought opportunities to ban and confiscate *Das Landvolk*.[165] However, he and the public prosecutors were frustrated to find that the movement's propaganda was almost always on the verge of the permissible and that the movement's hazy language often made it difficult to apply the law.[166]

Seeking to curtail *Das Landvolk's* income, the authorities had long terminated the publisher's commission to print the county gazette for Steinburg.[167] Then, on the grounds that it had frequently attacked state and communal institutions, Abegg requested all public bodies refrain from commissioning the paper to print official

---

162 Report from the Landeskriminalpolizeistelle, Abteilung IA, titled "Bericht" Altona, 8.6.1930, in both LAS 301/4697 and 309/22590; reports from the Landrat to the Regierungspräsident in Schleswig concerning "Landvolk-Notspende" Heide, 24.6.1930 and from the Landjägereiamt Bordesholm, Kreis Bordesholm, to the Landjägereiabteilung Bordesholm, Bordesholm, 27.7.1930, both in LAS 309/22590; report from the Abteilung IA titled "Bericht über die Landvolkbewegung" Altona, 15.9.1930, LAS 301/4697; report from the Landeskriminalpolizeistelle Altona, Abteilung IA, concerning "Landvolkbewegung" Itzehoe, 14.11.1929, in both LAS 301/4697 and 309/22783.

163 Thyssen stated that Luetgebrune was described as such. Thyssen, *Bauer und Standesvertretung*, p. 409.

164 See, for example, Regierungspräsident to the Oberstaatsanwalt in Flensburg concerning "Strafverfahren wegen der Vorgänge in Husum am 1.7.1929" Schleswig, 11.3.1930, LAS 309/22783; Regierungspräsident to the Minister des Innern in Berlin concerning "Vollstreckung der gegen den Landvolkführer Hamkens - Tetenbüll rechtskräftig erkannten Gefängnisstrafe wegen Beteiligung an dem nächtlichen Überfall auf das Amtsgerichtsgefängnis in Hohenwestedt" Schleswig, 2.4.1930, in both LAS 301/4697 and 309/22590; Regierungspräsident to the Generalstaatsanwalt in Kiel, Schleswig, 27.2.1930, in both LAS 301/4697 and 309/22783; Regierungspräsident to the Generalstaatsanwalt in Kiel concerning "Strafverbüßung Wilhelm Hamkens – Tetenbüll" Schleswig, 30.6.1931, LAS 309/22701.

165 Regierungspräsident to the Oberpräsident in Kiel, Schleswig, 17.7.1930, LAS 301/4697.

166 Regierungspräsident to the Preuß. Minister des Innern concerning "Politische Lage im Reg. Bez. Schleswig" Schleswig, 5.6.1929, LAS 309/22668; Oberstaatsanwalt to the Generalstaatsanwalt in Kiel concerning "Schreiben des Regierungspräsidenten in Schleswig vom 30.Dez. 1929 .... über Beschlagnahme von Tageszeitungen" Altona, 3.1.1930, LAS 301/4697.

167 Report from the Landeskriminalpolizeiamt Berlin concerning "Süderdithmarschen" Itzehoe, 26.3.1929, LAS 309/22668.

notices.[168] This endeavour was not entirely successful, however, for at the end of November 1929 the State Finance Office commissioned *Das Landvolk* to print some 30,000 tax assessment forms and even delivered the necessary paper.[169] Abegg's frustration at the State Finance Office's lack of judgement was evident from a marginal comment in the files, "Simply grotesque! Just as regrettable from the State Finance Office as it is immoral from the newspaper!"[170] Whether *Das Landvolk* had accepted the assignment out of cynical opportunism or sheer financial desperation is unclear.

While the earlier bans of the newspaper had dealt significant blows to its finances, endeavours to inhibit its collection of donations were less successful.[171] The authorities failed to find the legal grounds to block the registration of the Volksmacht.[172] Furthermore, charges against a total of nine Landvolk followers for breaching the charity law of February 15, 1917 by collecting donations without licences also proved futile. In three separate cases, sentences and charges were ultimately overturned because the courts determined the collections were to pay debts to Luetgebrune whereas the law from 1917 only applied to collections for charitable purposes.[173]

The Landvolkbewegung's overall decline was not simply a result of problems within the movement itself and the hostility of the authorities. Renewed hopes in an improvement in the farmers' financial circumstances also curbed support for the movement. Already in the spring of 1929, police reports associated less interest in the movement in certain areas of the province with signs of recovery in some

---

168 Regierungspräsident to the Oberpräsident in Kiel concerning "Amtliche Anzeigen in der Zeitung 'Das Landvolk' " Schleswig, 9.10.1929, LAS 301/4697.

169 Report from the Landeskriminalpolizeistelle Altona, Abteilung IA, concerning "Landvolkbewegung" Altona, 10.12.1929, in both LAS 301/4697 and 309/22783.

170 Preuß. Polizeipräsident Altona-Wandsbek to the Regierungspräsident in Schleswig concerning "Landvolkbewegung" Altona, 14.12.1929, LAS 309/22783; Regierungspräsident to the Präsident des Landesfinanzamtes in Kiel, Schleswig, 19.12.1929, LAS 309/22783 and Schleswig, 15.3.1930, in both LAS 301/4697 and 309/22783.

171 Report from the Landeskriminalpolizeiamt Berlin concerning "Süderdithmarschen" Itzehoe, 26.3.1929, LAS 309/22668.

172 See the note in the Regierungspräsident's files, Schleswig, 19.12.1929, LAS 309/22783; Regierungspräsident to the Oberpräsident in Kiel, Schleswig, 30.12.1929, in both LAS 301/4697 and 309/22783.

173 Indictment sent from the Oberstaatsanwalt to the Regierungspräsident in Schleswig, Altona, 27.5.1931, issues of verdicts of the Amtsgericht Meldorf from 15.7.1931 and of the Amtsgericht Itzehoe, 22.9.1931, and reports to the Regierungspräsident from the Landrat concerning "Sammlung für das Landvolk" Meldorf, 5.12.1931, from the Landrat concerning "Öffentliche Sammlung für die Landvolkbewegung" Meldorf, 4.7.1931, and from the Landrat concerning "Ausgang des Verfahrens gegen Karstens – Nesserdeich u. Gen. wegen verbotener Sammlung für das Landvolk" Meldorf, 15.4.1932, all in LAS 309/22701; *Schleswig-Holsteinische Tageszeitung* No. 197, 23.8.1931, p. 4; *Schleswiger Nachrichten* No. 195, 21.8.1931, p. 7.

branches of agriculture.[174] In 1930 the appointment of Schiele as Minister of Agriculture under Brüning gave many farmers cause for optimism and helped to bring relative peace to the countryside. Although the effects of policy, particularly in Schleswig-Holstein's livestock sector, remained to be seen, and though radical Landvolk followers were unwilling to acknowledge the concessions agriculture received as anything but "sedatives",[175] the promise of tariff increases quelled the farmers' rebelliousness.[176]

It was also in early 1930 that the political agricultural organisations re-emerged as a genuine force in the province. Despite so many setbacks and interminable tension between them, the Farmers' Association, the Farmers' League and the Rural League had never entirely abandoned the idea of unification. Although an attempt to merge the Farmers' League and Farmers' Association failed at the end of 1929, at the beginning of 1930 the Farmers' League and Rural League succeeded in forming the "Rural and Farmers' League" with the Reich Rural League as sole national organisation.[177]

Significantly, it was the radicalisation of the Farmers' League that had made fusion with the Rural League possible and the Regierungspräsident recognised that the Landvolkbewegung's influence had played a major role in this process. He noted the Farmers' League's cooperation with the movement, its support of the Landvolk's resistance to compulsory enforcement, participation in its criticisms of the authorities and involvement in the boycott of Neumünster. Moreover, the Farmers' League's avowals from the autumn of 1929 of its acceptance of the state but mistrust of the current state system, as well as its acknowledgment of its place within the "Volksgemeinschaft", were factors which had hindered a merger with the more moderate Farmers' Association, but which now made the Farmers' League amenable to the more extreme Rural League.[178]

---

174 See, for example, the report titled "Bericht über Glückstadt u. Umgegend sowie die Krempermarsch" Itzehoe, 27.3.1929, appended to the report from the Landeskriminalpolizeistelle to the Regierungspräsident in Schleswig, Flensburg, 2.4.1929, LAS 309/22921.

175 See, for example, *Das Landvolk* No. 67, 20.3.1930, p. 1.

176 *Frankfurter Zeitung* No. 17, 7.1.1930, LAS 301/4697; Uhse, *Söldner und Soldat*, p. 276. Also see chapter 2 above for developments in agricultural policy.

177 Reports from the Regierungspräsident to the Preuß. Minister des Innern in Berlin concerning "Landwirtschaftliche Organisationen" Schleswig, 16.1.1930 and from the Regierungspräsident to the Minister des Innern in Berlin concerning "Die landwirtschaftliche Bewegung in Schleswig-Holstein" Schleswig, 24.1.1930, both in LAS 309/22765. Also see Thyssen, *Bauer und Standesvertretung*, pp. 389-394.

178 Reports from the Regierungspräsident to the Oberpräsident in Kiel concerning "Entwicklung des landwirtschaftlichen Berufsorganisationswesens in Schleswig-Holstein" Schleswig, 22.1.1930 and from the Regierungspräsident to the Minister des Innern in Berlin concerning "Die landwirtschaftliche Bewegung in Schleswig-Holstein" Schleswig, 24.1.1930, both in LAS 309/22765.

The fusion of the Rural League and Farmers' League did not bode well for moderation with the majority of the province's farmers united under the nationalist Reich Rural League. Furthermore, the Regierungspräsident feared that continued association with the Landvolkbewegung would also lead to further radicalisation.[179] He judged the provincial Rural League's friendship with the movement so intimate he considered its newspaper the Landvolk's, and although he did subsequently observe that the Rural and Farmers' League treated the Landvolk with some caution, he nevertheless equated its approach as one of "benevolent toleration".[180]

However, by uniting the great majority of the province's farmers, the creation of the Rural and Farmers' League took away one of the principal attractions of the Landvolkbewegung. The Landvolk had profited as an alternative form of unity so long as agricultural-political organisations had remained divided. Quite rightly, Abegg expected the new, unified Rural and Farmers' League to draw support away from the movement even if he was not pleased with the consequence, for he would still be faced with a radical farmers' organisation.[181]

It must be noted here that only after the Farmers' League and Rural League agreed to amalgamate did Abegg truly recognise the need for a strong rival organisation to counterbalance the growth of the right-wing organisations. In January 1930 he envisaged the revival of the Farmers' Association as a means of hastening the decline of the Landvolkbewegung and hindering the advance of National Socialism in the countryside, and he informed the Prussian Interior Ministry of his intention to assist it inconspicuously, presumably with financial aid.[182] However, the optimism he expressed in the Farmers' Association's recovery as late as August 1930 was completely misplaced since it had already lost all significance in the province.[183]

Beside the Rural and Farmers' League, the other rising force in the province that contributed significantly to the decline of the Landvolkbewegung was, of course, the NSDAP. There are reports of the NSDAP drawing support away from the Landvolkbewegung from as early as the spring of 1929, particularly after the

179 Report from the Regierungspräsident to the Preuß. Minister des Innern in Berlin concerning "Die landwirtschaftliche Bewegung in Schleswig-Holstein" Schleswig, 24.1.1930, LAS 309/22765.

180 Regierungspräsident to the Regierungspräsident Dr. v. Bahrfeldt in Königsberg, Schleswig, 10.3.1930, LAS 309/22783.

181 Report from the Regierungspräsident to the Preuß. Minister des Innern in Berlin concerning "Landvolkbewegung" Schleswig, 23.12.1929, in both LAS 301/4697 and 309/22783.

182 Regierungspräsident to the Preuß. Minister des Innern in Berlin concerning "Die landwirtschaftliche Bewegung in Schleswig-Holstein" Schleswig, 24.1.1930, LAS 309/22765. Beyer read this as evidence that Abegg did actually subsidise the Farmers' Association. See Beyer, 'Die Agrarkrise und die Landvolkbewegung in den Jahren 1928-1932', p. 175 and 'Das Bauerntum Angelns während der großen Krise 1927 bis 1932', p. 145.

183 Report from the Regierungspräsident to the Preuß. Minister des Innern in Berlin, Schleswig, 19.8.1930, LAS 301/4697; Thyssen, *Bauer und Standesvertretung*, p. 394.

Wöhrden incident, but the party contributed to the movement's decline to a much greater degree after October.[184] The party profited more than any other rival group, though others, principally the Jungdeutscher Orden in South Tondern where it already had a strong supporter base, also gained followers at the Landvolk's expense.[185] Reports of Landvolk followers going over to the NSDAP support Gerhard Stoltenberg's view that the NSDAP took over as the dominant force shaping rural opinion first on the west coast and then increasingly in other areas of Schleswig-Holstein from the autumn and winter of 1929/1930.[186]

The NSDAP's membership in Schleswig-Holstein grew from 4,000 at the end of 1928 to approximately 8,000 by April 1929 and reached 10,400 by the beginning of December. Until this period the party's membership was concentrated in the province's western counties, with 3,000 registered in Dithmarschen alone. Thereafter, growth accelerated in the north and east.[187]

Unlike their followers, few Landvolk leaders joined the NSDAP until the movement had almost completely collapsed, despite all the similarities of ideology, and despite earlier moments of fraternity and several gestures on the part of Landvolk leaders to promote friendly relations. There were some notable changes of allegiance after this time, among them Peter Petersen, who continued his agitation in Plön as a speaker for the NSDAP when interest in the Landvolkbewegung waned.[188] One did not, of course, necessarily have to become a member of the party in order to support it. Be that as it may, only a relatively small number of Landvolk leaders and "trusted men" are to be found in the NSDAP's membership records of the Berlin Document Center (BDC). A search for 147 Landvolk leaders and "trusted people" (*Vertrauensleute*) in Schleswig-Holstein on the basis of names,

---

184 See chapter 5, pp. 171-172, chapter 7, p. 261 and chapter 8, pp. 281 and 285.

185 For the Landvolk and Jungdeutscher Orden in South Tondern, see the reports from the Kriminal- und Grenzkommissariat Niebüll to the Landrat in Niebüll, Niebüll, 16.1.1930 and from the Landrat to the Regierungspräsident in Schleswig, Niebüll, 24.1.1930, both in LAS 309/22783. Also see Koops, *Südtondern in der Zeit der Weimarer Republik*, p. 257. The Landvolkbewegung was also subject to greater competition from the Tannenbergbund after October 1929. See the report from the Landeskriminalpolizeiamt titled "Bericht" Altona, 7.10.1929, LAS 309/22784.

186 Stoltenberg, *Politische Strömungen im schleswig-holsteinischen Landvolk*, p. 141.

187 See LAS 309/22998; Stoltenberg, *Politische Strömungen im schleswig-holsteinischen Landvolk*, pp. 162-163.

188 For Peter Petersen's shift to the NSDAP and continuing efforts to encourage resistance to compulsory enforcement, see the reports from the Landrat to the Regierungspräsident in Schleswig, Plön, 22.2.1932 and Plön, 25.2.1932 and concerning "Zusammenziehung von Mitgliedern des Land- und Bauernbundes als Demonstration bei einer Zwangsversteigerung" Plön, 28.11.1932, LAS 309/22765. Also see Petersen, *Fliegender Sand*, p. 77. For the shift of erstwhile Landvolk supporters, Bender and Beyer, on the island of Fehmarn to supporting the NSDAP, see the report on meetings in Burg on Fehmarn from the Polizeiverwaltung, Burg a/F, 7.12.1929, LAS 309/22783; *Fehmarnscher Inselbote* No. 219, 18.9.1931, LAS 309/22701.

locations of residence, occupations and, where available, dates of birth, produced 44 likely matches.[189] Of these, two had joined the party and left it by 1926. Just nine joined the party from the beginning of 1928 to the end of 1930, that is during the period in which the Landvolkbewegung peaked and began to decline. By comparison, 11 became members in 1931, including six in November and December alone, while a further six joined in 1932, including five in the first three months of that year. 16 of the 44 found did not join until after the Reichstag elections in March 1933 or later. It would appear that to have played a leading role in the Landvolkbewegung – and therefore to have been involved in the enduring antagonism between the two movements[190] – remained a barrier to going over to the NSDAP. Importantly, the growing number who joined in the winter of 1931/1932 came only after the closure of *Das Landvolk* and after the Rural and Farmers' League and NSDAP had taken over the movement's obstruction of confiscations and compulsory auctions. Therefore, the defection of its leaders to the NSDAP cannot be considered a significant factor in explaining the decline of the Landvolkbewegung.

The publication of the party's one and only agricultural programme in March 1930 marked the NSDAP's growing interest in the rural population.[191] Neverthe-

---

189 BDC: 31XX: Zentralkartei and 3200: Ortskartei. A small number of matches may be subject to error due to the limitations of the search criteria and because the files of the BDC are believed to contain approximately 90 percent of what remains of the NSDAP's membership records. Those selected for comparison with the NSDAP membership lists in the BDC were taken chiefly from lists compiled by the Landeskriminalpolizeistelle Altona-Wandsbek, Abteilung IA. See the list concerning "Landvolkbewegung" Itzehoe, 20.3.1930 and the subsequent undated additions in both LAS 301/4697 and 309/22590.

190 Differences between the Landvolkbewegung and NSDAP did not abate. See, for example, the report from the Abteilung IA titled "Bericht über die Landvolkbewegung in Schleswig-Holstein" Altona, 28.1.1931, LAS 301/4698.

191 The NSDAP's agricultural programme extolled farmers for both their economic and "biological" importance as the main support of a "truly national German state"; emphasising the necessity of autarky and distinguishing the rural population as the bearers of the nation's health, the source of its youth and backbone of its defence forces. It blamed agriculture's lack of profitability on international capital and Weimar's deficient taxation and tariff policies as well as wholesale profits and excessive rates for electricity and fertilisers. These sectors were said to be mainly in the hands of Jews, who as moneylenders were also profiting from exorbitant interest for loans. The programme promised to restore profitability by redressing these problems, to end land speculation and prevent the subdivision of property on inheritance. It obliged farm owners to use their lands in the national interest, proposed the colonisation of Germany's eastern frontiers and for foreign policy to provide space for settlement. It promised assistance to professional organisations to improve production, demanded socially just contracts for farm labourers and the future elimination of foreign rural labourers, and the promotion of technical education, including support for "boys without means". Finally, the programme steadfastly promoted the idea that agriculture shared the fate of the entire nation, that the NSDAP would struggle not for a single profession but for the whole nation. G. Feder, *Das Programm der NSDAP und seine weltanschaulichen*

less, when explaining the party's success in the countryside, historians agree that its programme was not original, that its demands differed little to those of the Rural League and the DNVP, and that the substance of its other propaganda was not much different from that of other völkisch groups. In the main, its rural propaganda focused on anti-Semitism, anti-liberalism and anti-Bolshevism. It attempted to exploit the discontent among the rural population over the crisis and the divisions among the agricultural organisations, rather than tackling concrete technical and economic agricultural questions.[192] What set the NSDAP apart was its authoritarian structure, the special position of the party leader, its total demands on its members, a record untainted by participation in government and unrelenting activism.[193]

Explaining how the NSDAP won over the Landvolkbewegung's followers and generally profited at its expense must, nevertheless, highlight the similarities in their ideas and propaganda, for it was this overlap that made the transition possible. As has been seen at various points in the preceding chapters, Landvolk and National Socialist propaganda was so similar as to have sometimes caused the public and even the authorities to confuse one movement with the other, and for the NSDAP to accuse the Landvolk of appropriating its ideas.[194] Like the Landvolk's anti-Semitism, for example, that of the NSDAP was typically found coupled with attacks on other institutions, on international finance capital and on Marxism. And while the NSDAP called more frequently on its followers to boycott Jewish department stores, personal denunciations of local Jews in the province did not figure prominently in either movement's propaganda.[195] Since their ideas were so similar,

---

Grundgedanken (Munich, 1931), pp. 6-12. An early translation of this fifth German edition is to be found in G. Feder, *Hitler's Official Programme* (London, 1934), pp. 29-37.

192 Gies, *R. W. Darré*, pp. 51-60 and 'NSDAP und landwirtschaftliche Organisationen in der Endphase der Weimarer Republik', pp. 352-359.

193 The following state all these or a combination of these factors in respect of the NSDAP's differences to, both or either, other völkisch groups and political parties, in particular, the DNVP: Childers, *The Nazi Voter*, p. 218; Gerschenkron, *Bread and Democracy in Germany*, p. 146; Gessner, *Agrarverbände in der Weimarer Republik*, pp. 244-245; Noakes, *The Nazi Party in Lower Saxony*, p. 125; Rietzler, *"Kampf in der Nordmark"*, pp. 409-410 and 426; Stoltenberg, *Politische Strömungen im schleswig-holsteinischen Landvolk*, p. 144. Werner Angress also pointed to the NSDAP's "pure" and "barren" record but differed in suggesting that its agricultural programme was more "daring and comprehensive" and therefore more appealing than that of other groups. This judgement, however, is contrary to Gerschenkron's upon which it was drawn. See W. T. Angress, 'The Political Role of the Peasantry in the Weimar Republic', *The Review of Politics*, 21 (1959), pp. 546-547.

194 See chapter 3, p. 103, chapter 4, pp. 138-139, chapter 5, p. 188 and chapter 6, pp. 226 ff.

195 Paul, *Aufstand der Bilder*, pp. 87-90 and Paul, ' "Die Juden sind unser Unglück!" ', pp. 191-192. For prominent examples of the Landvolk's calls to avoid Jewish businesses, see the poster titled "*Deutsches Weihnachtsfest!*" appended to the report from the Landjägereiposten Kreis Bordesholm to the Landrat in Bordesholm, Wittorf b. Neumünster, 17.12.1928, LAS 309/22696 and the aims of the Landvolkbewegung as announced at the Landesthing in November 1929: transcript of the Landesthing from the Polizeiverwaltung in

the NSDAP was able to build on propaganda already familiar to the rural people when it expanded from the west coast across the countryside in 1929 and 1930.

What actually drew the Landvolk's supporters to the NSDAP were the party's better prospects of achieving its aims. The NSDAP often criticised the Landvolkbewegung for its lack of leadership and loose organisation, for its failure to collaborate with other occupational groups, and for the futility of its methods – the inability of its activism and above all the bombings – to effect a lasting improvement of the farmers' fortunes, or to eliminate the system. The NSDAP underlined its strengths in these areas as well as its determined strategy and parliamentary participation as promising greater success.[196] Indeed, according to the police and provincial authorities, the NSDAP persuaded a growing number of farmers that only it could rescue them from further misfortune, not only because of its programme but also because of its tighter organisation, its more pragmatic methods and rejection of the Landvolk's excesses.[197]

The NSDAP profited from the Landvolkbewegung's radicalisation of the rural population at elections, since the Landvolk refrained from active parliamentary participation. On occasion, individual Landvolk leaders did support particular parties. At the communal *Provinziallandtag* elections of November 1929, for instance, Paul Adam Ross ran as a candidate for the "Economic Bloc" in North Dithmarschen, while in Eiderstedt Hamkens told his followers only to vote for candidates who opposed the system, declaring on this occasion that the NSDAP was the only party that came into consideration.[198] Nevertheless, this advocacy represented but a fleeting respite in the otherwise strained relations between Hamkens and the NSDAP.

Rendsburg titled "Landesthing des Schleswig-Holsteinischen Landvolkes, am 10. November 1929 in Rendsburg im 'Schützenhof' nachmittags 3 Uhr" Rendsburg, 12.11.1929, in both LAS 301/4696 and 309/22590.

196 See chapter 3, p. 103, chapter 5, p. 189, chapter 6, p. 226 and chapter 7, pp. 244-245 and 253.

197 Reports from the Landeskriminalpolizeiamt Berlin, Abteilung IA, Heide, 21.3.1929, LAS 309/22668 and Berlin, 2.4.1929, in both LAS 301/4695 and 309/22668; report from the Regierungspräsident to the Preuß. Minister des Innern concerning "Politische Lage im Reg. Bez. Schleswig" Schleswig, 5.6.1929, LAS 309/22668; Regierungspräsident to the Regierungspräsident Dr. v. Bahrfeldt in Königsberg, Schleswig, 10.3.1930, LAS 309/22783.

198 Report from the Landeskriminalpolizeistelle Altona, Abteilung IA, concerning "Landvolkbewegung" Altona, 9.10.1929, in both LAS 301/4696 and 309/22784; reports titled "Bericht" Heide, 12.10.1929 and titled "Bericht" and concerning "Landvolkbewegung" Flensburg, 20.10.1929, sent by the Polizeipräsident, Landeskriminalpolizeiamt, Abteilung IA, to the Regierungspräsident in Schleswig, Altona, 21.10.1929, both in LAS 309/22784; report from the Landeskriminalpolizeistelle Flensburg titled "Bericht" and concerning "Landvolkbewegung" Husum, 8.11.1929, LAS 309/22783; reports from the Landrat to the Regierungspräsident in Schleswig concerning "Dorfgemeindevorsteher Paul Adam Ross" Heide, 12.10.1929 and Heide, 2.11.1929, LAS 309/22784; report from the Regierungspräsident to the Oberpräsident in Kiel concerning "Landvolkführer" Schleswig, 18.10.1929, in both LAS 301/4696 and 309/22784.

Moreover, the Landvolkbewegung never officially avowed its support for any particular political party at any elections of any kind.

Other parties, foremost the DNVP and CNBLP, may have sought agricultural votes too, but ultimately the NSDAP was left as the party most likely to deliver the fundamental change the Landvolk demanded. It should be remembered that the NSDAP only polled four percent of the vote at the Reichstag election in 1928, and consequently the Provinziallandtag elections provided a clear sign of the party's growing strength, with 10.3 percent across the province.[199] Following the publication of its agricultural programme, continuing expansion of its membership and organisation, and a rigorous campaign which included some 1,000 meetings and marches throughout the province, the Reichstag election of September 1930 heralded National Socialism's breakthrough as a major political force in Schleswig-Holstein and across Germany.[200]

Enjoying vast nationwide gains to become the second strongest party behind the SPD, the NSDAP polled 18.3 percent of the vote and no less than 27 percent in Schleswig-Holstein.[201] Its success in the province was due in large part to its advances in the countryside where it polled 35.1 percent of the vote. Particularly strong in the rural areas of the Geest with 45.9 and in the marshes 41.2 percent, it gained chiefly at the cost of the DNVP and the DVP.[202] Their shares of the vote fell sharply from 1928 from 23 to 6.1 and 13.7 to 7.3 percent respectively, while the DDP's decline continued under its new name, the *Deutsche Staatspartei* (German State Party, DSP), from 5.7 to 4.7 percent. Support for the left, by contrast, was relatively stable as a whole, the SPD's share of the vote falling from 35.3 to 29.8 percent and the KPD's rising from 7.9 to 10.6 percent.

The NSDAP's dramatic election success in 1930 served to bolster the promise of the party's parliamentary course as opposed to the Landvolk's insistence on remaining outside the system in order to defeat it. Increasingly, the prospects of the broad-based party contrasted sharply with the growing scepticism over the primarily rural movement's methods and want of a formal programme.

A further important way in which the NSDAP consolidated its position in rural areas was its systematic nationwide strategy to establish an Agricultural Political Apparatus (*Agrarpolitischer Apparat*). Its development from the middle of 1930 came too late to be considered a major factor in the party's election success that

---

199 *Statistisches Jahrbuch für den Freistaat Preußen,* vol. 26 (Berlin, 1930), p. 354; F. Omland, '"Die Nordmark steht zu Hitler!" Wählerwanderungen und Wählerherkunft der NSDAP 1924-1933' *Informationen zur Schleswig-Holsteinischen Zeitgeschichte*, 48 (2007), pp. 26-28.

200 Stoltenberg, *Politische Strömungen im schleswig-holsteinischen Landvolk*, pp. 162-163; Gessner, *Agrarverbände in der Weimarer Republik*, p. 243.

201 *Statistisches Jahrbuch für das Deutsche Reich* (Berlin, 1931), pp. 546-547.

202 These figures apply to rural communities in Schleswig-Holstein with a population under 2,000. See Heberle, *From Democracy to Nazism*, p. 99.

year.[203] Nevertheless, it constituted a significant step in the party's endeavours to secure the farmers' support and underpinned its continuing advance in the countryside.

Appointed in May to advise the Reich leadership on agriculture, Richard Walter Darré, otherwise known for his racial theories, masterminded the expansion of the Agricultural Political Apparatus).[204] The original idea, however, belonged to Kenstler. Though he had become involved in the Landvolkbewegung from the autumn of 1929, as one of the main leaders of the Artam movement, Kenstler also had close ties to the NSDAP in Thuringia. Before he recognised the Reich leadership's reservations toward the Landvolkbewegung, Kenstler had hoped that the crisis and the Landvolkbewegung might be used to effect a revolution that would allow Germany's "re-agrarianisation" and he supposed the NSDAP might offer a suitable medium to transform the Landvolk into a Reich-wide movement. To this end, he came up with the idea of a centralised agricultural apparatus for the party and he developed the concept with the editor of the party's Gau newspaper in Thuringia, Ziegler, and with Darré. Some months later, when the party leadership commissioned Darré as a consultant on agricultural matters, he put Kenstler's idea into practice.[205]

Though it ended up aiding the NSDAP's rural expansion, the initial impetus for the creation of the Agricultural Political Apparatus, therefore, came from the Landvolkbewegung. The Landvolkbewegung, however, did not serve as a model for its organisation. On the contrary, as a base for the party's rural agitation, it comprised a network of expert advisers which paralleled the party's centralised structure at Gau, county and local levels and was responsible both to Darré and regional political leaders.[206]

With the establishment of the Agricultural Political Apparatus in mid-1930, the NSDAP began to use its tight organisation to infiltrate and exercise influence over the political agricultural associations, that is, in place of creating an agricultural organisation of its own. Acknowledging the Rural Leagues as the most suitable for this purpose, Darré encouraged party members to join them, and to run for office, in order to conquer them position by position and eliminate opposition.[207]

In Schleswig-Holstein this signified reinforcement of the party's earlier strategy rather than an essentially new course. Gauleiter Lohse had already voiced his

203 Gustavo Corni, for example, did not point to the Agricultural Political Apparatus as a factor in the NSDAP's success in 1930: Corni, *Hitler and the Peasants*, p. 25.

204 Corni, *Hitler and the Peasants*, pp. 21-23; Gies, *R. W. Darré*, pp. 40-41.

205 Bramwell, *Blood and Soil*, pp. 76-78; Gies, *R. W. Darré*, pp. 37-39; Gies, 'NSDAP und landwirtschaftliche Organisationen in der Endphase der Weimarer Republik', *Vierteljahrshefte für Zeitgeschichte*, 15 (1967), pp. 343-344.

206 Bramwell, *Blood and Soil*, pp. 76-78 and 84-86; Bergmann, *Agrarromantik und Großstadtfeindschaft*, pp. 289-297; Gies, *R. W. Darré*, pp. 39-39 and 43-60; Gies, 'NSDAP und landwirtschaftliche Organisationen in der Endphase der Weimarer Republik', pp. 343-344.

207 Gessner, *Agrarverbände in der Weimarer Republik*, p. 246.

resolve against establishing a further rival agricultural organisation, and since April 1929 the party had openly attempted to put pressure on the Farmers' League.[208] When the Farmers' League amalgamated with the Rural League at the beginning of 1930, the NSDAP continued this strategy with the sole difference that it now courted an organisation that encompassed the vast majority of the province's farmers.

In March 1930 the Regierungspräsident predicted that the party's proffer of friendship would compel the Rural and Farmers' League to respond with goodwill in order to avoid risking the party's opposition.[209] These fears were justified. At the end of the year Lohse notified members that the Rural and Farmers' League was the only acceptable professional agricultural organisation and, once the Reich Rural League had agreed to support a course of right-wing opposition, Hitler's call followed in July 1931.[210]

By this stage, the NSDAP's strategy had already borne fruit. The Rural and Farmers' League's chairmen, first Tönnsen and then Köhler, joined the party and they were soon followed by other prominent agricultural leaders, including the leader of the rallies of January 1928, Otto Johannsen.[211] The victories of the NSDAP's endorsed Rural and Farmers' League candidates at the half-elections for Prussia's agricultural chambers in October 1931 signalled the party's ever-growing influence.[212] Significantly, by the spring of 1932 this extended to securing the support of the Rural and Farmers' League for the National Opposition at the Reich President, Prussian Landtag and Reichstag elections.[213]

Having recognised that the creation of a further political agricultural organisation in the province would only cause greater disunity and rivalry, the NSDAP had

---

208 See chapter 5, pp. 190-191.

209 Report from the Regierungspräsident to the Minister für Landwirtschaft, Domänen und Forsten in Berlin concerning "Schleswig-Holsteinischer Land- und Bauernbund" Schleswig, 24.3.1930, in both LAS 309/22765 and 301/4697.

210 *Schleswig-Holsteinische Tageszeitung* No. 304, 28.12.1930, LAS 309/22765; *Schleswig-Holsteinischer Bauernbund* No. 27, 5.7.1931, p. 3. According to Gessner, Hitler made such a call as early as March 1931 at a rally in Munich. Gessner, *Agrarverbände in der Weimarer Republik*, p. 247.

211 For Tönnsen, see the report from the Regierungspräsident to the Minister für Landwirtschaft, Domänen und Forsten in Berlin concerning "Schleswig-Holsteinischer Land- und Bauernbund" Schleswig, 24.3.1930, in both LAS 309/22765 and 301/4697. For Köhler, see Stoltenberg, *Politische Strömungen im schleswig-holsteinischen Landvolk*, p. 169 and Thyssen, *Bauer und Standesvertretung*, p. 397. Otto Johannsen joined the party in March 1931. See the *Schleswig-Holsteinische Tageszeitung* No. 76, 31.3.1931, p. 5.

212 *Schleswig-Holsteinischer Bauernbund* No. 42, 25.10.1931, p. 1-4 and No. 44, 1.11.1931, p. 2; *Schleswig-Holsteinische Tageszeitung* No. 253, 28.10.1931, p. 5 and No. 258, 3.11.1931, p. 4.

213 *Schleswig-Holsteinischer Bauernbund* No. 50, 18.12.1932, p. 1, No. 9, 6.3.1932, p. 1, No. 10, 13.3.1932, p. 1, No. 12, 27.3.1932, p. 1, No. 14, 10.4.1932, p. 1 and No. 16, 24.4.1932, p. 1, No. 17, 1.5.1932, p. 2 and No. 30, 31.7.1932, p. 1.

instead cleverly sided with the strongest and most promising existing association, enhancing the party's rural influence and securing the support that facilitated its future electoral success. The fusion of the Farmers' League and the Rural League had all but eliminated the appeal of the Landvolkbewegung as an alternative to the unity organisation, while the NSDAP's dramatic growth surpassed the Landvolkbewegung thereby reducing its attraction as a centre of opposition to the system.

From the autumn of 1931, as discussed earlier in this chapter, the Rural and Farmers' League and the NSDAP removed from the Landvolkbewegung its primary purpose: organising actions at confiscations and compulsory auctions. Under National Socialist and Rural and Farmers' League leadership, the Crisis and Shared Fate Communities were so strong and effective as to render the Landvolk redundant. Only in the province's north-west did pockets of the Landvolkbewegung remain, but these struggled to survive.

At the end of 1931 the NSDAP became openly hostile to the Landvolk and went on to crush it. In South Tondern Bossen had tried to keep the movement going before being compelled in November 1931 to commence his prison term.[214] Writing to Claus Heim, he described the NSDAP's conquest of the Landvolkbewegung as "the saddest event in Germany" and complained that the party had "seized" many of its "best men" and taken advantage of their "educational work."[215] In localities where the movement remained, National Socialists disrupted its meetings and even brawled with its followers.[216]

In Eiderstedt, Hamkens endured as chairman of the Unity Organisation and continued to aid his colleagues in their dealings with creditors and the authorities.[217] The county administrator blamed his agitation for the persistent unwillingness of the populace to pay taxes, but also noted that he had lost much prestige. The NSDAP and the Rural and Farmers' League impeded the development of the Landvolk's new Village Communities, by restraining their members from taking part in Village Communities' activities. Indeed, at one meeting where Hamkens was advertised to appear, an audience voted against him speaking.[218]

---

214 Oberstaatsanwalt to the Regierungspräsident in Schleswig, Altona, 28.11.1931 and 29.11.1931, LAS 309/22701.

215 Copy of an undated letter from E. Bossen, Neulandshof bei Leck, to Claus Heim, LAS 309/22701.

216 Copy of a report from the Landjägereiposten Ladelund, Kreis Südtondern, to the Landrat in Niebüll, Ladelund, 10.11.1931, LAS 309/22701; report from the Landjägereiposten Sande, Kreis Südtondern, to the Landrat in Niebüll, Sande b. Stedesand, 10.11.1931, LAS 309/22924.

217 For a late example, see Hamkens' letter in the *Husumer Nachrichten* No. 118, 22.5.1933, p. 7.

218 Report from the Landrat des Kreises Eiderstedt to the Regierungspräsident in Schleswig concerning "Landvolkbewegung im Kreise Eiderstedt" Tönning, 22.12.1931, LAS 309/22924.

The Landvolk rejected the Rural and Farmers' League's Crisis and Shared Fate Communities but it now lacked the strength to resist them.[219] In Rendsburg in January 1932, a meeting of 60 of the movement's "trusted men" resolved to continue to hold meetings of the Village Communities. However, they almost unanimously turned down a suggestion to oppose the NSDAP openly, nor did they dare to campaign against the Rural and Farmers' League.[220] The Landvolkbewegung survived at the local level but was no longer a significant power in the province. In fact, by the end of February 1932 it ceased to be an object of regular monitoring and reports for the provincial authorities.

In the Reich elections of July 31, 1932, the NSDAP gained 51 percent of the vote in Schleswig-Holstein, including 63.8 percent in rural areas. In the countryside the party gained majorities in all but the eastern counties where farm labourers continued to support the Social Democrats. Consistent with the nationwide trend, the SPD's vote remained relatively stable, but the DNVP suffered severe losses and the DSP, DVP and Economic Bloc were all but eliminated. At the following elections in November, the NSDAP experienced a small set back in the province, but with Hitler as Reich Chancellor, the party managed to consolidate its position with 53.2 percent in Schleswig-Holstein on March 5, 1933.[221]

Since the vast majority of Schleswig-Holstein's farmers were already united in the NSDAP-friendly Rural and Farmers' League, the subsequent elimination of opposition within agriculture in the province was swift. Within two months, objections to National Socialist dominance from the DNVP were silenced and the Farmers' Association was forced to dissolve. At the same time, the NSDAP pressured the Rural and Farmers' League to replace leaders who were not party members.[222]

Only Hamkens and Heim remained as possible rallying points for discontent, Hamkens as leader of the Unity Organisation in Eiderstedt, which refused to join the Rural and Farmers' League, and Heim in North Dithmarschen, who on his release from prison had established a new weekly Landvolk newspaper. When Hamkens refused to yield to the demands of Struve, the freshly appointed provincial agricultural leader (*Landesbauernführer*), Struve called a meeting of Eiderstedt's Unity Organisation at the end of May. Insisting it join the Rural and Farmers' League, Struve called for Hamkens' resignation and blocked a motion of confi-

219 Report from the Abteilung IA concerning "bevorstehende Aktionen des Land- und Bauernbundes in der Provinz Schleswig-Holstein" Altona, 10.10.1931, LAS 301/4698.

220 Report from the Landrat des Kreises Eiderstedt to the Regierungspräsident in Schleswig concerning "Landvolkbewegung im Kreise Eiderstedt" Tönning, 22.12.1931, LAS 309/22924; report from the Abteilung IA concerning "Landvolkbewegung" Altona, 23.2.1932, LAS 309/22701.

221 *Statistisches Handbuch für Schleswig-Holstein*, p. 537. Also see Heberle, *From Democracy to Nazism*, pp. 95 and 99.

222 Stoltenberg, *Politische Strömungen im schleswig-holsteinischen Landvolk*, pp. 186-193; Thyssen, *Bauer und Standesvertretung*, pp. 401-402 and 412-414.

dence on his behalf, upon which Hamkens defiantly closed the meeting. Later, Hamkens – and his entire board – resigned.[223]

Dusendüwelswarf

Nr. 23, 1933 | Im Geiste der Landvolkbewegung herausgegeben von Claus Heim. | 1. Juli

Entweder – oder!

Wir verweisen auf das Buch von Weigle „Zwischen Gott und Mammon ein Wegweiser in die neue deutsche Volkswirtschaft" Hammer-Verlag, Leipzig 1932, Preis 1,85. Die Schriftleitung.

Der Bauer weiß, daß er auf seinem Acker entweder gutes Korn bauen oder Unkraut wachsen lassen kann. Der Gärtner weiß, daß an seinen Apfelbäumen entweder gesunde Früchte oder Blutläuse wachsen können; beides zusammen geht auf die Dauer nicht. Ebenso kann man immer wieder beobachten, daß gewisse Rank- und Schlinggewächse schließlich auch den stärksten Baum umbringen. Der harmlose Gartenfreund kann wohl seine Freude daran haben, wie der Efeu am Baum emporklimmt und Aeste und Zweige üppiger und üppiger umrankt. Der Gärtner aber weiß, daß

sundem Menschenverstand die Gegenwart erlebt, erkennt auch ohne großes Geschichtswissen das Uebel, das ganz allein all das Elend über uns gebracht hat. Aber es tut bitter not, diesen Schmarotzer immer wieder bei seinem richtigen Namen zu nennen und nach dem Gärtner zu rufen, der ihn — in klarer Erkenntnis daß die Stunde des Entweder — oder da ist — mit Stumpf und Stiel ausrodet.

Entweder – oder! Es gibt keine Möglichkeit mehr für Volk und Kapital, von Baum und Schmarotzer. Es heißt jetzt: entweder Ernährung oder Währung, entweder Wirtschaft oder Bürgschaft. Entweder Leben, Weib und Kind, Haus, Hof, Acker, Vieh und alles, was des Menschen ist, — oder Hypothek und Zinsen, Aktie und Dividende, Pacht, Miete, Steuer und alles, was des

in ein Bekenntnis zur völkischen Idee gekleidet haben, müssen bei allen Aufgaben, deren Lösung noch aussteht, die eherne Naturnotwendigkeit des Entweder — oder immer wieder und wieder und überall erkennen lernen und durch Betätigung bestätigen. Machen wir uns ein paar brennende Fragen doch einmal klar an diesem so einfachen und lehrreichen Beispiel aus der Natur: entweder Baum oder Schmarotzer, ohne daß wir uns auf die Kaugummiworte aus „ismus" verlassen; wie Kapitalismus, Sozialismus, Nationalismus, Marxismus etc.

Was hat es denn da z. B. für eine Bewandtnis mit diesem „Marxismus"? Der Marxismus ist ein Krebsgeschwür am deutschen Volksbaum. Wie ein solches am Baum im Garten entsteht, haben wir schön gesehen: ein aus natürlichem

The Dusendüwelswarf was published from January to September 1933[224]

This left Heim's endeavours to revive the Landvolkbewegung with his newspaper, the *Dusendüwelswarf*, as the final remnant of independence in rural Schleswig-Holstein. It had commenced publication in January 1933. Though appropriately named after the symbol of the farmers' freedom, the obelisk marking the victory of Dithmarschen's peasant army over the forces of the Danish king and Gottorf Duke in 1500, the *Dusendüwelswarf* could continue for a short time. Völkisch, nationalistic, anti-capitalist, anti-Semitic, anti-liberal, anti-Catholic, anti-pacifist, anti-Communist and anti-materialistic, many of the articles published in all 36 issues of the newspaper were compatible with National Socialist thinking.[225] Yet the *Dusendüwelswarf* also expressed disappointment in the new government's at-

---

223 For reports of the meeting in Garding, see the *Eiderstedter Nachrichten* No. 8883, 1.6.1933, p. 2 and *Husumer Nachrichten* No. 125, 31.5.1933, p. 12.

224 *Dusendüwelswarf* No. 23, 1.7.1933, p. 1.

225 For the survey of comment and criticism offered here, see the *Dusendüwelswarf* No. 1, 28.1.1933- No. 36, 30.9.1933.

tempts to control farmers' debts and was disapproving of its reluctance in dealing with capitalism and in freeing farmers from "interest servitude". Further, it resisted National Socialist centralism in its calls for stronger village administration and local initiative, and it dared to warn against the suppression of freedom of thought and expression. Such liberties soon proved too much for the new authorities. Heim and his newspaper, like the visitors to his farm and their correspondence, were closely monitored after the seizure of power.[226] An article decrying that farms lost at compulsory auctions had not been returned to their owners provided a convenient pretext to ban the *Dusendüwelswarf* on October 1 for six months on the grounds it was "hostile to the state".[227] Effectively, this put an end to the newspaper and Heim, like Hamkens, subsequently withdrew from public life.

Signs of the Landvolkbewegung's decline were visible from October 1929 but the movement's activities, and attendances at them, decreased dramatically in the first months of the new year, accompanied by falling subscriptions to *Das Landvolk* and mounting financial difficulties. It was in the winter of 1929/1930 that the NSDAP became the dominant rural political force in the province. Although the Landvolkbewegung carried on after this time, the level of support it received was subject to various regional fluctuations and much diminished overall. Renewed interest in the "tax struggle" signalled a partial revival in 1931 but this was cut short as other groups, principally the Rural and Farmers' League and the NSDAP, adopted Landvolk tactics. Only remnants of the Landvolkbewegung survived by early 1932, yet Hamkens and Heim both doggedly endeavoured to continue the movement into 1933.

The explanation of the movement's decline cannot be reduced to any one decisive factor, nor should any single factor be accorded undue weight; a multi-faceted explanation is required. The most important internal factors included the reactions of the public and of the movement's leaders to the bombings and the bombers' arrests; the Landvolk's failure to secure a lasting improvement of the farmers' fortunes, to clarify its practical aims and to formulate a coherent programme to balance its predominantly negative criticisms; the loss of many key leaders due to im-

---

226 Landrat Meldorf to the Regierungspräsident in Schleswig concerning "Claus Heim. St. Annen – Oesterfeld" Meldorf, 20.7.1933 and Meldorf, 22.7.1933, both in LAS 309/22764; copy of a letter dated Heimhof, St. Annen-Osterfeld, 11.7.1933, sent to the Regierungspräsident in Schleswig from the Staatspolizeistelle in Altona, 9.8.1933, in both LAS 301/4699 and 309/22764; Landrat Meldorf to the Regierungspräsident in Schleswig concerning "Archivarin Helene Höhnk [sic]-Marne" Meldorf, 11.8.1933, LAS 309/22764; letters from Helene Hönck to Marie Peters in Wesselburen, Marne, 7.9.1933 and from the Abteilung IA 1 to Bürgermeister Herwig in Wesselburen, Flensburg, 4.8.1933, LAS 301/4699. Also see the extract from Hönck's manuscript sent from the Staatspolizeistelle in Altona to the Geheimes Staatspolizeiamt in Berlin, 14.12.1933, LAS 301/4699.

227 See the article, "Die verlorene Scholle", and the notice of the ban in a supplement to the final issue, *Dusendüwelswarf* No. 36, 30.9.1933.

prisonment and, from mid-1930, to a series of disputes and fractures, that were particularly harmful because there was no organisational structure that could sustain the movement in their absence; and, falling income as the movement's popularity waned, exacerbated by rumours of misappropriation, in conjunction with the substantial costs of legal proceedings and operation of *Das Landvolk*. An important external factor was the bans of the newspapers which also worsened the movement's financial difficulties while depriving it of its principal mouthpiece in the critical period of potential revival in the summer and autumn of 1931. The further key external factors were the farmers' growing hopes in the first half of 1930 in an improvement in their fortunes, bolstered, in particular, by Schiele's appointment as Reich Minister for Agriculture; the creation of the Rural and Farmers' League which fulfilled the long-held desire for unity of the majority of the province's farmers; and, the loss of support from the Landvolkbewegung to the NSDAP as the party expanded across the province and then proved its greater potential as a force against the Republic working within the system by way of its election successes. The Rural and Farmers' League, with the NSDAP's support, put an end to the Landvolkbewegung's revival, not only by adopting its methods in dealing with confiscations and compulsory auctions, but also by appropriating its campaign for the release of Claus Heim. Yet it was not until 1933 that the NSDAP finally silenced Hamkens and Heim, compelling Hamkens to yield to the Rural and Farmers' League, and by forcing the closure of Heim's newspaper.

Conclusion

The principal concern of the previous chapters has been to determine and explain the origins, course and decline of the Landvolkbewegung, together with the profound consequences of the movement's radicalising effects in Schleswig-Holstein's countryside. In short, the chief object throughout has been to elucidate how and why the province's rural population rebelled. The Landvolkbewegung mobilised the rural people and endeavoured to unify them in rebellion. Its protest meetings and marches, publications, civil disobedience and terrorism destabilised the established agricultural organisations and representation, shook confidence in the community and county administrations, placed great strain on the provincial authorities, and demanded the attention of the governments in Berlin. Consequently, both the authorities' efforts to contend with the Landvolk and the Landvolk's reactions to the authorities' measures called for thorough analysis. While a careful differentiation between the various phases of the farmers' protests is necessary, on the whole, it can be said that the movement's development was characterised by growing militancy in its actions and increasing hostility in expression. The abundant and diverse sources examined in this text have illustrated the Landvolk's manifold protest forms, its ideology, tactics and strategy, and the intricacies of its organisational and leadership structures. It is on account of these that the Landvolkbewegung must be recognised as a distinct and distinctive regional phenomenon as well as a remarkable rural protest movement. The Landvolkbewegung must also be identified within a wider shift to the political right in the province. The farmers' movement represented a sudden and dramatic stage in that process. More broadly, the history of the Landvolkbewegung furthers our knowledge of Weimar's radical right, of the pressures on the Republic and the mobilisation of opposition to it, as well as of the rise of the NSDAP in Schleswig-Holstein. The Landvolkbewegung's significance rests upon all of these grounds and not solely on the movement's complex relationship with the NSDAP and contribution to its success.

From the autumn of 1928 the Landvolkbewegung developed along Schleswig-Holstein's west coast as farmers banded together to block confiscations and compulsory auctions and held a number of impressive meetings. Expanding throughout the province in 1929, the rallies and bombings of the summer marked the movement's climax. By the end of the year, however, the Landvolkbewegung had lost momentum. Though the trials, meetings and the boycott of Neumünster continued, from 1930 it fell into a protracted decline.

The volume of rural unrest in the province, however, did not simply decrease thereafter.[1] In the context of the Landvolkbewegung's decline, a second phase of obstructive actions commenced in the autumn of 1931, largely conducted under the

1 This was Bergmann and Megerle's conclusion. See Bergmann and Megerle, 'Protest und Aufruhr der Landwirtschaft', pp. 228-234.

leadership of the Rural and Farmers' League and the NSDAP. Though historians have noted the creation of the Crisis and Shared Fate Communities in Schleswig-Holstein and in other regions of Germany, the resultant obstructive actions have been largely overlooked, even in the regional literature.[2] These merit greater recognition, and further research is essential to determine their nature and extent beyond Schleswig-Holstein.

As discussed, assessing the Landvolkbewegung's size and strength is difficult, because the movement did not keep official records. Yet the level of support appeared to peak in October 1929 when meetings attracted audiences typically between 300 and 500 strong and *Das Landvolk* printed 12,000 copies daily, making it one of the most widely read papers in the province. In its composition, the movement was almost entirely rural and in fact always found it difficult to attract the support of occupations outside of farming. The movement was overwhelmingly male. Women attended meetings but were always considerably fewer in number than men and played no active role in the movement. Unfortunately, there is no data available to indicate whether the Landvolk's appeal was stronger among younger or older farmers.

The Landvolkbewegung's avoidance of organisational structure set it apart from Weimar's many parties and associations. The movement never introduced statutes, membership lists or other forms of bureaucratic structure or procedure. Consequently, it was highly dependent on a small number of energetic and determined leaders. Many came from the veteran, nationalist and agricultural organisations, which were essential in establishing the connections upon which the movement formed. Of greater importance overall, however, were networks of "trusted men" and familial, neighbourly and occupational ties that facilitated the coordination of actions, meetings and fund-raising. Nevertheless, the movement could not function without some forms of organisation. The Defence Association vigilante network maintained contact among the farmers in the winter of 1928/1929, the Crisis Assistance provided a further link between the movement's followers and its "trusted men" from the spring of 1929, and from November the Volksmacht Association provided financial support for the newspaper. Though it was not an administrative body, *Das Landvolk* was of particular significance because it offered the movement's followers a focal point with which they could identify and a centre for information and communication.

The agricultural crisis was the central factor leading to unrest in Schleswig-Holstein's countryside, but the direction of the Landvolkbewegung and the nature of its protests were shaped by the collision of economic factors with longer-term political and social structures and processes, as well as by the unique historical experience in the province. Of particular importance were regional patriotism, nationalism and the rejection of socialism, which were entrenched in the liberal parties that dominated politics in Schleswig-Holstein before the First World War. Like-

2 See chapter 8.

wise, anti-Semitism and völkisch thought played a significant role, and had also been a part of the province's political culture since the late nineteenth century. During the Weimar period the forces of democracy in the countryside were not consolidated, rural voter loyalties were weak and until 1924 the parties of the right increasingly gained electoral support. This development was promoted, above all, by the rejection of the revolution and of social democracy, by the association of the Republic with the humiliation of defeat and the Treaty of Versailles as well as the prolongation of the "controlled economy", inflation and the implementation of tariffs in the face of continuing economic uncertainty. Long before the onset of the crisis, farmers had already begun to turn to the right and there were already signs that rural circles were receptive to right-wing interpretations of and solutions to their difficulties.

The pronounced confidence and defiant natures of the farmers on Schleswig-Holstein's west coast were additional factors explaining why rural protest arose there first. These were products of the former affluence, the social conditions and unique histories of independence and local administration of the region. The inspiration the Landvolkbewegung drew from these histories is evident in the vocabulary, symbols and forms of its protests.

The agricultural crisis of the late 1920s struck Schleswig-Holstein, and especially its west coast, with particular severity. Although the farmers in the region had long been accustomed to prosperity, they were very vulnerable to crises given the lack of diversity of their products and the large capital investments required for commercial livestock farming. When profitability declined from 1927 as a consequence of falling prices and rising production costs, they were caught with perilous levels of debt at high rates of interest and a tax burden greater than Reich and province averages. Additional regional-specific afflictions, including animal disease and a poor harvest, aggravated the difficulties faced by farmers in meeting their financial obligations and produced the intense discontent that led to the unprecedented mobilisation in the countryside.

The great rallies of January 28, 1928 were both an appeal to the Reich government for fundamental reform of trade and tax policy and an attempt from below to compel the province's disparate agricultural associations to fuse, in order to provide a more effective lobby. Yet the rallies raised the farmers' expectations to unreasonable levels. The Reich government's measures fell short of the farmers' inflated hopes and ultimately this undermined their confidence in the state. Moreover, the rallies put the agricultural associations under immense pressure to pursue nothing less than complete amalgamation, rather than simply to seek greater cooperation as they did outside Schleswig-Holstein.

The unity endeavours left the province's farmers frustrated and with an organisational constellation badly shaken but no less fragmented than before the rallies. This proved especially devastating in Schleswig-Holstein because it was the agricultural associations, rather than any particular political party, to which the farmers looked for the representation of their interests. The middle-class parties lacked rural organisational structure and the active participation of farmers, and in their voting behaviour the rural population's ties to them were already fragile. This explains why the rallies and the failure of the unity endeavours, by unsettling the farmers' allegiances to their agricultural associations, created such a volatile situation.

The next major turning point came as the postponement of tax payments ended in October 1928 and bailiffs began to recover outstanding sums. Complaints about taxes had featured strongly at the rallies of January 28, 1928, but relief measures had given the farmers a reprieve and temporarily curbed unrest. The tax issue, however, resurfaced and quickly triggered new protests and the development of the Landvolkbewegung from the autumn of 1928.

The Landvolkbewegung moved to action by inciting tax-strikes and obstructing confiscations and compulsory auctions. It retained the spirit of January 28, 1928 by seeking the unity of all the province's farmers. Nonetheless, one must differentiate between the Landvolkbewegung from the autumn of 1928 and the earlier rallies, not only because of the Landvolk's adoption of direct action but also because of its vehement opposition to what it described as Weimar's "Jewish parliamentary system". It is not true, as Stoltenberg and others since have argued, that economic motives were "immediately" mixed with political opposition from the demonstrations of January 28, 1928.[3] At that time the majority of the province's farmers still believed that reform of trade and tax policy was the solution to the crisis. They were working within the system and were largely cooperative with it. Even the most extreme demands made at some of the rallies, such as the calls for the refutation of Germany's sole war guilt and for a revision of reparation payments, were moderate compared with the Landvolkbewegung's later outright rejection of "the system". The radicalisation of Schleswig-Holstein's rural population actually began later and was, therefore, much swifter and more spectacular than has previously been acknowledged.

The Landvolkbewegung was significant for its radicalisation of Schleswig-Holstein's rural population, for channelling the farmers' discontent over the agricultural crisis into extreme opposition to the Weimar Republic. From the autumn of 1928 the Landvolk interpreted agriculture's difficulties as a consequence of the failings of the Weimar system and created a deep division between the farmers and the authorities. Moreover, the process of radicalisation was characterised not only by the ever more extreme nature of the movement's methods, but also by a shift in the purpose of the movement's actions and its propaganda. In short, the movement's focus altered from the agricultural crisis to defiance of the authorities.

The Landvolk's meetings were especially important in this respect because they fostered the idea that the farmers' actions were part of a struggle against "the system" itself. Within a month of Hamkens and Heim's public rejections of tax payments, the tax issue led to conflict with the authorities at Lunden and Beiden-

3 See chapter 2.

fleth. When their expectations for further leniency were dashed, farmers turned to what they termed "self-help", collaborating with their neighbours to block compulsory tax collection. It was these actions that created the division between the farmers and the authorities, uniting the former against the bailiffs, community chairmen and county administrators who conducted and authorised the confiscations and compulsory auctions. Lunden and Beidenfleth were, of course, important as models for further action. Yet essential for the process of radicalisation was that the Landvolk reasoned at meetings that their actions did not simply oppose tax payments, but also protected their property from expropriation, which they maintained was conducted under the guise of legality by authorities serving as the "lackeys" of international capital and the Weimar system. Going much further than the complaints of the agricultural associations that Weimar paid insufficient attention to their interests, the Landvolk claimed that the Republic was deliberately allowing agriculture's destruction because it was serving "foreign" interests.

Alongside these condemnations, the Landvolk adopted an uncompromising stance, separating friends and enemies. The latter were not only the authorities and those who conducted, supported or took part in confiscations and compulsory auctions, but also anyone who dared to thwart or criticise the movement. Among the former, the Landvolk strove to reinforce unity. The Landvolk's meetings were generally characterised by high levels of enthusiasm and approval, at least until late 1929, and the movement's leaders astutely bolstered cohesion with gestures and acts of camaraderie and accord. This was achieved, for instance, by the singing of patriotic songs, asking audiences rhetorical questions, or calling on audiences to raise their fists to display their united resolve. Those who supported the movement were assured of assistance, both legitimate and illegal, in dealing with their financial difficulties, while those who disapproved of the movement or dared to bid at auctions were subject to denunciation, intimidation and ostracism.

The Landvolk's trials further deepened the division between the authorities and farmers. Convinced that they were entitled to the state's assistance, farmers were incensed at being treated as criminals and they used the courts to publicise their grievances and criticise the authorities and the system. Meetings and marches of solidarity for the convicted, as well as receptions for those released, fostered the farmers' identification with the movement while demonstrating their opposition to the authorities. Then, when the rallies and marches led to confrontations with the police, this served to underline the divide between the farmers and the authorities once again. The spread of conflict into the courts and onto the streets reflected the growing extent of radicalisation, as did the expansion of protests from the west coast across the entire province in 1929. At the same time, increasing militancy and violence marked the growing intensity of the Landvolk's radicalisation.

The Landvolkbewegung broke away from the rallies of January 28, 1928 by turning to direct action and extreme opposition, yet these responses remained an extension of the earlier protests in that they sought to contend with the agricultural crisis and to unify the province's farmers. As the movement radicalised, however, there was a further shift as defiance of the authorities and opposition to the system became ever more central. While obstructive actions at confiscations and auctions were quite palpably a form of crisis relief, later undertakings such as the Crisis Assistance clearly served a dual purpose, seeking both to institutionalise tax obstruction and gain greater control over local and provincial affairs. Similarly, although Heim's first project, the Defence Association, was devised to protect farm property, the subsequent bombings were an attempt to draw attention to the farmers' plight, intimidate the authorities and destabilise "the system". Increasingly, challenging the legitimacy of the authorities became an end in itself. The raid on the courthouse prison in Hohenwestedt, the boycott of Neumünster, not to mention pranks such as the abduction of the director of the agricultural school in Garding, were only marginally connected to the agricultural crisis.

As the agricultural crisis, too, fell into the background, even tax obstruction was reduced to just one element in the movement's calls for rejection of cooperation with "the system". Ever more, the movement's energy was geared toward generating and spreading such opposition, and this was also seen at Landvolk meetings. Discussion of the crisis, of the farmers' hardships, of the necessary changes to agricultural and trade policy soon diminished, as the movement's leaders declared that the solution to their difficulties was the elimination of the Weimar system.

Landvolk leaders understood that it was their task at meetings to awaken their colleagues to the true nature of their plight. As has been amply demonstrated, the movement's leaders across the province all embraced and adhered to a body of anti-democratic, anti-socialist, anti-liberal, anti-Semitic, anti-capitalistic, nationalistic and völkisch ideas. Setting out to indoctrinate their audiences, from early 1929 their meetings developed a routine, the centrepiece of which was an address delivered by the chief speaker, usually Hamkens, explaining the agricultural crisis and indeed all ills afflicting Germany. In short, they traced all developments from the First World War, from defeat, revolution and the birth of the Republic to reparations, inflation, trade agreements, debts and taxation as leading to the agricultural crisis by the design of "international powers" seeking to seize the farmers' property, to enslave and ultimately destroy Germany.

The movement's struggle against "the system" gained greater momentum with the founding of *Das Landvolk*. Quite frank in its intentions, the Landvolkbewegung established the newspaper as an instrument of propaganda, employing Bruno von Salomon as editor and other nationalist revolutionary intellectuals to contribute articles. *Das Landvolk* seized every opportunity to deride the authorities and the "administrative apparatus". As its circulation increased during 1929, *Das Landvolk* was able to stir up opposition and spread extreme nationalism, anti-Semitism and anti-system sentiment ever more effectively.

So pervasive was the Landvolkbewegung's abhorrence of "the system" that it shaped the reasoning behind the movement's rejection of organisational structure.

Landvolk leaders argued that the absence of organisation facilitated cooperation with other groups, allowed all farmers to feel part of the movement and made it difficult for the authorities to prohibit its activities. They also maintained that it encouraged followers to assume responsibility and would ensure the movement did not succumb to bureaucratic ossification. Moreover, they made it very clear that the absence of an organisational apparatus was essential in order to combat Weimar successfully. They insisted that it was the methods and structures of the parties and associations that exposed them to the factionalism of the Republic, engendered "brotherly struggle" and made them instruments of "international powers" and "Jewish high-finance".

Anti-Semitism played a central and prominent role in the Landvolkbewegung's ideology, an aspect paid insufficient attention in previous studies of the movement. Manifest from the movement's emergence in the autumn of 1928, anti-Semitism featured repeatedly in almost every issue of *Das Landvolk*, at meetings, rallies and courthouses. All Landvolk leaders employed it. There is no reliable gauge for the reception of anti-Semitism among the movement's followers and, of course, no means of determining the extent to which they actually embraced anti-Semitic ideas. Reports on Landvolk meetings show, however, that speeches containing anti-Semitic content invariably received applause and sometimes incited anti-Semitic cries or comments from audiences, and this does point to the widespread acceptance of, if not approval for, these views.

The Landvolkbewegung's anti-Semitism was economic, political, cultural and racist. Though used in a variety of contexts, it was most often linked with criticisms of international capital, the Weimar system and, to a slightly lesser extent, the press sympathetic to the Republic. Nonetheless, anti-Semitism's most important function in the Landvolkbewegung was its integral position in explaining the crisis. It was the essential link between parliamentary democracy and international capital, since Jews or "international powers" were said to be using these means to destroy agriculture and enslave the German people. Anti-Semitism, therefore, served to provide a readily comprehensible explanation for the complex causes of the farmers' misfortunes.

The Landvolkbewegung was intensely anti-Semitic, attributing its difficulties to "Jewish influences" and the domination of a "foreign race". Yet given that the movement was not interested in practical politics, it is perhaps unsurprising that it was not concerned with matters such as assimilation or the place of Jews in public life. Furthermore, the Landvolk's anti-Semitism was not directed at local individuals. The closest was Regierungspräsident Abegg in Schleswig, while all others were ministers in Berlin. Even travelling cattle traders, who one might expect to have been depicted as profiteers, were not targeted. In Schleswig-Holstein bidders at compulsory auctions were sometimes labelled "Jews" but here the word had an expletive function. Jews of the region were not branded as profiteers, let alone deemed agents of other evils. The Landvolk did make numerous calls for its followers not to support Jewish businesses. However, these calls were always general and did not concern specific local ones.

The farmers' interactions with Jews or fears of economic rivalry do not explain the appeal and depth of anti-Semitism in the Landvolkbewegung. The Landvolk's anti-Semitism bore no relation to the farmers' experiences of Jews in the province. The abstractions of the movement's anti-Semitism were founded upon prejudices and fallacies which can be traced back to the political anti-Semitism of the late nineteenth century. These were bound into broader conservative-nationalist ideology and adapted to the new social and political conditions of the Republic by the veteran and paramilitary organisations, the DNVP and, above all, the Rural League. Drawing on their experiences in these organisations, the Landvolk leaders reactivated old concepts and tailored them to the agricultural crisis.

The Landvolkbewegung must be ascribed a significant role in disseminating anti-Semitic views in rural Schleswig-Holstein. Further research is required, however, to build on the pioneering work conducted by Gerhard Paul and Peter Wulf in order to determine the nature and strength of anti-Semitism in the countryside for the entire Weimar period.[4]

The counterpart to anti-Semitism was völkisch thought. By placing an emphasis on the concept of Volk, the Landvolkbewegung offered farmers a sense of belonging stronger than that of the shared experience of crisis, together with the unity that had long been desired from the divided agricultural associations. In addition, the racial definition, at least in theory, allowed Germans outside of agriculture to join the movement. It also presented the farmers with idealistic grounds, rather than self-interest, for the preservation of their property, a reason to believe that their struggle was one of protecting German soil in the national interest. Convinced that their occupation was not just of fundamental value to the economy but to the German nation at large, völkisch thought gave the farmers' further claim to special entitlement. It elevated agriculture's importance in a racial sense, especially when combined with notions of healthy rural living and the purity of blood of Schleswig-Holstein's old tribes. This thinking paved the way for a small number of fanatics who, from late 1929, used the movement to air their pseudo-scientific understanding of race and of history as a racial struggle.

Connections to other groups and the involvement of individuals from them were manifold, but the Landvolkbewegung always remained an independent farmers' movement. Its open organisational form welcomed the participation and contribution of members of the Stahlhelm, DNVP, Rural League, Tannenbergbund, Artam movement and of Ehrhardt's circle without ever requiring them to relinquish their allegiances. Some who played leading roles, like Weschke, Muthmann and Kenstler, were not farmers, but were closely connected to agriculture by qualification or occupation. Still others, central in the Defence Association or in the preparation and execution of the bombings such as Volck, Nickels, Kaphengst and

---

4 Paul, ' "Die Juden sind unser Unglück!" '; Wulf, ' "Jüdische Weltherrschaftspläne".'

Schmidt, were strangers to farming. Bruno von Salomon was an important figure at *Das Landvolk*, and he and other publicists made a significant contribution to the radicalisation of the movement. Hans Beyer, however, made far too much of their influence as did Otto-Ernst Schüddekopf, who held them responsible for politicising the farmers' economic protests.[5] The farmers had begun to radicalise before *Das Landvolk* commenced publication and national revolutionary publicists did not turn the movement, or even the newspaper, into their instrument. The Landvolkbewegung remained under the farmers' control. Farmers carried out the movement's obstructive actions, Heim was the undisputed leader of the bombings and he, Hamkens and many other farmers remained active as speakers and organisers and oversaw the management and content of *Das Landvolk*. Various right-wing parties and organisations as well as the agricultural associations hoped to use the Landvolkbewegung for their own ends. However, when outside groups sought to gain greater influence within the movement – such as the Tannenbergbund members' attempt to take control of *Das Landvolk* – their endeavours caused division and eroded support, and failed completely. It is mistaken, therefore, to argue that organisations or individuals outside the Landvolkbewegung steered or took over the movement just as it is quite incorrect to contend that they were responsible for the Landvolkbewegung's radicalisation.

The extent of the KPD's role in the Landvolkbewegung should also be considered carefully. The KPD became interested in the farmers' struggle after Uhse and Bruno von Salomon joined the party. Communists took part in some obstructive actions in 1931 and 1932, directed more propaganda toward the farmers in this period and supported the campaign for Claus Heim's release. In 1932 the KPD managed to establish a number of committees in Schleswig-Holstein's south-west and found some pockets of electoral support among the small farmers there. Ehrenfried Friedrich, however, overstated the party's influence in the countryside and the extent to which the party won over the Landvolk's followers. In prison, Heim was cordial toward visitors from the KPD and declared his readiness to cooperate with all groups. Even so, he always insisted on the Landvolk's independence. The movement did not support the party and there is no evidence that the KPD ever gained any influence within the movement.

It must be acknowledged that the authorities faced great difficulties in dealing with the Landvolkbewegung. The state's assistance for agriculture had not met the farmers' high expectations and the level of discontent in the countryside was considerable. Regierungspräsident Abegg, in particular, was in an unfortunate position facing Landvolk leaders who seemed committed to defiance and who were prejudiced against him because he was Jewish and foreign to the province. Furthermore, laws and regulations limited the authorities' options in tackling the movement, for instance in the imposition of bans and the relocation of auctions. The Landvolk's elusive language often made it difficult to press charges and on some occasions

5 Schüddekopf, *Linke Leute von rechts*, pp. 306-307.

sympathetic courts passed mild sentences so that trials had deficient deterrent effect. Mistakes and poor communication and coordination on the part of the provincial, county, justice and prison authorities, the police and finance offices sometimes made the task more difficult still.

Nevertheless, this should not obscure the fact that the authorities' responses to the Landvolkbewegung were unsatisfactory. Nils Werner argued correctly that Abegg hid behind the courts when he should have sought to earn the farmers' respect by confronting them.[6] This was merely one failing in the provincial authorities' handling of the movement. The authorities were indeed largely dependent on police intervention and legal action in dealing with the Landvolk's insubordination. The important point is, however, that the authorities' response to the movement did not simply fail to curb unrest in the countryside but actually led to further radicalisation. However merited they were, bans, warnings, arrests and trials enraged the farmers, produced greater antagonism and ultimately exacerbated the divide between the farmers and the authorities. Moreover, it is essential to recognise that it did so because there was no counterbalance to suppression. Unlike his predecessor, Johannsen, who had been proactive in his endeavours to gain relief for agriculture, Abegg showed little initiative on the farmers' behalf and failed to gain their confidence. His gestures and words of concern were unconvincing because they were not supported by genuine commitment to the farmers' welfare. Indeed, Abegg's lack of understanding for the farmers' plight and his aloofness to the province's rural population played into the hands of Landvolk leaders who disparaged him as a stranger to the province.

Abegg rightly recognised that the Landvolk leaders were deliberately creating trouble. However, he was so convinced that they were not in distress but simply being provocative that he failed to acknowledge their hardship, even ignoring reports to the contrary, reports he himself had commissioned. His determination to uphold the state's authority, and sometimes his own reputation, led him to clumsy and even dishonest conduct on a number of occasions. Furthermore, the omissions of details in clarifications and investigations, the deliberate exaggerations of the danger the farmers and their actions posed, the interference in the course of justice and the attempts to conceal evidence that gave the appearance of further transgressions backfired when exposed. These actions damaged his credibility, reinforced the impression that the farmers were being treated in a heavy-handed fashion and ultimately undermined the state's authority.

Constructive counter measures were required but Abegg's decision to support the moderate Farmers' Association came too late in mid-1930 and his idea of a calming propaganda campaign, highlighting the Republic's achievements, proved too cumbersome for the provincial press and was abandoned. Additionally, Abegg failed to engage with the farmers. He expected the agricultural associations to support him in condemning the Landvolk's actions but offered them little in return.

---

6 See the introduction, p. 8.

Abegg's negotiations with them over the measures to alleviate the crisis were token and so too was the farmers' participation in the "tax commissions" after he diluted their powers. Abegg failed to comprehend that the farmers' unrest and their turn to "self-help" was partly a product of their feeling helpless to influence their circumstances. Abegg did not understand that the farmers' involvement was required in order to restore the state's authority.

Arguably, the authorities' measures did help to stop the Landvolkbewegung. Legal proceedings, for example, led to the imprisonment of its leaders, and bans of *Das Landvolk* hastened its liquidation. All the same, the authorities' strategy failed. The farmers' experience of their provincial administration was essentially one of suppression, little understanding and even less participation. Since for the farmers the provincial administration represented the Weimar system, that experience fed anti-Republican sentiment. Perhaps it is not surprising that the farmers adopted direct action when their hopes in Weimar waned. Economic difficulties, however, were only part of the problem. Genuine participation was required if there were to be a chance of overcoming the Landvolkbewegung's contention that Weimar was not their Republic but a foreign system. For the farmers, Weimar democracy spelt powerlessness.

Though adamant in its determination to preserve the farmers' property and to oppose international capital and the "Jewish parliamentary system", the Landvolkbewegung failed to produce a concept that could lead to a practical or attainable solution to the problems it had identified as afflicting agriculture and Germany. Its attempt to clarify its aims in November 1929 confirmed that it envisioned a Volksgemeinschaft for Germany's future. But beyond cursory assertions that it be authoritarian, the movement never ventured to formulate a model for an alternative state structure nor, more importantly, did it put forward any plan, programme or intermediate objectives that showed how a Volksgemeinschaft could be established in Weimar's place. The Landvolkbewegung rejected a programme just as it rejected organisational structure. But, in any case, by the end of 1929 the movement had exhausted its own possibilities.

There was an immense gap between the Landvolkbewegung's wish to eliminate the system and its actions and options. These were limited anyway by the movement's refusal to work within the system. Having despaired in obtaining relief from Weimar, the movement left the business of lobbying for assistance and policy reform to the agricultural associations, and it rejected parliamentary participation. A lack of ambition to seize power further limited the movement's options. Vigilante forces set up to defend farm property were never organised, or indeed strong enough, for the movement's leaders to contemplate any form of offensive action, and there is no evidence to suggest they ever considered it a possibility.

Yet the alternatives that remained were neither suited nor sufficient to destroy the Republic. At meetings Hamkens declared that the system would collapse if the farmers refused to cooperate with it, but in practical terms this amounted to little more than the obstruction of compulsory tax collection and civil disobedience, chiefly in the form of the defiance of police orders. There was no mass basis upon which passive resistance could possibly succeed in overthrowing the state's authority since the movement was confined largely to one province and a single occupation.

It is also difficult to consider the Crisis Assistance committees as an attempt to depose of "the system". The committees were unquestionably an expression of the desire for local autonomy that was rooted deep within the west coast consciousness, but they did not seek to assume authority beyond the local level and were never expected to do more than serve alongside the existing institutions as an "additional government". Nevertheless, even if one insisted that this was a veiled attempt at revolution, it would simply highlight how far the Landvolkbewegung was from finding a realistic strategy. Except for their function as a system of tax obstruction, the Crisis Assistance committees were confined to just three counties, were poorly organised, with little public support and were consequently soon driven underground by the provincial authorities.

While Heim's objectives in resorting to bombings were manifold, it appears he anticipated the attacks would generate such widespread fear as to paralyse political life. But even if one were to suppose that a bombing campaign greater in scale could have produced such an intimidating effect, the fall of "the system" would be no corollary. It seems Heim had not thought his plans through, and Schleswig-Holstein's rural population widely agreed that bombings were not a realistic method of eliminating the Republic.

In essence, the Landvolkbewegung's demagogy, obstructive actions, demonstrations and bombings stirred up resentment, weakened the authorities' standing and strained law enforcement. Though effective in encouraging the desire for radical change, the movement could produce neither a concept that might possibly destroy Weimar nor come up with a clear vision of what might replace it.

Doubts over its capacity to effect improvement and over its lack of a programme and established goals were among numerous factors that contributed to the movement's decline from the end of 1929. Further significant factors included the rejection of bombs as violent and ineffective, financial strains exacerbated by rumours of misappropriation, the imprisonment of the movement's principal leaders and bans of its newspapers, as well as tension between the leaders that led to divisions and departures and drove away followers. In 1930 a short period of renewed hope in agriculture's fortunes and in favourable policy changes also diminished the movement's attraction. So too did competition from other groups, above all from the NSDAP and Rural and Farmers' League. The Rural and Farmers' League satisfied the rural population's desire for a unified agricultural organisation and deprived the Landvolkbewegung of its appeal as an alternative source of unity. Significantly, it was the radicalisation of the province's farmers, and their shift to uncompromising opposition of the Republic, that ultimately served to bridge the differences in interests that had separated the associations for so long.

There can be no question that part of the significance of the Landvolkbewegung was its substantial contribution to the rise of National Socialism in Schleswig-Holstein. The NSDAP was the greatest beneficiary of the Landvolk's radicalisation of the countryside. Nonetheless, there was a great deal more to the relationship between the two movements than the NSDAP taking advantage of the Landvolkbewegung's spread of right-wing propaganda and hostility to the Weimar system.

Before the farmers began to mobilise themselves, the NSDAP's growth in the countryside was limited. Gauleiter Lohse was unusual among the party's leaders for his early efforts in directing propaganda toward agriculture and in attempting to develop the party's organisational structure accordingly, and this put the NSDAP in a good position to capitalise on the increasing unrest in the countryside. Even so, it was only after the rallies of January 28, 1928 that the party really began to expand beyond South Dithmarschen and that strong campaigning led to promising results in some counties on the west coast at the Reichstag election of May 1928.

The NSDAP grew throughout 1928. It was the Landvolkbewegung, however, with its actions and boisterous meetings, that was the dynamic movement in the countryside from the autumn of 1928 and throughout 1929. During this period, the authorities held the Landvolkbewegung largely responsible for the radicalisation in the countryside, while other groups, including the NSDAP, were seen more as responding to the farmers' unrest. Though the provincial authorities soon recognised that the NSDAP posed a greater long-term danger, because of its activism, the Landvolkbewegung was of greater concern in 1929. Not until early 1930, when the Landvolkbewegung had begun to decline, did the NSDAP take over as the dominant rural political force in the province.

The NSDAP began to win over Landvolk supporters from the spring of 1929, especially after the Wöhrden incident. It then gained their support in much larger numbers from the winter of 1929/1930. The records of the Berlin Document Center, however, suggest, that only a small number of leaders and "trusted men" ever joined the party. Among the few who did, most did not join before late 1931, when the Landvolkbewegung was truly in decline, or until after the seizure of power.

Where the NSDAP stood to gain most from the radicalisation of the rural population was at elections. The Landvolkbewegung rejected parliamentary participation as cooperation with "the system", leaving the NSDAP as the party most likely to deliver the fundamental change it demanded. Though the NSDAP did not achieve its extraordinary result of 51 percent until July 1932, it was at the Reichstag election of September 1930 that it made its decisive electoral breakthrough, gaining 27 percent in Schleswig-Holstein, including no less than 35.1 percent in the countryside.

The Landvolk's dissemination of extreme right-wing propaganda and its promotion of radical opposition to Weimar among the rural population played a major role in facilitating the NSDAP's growth in the province. The Landvolk's anti-Semitism and anti-capitalism, its anti-democratic and anti-system sentiment made it easy for the NSDAP to build on propaganda already familiar in the countryside as it expanded from the west coast across the countryside in 1929 and 1930. Furthermore, their common ideas aided the farmers' transition from the Landvolkbewegung to the NSDAP.

It must be emphasised that the NSDAP was calculating, and actively sought to profit from the Landvolkbewegung. Among the few historians to have recognised this, Rudolf Rietzler observed that the party deliberately held its activities in those areas where the farmers' unrest was strongest, and that it profited in areas where the Landvolkbewegung had found support.[7] Yet the NSDAP did more than this. Not only did it adopt Landvolk vocabulary and time campaigns in its newspaper for maximum effect, it also cleverly exploited the Landvolk's actions, using the excitement they had caused, to lure crowds to its own promotional meetings. When the Landvolk held successful meetings, the NSDAP claimed these were the fruit of its own hard work, and in 1930, when many of the Landvolk leaders were in prison, the party took advantage of their absence, establishing locals among former Landvolk followers.

Though the Landvolkbewegung would have welcomed cooperation, its emergence in the autumn of 1928 immediately led to tension with the NSDAP because the party saw the Landvolkbewegung as a rival. The public, press and even the police sometimes confused the two movements as a result of the similarities of their ideologies. The NSDAP's anti-Semitism, for instance, was much like the Landvolk's in that it was generally used in combination with attacks on other institutions and organisations such as Marxism and "high finance".[8] Both movements, as well as the provincial authorities, recognised that little separated them ideologically and for a time the provincial authorities even speculated whether there might in fact be closer organisatorial ties between them. Leaders of both movements always denied any links existed, although there were in fact many instances of fraternisation and cooperation at the local level.

The National Socialist Gau and Reich leaderships refused to reciprocate the Landvolkbewegung's friendly stance. They feared that association with the Landvolk's unruly and illegal activities would frighten away potential supporters and, above all, give the authorities a pretext to ban the party. Though it was never articulated, it would also appear that the party could not tolerate the Landvolk's activism because it weakened its own dynamic image. Privately, some prominent National Socialists approved of the Landvolk's greater activism. But abiding by the Reich leadership's legal course, the party in Schleswig-Holstein resisted pressure to be-

7 Rietzler, *"Kampf in der Nordmark"*, p. 426.

8 See chapter 8, p. 318.

come more extreme. Hitler explicitly prohibited his members from participation in the Landvolkbewegung and the party expelled those who violated these instructions too openly. The Landvolkbewegung was angered by what it saw as the condemnation of its activism but it paid little heed to the NSDAP in its choice of strategy.

The Landvolkbewegung presented a special challenge for the NSDAP since its rejection of parliamentary politics and organisation gave it the appearance of a most uncompromising opponent of the Republic. The party defended its parliamentary aspirations and took every opportunity to undermine the farmers' movement, attacking its actions at confiscations and the bombings, in particular, as ineffective. It highlighted its lack of organisation, underlined its inability to influence agricultural policy, drew public attention to moments of tension within the movement and noted its failure to attract the support of other occupational groups.

At the same time, the party sympathised with the farmers. The *Schleswig-Holsteinische Tageszeitung* supported Landvolk followers on trial as victims of a system that had driven them to despair but it stopped short of siding with the movement itself. Seeking to win over the farmers, the NSDAP claimed that the Landvolkbewegung's ideas merely mirrored its own. Again and again the party maintained that only it truly represented agriculture, that only it had the programme, organisational structure and leader capable of saving Germany.

Rumours that the NSDAP was involved in the bombings put great pressure on it, but once free of suspicion, it stood to gain as public opinion turned and wide sections of the rural population expressed approval of the bombers' motives but continued to object to their methods as inappropriate and ineffective. As scepticism grew over the primarily rural movement's methods and want of strategy, the NSDAP appeared a more promising alternative with its programme, tight organisation, pragmatic methods, determination to work within the system and claims to represent all classes and occupations. The election success of 1930 confirmed the party's prospects and it reinforced its growing influence in the countryside with the development of the Agricultural Political Apparatus and its infiltration of the Rural and Farmers' League.

Where the Landvolkbewegung's terrorist acts, wish to control local affairs and rejection of party-politics invite comparison with anarchism or syndicalism, it is at the movement's desire for an authoritarian state that the resemblance stops. However, it is just as inconceivable that the Landvolk imagined an authoritarian alternative to Weimar as totalitarian. The Landvolkbewegung was not organised on the same insistent leadership principle (*Führerprinzip*) as the NSDAP rather it permitted its followers great freedom of action.[9] As Kurt Sontheimer explained in *Antidemokratisches Denken in der Weimarer Republik*, an authoritarian state was bound to become totalitarian under the conditions of industrial mass society, but the

9 In this respect, Beyer's assessment that the Landvolkbewegung opposed National Socialism is accurate: Beyer, 'Die Agrarkrise und die Landvolkbewegung', p. 173 and 'Landbevölkerung und Nationalsozialismus', p. 74.

idea itself, he argued, allowed for decentralisation and self-administration, for local and professional concerns to be run by locals.[10] Certainly this is the way the Landvolk would have had it. The unique history of the west coast was reflected in the Landvolk's protests not just in the farmers' cries of "Rather dead than a slave", but also in their defiance in dealing with the police and declared aversion to "non-native" administration in the province. Similarly, the Crisis Assistance committees were a sign both of the Landvolk's antipathy to central Prussian control and confidence in their ability to manage their own affairs.

When Rudolf Heberle wrote that Schleswig-Holstein's farmers ended up voting for a party contradictory to its political tradition, he was reflecting on the path the province's rural population took from supporting the liberal parties at the beginning of the Weimar Republic to supporting the NSDAP from 1930. The strength of this liberal tradition has since been qualified and its nature clarified. By voting for the NSDAP, however, Schleswig-Holstein's rural people went against a far more deeply rooted tradition. As much as the Landvolk desired a strong state, National Socialist Gleichschaltung did not accord with their independent character and faith in local autonomy. This is not to suggest that the independent mindedness of the west coast farmers led to a diminution of support there for the NSDAP after the seizure of power. Yet in 1933 Hamkens' symbolic refusal to comply with the party's provincial agricultural leader and Heim's calls for stronger village administration and criticisms of the suppression of freedom of thought showed that they, at least, never intended to submit to National Socialism. Though their wish for the destruction of the Weimar system was fulfilled, still Hamkens and Heim were not quite at home in the Third Reich.

---

10 K. Sontheimer, *Antidemokratisches Denken in der Weimarer Republik* (Munich, 1978), pp. 203-209.

Bibliography

UNPUBLISHED SOURCES

1. Bundesarchiv (Berlin)

Berlin Document Center:
31XX: Zentralkartei
3200: Ortskartei

2. Geheimes Staatsarchiv Preußischer Kulturbesitz (Berlin)

Files of the Prussian Ministry of Justice – Rep 84a
55359
55362-55365
55367
55372
55374-55375
55379
55381-55382
55389
55397-55398
55410
55430
55434

3. Landesarchiv Schleswig-Holstein (Schleswig)

Files of the Oberpräsidium of the Province Schleswig-Holstein in Kiel:
301/3518
301/4125
301/4589
301/4695-4705

Files of the Regierungspräsident in Schleswig:
309/22569
309/22576
309/22590
309/22668
309/22696
309/22701

309/22759
309/22761
309/22764-22765
309/22783-22784
309/22921
309/22924
309/22998
309/23017

4. Stadtarchiv Neumünster

MA 5360 (I):

Letter from Claus Heim to Johannes Rathjen, Celle, 2.2.1932.
Letter from Fritz Rehling to Johannes Rathjen, Mülheim (Ruhr), 25.4.1935.
Letter from Hans Fallada (Rudolf Ditzen) to "Ibeth", Neumünster, 20.11.1929.

5. Zeitzeichen Archiv (Kiel)

Interview with Peter Petersen (1990)
Interview with Mrs Hamkens (1990)

PUBLISHED SOURCES

a. The Press

Bargteheider Zeitung
Ein Blatt am Montag
Blut und Boden
Bramstedter Nachrichten
Deutsche Zeitung
Deutscher Landvolk-Kampfbund
Dusendüwelswarf
Eckernförder Zeitung
Eiderstedter Nachrichten
Fehmarnscher Inselbote
Flensborg Avis
Flensburger Nachrichten
Frankfurter Zeitung
Germania
Grüne Front
Hamburger Echo
Hamburger Fremdenblatt
Hamburger Nachrichten
Hamburger Volkszeitung
Heider Anzeiger
Holsteinischer Courier
Husumer Nachrichten
Itzehoer Nachrichten
Kämpfendes Landvolk
Kaltenkirchener Zeitung
Kieler Neueste Nachrichten
Kieler Zeitung
Kölnische Zeitung
Landbote für Schleswig-Holstein
Das Landvolk
Der Landvolkkampf
Lecker Anzeiger
Marner Zeitung
Mitteilungsblatt der Kreisbauernschaft-Segeberg
Der Nationale Sozialist
Neue Preußische Kreuz Zeitung
Norddeutsche Zeitung
Nordischer Kurier
Ostholsteinisches Tageblatt
Preetzer Zeitung
Rendsburger Tageblatt (Schleswig-Holsteinische Landeszeitung)
Der Schleswiger
Schleswiger Nachrichten
Schleswig-Holsteinische Tageszeitung
Schleswig-Holsteinische Volkszeitung
Schleswig-Holsteinischer Bauernbund
Segeberger Kreis- und Tageblatt
Soltauer Nachrichten
Tempo
Völkischer Beobachter
Volkszeitung
Vorwärts
Vossische Zeitung
Die Weltbühne
Westküste
Wilstersche Zeitung

b. Statistical and Documentary Material

G. Abramowski (ed.), *Akten der Reichskanzlei: Die Kabinette Marx III und IV*, vol. 2 (Boppard/Rhein, 1988).

L. Ebermayer, A. Lobe and W. Rosenberg (eds), *Kommentar zum Reichs-Strafgesetzbuch* (Berlin and Leipzig, 1925).

K. D. Erdmann (ed.), *Akten der Reichskanzlei Weimarer Republik: Das Kabinett Müller II*, vol. 2 (Boppard/Rhein, 1970).

G. Feder, *Das Programm der NSDAP und seine weltanschaulichen Grundgedanken* (Munich, 1931).

Gemeindelexikon für den Freistaat Preußen, Band IX: Provinz Schleswig-Holstein, Preußisches Statistisches Landesamt (Berlin, 1930).

Geschäftsbericht der Landwirtschaftskammer für die Provinz Schleswig-Holstein (Kiel, 1921/1922 - 1929/1930).

E. R. Huber (ed.), *Dokumente zur deutschen Verfassungsgeschichte*, vol. 4 (Stuttgart, Berlin and Cologne, 1992).

Landwirtschaftliche Buchführungsergebnisse, Enquete-Ausschuß, Verhandlungen und Berichte des Unterausschusses für Landwirtschaft (II. Unterausschuß), vol. 4 (Berlin, 1929).

Nationalsozialistisches Jahrbuch, published in collaboration with the Hauptparteileitung der NSDAP (Munich, 1927 and 1929).

Reichsgesetzblatt (Berlin, 1923).

'Sammlung der Drucksachen des Preußischen Landtags, 3. Wahlperiode 1. Tagung begonnen am 8. Juni 1928', vol. 15 (Berlin, 1932)

K. Schaap (ed.), *Oldenburgs Weg ins "Dritte Reich": Quellen zur Regionalgeschichte Nordwest-Niedersachsens* (Oldenburg, 1983).

W. Sellert and H. Rüping, *Studien- und Quellenbuch zur Geschichte der deutschen Strafrechtspflege*, vol. 2 (Aalen, 1994).

Statistisches Handbuch für Schleswig-Holstein, Statistisches Landesamt Schleswig-Holstein (Kiel, 1951).

Statistisches Jahrbuch für den Freistaat Preußen, Statistisches Landesamt Preußen (Berlin, 1926 - 1931).

Statistisches Jahrbuch für das Deutsche Reich, Statistisches Reichsamt (Berlin, 1919, 1921, 1924 and 1926 - 1931).

Die Verschuldungs- und Kreditlage der deutschen Landwirtschaft in ihrer Entwicklung von der Währungsbefestigung bis Ende 1928. Enquete-Ausschuß, Verhandlungen und Berichte des Unterausschusses für Landwirtschaft (II. Unterausschuß), vol. 12 (Berlin, 1930).

c. Memoirs, Speeches, etc.

A. Bartels, *Die Dithmarscher: historischer Roman in vier Büchern* (Hamburg, 1927).

S. O. Berggöte (ed.), *Ernst Jünger: Politische Publizistik 1919 bis 1933* (Stuttgart, 2001).

B. Dusik (ed.), *Hitler: Reden, Schriften, Anordnungen: Februar 1925-Januar 1933*, vol. 2 (2) (Munich, London, New York and Paris, 1992).

B. Dusik and K. A. Lankheit (eds), *Hitler: Reden, Schriften, Anordnungen: Februar 1925-Januar 1933*, vol. 3 (1) (Munich, New Providence, London and Paris, 1994).

H. Fallada, *A Small Circus* (London, 2012)

H. Fallada, *Bauern, Bonzen und Bomben* (Hamburg, 1997).
H. Fallada, *Bauern, Bonzen und Bomben* (adapted and directed by E. Monk, Studio Hamburg 1972).
C. Heim, 'Claus Heim berichtet. Erstveröffentlichung seiner Erinnerungen', *Bauernruf. Land-Rundschau für Politik und Agrarwirtschaft* No. 23, 26.11.1965 – No. 15, 5.8.1966, LAS EI/4434.
J. Goebbels, *Tagebücher: 1924-1945, 1. Einführung, 1924-1929* (Munich, 1992).
A. E. Günther, 'Die Schwarze Fahne', *Deutsches Volkstum*, 12 (1930), pp. 335-342.
G. Ipsen, *Das Landvolk. Ein soziologischer Versuch* (Hamburg, 1933).
K. A. Lankheit (ed.), *Hitler: Reden Schriften Anordnungen: Februar 1925 bis Januar 1933*, vol. 3 (2) (Munich, 1994).
W. Luetgebrune, *Neu-Preußens Bauernkrieg: Entstehung und Kampf der Landvolkbewegung* (Hamburg, 1931).
E. Niekisch, *Erinnerungen eines deutschen Revolutionärs*, vol. 1 (Cologne, 1974).
F. W. v. Oertzen, *Im Namen der Geschichte! Politische Prozesse der Nachkriegszeit* (Hamburg, 1934).
P. Petersen, *Fliegender Sand* (Norderstedt, 1985).
P. Petersen, *Telse Kampen. Aus Dithmarschens großer Zeit* (Neumünster, 1925).
E. v. Salomon, *Der Fragebogen (The Questionnaire)* (New York, 1955).
E. v. Salomon, *Der Fragebogen* (Reinbek, 1961).
E. v. Salomon, *Deutschland deine Schleswig-Holsteiner* (Hamburg, 1971).
E. v. Salomon, *Die Stadt* (Berlin, 1932).
R. Schapke, *Aufstand der Bauern* (Leipzig, 1933).
J. Schimmelreiter (pseudonym), *Unter der schwarzen Bauernfahne. Die Landvolkbewegung im Kampf für Deutschlands Befreiung* (Munich, 1929).
C. Severing, *Mein Lebensweg*, vol. 2 (Cologne, 1950).
L. D. Stokes, *Kleinstadt und Nationalsozialismus: ausgewählte Dokumente zur Geschichte von Eutin, 1918-1945* (Neumünster, 1984).
G. O. Tönnies, *Die Auflehnung der Nordmarkbauern. Eine politische Heimatschrift* (Flensburg, 1930).
B. Uhse, *Söldner und Soldat* (Paris, 1935).
H. Volck, *Rebellen um Ehre. Mein Kampf für die nationale Erhebung 1918-33* (Gütersloh, 1936).
C. Vollnhals (ed.), *Hitler: Reden, Schriften, Anordnungen: Februar 1925-Januar 1933*, vol. 1 (Munich, London, New York and Paris, 1992).
G. Zarnow, *Gefesselte Justiz*, vol. 1 and 2 (Munich, 1932).

SECONDARY SOURCES

W. Abel, *Agrarkrisen und Agrarkonjunktur: Eine Geschichte der Land- und Ernährungswirtschaft Mitteleuropas seit dem hohen Mittelalter* (Hamburg and Berlin, 1978).

W. Abel, *Geschichte der deutschen Landwirtschaft vom frühen Mittelalter bis zum 19. Jahrhundert* (Stuttgart, 1978).
F. Aereboe, *Allgemeine landwirtschaftliche Betriebslehre* (Berlin, 1918).
F. Aereboe, *Der Einfluß des Krieges auf die Landwirtschaftliche Produktion in Deutschland* (Stuttgart, Berlin and Leipzig, 1927).
F. Aereboe (ed.), *Wirtschaftslehre des Landbaues* (Berlin, 1930).
W. T. Angress, 'The Political Role of the Peasantry in the Weimar Republic', *The Review of Politics*, 21 (1959), pp. 530-549.
C. Applegate, *A Nation of Provincials: The German Idea of Heimat* (Berkeley, Los Angeles and Oxford, 1990).
K. J. Bade, 'Arbeitsmarkt, Bevölkerung und Wanderung in der Weimarer Republik' in M. Stürmer (ed.), *Die Weimarer Republik* (Königstein, 1980), pp. 160-187.
A. Bantelmann, A. Panten, R. Kuschert and T. Steensen, *Geschichte Nordfrieslands* (Heide, 1995).
S. Baranowski, *The Sanctity of Rural Life: Nobility, Protestantism, and Nazism in Weimar Prussia* (New York and Oxford, 1995).
H. Becker, *Handlungsspielräume der Agrarpolitik in der Weimarer Republik zwischen 1923 und 1929* (Stuttgart, 1990).
V. R. Berghahn, *Der Stahlhelm Bund der Frontsoldaten 1918-1935* (Düsseldorf, 1966).
J. Bergmann and K. Megerle, 'Protest und Aufruhr der Landwirtschaft in der Weimarer Republik (1924-1933). Formen und Typen der politischen Agrarbewegung im regionalen Vergleich' in J. Bergmann, J. Brockstedt, R. Fremdling et al., *Regionen im historischen Vergleich. Studien zu Deutschland im 19. und 20. Jahrhundert* (Opladen, 1989), pp. 200-287.
K. Bergmann, *Agrarromantik und Großstadtfeindschaft* (Meisenheim/Glan, 1970).
R. Bessel, *Germany After the First World War* (Oxford, 1993).
R. Bessel, 'Why did the Weimar Republic Collapse?', in I. Kershaw (ed.), *Weimar and Hitler: Why Did German Democracy Fail?* (New York, 1990), pp. 120-152.
H. Beyer, 'Die Agrarkrise und die Landvolkbewegung in den Jahren 1928-1932: ein Beitrag zur Geschichte "revolutionärer" Bauernbewegungen zwischen den beiden Weltkriegen', *Archiv für Agrargeschichte der holsteinischen Elbmarschen*, 5 (1983), pp. 156-187.
H. Beyer, 'Die Landvolkbewegung Schleswig-Holsteins und Niedersachsens 1928-1932', *Jahrbuch Kreis Eckernförde*, 15 (1957), pp. 173-202.
H. Beyer, 'Das Bauerntum Angelns während der großen Krise 1927 bis 1932', *Jahrbuch des Angler Heimatvereins*, 26 (1962), pp. 131-161.
H. Beyer, 'Landbevölkerung und Nationalsozialismus in Schleswig-Holstein', *Zeitschrift für Agrargeschichte und Agrarsoziologie*, 11 (1963), pp. 69-74.

H. Beyer, 'Der Friese Cornelius Petersen und "Bondens Selvstyre"', *Zeitschrift für Agrargeschichte und Agrarsoziologie*, 10 (1962), pp. 212-230.

H. Beyer, 'Schleswig-Holsteins Landvolk im Zwielicht', *Die Landwirtschaftliche Berufsschule*, 18 (1968), pp. 13-15.

C. H. Bill, 'Organisation und Finanzierung der Landvolkbewegung in Schleswig-Holstein 1928-1933', *Informationen zur Schleswig-Holsteinischen Zeitgeschichte,* 33/34 (1998), pp. 33-52.

F. Blaich, 'Der "Winzersturm von Bernkastel" Ursachen und Auswirkungen eines Steuerstreiks in der Weimarer Republik', *Zeitschrift für Agrargeschichte und Agrarsoziologie*, 33 (1985), pp. 2-26.

M. Bloch, 'The Rise of Dependent Cultivation and Seigniorial Institutions' in M. Postan (ed.), *The Cambridge Economic History of Europe, vol. 1: The Agrarian Life of the Middle Ages* (Cambridge, 1966), pp. 235-289.

O. Böckmann, *Landwirtschaft im Oldenburger Münsterland 1919-1933* (Vechta, 2000).

F. Bösch, 'Von der Einwohnerwehr zur Volkspartei' in D. Münkel (ed.), *Der lange Abschied vom Agrarland. Agrarpolitik, Landwirtschaft und ländliche Gesellschaft zwischen Weimar und Bonn* (Göttingen, 2000), pp. 227-248.

K. D. Bracher, *The German Dictatorship. The Origins, Structure and Consequences of National Socialism* (London, 1970).

J. Bracker, 'Die Einwohnerwehren in Schleswig-Holstein', *Zeitschrift der Gesellschaft für Schleswig-Holsteinische Geschichte*, 99 (1974), pp. 255-269.

A. Bramwell, *Blood and Soil: Richard Walther Darré and Hitler's 'Green Party'* (Bourne End, 1985).

O. Brandt, *Geschichte Schleswig-Holsteins: Ein Grundriß* (Kiel, 1981).

T. Bredohl, 'Some thoughts on the Political Opinions of Hans Fallada: A Response to Ellis Shookman', *German Studies Review*, 15 (1992), pp. 525-545.

J. Brockstedt, 'Konjunkturen in der Wirtschaftsentwicklung Schleswig-Holsteins 1864-1913, mit einem Überblick der Periode 1773-1864' in J. Brockstedt (ed.), *Wirtschaftliche Wechsellagen in Schleswig-Holstein vom Mittelalter bis zur Gegenwart* (Neumünster, 1991), pp. 221-263.

K. Bullivant, 'The Conservative Revolution' in A. Phelan (ed.), *The Weimar Dilemma: intellectuals in the Weimar Republic* (Manchester, 1985), pp. 47-70.

W. Carr, *Schleswig-Holstein 1815-1848* (Manchester, 1963).

T. Childers, *The Nazi Voter: The Social Foundations of Fascism in Germany, 1919-1933* (Chapel Hill and London, 1983).

T. Childers, 'Inflation, Stabilization, and Political Realignment in Germany 1924 to 1928' in G. Feldman, C.-L. Holtfrerich, G. Ritter and P. Witt (eds), *Die Deutsche Inflation: Eine Zwischenbilanz* (Berlin and New York, 1982), pp. 409-431.

T. Childers, 'Interest and Ideology: Anti-System Politics in the Era of Stabilization 1924-1928' in G. Feldman (ed.), *Die Nachwirkungen der Inflation auf die deutsche Geschichte 1924-1933* (Munich, 1985), pp. 1-20.

J. Christoph, *Die politischen Reichsamnestien 1918-1933* (Frankfurt/M, 1988).

O. Clausen, 'Der Aufmarsch des Landvolkes 1928', *Schleswig-Holstein*, 30 (1978), p. 18.

P. Clavin, *The Great Depression in Europe, 1929-1939* (New York, 2000).

G. Coppin, *Landvolkbewegung und Nationalsozialismus im schleswig-holsteinischen Marschgebiet 1928-1932* (Lille, 1975).

G. Corni, *Hitler and the Peasants. Agrarian Policy of the Third Reich 1930-1939* (Oxford, 1990).

G. Corni and H. Gies, *"Blut und Boden": Rassenideologie und Agrarpolitik im Staat Hitlers* (Idstein, 1994).

E. Danker, *Die Verschuldung der schleswig-holsteinischen Landwirtschaft in ihrer regionalen Bedingtheit unter besonderer Berücksichtigung der bäuerlichen Bezirke* (Kiel, 1931).

U. Danker, 'Die Zerstörung der Weimarer Demokratie in Schleswig-Holstein' in G. Wewer (ed.), *Demokratie in Schleswig-Holstein* (Opladen, 1998), pp. 213-228.

U. Danker, '"Von heute ab zahle ich keine Steuern mehr!" Steuerstreik, Stadtboykott und Bomben; die schleswig-holsteinische Landvolkbewegung 1928-1930' in U. Danker (ed.), *Die Jahrhundertstory* (Flensburg, 1998), pp. 48-67.

R. E. Dickinson, *The Regions of Germany* (London, 1945).

J. Diehl, *Paramilitary Politics in Weimar Germany* (Bloomington and London, 1977).

A. Dopsch, 'Agrarian Institutions of the Germanic Kingdoms from the fifth to the ninth century' in M. Postan (ed.), *The Cambridge Economic History of Europe, vol. 1: The Agrarian Life of the Middle Ages* (Cambridge, 1966), pp. 180-204.

J. Falter, *Hitlers Wähler* (Munich, 1991).

J. E. Farquharson, *The Plough and the Swastika: The NSDAP and Agriculture in Germany 1928-1945* (London and Beverly Hills, 1976).

J. E. Farquharson, 'The Agrarian Policy of National Socialist Germany' in R. G. Moeller (ed.), *Peasants and Lords in Modern Germany* (London, 1986), pp. 233-259.

G. Feldman (ed.), *Die Nachwirkungen der Inflation auf die deutsche Geschichte 1924-1933* (Munich, 1985).

T. Fink, *Geschichte des schleswigschen Grenzlandes* (Copenhagen, 1958).

C. Fischer, *The Ruhr Crisis, 1923-1924* (Oxford, 2003).

L. Fischer, *Haubarge: Eine Bauernhausform hat abgewirtschaftet?* (Bredstedt, 1987).

J. Flemming, 'Die Bewaffnung des "Landvolks". Ländliche Schutzwehren und agrarischer Konservatismus in der Anfangsphase der Weimarer Republik', *Militärgeschichtliche Mitteilungen*, 26 (1979), pp. 7-36.

J. Flemming, *Landwirtschaftliche Interessen und Demokratie: Ländliche Gesellschaft, Agrarverbände und Staat 1890-1925* (Bonn, 1978).

P. Fritzsche, *Rehearsals for Fascism: Populism and Political Mobilization in Weimar Germany* (New York and Oxford, 1990).

E. Friedrich, *Ursachen, Verlauf und agrarhistorische Bedeutung der Landvolkbewegung in Schleswig-Holstein von 1928-1930* (PhD diss., Leipzig, 1967).

P. Gay, *Weimar Culture. The Outsider as Insider* (London, 1969).

D. Geary, 'Employers, Workers, and the Collapse of the Weimar Republic', in I. Kershaw (ed.), *Weimar: Why did German Democracy Fail?* (New York, 1990), pp. 92-119.

A. Gerschenkron, *Bread and Democracy in Germany* (Berkeley and Los Angeles, 1943).

D. Gessner, *Agrarverbände in der Weimarer Republik: Wirtschaftliche und soziale Voraussetzungen agrarkonservativer Politik vor 1933* (Düsseldorf, 1976).

D. Gessner, 'The Dilemma of German Agriculture during the Weimar Republic' in R. Bessel and E. J. Feuchtwanger (eds), *Social Change and Political Development in Weimar Germany* (London, 1981).

H. R. Gies, *Richard Walther Darré und die nationalsozialistische Bauernpolitik in den Jahren 1930 bis 1933* (Frankfurt/M, 1966).

H. Gies, 'Die nationalsozialistische Machtergreifung auf dem agrarpolitischen Sektor', *Zeitschrift für Agrargeschichte und Agrarsoziologie*, 16 (1968), pp. 210-232.

H. Gies, 'NSDAP und landwirtschaftliche Organisationen in der Endphase der Weimarer Republik', *Vierteljahrshefte für Zeitgeschichte*, 15 (1967), pp. 341-376.

H. W. Gondesen, 'Zur Entstehung der schleswig-holsteinischen Landvolkbewegung', *Jahrbuch für die Schleswigsche Geest*, 34 (1986), pp. 8-32.

H. W. Gondesen, 'Die Landvolkbewegung 1929', *Jahrbuch für die Schleswigsche Geest*, 35 (1987), pp. 48-73.

H. W. Gondesen, '1930 und 1931, Niedergang der Landvolkbewegung, Aufstieg der NSDAP', *Jahrbuch für die Schleswigsche Geest*, 36 (1988), pp. 155-176.

H. W. Gondesen, 'Ende der Weimarer Republik, Ende des Landvolks', *Jahrbuch für die Schleswigsche Geest*, 37 (1989), pp. 7-27.

U. Greve, 'Revolte der Verzweifelten: Aufstand; bäuerliche Protestbewegung 1928', *Logo Schleswig-Holstein*, 8 (1989), pp. 34-35.

J. H. Grill, 'The Nazi Party's Rural Propaganda Before 1928', *Central European History*, 15 (1982), pp. 149-185.

H. Haasis, *Spuren der Besiegten* (Reinbek, 1984).

G. Hamkens, *Die Familie Hamkens aus Eiderstedt: eine Chronik* (Lübeck, 1972).

H. Hannover and E. Hannover-Drück, *Politische Justiz, 1918-1933* (Bornheim-Merten, 1987).

N. Hansen, 'Aufbruch in eine neue Zeit: Dithmarschen 1864-1918' in M. Gietzelt (ed.), *Geschichte Dithmarschens* (Heide, 2000), pp. 257-298.

O. Hauser, *Preußische Staatsräson und nationaler Gedanke* (Neumünster, 1960).

H. Haushofer, *Die deutsche Landwirtschaft im technischen Zeitalter* (Stuttgart, 1963).

R. Heberle, *From Democracy to Nazism: A Regional Case Study* (Baton Rouge, 1945).

R. Heberle, *Landbevölkerung und Nationalsozialismus: Eine soziologische Untersuchung der politischen Willensbildung in Schleswig-Holstein 1918-1932* (Stuttgart, 1963).

R. Heberle, 'Die Politische Haltung des Landvolkes in Schleswig-Holstein 1918-1932', *Volksspiegel, Zeitschrift für deutsche Soziologie und Volkswissenschaft*, 1 (1934), pp. 166-172.

R. Heberle, 'The Ecology of Political Parties: A Study of Elections in Rural Communities in Schleswig-Holstein. 1918-1932', *American Sociological Review*, 9 (1944) pp. 401-414.

A. Heggen 'Bauern, Bomben und Bonzen – Hans Fallada und der Landvolkprozess in Neumünster 1929' in A. Heggen (ed.) *Projekt Zeitgeschichte Neumünster* (Neumünster, 2006), pp. 42-66.

S. Heim, *Die Landvolkbewegung in Schleswig-Holstein 1928/1929: Eine Analyse ihrer sozioökonomischen Entstehungsbedingungen und politischen Aktionsformen* (Examensarbeit, Hamburg 1980).

P. Heinacher, *Der Aufstieg der NSDAP im Stadt- und Landkreis Flensburg* (Flensburg, 1986).

F.-W. Henning, *Landwirtschaft und ländliche Gesellschaft in Deutschland Band 2: 1750 bis 1976* (Paderborn, 1978).

R. Heydeloff, 'Staranwalt der Rechtsextremisten: Walter Luetgebrune in der Weimarer Republik', *Vierteljahrshefte für Zeitgeschichte*, 32 (1984), pp. 373-421.

G. Hoch, 'Artamanen in Schleswig-Holstein' in E. Hoffmann and P. Wulf (eds), "*Wir bauen das Reich.*" *Aufstieg und erste Herrschaftsjahre des Nationalsozialismus in Schleswig-Holstein* (Neumünster, 1983), pp. 137-148.

G. Hoch, *Das Scheitern der Demokratie in Schleswig-Holstein: Das Beispiel der Region Kaltenkirchen/Henstedt-Ulzburg 1870-1933* (Kiel, 1988).

J. B. Holt, *German Agricultural Policy 1918-1934* (Chapel Hill, 1936).

C.–L. Holtfrerich, *Die deutsche Inflation 1914-1923* (Berlin, 1980).

C.–L. Holtfrerich, Economic Policy Options and the End of the Weimar Republic', I. Kershaw (ed.), *Weimar: Why did German Democracy Fail?* (New York, 1990), pp. 58-91.

K. R. Homes, 'The Forsaken Past. Agrarian Conservatism and National Socialism in Germany', *Journal of Contemporary History*, 17 (1982), pp. 671-688.
R. Hopcroft, *Regions, Institutions, and Agrarian Change in European History* (Michigan, 1999).
E. R. Huber, *Deutsche Verfassungsgeschichte seit 1789 Band V: Weltkrieg, Revolution und Reichserneuerung 1914-1919* (Stuttgart, Berlin, Cologne and Mainz, 1978).
E. R. Huber, *Deutsche Verfassungsgeschichte seit 1789 Band VII: Ausbau, Schutz und Untergang der Weimarer Republik* (Stuttgart, Berlin, Cologne and Mainz, 1984).
P. Hüttenberger, *Die Gauleiter: Studie zum Wandel des Machtgefüges in der NSDAP* (Stuttgart, 1969).
H. James, *The German Slump: Politics and Economics 1924-1936* (New York, 1988).
H. James, 'Economic Reasons for the Collapse of the Weimar Republic' in I. Kershaw (ed.), *Weimar: Why did German Democracy Fail?* (New York, 1990), pp. 30-57.
G. Jasper, 'Justiz und Politik in der Weimarer Republik', *Vierteljahrshefte für Zeitgeschichte*, 30 (1982), pp. 167-205.
W. Jochmann, *Gesellschaftskrise und Judenfeindschaft in Deutschland 1870-1945.* (Hamburg, 1988).
F. Johannsen, 'Zur Erinnerung an Wilhelm Hamkens, Tetenbüll', *Zwischen Eider und Wiedau, Heimatkalender für Nordfriesland* (Husum, 1986), pp. 179-182.
L. E. Jones, 'Crisis and Realignment: Agrarian Splinter Parties in the Late Weimar Republic, 1928-1933' in R. G. Moeller (ed.), *Peasants and Lords in Modern Germany* (London, 1986), pp. 198-232.
L. E. Jones, *German Liberalism and the Dissolution of the Weimar Party System 1918-1933* (Chapel Hill and London, 1988).
M. H. Kater, 'Die Artamanen – Völkische Jugend in der Weimarer Republik', *Historische Zeitschrift*, 213 (1971), pp. 577-638.
M. H. Kater, *The Nazi Party: A Social Profile of Members and Leaders, 1919-1945* (Oxford, 1983).
I. Kershaw (ed.), *Weimar: Why did German Democracy Fail?* (New York, 1990).
M. J. Klein, *Ernst von Salomon: Eine politische Biographie mit einer vollständigen Bibliographie* (Limburg/Lahn, 1994).
G. Koch-Facompré, 'Die Landvolkbewegung', *Dithmarschen - Zeitschrift für Landeskunde und Landschaftspflege* (1991), pp. 77-84.
J. Kocka, *Klassengesellschaft im Krieg* (Frankfurt/M, 1988).
W. Koops, *Südtondern in der Zeit der Weimarer Republik (1918-1933)* (Neumünster, 1993).
W. Koops, 'Die Bombe von Niebüll: Sprengstoffanschlag als Bauern-Protest', *Nord-Friesland*, 111 (1995), pp. 18-20.

W. Kopitzsch, 'Politische Gewalttaten in Schleswig-Holstein in der Endphase der Weimarer Republik' in E. Hoffmann and P. Wulf (eds), "*Wir bauen das Reich." Aufstieg und erste Herrschaftsjahre des Nationalsozialismus in Schleswig-Holstein* (Neumünster, 1983), pp. 19-39.

R. Kuczynski, *Deutsche Anleihen im Ausland 1924-1928* (Berlin, 1929).

R. Kuschert, *Landesherrschaft und Selbstverwaltung in der Landschaft Eiderstedt unter den Gottorfern (1544 bis 1713)* (Neumünster, 1981).

W. Lammers, *Die Schlacht bei Hemmingstedt: Freies Bauerntum und Fürstenmacht im Nordseeraum* (Heide, 1982).

G. I. Langmuir, *Toward a Definition of Antisemitism* (Los Angeles, 1990).

M. Le Bars, *Le Mouvement Paysan Dans Le Schleswig-Holstein 1928-1932* (Berne, 1986).

M. Le Bars, 'Die Landvolkbewegung in Schleswig-Holstein: Geschichte und Literatur' in G. Müller-Waldeck and R. Ulrich (eds), *Hans Fallada: Beiträge zu Leben und Werk* (Rostock, 1995), pp. 67-99.

S. Lehmann, *Kreisleiter der NSDAP in Schleswig-Holstein: Lebensläufe und Herrschaftspraxis einer regionalen Machtelite* (Bielefeld, 2007).

G. Liebchen, *Ernst Jünger. Seine literarischen Arbeiten in den zwanziger Jahren. Eine Untersuchung zur gesellschaftlichen Funktion von Literatur* (Bonn, 1977).

C. P. Loomis and J. A. Beegle, 'The Spread of German Nazism in Rural Areas', *American Sociological Review*, 11 (1946), pp. 724-734.

K-J. Lorenzen-Schmidt, 'Neuorientierung auf den deutschen Wirtschaftsraum – Wirtschaftliche Entwicklung 1864-1918' in U. Lange (ed.), *Geschichte Schleswig-Holsteins* (Neumünster, 1996), pp. 385-399.

K.-J. Lorenzen-Schmidt, 'Die Landvolkbewegung in Schleswig-Holstein' in J. Diedrichs and H.-H. Weibe, *Schleswig-Holstein unter dem Hakenkreuz* (Bad Segeberg, 1985), pp. 43-51.

K.-J. Lorenzen-Schmidt, 'Kehrseite der Modernisierung: die Krise in Landwirtschaft und Industrie' in G. Paul, U. Danker and P. Wulf (eds), *Geschichtsumschlungen Schleswig-Holstein 1848-1948* (Bonn 1996), pp. 185-191.

A. Mayhew, *Rural Settlement and Farming in Germany* (London, 1973).

W. C. McNeil, *American Money and the Weimar Republic. Economics and Politics on the Eve of the Great Depression* (New York, 1986).

S. Meinl, *Nationalsozialisten gegen Hitler: Die nationalsozialrevolutionäre Opposition um Friedrich Wilhelm Heinz* (Berlin, 2000).

S. Meinl and D. Krüger, 'Der Politische Weg von Friedrich Wilhelm Heinz: Vom Freikorpskämpfer zum Leiter des Nachrichtendienstes im Bundeskanzleramt', *Vierteljahrshefte für Zeitgeschichte*, 42 (1994), pp. 36-69.

S. Merkenich, *Grüne Front gegen Weimar: Reichs-Landbund und agrarischer Lobbyismus 1918-1933* (Düsseldorf, 1998).

R. G. Moeller, 'Winners as Losers in the German Inflation: Peasant Protest over the Controlled Economy, 1920-1923' in G. Feldman, C.-L. Holtfrerich, G. Ritter and P. Witt (eds), *Die Deutsche Inflation: Eine Zwischenbilanz* (Berlin and New York, 1982), pp. 255-288.

R. G. Moeller, *German Peasants and Agrarian Politics, 1914-1924: The Rhineland and Westphalia* (Chapel Hill and London, 1986).

R. G. Moeller, 'Economic dimensions of Peasant Protest in the Transition from Kaiserreich to Weimar' in R. G. Moeller (ed.), *Peasants and Lords in Modern Germany: Recent studies in Agricultural History* (Sydney, 1986), pp. 140-167.

R. Möller, *Eine Küstenregion im politisch-sozialen Umbruch (1860-1933). Die Folgen der Industrialisierung im Landkreis Steinburg* (Hamburg, 2007).

R. Möller, 'Zur wirtschaftlichen Lage der Landwirtschaft und zur kommunalen Agrarkrisenbewältigungspolitik im Kreis Steinburg 1927-1930', *Archiv für Agrargeschichte der Holsteinischen Elbmarschen,* 1 (1988), pp. 1-5.

R. Möller, 'Zur Landvolkbewegung und ihren ökonomischen Entstehungsbedingungen', *Archiv für Agrargeschichte der Holsteinischen Elbmarschen,* 2 (1988), pp. 48-57.

A. Mohler, *Die konservative Revolution in Deutschland* (Darmstadt, 1972).

A. Mohler, *Die konservative Revolution in Deutschland 1918-1932. Ergänzungsband* (Darmstadt, 1989).

M. Müller, *Die Christlich-Nationale Bauern- und Landvolkpartei, 1928-1933* (Düsseldorf, 2001).

T. Nipperdey, 'Die deutsche Landbevölkerung und die Politik 1918-1933', *Neue Politische Literatur*, 9 (1964), pp. 390-400.

N. Nissen, *Landwirtschaft im Wandel* (Heide, 1989).

J. Noakes, *The Nazi Party in Lower Saxony 1921-1933* (London, 1971).

M. Oddey, *Unter Druck gesetzt. Presse und Nationalsozialismus in Schleswig-Holstein. Struktur – Wahrnehmung – Herrschaftsakzeptanz* (Eutin, 2006).

F. Omland, '"Die Nordmark steht zu Hitler!" Wählerwanderungen und Wählerherkunft der NSDAP 1924-1933' *Informationen zur Schleswig-Holsteinischen Zeitgeschichte*, 48 (2007), pp. 4-67.

F. Omland, '"Gegen das System!" Zum Aufstieg der NSDAP bei den Wahlen in Schleswig-Holstein 1924-1933' in K. Dohnke, R. Dopheide, T. Jacobs and F. Omland (eds) *"Siegeszug in der Nordmark" Schleswig-Holstein und der Nationalsozialismus 1925-1950. Schlaglichter – Studien – Rekonstrucktionen* (Kiel 2008), pp. 28-53.

E. Opitz, 'Dithmarschen 1773-1867' in M. Gietzelt (ed.), *Geschichte Dithmarschens* (Heide, 2000), pp. 217-254.

D. Orlow, *The History of the Nazi Party, 1919-1933* (Trowbridge and London, 1971).

D. Orlow, *Weimar Prussia, 1918-1925. The Unlikely Rock of Democracy* (Pittsburgh, 1986).

D. Orlow, *Weimar Prussia 1925-1933. The Illusion of Strength* (Pittsburgh, 1991).

J. Osmond, *Rural Protest in the Weimar Republic: The Free Peasantry in the Rhineland and Bavaria* (New York, 1993).

J. Osmond, 'Peasant Farming in South and West Germany during the War and Inflation 1914 to 1924: Stability or Stagnation' in G. Feldman, C.-L. Holtfrerich, G. Ritter and P. Witt (eds), *Die Deutsche Inflation: Eine Zwischenbilanz* (Berlin and New York, 1982), pp. 289-307.

A. Otto-Morris, '"Only united can we escape certain ruin": rural protest at the close of the Weimar Republic', *Rural History*, 20, 2 (2009), pp. 187-208.

A. Otto-Morris, "Bauer, wahre dein Recht!" Landvolkbewegung und Nationalsozialismus 1928/30" in K. Dohnke, R. Dopheide, T. Jacobs and F. Omland (eds) *"Siegeszug in der Nordmark" Schleswig-Holstein und der Nationalsozialismus 1925-1950. Schlaglichter – Studien – Rekonstrucktionen* (Kiel 2008), pp. 54-73.

A. Panzer, *Das Ringen um die deutsche Agrarpolitik* (Kiel, 1970).

A. Panzer, 'Parteipolitische Ansätze der deutschen Bauernbewegung bis 1933' in H. Gollwitzer (ed.), *Europäische Bauernparteien im 20. Jahrhundert* (Stuttgart and New York, 1977).

G. Paul, *Aufstand der Bilder. Die NS-Propaganda vor 1933* (Bonn, 1990).

G. Paul, '"Die Juden sind unser Unglück!" Zum Antisemitismus in einer frühen Hochburg der NSDAP' in G. Paul, and M. Gills-Carlebach (eds), *Menora und Hakenkreuz. Zur Geschichte der Juden in und aus Schleswig-Holstein, Lübeck und Altona (1918-1998)* (Neumünster, 1998), pp. 183-204.

V. Pauls, *Landesherrschaft und Selbstverwaltung in Eiderstedt* (Garding, 1932).

P. Petersen, '60 Jahre Schwarze Bauernfahne mit Pflug und Schwert', *Bauernblatt für Schleswig-Holstein*, 28 (1989), pp. 21-22.

P. Petersen, 'Die schleswig-holsteinische Landvolkbewegung 1927-1930', *Jahrbuch für die Schleswigsche Geest*, 32 (1984), pp. 181-191.

D. Petzina, *Die deutsche Wirtschaft in der Zwischenkriegzeit* (Wiesbaden, 1977).

D. Petzina, 'Was There a Crisis Before the Crisis? The State of the German Economy in the 1920s' in J. v. Kruedener (ed.), *Economic Crisis and Political Collapse: the Weimar Republic, 1924-1933* (New York, Oxford and Munich, 1990), pp. 1-19.

D. Peukert, *The Weimar Republic. The Crisis of Classical Modernity* (New York, 1993).

O. Poppinga, *Bauern und Politik* (Frankfurt/ M, 1975).

F. Priewe, 'Der Aufstand des Landvolkes', *Schleswig-Holstein*, 32 (1980), pp. 19-21.

H. - J. Puhle, *Agrarische Interessenpolitik und preußischer Konservatismus im wilhelminischen Reich (1893-1914)* (Hanover, 1967).

H. - J. Puhle, *Politische Agrarbewegungen in kapitalistischen Industriegesellschaften* (Göttingen, 1975).
H. - J. Puhle, *Von der Agrarkrise zum Präfaschismus. Thesen zum Stellenwert der agrarischen Interessenverbände in der deutschen Politik am Ende des 19. Jahrhunderts* (Wiesbaden, 1972).
W. Pyta, *Dorfgemeinschaft und Parteipolitik 1918-1933* (Düsseldorf, 1996).
H. Reif, 'Antisemitismus in den Agrarverbänden Ostelbiens während der Weimarer Republik' in H. Reif (ed.), *Ostelbische Agrargesellschaft im Kaiserreich und in der Weimarer Republik* (Berlin, 1994), pp. 379-411.
G. Reinhardt, *Preußen im Spiegel der öffentlichen Meinung Schleswig-Holsteins 1866-1870* (Neumünster, 1954).
R. Rietzler, *"Kampf in der Nordmark" Das Aufkommen des Nationalsozialismus in Schleswig-Holstein (1919-1928)* (Neumünster, 1982).
R. Rietzler, 'Gegründet 1928/29: Die Schleswig-Holsteinische Tageszeitung: Erste Gau-Tageszeitung der NSDAP' in E. Hoffmann and P. Wulf, *"Wir bauen das Reich" Aufstieg und erste Herrschaftsjahre des Nationalsozialismus in Schleswig-Holstein.* (Neumünster, 1983), pp. 117-133.
K.-G. Riquarts, *Der Antisemitismus als politische Partei in Schleswig-Holstein und Hamburg 1871-1914* (PhD diss., Kiel, 1975).
H. Rüping and G. Jerouschek, *Grundriß der Strafrechtsgeschichte* (Munich, 2002).
M. Sadek, 'Bauern, Bonzen und Bomben – Realität und Roman' in G. Müller-Waldeck and R. Ulrich (eds), *Hans Fallada: Beiträge zu Leben und Werk* (Rostock, 1995), pp. 100-116.
M. Schäfer, *Die schleswig-holsteinischen Bauernunruhen 1928-1932. Eine Darstellung unter besonderer Berücksichtigung der Bestände der Kieler Bibliotheken* (Hausarbeit, Hamburg, 1974).
A. Scharff, *Geschichte Schleswig-Holsteins* (Freiburg, 1991).
O.-E. Schüddekopf, *Linke Leute von rechts* (Stuttgart, 1960).
H.- P. Schwarz, *Der konservative Anarchist. Politik und Zeitkritik Ernst Jüngers* (Freiburg, 1962).
H. Schwilk, *Ernst Jünger: Ein Jahrhundertleben. Die Biographie* (Munich and Zurich, 2007).
M. Sering, *Die deutsche Landwirtschaft unter volks- und weltwirtschaftlichen Gesichtspunkten* (Berlin, 1932).
E. Shookman, 'Making History in Hans Fallada's Bauern, Bonzen und Bomben: Schleswig-Holstein, Nazism, and the Landvolkbewegung', *German Studies Review*, 13 (1990), pp. 461-480.
E. Shookmann, 'Figuring Out Hans Fallada: A Reply to Thomas Bredohl', *German Studies Review*, 17 (1994), pp. 79-81.
A. Skalweit, *Die deutsche Kriegsernährungswirtschaft* (Stuttgart, Berlin and Leipzig, 1927).

C. M. Sörensen, *Politische Entwicklung und Aufstieg der NSDAP in den Kreisen Husum und Eiderstedt 1918-1933* (Neumünster, 1995).

K. Sontheimer, *Antidemokratisches Denken in der Weimarer Republik* (Munich, 1978).

P. D. Stachura, 'Der kritische Wendepunkt? Die NSDAP und die Reichstagswahlen von 1928', *Vierteljahrshefte für Zeitgeschichte*, 26 (1978), pp. 66-99.

P. D. Stachura, *Gregor Strasser and the Rise of Nazism* (London, 1983).

T. Steensen, *Geschichte Nordfrieslands von 1918 bis in die Gegenwart* (Bredstedt, 2008), pp. 49-53.

T. Steensen, ' "In Land und Stadt das Heimatblatt" Zur Zeitungslandschaft in Schleswig-Holstein – insbesondere Nordfriesland – während der Weimarer Republik' in W. Paravicini (ed.), *Mare Balticum: Beiträge zur Geschichte des Ostseeraums in Mittelalter und Neuzeit* (Sigmaringen, 1992), pp. 391-412.

D. Stegmann, *Politische Radikalisierung in der Provinz. Lageberichte und Stärkemeldungen der Politischen Polizei und der Regierungspräsidenten für Osthannover 1922-1933* (Hanover, 1999).

H. Stern, 'The Organisation Consul', *The Journal of Modern History*, 35 (1963), pp. 20-32.

F. Stoehr, *Die Landvolkbewegung in Schleswig-Holstein* (Hausarbeit, Kiel 1986).

G. Stoltenberg, *Politische Strömungen im schleswig-holsteinischen Landvolk 1918-1933. Ein Beitrag zur politischen Meinungsbildung in der Weimarer Republik* (Düsseldorf, 1962).

G. Stoltenberg, 'Der Politische Umbruch in Schleswig Holstein, 1928-1930' Neumünster, 13.11.1979, StaNms MA 5360 (III).

W. Theine, *Westfälische Landwirtschaft in der Weimarer Republik* (Paderborn, 1991).

J. W. Thomsen, *Landleben in der Weimarer Republik* (Heide, 1989).

T. Thyssen, 'Schleswig-Holsteinisches Bauerntum zwischen den beiden Weltkriegen', *Jahrbuch des Angler Heimatvereins*, 27 (1963), pp. 150-153.

T. Thyssen, *Bauer und Standesvertretung. Werden und Wirkung des Bauerntums in Schleswig-Holstein seit der Agrarreform* (Neumünster, 1958).

T. A. Tilton, *Nazism, Neo-Nazism and the Peasantry* (Bloomington, 1975).

E. Topf, *Die Grüne Front* (Berlin, 1933).

M. Tönnsen, 'Rückschau und Erinnerungen. Betrachtungen und Erlebnisberichte zur Bauernbewegung nach dem ersten Weltkrieg bis zur Machtübernahme', *Jahrbuch des Angler Heimatvereins*, 28 (1964), pp. 130-138.

F. Trende, *Die Schlacht bei Hemmingstedt. Ein deutscher Mythos zwischen Politik, Poesie und Propaganda* (Heide, 2000).

H. A. Turner (ed.) *Nazism and the Third Reich* (New York, 1972).

W. L. Urban, *Dithmarschen: A Medieval Peasant Republic* (Lewiston 1991).

S. Volkov, 'Anti-Semitism as A Cultural Code: Reflections on the History and Historiography of Antisemitism in Imperial Germany', *Leo Baeck Institute Year Book*, 23 (1978), pp. 25-46.

R. G. L. Waite, *Vanguard of Nazism: The Free Corps Movement in Postwar Germany, 1918-1923* (New York 1969).

A. Walter, *Die Entwicklung der schleswig-holsteinischen Landvolkbewegung im Spiegel ihres Zentralorgans "Das Landvolk" (1928-1932)* (Hamburg, 1995).

N. Werner, *Die Prozesse gegen die Landvolkbewegung in Schleswig-Holstein 1929-1932* (Frankfurt, 2001).

A. W. Winkler, 'Die deutsche Gesellschaft der Weimarer Republik und der Anti-Semitismus - Jude als "Blitzableiter" ' in W. Benz and W. Bergmann (eds), *Vorurteile und Völkermord. Entwicklungslinie des Anti-Semitismus* (Freiburg, 1997), pp. 341-363.

K. W. Wippermann, *Politische Propaganda und staatsbürgerliche Bildung. Die Reichszentrale für Heimatdienst in der Weimarer Republik* (Cologne, 1976).

K. A. Wittvogel, 'Bauern, Bonzen, Faschisten: die Geheimnisse von Neumünster', *Die Linkskurve*, 2 (1970), pp. 28-32.

P. Wulf, *Die politische Haltung des schleswig-holsteinischen Handwerks 1928-1932* (Kiel, 1967).

P. Wulf, '"Juda ist überall". Antisemitismus in Schleswig-Holstein in der Zeit der Weimarer Republik' in R. Wenzel (ed.), *Ausgegrenzt – Verachtet – Vernichtet. Zur Geschichte der Juden in Schleswig-Holstein.* (Kiel, 1994), pp. 71-81.

P. Wulf, '"Jüdische Weltherrschaftspläne" Anti-Semitismus in bürgerlichen und bäuerlichen Parteien und Verbänden in Schleswig-Holstein zu Beginn der Weimarer Republik', *Zeitschrift der Gesellschaft für Schleswig-Holsteinische Geschichte*, 128 (2003), pp. 172-176.

P. Wulf, 'Die Not hat uns zusammengeschmiedet: die Landvolkbewegung Ende der 20er Jahre' in G. Paul, U. Danker and P. Wulf (eds), *Geschichtsumschlungen Schleswig-Holstein 1848-1948* (Bonn, 1996), pp. 192-199.

H. Zimmermann, 'Die Einwohnerwehren. Selbstschutzorganisationen oder konterrevolutionäre Kampforgane?', *Zeitschrift der Gesellschaft für Schleswig-Holsteinische Geschichte*, 128 (2003), pp. 184-211.

Z. Zofka, Die Ausbreitung des Nationalsozialismus auf dem Lande: Eine regionale Fallstudie zur politischen Einstellung der Landbevölkerung in der Zeit des Aufstiegs und der Machtergreifung der NSDAP 1928-1936 (Munich, 1979).

Index

KIELER WERKSTÜCKE

Reihe A: Beiträge zur schleswig-holsteinischen und skandinavischen Geschichte
Hrsg. von Oliver Auge und Thomas Riis

Band 1 Kai Fuhrmann: Die Auseinandersetzung zwischen königlicher und gottorfischer Linie in den Herzogtümern Schleswig und Holstein in der zweiten Hälfte des 17. Jahrhunderts. 1990.

Band 2 Ralph Uhlig (Hrsg.): Vertriebene Wissenschaftler der Christian-Albrechts-Universität zu Kiel (CAU) nach 1933. Zur Geschichte der CAU im Nationalsozialismus. Eine Dokumentation, bearbeitet von Uta Cornelia Schmatzler und Matthias Wieben. 1991.

Band 3 Carsten Obst: Der demokratische Neubeginn in Neumünster 1947 bis 1950 anhand der Arbeit und Entwicklung des Neumünsteraner Rates. 1992.

Band 4 Thomas Hill: Könige, Fürsten und Klöster. Studien zu den dänischen Klostergründungen des 12. Jahrhunderts. 1992.

Band 5 Rüdiger Wurr / Udo Gerigk / Uwe Törper / Alfred Sielken: Türkische Kolonie im Wandel. Ausländersozialarbeit und Ausländerpädagogik in Schleswig-Holstein (Bandhrsg.: Kai Fuhrmann und Ralph Uhlig). 1992.

Band 6 Torsten Mußdorf: Die Verdrängung jüdischen Lebens in Bad Segeberg im Zuge der Gleichschaltung 1933-1939 (Bandhrsg.: Kai Fuhrmann und Ralph Uhlig).1992.

Band 7 Thorsten Afflerbach: Der berufliche Alltag eines spätmittelalterlichen Hansekaufmanns. Betrachtungen zur Abwicklung von Handelsgeschäften. 1993.

Band 8 Ralph Uhlig: *Confidential Reports* des Britischen Verbindungsstabes zum Zonenbeirat der britischen Besatzungszone in Hamburg (1946-1948). Demokratisierung aus britischer Sicht. 1993.

Band 9 Broder Schwensen: Der Schleswig-Holsteiner-Bund 1919-1933. Ein Beitrag zur Geschichte der nationalpolitischen Verbände im deutsch-dänischen Grenzland. 1993.

Band 10 Matthias Wieben: Studenten der Christian-Albrechts-Universität im Dritten Reich. Zum Verhaltensmuster der Studenten in den ersten Herrschaftsjahren des Nationalsozialismus. 1994.

Band 11 Volker Henn / Arnved Nedkvitne (Hrsg.): Norwegen und die Hanse. Wirtschaftliche und kulturelle Aspekte im europäischen Vergleich. 1994.

Band 12 Jürgen Hartwig Ibs: Die Pest in Schleswig-Holstein von 1350 bis 1547/48. Eine sozialgeschichtliche Studie über eine wiederkehrende Katastrophe. 1994.

Band 13 Martin Höffken: Die "Kieler Erklärung" vom 26. September 1949 und die "Bonn-Kopenhagener Erklärungen" vom 29. März 1955 im Spiegel deutscher und dänischer Zeitungen. Regierungserklärungen zur rechtlichen Stellung der dänischen Minderheit in Schleswig- Holstein in der öffentlichen Diskussion. 1994.

Band 14 Erich Hoffmann / Frank Lubowitz (Hrsg.): Die Stadt im westlichen Ostseeraum. Vorträge zur Stadtgründung und Stadterweiterung im Hohen Mittelalter. Teil 1 und 2. 1995.

Band 15 Claus Ove Struck: Die Politik der Landesregierung Friedrich Wilhelm Lübke in Schleswig-Holstein (1951-1954). 1997.

Band 16 Hannes Harding: Displaced Persons (DPs) in Schleswig-Holstein 1945-1953. 1997.

Band 17 Olav Vollstedt: Maschinen für das Land. Agrartechnik und produzierendes Gewerbe Schleswig-Holsteins im Umbruch (um 1800-1867). 1997.

Band 18 Jörg Philipp Lengeler: Das Ringen um die Ruhe des Nordens. Großbritanniens Nordeuropa-Politik und Dänemark zu Beginn des 18. Jahrhunderts. 1998.

Band 19 Thomas Riis (Hrsg.): Tisch und Bett. Die Hochzeit im Ostseeraum seit dem 13. Jahrhundert. 1998.

Band 20 Alf R. Bjercke: Norwegische Kätnersöhne als königliche Dragoner. Eine Abhandlung über den Dragonerdienst in Norwegen und die Grenzwache in Schleswig-Holstein 1758-1762. 1999.

Band 21 Niels Bracke: Die Regierung Waldemars IV. Eine Untersuchung zum Wandel von Herrschaftsstrukturen im spätmittelalterlichen Dänemark. 1999.

Band 22 Lutz Sellmer: Albrecht VII. von Mecklenburg und die Grafenfehde (1534-1536). 1999.

Band 23 Ernst-Erich Marhencke: Hans Reimer Claussen (1804-1894). Kämpfer für Freiheit und Recht in zwei Welten. Ein Beitrag zu Herkunft und Wirken der "Achtundvierziger". 1999.

Band 24 Hans-Otto Gaethke: Herzog Heinrich der Löwe und die Slawen nordöstlich der unteren Elbe. 1999.

Band 25 Henning Unverhau: Gesang, Feste und Politik. Deutsche Liedertafeln, Sängerfeste, Volksfeste und Festmähler und ihre Bedeutung für das Entstehen eines nationalen und politischen Bewußtseins in Schleswig-Holstein 1840-1848. 2000.

Band 26 Joseph Ben Brith: Die Odyssee der Henrique-Familie (Bandhrsg.: Björn Marnau und Ralph Uhlig). 2001.

Band 27 Karl-Otto Hagelstein: Die Erbansprüche auf die Herzogtümer Schleswig und Holstein 1863/64. 2003.

Band 28 Annegret Wittram: Fragmenta. Felix Jacoby und Kiel. Ein Beitrag zur Geschichte der Kieler Christian-Albrechts-Universität. 2004.

Band 29 Sönke Loebert: Die dänische Vergangenheit Schleswigs und Holsteins in preußischen Geschichtsbüchern. 2008.

Band 30 Hans Gerhard Risch: Der holsteinische Adel im Hochmittelalter. Eine quantitative Untersuchung. 2010.

Band 31 Silke Hinz: Hochzeit in Kiel. Wandel im Hochzeitsgeschehen von 1965 bis 2005. 2011.

Band 32 Sönke Loebert / Okko Meiburg / Thomas Riis: Die Entstehung der Verfassungen der dänischen Monarchie (1848-1849). 2012.

Band 33 Franziska Nehring: Graf Gerhard der Mutige von Oldenburg und Delmenhorst (1430-1500). 2012.

Band 34 Simon Huemer: Studienstiftungen an der Christian-Albrechts-Universität zu Kiel. Private Bildungsförderung zwischen Stiftungsnorm und Stiftungswirklichkeit. 2013.

Band 35 Marina Loer: Die Reformen von Windesheim und Bursfelde im Norden. Einflüsse und Auswirkungen auf die Klöster in Holstein und den Hansestädten Lübeck und Hamburg. 2013.

Band 36 Alexander Otto-Morris: Rebellion in the Province: The Landvolkbewegung and the Rise of National Socialism in Schleswig-Holstein. 2013.

Reihe B: Beiträge zur nordischen und baltischen Geschichte
Hrsg. von Hain Rebas

Band 1 Rainer Plappert: Zwischen Zwangsclearing und Entschädigung. Die politischen Beziehungen zwischen der Bundesrepublik Deutschland und Schweden im Schatten der Kriegsfolgefragen 1949-1956. 1996.

Band 2 Volker Seresse: Des Königs "arme weit abgelegenne Vntterthanen". Oesel unter dänischer Herrschaft 1559/84-1613. 1996.

Band 3 Ingrid Bohn: Zwischen Anpassung und Verweigerung. Die deutsche St. Gertruds Gemeinde in Stockholm zur Zeit des Nationalsozialismus. 1997.

Band 4 Saskia Pagell: Souveränität oder Integration? Die Europapolitik Dänemarks und Norwegens von 1945 bis 1995. 2000.

Band 5 Ulrike Hanssen-Decker: Von Madrid nach Göteborg. Schweden und der EU-Beitritt Estlands, Lettlands und Litauens, 1995-2001. 2008.

Reihe C: Beiträge zur europäischen Geschichte des frühen und hohen Mittelalters
Hrsg. von Hans Eberhard Mayer

Band 1 Martin Rheinheimer: Das Kreuzfahrerfürstentum Galiläa. 1990.

Band 2 Oliver Berggötz: Der Bericht des Marsilio Zorzi. Codex Querini-Stampalia IV 3 (1064). 1990.

Band 3 Thomas Eck: Die Kreuzfahrerbistümer Beirut und Sidon im 12. und 13. Jahrhundert auf prosopographischer Grundlage. 2000.

Reihe D: Beiträge zur europäischen Geschichte des späten Mittelalters
Hrsg. von Werner Paravicini

Band 1 Holger Kruse, Werner Paravicini, Andreas Ranft (Hrsg.): Ritterorden und Adelsgesellschaften im spätmittelalterlichen Deutschland. Ein systematisches Verzeichnis. 1991.

Band 2 Werner Paravicini (Hrsg.): Hansekaufleute in Brügge. Teil 1: Die Brügger Steuerlisten 1360-1390, hrsg. von Klaus Krüger. 1992.

Band 3 Les Chevaliers de l'Ordre de la Toison d'or au XVe siècle. Notices bio-bibliographiques publiées sous la direction de Raphaël de Smedt. 1994. 2. Auflage 2000.

Band 4 Werner Paravicini (Hrsg.): Der Briefwechsel Karls des Kühnen (1433-1477). Inventar. Redigiert von Sonja Dünnebeil und Holger Kruse. Bearbeitet von Susanne Baus u.a. Teil 1 und 2. 1995.

Band 5 Werner Paravicini (Hrsg.): Europäische Reiseberichte des späten Mittelalters. Eine analytische Bibliographie. Teil 1: Deutsche Reiseberichte, bearb. von Christian Halm. 1994. 2., durchgesehene und um einen Nachtrag ergänzte Auflage 2001.

Band 6 Rainer Demski: Adel und Lübeck. Studien zum Verhältnis zwischen adliger und bürgerlicher Kultur im 13. und 14. Jahrhundert. 1996.

Band 7 Anne Chevalier-de Gottal: Les Fêtes et les Arts à la Cour de Brabant à l'aube du XVe siècle. 1996.

Band 8 Stephan Selzer: Artushöfe im Ostseeraum. Ritterlich-höfische Kultur in den Städten des Preußenlandes im 14. und 15. Jahrhundert. 1996.

Band 9 Werner Paravicini (Hrsg.): Hansekaufleute in Brügge. Teil 2. Georg Asmussen: Die Lübecker Flandernfahrer in der zweiten Hälfte des 14. Jahrhunderts (1358-1408). 1999.

Band 10 Jean Marie Maillefer: Chevaliers et princes allemands en Suède et en Finlande à l'époque des Folkungar (1250-1363). Le premier établissement d'une noblesse allemande sur la rive septentrionale de la Baltique. 1999.

Band 11 Werner Paravicini, Horst Wernicke (Hrsg.): Hansekaufleute in Brügge. Teil 3. Prosopographischer Katalog zu den Brügger Steuerlisten 1360-1390. Bearbeitet von Ingo Dierck, Sonja Dünnebeil und Renée Rößner. 1999.

Band 12 Werner Paravicini (Hrsg.): Europäische Reiseberichte des späten Mittelalters. Eine analytische Bibliographie. Teil 2: Französische Reiseberichte, bearbeitet von Jörg Wettlaufer in Zusammenarbeit mit Jacques Paviot. 1999.

Band 13 Nils Jörn, Werner Paravicini, Horst Wernicke (Hrsg.): Hansekaufleute in Brügge. Teil 4. Beiträge der Internationalen Tagung in Brügge April 1996. 2000.

Band 14 Werner Paravicini (Hrsg.): Europäische Reiseberichte des späten Mittelalters. Eine analytische Bibliographie. Teil 3. Niederländische Reiseberichte. Nach Vorarbeiten von Detlev Kraack bearbeitet von Jan Hirschbiegel. 2000.

Band 15 Werner Paravicini (Hrsg.): Hansekaufleute in Brügge. Teil 5. Renée Rößner: Hansische Memoria in Flandern. Alltagsleben und Totengedenken der Osterlinge in Brügge und Antwerpen (13. bis 16. Jahrhundert). 2001.

Band 16 Werner Paravicini (Hrsg.): Hansekaufleute in Brügge. Teil 6. Anke Greve: Hansische Kaufleute, Hosteliers und Herbergen im Brügge des 14. und 15. Jahrhunderts. 2011.

Reihe E: Beiträge zur Sozial- und Wirtschaftsgeschichte
Hrsg. von Gerhard Fouquet

Band 1 Thomas Hill / Dietrich W. Poeck (Hrsg.): Gemeinschaft und Geschichtsbilder im Hanseraum. 2000.

Band 2 Gabriel Zeilinger: Die Uracher Hochzeit 1474. Form und Funktion eines höfischen Festes im 15. Jahrhundert. 2002.

Band 3 Sascha Taetz: Richtung Mitternacht. Wahrnehmung und Darstellung Skandinaviens in Reiseberichten städtischer Bürger des 16. und 17. Jahrhunderts. 2004.

Band 4 Harm von Seggern / Gerhard Fouquet / Hans-Jörg Gilomen (Hrsg.): Städtische Finanzwirtschaft am Übergang vom Mittelalter zur Frühen Neuzeit. 2007.

Band 5 Gerhard Fouquet (Hrsg.): Die Reise eines niederadeligen Anonymus ins Heilige Land im Jahre 1494. 2007.

Band 6 Sven Rabeler: Das Familienbuch Michels von Ehenheim (um 1462/63-1518). Ein niederadliges Selbstzeugnis des späten Mittelalters. Edition, Kommentar, Untersuchung. 2007.

Band 7 Gerhard Fouquet / Gabriel Zeilinger (Hrsg.): Die Urbanisierung Europas von der Antike bis in die Moderne. 2009.

Band 8 Dietrich W. Poeck: Die Herren der Hanse. Delegierte und Netzwerke. 2010.

Band 9 Carsten Stühring: Der Seuche begegnen. Deutung und Bewältigung von Rinderseuchen im Kurfürstentum Bayern des 18. Jahrhunderts. 2011.

Band 10 Sina Westphal: Die Korrespondenz zwischen Kurfürst Friedrich dem Weisen von Sachsen und der Reichsstadt Nürnberg. Analyse und Edition. 2011.

Band 11 Ulf Dirlmeier: Menschen und Städte. Ausgewählte Aufsätze. Herausgegeben von Rainer S. Elkar, Gerhard Fouquet und Bernd Fuhrmann. 2012.

Reihe F: Beiträge zur osteuropäischen Geschichte
Hrsg. von Rudolf Jaworski und Ludwig Steindorff

Band 1 Peter Nitsche (Hrsg.), unter Mitarbeit von Ekkehard Klug: Preußen in der Provinz. Beiträge zum 1. deutsch-polnischen Historikerkolloquium im Rahmen des Kooperationsvertrages zwischen der Adam-Mickiewicz-Universität Poznań und der Christian-Albrechts-Universität zu Kiel. 1991.

Band 2 Rudolf Jaworski (Hrsg.): Nationale und internationale Aspekte der polnischen Verfassung vom 3. Mai 1791. Beiträge zum 3. deutsch-polnischen Historikerkolloquium im Rahmen des Kooperationsvertrages zwischen der Adam-Mickiewicz-Universität Poznań und der Christian-Albrechts-Universität zu Kiel, unter Mitarbeit von Eckhard Hübner. 1993.

Band 3 Peter Nitsche (Hrsg.): Die Nachfolgestaaten der Sowjetunion. Beiträge zur Geschichte, Wirtschaft und Politik. Herausgegeben unter Mitarbeit von Jan Kusber. 1994.

Band 4 Stephan Conermann / Jan Kusber (Hrsg.): Die Mongolen in Asien und Europa. 1997.

Band 5 Randolf Oberschmidt: Rußland und die schleswig-holsteinische Frage 1839-1853. 1997.

Band 6 Rudolf Jaworski / Jan Kusber / Ludwig Steindorff (Hrsg.): Gedächtnisorte in Osteuropa. Vergangenheiten auf dem Prüfstand. 2003.

Band 7 Ulrich Kaiser: Realpolitik oder antibolschewistischer Kreuzzug? Zum Zusammenhang von Rußlandbild und Rußlandpolitik der deutschen Zentrumspartei 1917-1933. 2005.

Band 8 Annelore Engel-Braunschmidt / Eckhard Hübner (Hrsg.): Jüdische Welten in Osteuropa. 2005.

Band 9 Martin Aust / Ludwig Steindorff (Hrsg.): Russland 1905. Perspektiven auf die erste Russische Revolution. 2007.

Reihe G: Beiträge zur Frühen Neuzeit
Hrsg. von Olaf Mörke

Band 1 Rolf Schulte: Hexenmeister. Die Verfolgung von Männern im Rahmen der Hexenverfolgung von 1530-1730 im Alten Reich. 2000. 2., ergänzte Auflage 2001.

Band 2 Jan Klußmann: Lebenswelten und Identitäten adliger Gutsuntertanen. Das Beispiel des östlichen Schleswig-Holsteins im 18. Jahrhundert. 2002.

Band 3 Daniel Höffker / Gabriel Zeilinger (Hrsg.): Fremde Herrscher. Elitentransfer und politische Integration im Ostseeraum (15.-18. Jahrhundert). 2006.

Band 4 Volker Seresse (Hrsg.): Schlüsselbegriffe der politischen Kommunikation in Mitteleuropa während der frühen Neuzeit. 2009.

Band 5 Björn Aewerdieck: Register zu den Wunderzeichenbüchern Job Fincels. 2010.

Band 6 Tatjana Niemsch: Reval im 16. Jahrhundert. Erfahrungsräumliche Deutungsmuster städtischer Konflikte. 2013.

Reihe H: Beiträge zur Neueren und Neuesten Geschichte
Hrsg. von Christoph Cornelißen

Band 1 Lena Cordes: Regionalgeschichte im Zeichen politischen Wandels. Die Gesellschaft für Schleswig-Holsteinische Geschichte zwischen 1918 und 1945. 2011.

Band 2 Birte Meinschien: Michael Freund. Wissenschaft und Politik (1945-1965). 2012.

Band 3 Stefan Bichow: Die Universität Kiel in den 1960er Jahren. Ordnungen einer akademischen Institution in der Krise. 2013.

www.peterlang.de

www.ingramcontent.com/pod-product-compliance
Ingram Content Group UK Ltd.
Pitfield, Milton Keynes, MK11 3LW, UK
UKHW041830100926
8556IPUK00002B/102

9 783631 581940